Recent Advances in Medical Thermology

Recent Advances in Medical Thermology

Edited by

E. Francis J. Ring

Royal National Hospital for Rheumatic Diseases
Bath, United Kingdom

and

Barbara Phillips

St. Martin's Hospital
Bath, United Kingdom

PLENUM PRESS • NEW YORK AND LONDON

Library of Congress Cataloging in Publication Data

International Congress of Thermology (3rd: 1982: Bath, Avon)
Recent advances in medical thermology.

"Proceedings of the Third International Congress of Thermology, held March 29–April 2, 1982, in Bath United Kingdom."—T.p. verso.
Includes bibliographical references and index.
1. Medical thermography—Congresses. 2. Thermography—Congresses. 3. Body temperature—Regulation—Congresses. I. Ring, E. F. J. II. Phillips, Barbara (Barbara Helen) III. Title. [DNLM: 1. Body temperature—Congresses. 2. Heat—Therapeutic use—Congresses. 3. Thermography—Congresses. W3 IN579 3rd 1982r/ WB 469 I61 1982r]
RC78.7.T5I55 1982 616.07′2 84-3366
ISBN 978-1-4684-7699-6 ISBN 978-1-4684-7697-2 (eBook)
DOI 10.1007/978-1-4684-7697-2

Proceedings of the Third International Congress of Thermology,
held March 29–April 2, 1982, in Bath, United Kingdom

Softcover reprint of the hardcover 1st edition 1984

A Division of Plenum Publishing Corporation
233 Spring Street, New York, N.Y. 10013

FOREWORD

I am delighted to have been invited to Bath for the opening of this Third International Congress of Thermology. The connection between the Congress and the City of Bath is significant. The properties of sunlight have been recognized throughout the centuries. Indeed, many ancient religions were based on the worship of the sun gods. The study of radiant heat was pioneered by Sir William Herschel, whose experiments led him to the study of heat and ultimately of infrared radiation. His son, John, furthered these experiments and formed an image by evaporating alcohol with carbon. In modern technology, infrared radiation plays a vital role in a wide range of applications. Thermal imaging is widely used in the manufacturing industries, especially plastics, glass and paper. The motor industry, for example, employs thermography in the design of windscreens and tire development. Chemical plants and refineries also use it in the important control of expensive energy losses. The communications industry makes extensive use of thermal imaging since overheating and cracks in insulation may cause the unscheduled shut-down of expensive equipment. There is now a special thermal imaging system for the examination of very large scale integrated circuits to help in the development of diagnostic tools for examining circuits which now have features of 1 micron in size, making the conventional method of mechanical probing impossible. This revolution in probing will enable us to maintain the high levels of quality control which are essential in the communications industry.

In medicine, the connection between temperature and disease has, of course, been recognized for centuries, and for a long time the thermometer was the principal tool for medical thermography, but the last twenty years have brought about a technical revolution. Thermal imaging has gone beyond the natural limits of human sight and allowed us to reach into the invisible world of infrared radiaiton. At the sametime, other techniques for viewing the human body have developed, using X-rays, gamma rays, ultrasound and now magnetic fields. But infrared imaging has the great advantage that the doctor realizes the information through a natural source of energy and it is, therefore, completely safe to use it for examination.

This conference brings together experts from a wide range of technical and medical disciplines. Medical research has increasingly become a team exercise. Disease and sickness know no frontiers and medical science cannot meet the world's needs without a continuing committment to exchange and fertilize ideas and concepts. The understanding of body temperature from childhood to old age, and the effects of disease, is still developing, and the purpose of this International Congress is to compare and discuss recent progress in these subjects. Three main areas of medicine - diseases of the circulation, arthritis and cancer - form special parts of the Congress programme and a forum is provided for doctors, scientists and engineers to collaborate in furthering the knowledge of medical thermography.

In opening this Congress, I wish you all the best of success and would ask you to give some thought to the immortal words of that great poet Milton:

"Accuse not nature, she hath done her part;
Do thou but thine, . . ."

His Royal Highness Prince Michael of Kent

Opening Address
Third International Congress of Thermology
Bath, United Kingdom

PREFACE

Increasing interest in human body temperature has derived from the development of improved technology. Thermal imaging in particular has opened up many areas for study in the last 20 years. This book contains a collection of papers based on the Third International Congress of Thermology. The Congress, opened by His Royal Highness the Duke of Kent, was held in Bath, United Kingdom, in April 1982.

A comprehensive review of new thermal technology applied to clinical medicine was given. The main themes of thermal physiology, instrumentation and technology, and their applications in vascular, locomotor and malignant diseases were represented. As this Congress is held every four years under the auspices of the European Association of Thermology, it forms a valuable basis for evaluating techniques of research interest. It also indicates those techniques which are more applied to clinical diagnosis. Since the majority of these procedures are non-invasive, an increasing trend to monitoring disease in response to treatment is shown.

Acknowledgements are especially due to those colleagues who applied their expertise in the scientific programme and the review of the many manuscripts. They are:

Dr R P Clark	Physiology
Dr C H Jones	Instrumentation and Technique
Dr E D Cooke	Vascular diseases
Dr D P Page Thomas	Locomotor diseases
Dr A M Stark	Malignant diseases.

We are particularly grateful to Dr R P Clark and Mr M Goff for their sustained assistance in preparing the many figures and illustrations found in this book. A considerable proportion of the work has been carried by Miss J Elswood and Miss F A H Davies in typing and preparing the many and varied manuscripts.

We are grateful to His Royal Highness Prince Michael of Kent for opening the Congress and to the Congress Chairman Mr K Lloyd-Williams of Bath.

Francis J Ring
Barbara Phillips

Royal National Hospital for
Rheumatic Diseases and
Royal United Hospital, Bath

CONTENTS

THERMAL PHYSIOLOGY

TECHNICAL DEVELOPMENTS AND INSTRUMENTATION

VASCULAR AND ENDOCRINE DISEASES

LOCOMOTOR DISEASES

MALIGNANT DISEASES

HYPERTHERMIA

THERMAL PHYSIOLOGY - INTRODUCTION

R P Clark

Clinical Research Centre
Watford Road
Harrow, UNITED KINGDOM

Human thermal physiology forms the basis of thermology. With this in mind, the physiology sessions at the Congress set out to provide a scientific framework to consider the basic physical and physiological principles that determine the thermal parameters in the healthy human, and against which pathological abnormalities may be judged.

If thermology is to establish itself as a valuable technique in medicine it is vital that further effort be put into extending our understanding of the basic processes involved. In many areas where thermology complements diagnosis or assessment, our appreciation of deviation from the "normal" is often poorly understood.

Because infrared thermography is the most widely used technique in thermology, much emphasis in the physiology sessions was placed on the role of skin temperature in thermoregulation.

Skin temperature patterns were discussed in relation to age, sex, exercise, obesity, etc. Differing, clearly developing patterns exist from infancy through childhood to adulthood. Although there is considerable individual variation in absolute temperatures between subjects, there are many overall similarities in pattern between comparable groups.

The modification seen in reponse to hot or cold stress on "normal" skin temperature patterns underlined the importance of choosing suitable environments and patient equilibration times for thermographic examination.

Until fairly recently, the description of thermal images produced by infrared thermography has relied on subjective assessment. Although analog analyses have been relatively successful, the real breakthrough has been with the availability of computer analysis and digital recording. It is now possible to quantify numerous thermal images in a way that was not practicable previously. This, coupled with the high sensitivity and stability of modern infrared scanning equipment, places a powerful tool in the hands of the physiologist and clinician. Digital techniques now allow for the possibility of standardization in the description of thermal images, which will facilitate the interchange of information between different groups of investigators.

Several papers were presented dealing with the mathematical analysis of heat transfer between the body and the surroundings. These drew particular attention to the influences of "limited" or "man-made" environments and to the problems of, for example, how reflected radiation may influence infrared measurements of skin temperature both indoors and outside.

The influence of centrally and peripherally acting drugs on temperature regulation was described, and the principles involved in thermoreception and temperature sensation were discussed. Topics such as temperature changes found in skin diseases and after tissue injury added to the comprehensive overview of mechanisms involved in the thermal balance of the body.

In discussion, various speakers were asked about the thermogenic importance of brown fat, particularly in regard to obesity, where it has recently attracted some following. In reply, it was felt that brown fat could not be ignored in the infant as it was the only thermogenic retaliation that the neonate had in response to cold stress. However, this role had not been convincingly demonstrated. In normal circumstances babies are rarely exposed to temperatures below 25°C for any length of time and at such temperatures any brown fat thermogenic mechanism is not "switched-on".

There is a smaller percentage of brown fat in relation to total body mass in childhood and adulthood compared with the neonate. Within the normal range of environmental temperatures in which people habitually live there was little, if any, evidence of an important role for brown fat in heat production.

Mechanisms based on differences in heat output between lean and obese subjects had been attributed to the presence or absence of brown fat, but some early claims had not been substantiated and these ideas have, in the main, been dropped.

Brown fat could be important if it were shown to be thermogenically active in the elderly, particularly when physical and physiological mechanisms of heat production were depressed. If it could be found in deep tissue, around the vessels of the heart, for example, this could be a useful source of heat production. However, the examination of post-mortem specimens from people dying in the seventh to eighth decade of life had been found to be completely devoid of brown fat when examined by light and electron microscopy.

The general conclusion to this discussion was that brown fat was no longer an important issue in human thermoregulation, at least in normal environments.

One of the areas where more research is indicated concerns temperature regulation during sleep. Skin temperature patterns, their rhythmic changes, transepidermal water loss and other variables are modified in sleep. The changes are not the same for adults and infants.

One problem outlined during discussion was concerned with the courses of infectious diseases in different "limited" environments. In going from hot to cold regions the ambient temperatures can modify the human thermoregulatory response and this in turn affects the manifestation of symptoms of some diseases. Any infections that change circulatory responses or hemoglobin levels, for example, may well be expected to influence physiological responses to changes in temperature. There was certainly some evidence that schistosomiasis in the Sudan modified the patient's exercise and thermoregulatory capacity, when tested by standard methods.

The wide ranging topics covered in the physiology sessions provide scope for broad collaboration between physiologists and clinicians for the development of thermology in general, and for infrared thermography in particular.

HUMAN SKIN TEMPERATURE AND ITS RELEVANCE IN PHYSIOLOGY AND CLINICAL ASSESSMENT

R P Clark

Laboratory for Aerobiology,
Clinical Research Centre,
Harrow, UNITED KINGDOM

INTRODUCTION

The temperature of the human body has been widely used as an index of illness since antiquity. The hand of an experienced physician laid upon the skin could provide much useful information about the temperature of a patient and the course of an illness. Eventually, more objective assessment was possible and the clinical thermometer (much as we know it today) was developed during the last century. This instrument rapidly became firmly established as this extract from a work of 1882 shows - "It (the thermometer) is now in daily, nay hourly, use in every hospital in London, and ranks in importance with the stethoscope. A doctor without his thermometer is like a sailor without his compass".[1]

The clinical thermometer is so useful because body temperature (by which is meant the temperature of the deep tissues) is fairly constant at around 37 ± 1°C for healthy subjects; Figure 1 shows the variation in health and disease. Normal temperature may be elevated during exercise to become as high as 40°C, but if values such as this occur during rest then the subject will be ill, have a fever and be described as being hyperthermic. At body temperature above 45-46°C death results.

At the other end of the scale, when body temperature falls below 35°C the subject will be described as being hypothermic, and this can occur in cold exposure and with surgical hypothermia. Temperatures below about 32-33°C again result in death.

Body temperature is a consequence of the heat energy that is continually being produced as a by-product of metabolism. A simple, but useful, concept in considering temperature in healthy subjects, particularly in calculations of heat exchange with the environment, regards

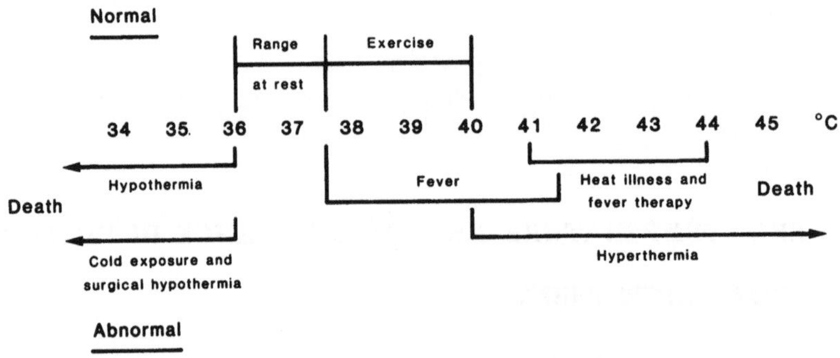

Fig. 1. Body temperature variation.

the body as having an inner core where heat production is centered. This will be the hottest region, and temperatures will fall progressively with increasing distance from the "core", the cooler and more superficial tissues being regarded as a surrounding "shell". This concept is discussed in more detail in references 2 and 3. The core temperature can be measured at several sites (rectum, esophagus, aural canal, or mouth). The values differ slightly from one another, but generally change consonantly with each other. The core temperature is virtually independent of environmental conditions over a wide range. It is, however, very sensitive to changes in work rate,[4] whereas skin temperature is markedly affected by environmental temperature and air movement.

In order that the core does not overheat, the energy produced must be exchanged with the environment. Heat loss to the surroundings must not be at such an excessive rate that the body core drops in temperature; a balance is maintained when the rate of heat production is equal to the heat lost by radiation, convection, conduction and evaporation. For a nude standing subject at rest in warm conditions (25°C, say) some 60% of the heat loss will be by radiation. About 30% is lost by convection and the rest by conduction (at body areas directly in contact with other surfaces such as the feet) and insensible water loss through the skin. During exercise, heat production can reach 6-7 times the resting level and the loss from the evaporation of sweat increases dramatically to become the dominant mode of heat exchange. All of this energy transfer occurs at the skin, which is the largest organ of the body and can be described as the interface between the deeper tissues and the environment. One consequence of its role in heat exchange is that a study of temperature distribution over the skin surface can give an insight into many physiological mechanisms concerned with thermoregulation and metabolism.[5] In pathological conditions (locomotor, vascular and malignant diseases, for example) skin temperature patterns often provide valuable diagnostic information, and can be important in choosing and assessing subsequent therapy.

In all such investigations it is necessary to measure accurately skin temperature and its distribution. This can be done in a number of ways, for example, by the use of thermistor and thermocouple probes. However, such methods cannot assess overall skin temperature patterns as comprehensively as the more sophisticated technique of infrared thermography. This is a non-invasive, no-touch method for the determination of surface temperature from the emission of infrared energy from the body surface.[6]

It was in 1800 that William Herschel discovered the infrared portion of the spectrum and in 1840, his son, John, produced the first infrared image. However, it was not until 1956 that Lawson produced the first medical thermogram. Since then there has been a continuous development in equipment suitable for physiological and clinical use. Figure 2 illustrates chronologically some of the landmarks in thermology, from Hippocrates to the present day.

Modern thermographic equipment for assessing temperature patterns over the human body intercepts electromagnetic energy emission in wavelengths between 2-6 microns using a special scanner. The infrared energy is converted into temperature levels, which may be displayed as a TV image in monochrome or color, where particular shades or colors indicate specific surface temperatures. The latest equipment can be combined with digital tape recording of thermal images. Combined with computer analysis, it provides the investigator with a powerful tool to quantify skin temperature patterns and variations in health and disease.

SKIN TEMPERATURE PATTERNS IN HEALTHY SUBJECTS

The starting point in human skin temperature investigations is the systematic study of overall temperature levels and patterns in normal healthy subjects of different age and physique. Despite considerable individual variability in absolute skin temperature in specific regions and the body as a whole, there are many common features in overall patterns for similar groups of subjects. These are dependent on a number of internal and external factors.

Skin temperature patterns develop with age from infancy to adulthood. In neonates,[7] the patterns appear to be due mainly to near uniform heat conduction through the tissues. They are not significantly influenced by regional differences in the cutaneous circulation. Direct conduction from heat-producing structures beneath the surface is not dominant. In childhood, very clear and definite patterns develop from the neonatal distribution. Regional differences in skin blood flow, with the insulating and heat conducting properties of subcutaneous structures, are important in determining the skin temperature patterns. Changes continue through adolescence into adulthood; the adult pattern, once established, can remain virtually unchanged in healthy subjects to the eighth decade of life and beyond.

Obese and lean subjects have important differences in temperature patterns. In the obese, the effect of direct heat conduction through the tissues is modified by a layer of subcutaneous fat.[8]

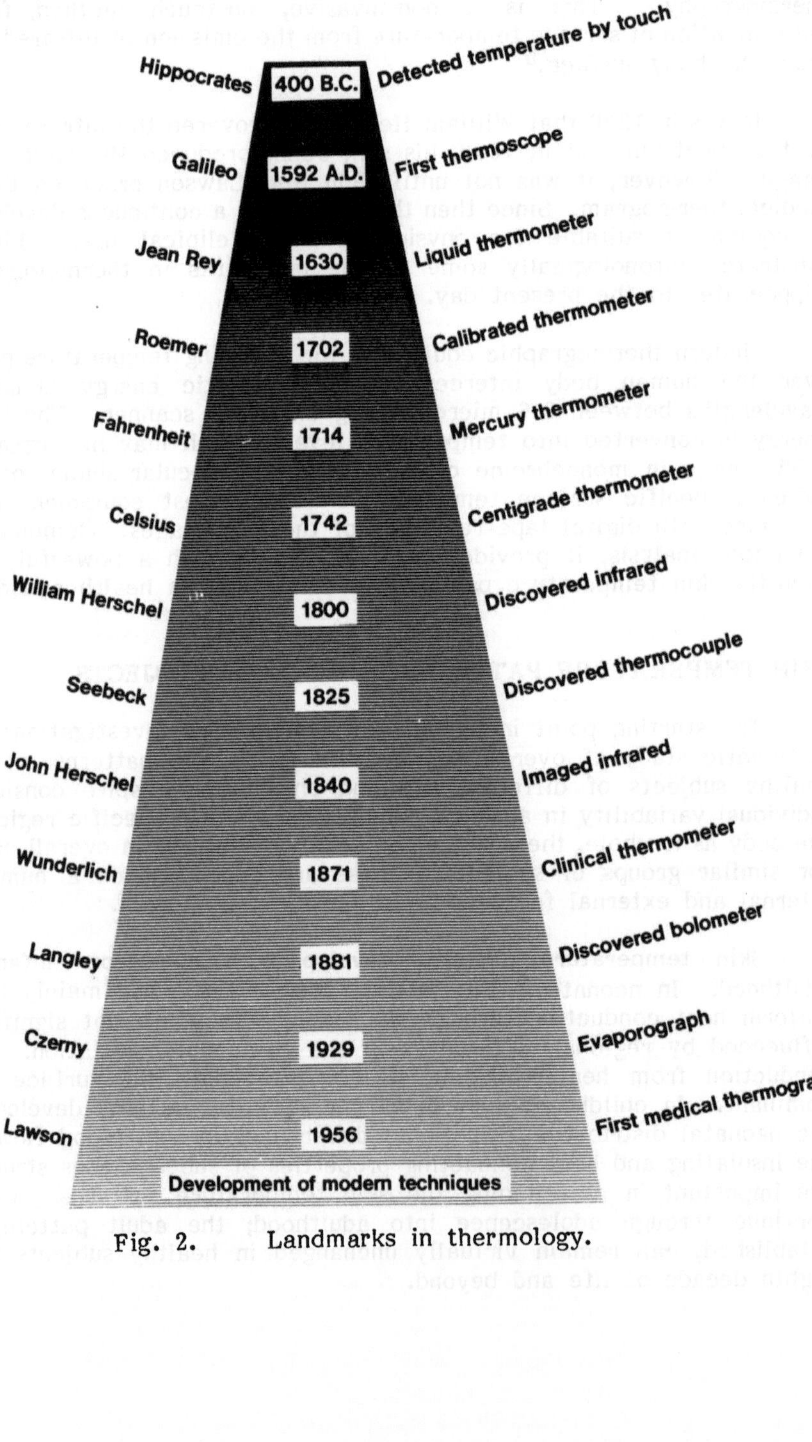

Fig. 2. Landmarks in thermology.

ENVIRONMENTAL CONDITIONS

Ambient temperature and air movement are important in determining skin temperature patterns and in some cases their influence may be dominant. In a cool environment, direct conduction of heat through the subcutaneous tissues has a marked effect, particularly where skin blood flow is reduced owing to vasconstriction. The resultant range of temperature is greater than 10°C over the skin at ambient temperatures between 10-20°C. In warm and hot environments, the difference between skin and air temperatures decreases and consequently the heat loss is reduced. This may be reversed when the body is subjected to high solar energy gain. In these conditions skin temperature is markedly influenced by evaporative cooling, and the range over the skin reduces to some 3-4°C at ambient temperatures above 35°C. Figure 3 illustrates some of these points; 3a is a thermogram of a subject in cool conditions at 10°C and the large range of temperature over the body surface is clearly seen. For example, the difference in temperature in the orbit of the eye and the skin of the forehead represents a very steep surface gradient (ie change of temperature across the skin). The skin of the axilla in this picture is at a high temperature due to direct heat conduction from the underlying blood vessels.

Figure 3b shows a thermogram of the same subject with the ambient temperature raised to 39°C. The reduced range over the skin is clearly seen, particularly over the face. The hair, which is a passive surface, takes up heat directly from the atmosphere. Because it does not have the ability to thermoregulate, it becomes a very warm area. The skin of the axilla, which was one of the hottest areas in the cold conditions, is now the coolest region due to the evaporation of sweat.

The influence of environmental factors has to be well understood in studies on thermal physiology, and they are most important when deciding on suitable ambient temperatures for thermographic investigations. Thermography is generally carried out at room temperatures between 20-27°C. The choice depends on the extent of the body area under investigation, the length of time of exposure, and the required state of the cutaneous vasculature in terms of dilatation or constriction. Air movement must also be carefully considered as cooling by forced convection produces large changes in skin temperature.

EFFECT OF EXERCISE

Energy expenditure in exercise can dramatically affect the skin temperature distribution. Muscular activity can produce an increase in the direct conduction of heat from the active muscles to the skin surface and this is often accompanied by a redistribution of skin blood flow. These combined effects greatly modify the skin temperature pattern from that observed at rest.[9]

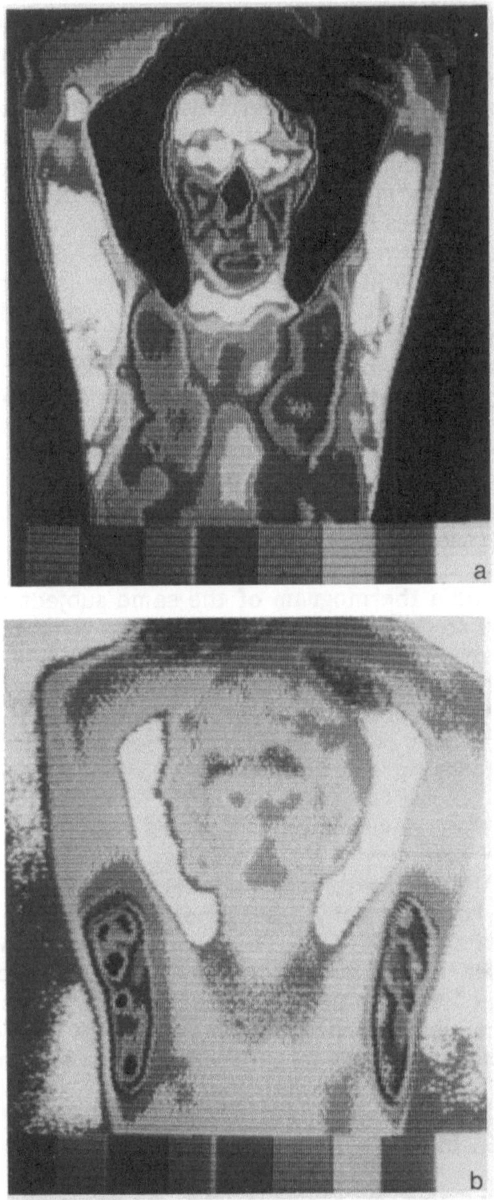

Fig. 3. Skin temperature patterns in cool conditions (10°C) (a) are quite different to those found in a hot environment (37°C). (b) The skin of the axilla is one of the warmest regions in the cold, but in hot conditions evaporative cooling makes this the coolest area.

EFFECT OF FOOD AND DRINK

Food and drink can significantly alter skin temperature. This mainly occurs by heat transfer changes brought about by gustatory sweating and by the effects of increased metabolic activity following a meal.[10,11,12,13]

TEMPERATURE CYCLING

In many physiological investigations a "steady state" is often assumed to exist when equilibrium between body heat production and environmental heat loss leads to a steady skin temperature pattern. Recent studies of subjects in well-controlled environments, however, have shown that the body surface temperature is in a continual state of rhythmic change. The changes can be observed over the arms, hands, trunk and head, but may not all be in phase or of the same amplitude.[14] Studies aimed at recording and quantifying these effects have particular relevance to pathological conditions, where there may be abnormal skin temperatures and patterns owing to defects in the nervous control of the peripheral circulation.

POTENTIAL FOR NUMERICAL ANALYSIS

Until recently, the interpretation of complex thermograms has often relied on visual inspection and qualitative description. Attempts at simple numerical analysis using analogue electrical techniques have also been made. The results of studies obtained using these methods have helped in establishing the value of thermography in physiology and medicine. However, in many thermographic investigations there is often a surfeit of information which is difficult to use or interpret. A thermographic image could be described as being composed of an infinite number of spot temperature measurements. In order to see trends or changes in any group of patients or subjects, it is important to be able to simplify such large amounts of information and to produce indices which can quantify the particular aspects of temperature pattern being investigated.

It is particularly difficult to produce a numerical index for a thermogram which can describe, for example, the difference between two images which may be quite obvious by visual inspection. If an index is to be of any value it has to be both accurate and sensitive to small differences which are instantly detectable by the naked eye.

It is only with the advent of equipment to digitize and store thermographic images on magnetic tape for subsequent retrieval and computer analysis, that the useful numerical interpretation of large numbers of thermograms has become a reality.[15,16] An example of suitable equipment is shown in the block diagram of Figure 4. Figure 5a shows a computer calibrated and processed thermal image of a hand with the gray levels (color in the original thermogram) calibrated to specific temperatures. A cursor under operator control can be positioned at any point on the image and the X and Y co-ordinates of the cursor displayed. The spot temperature at this position (at the center of the cross wires) can be immediately displayed and in this case is seen as 33.9°C. Any feature on the thermogram can be identified in terms of

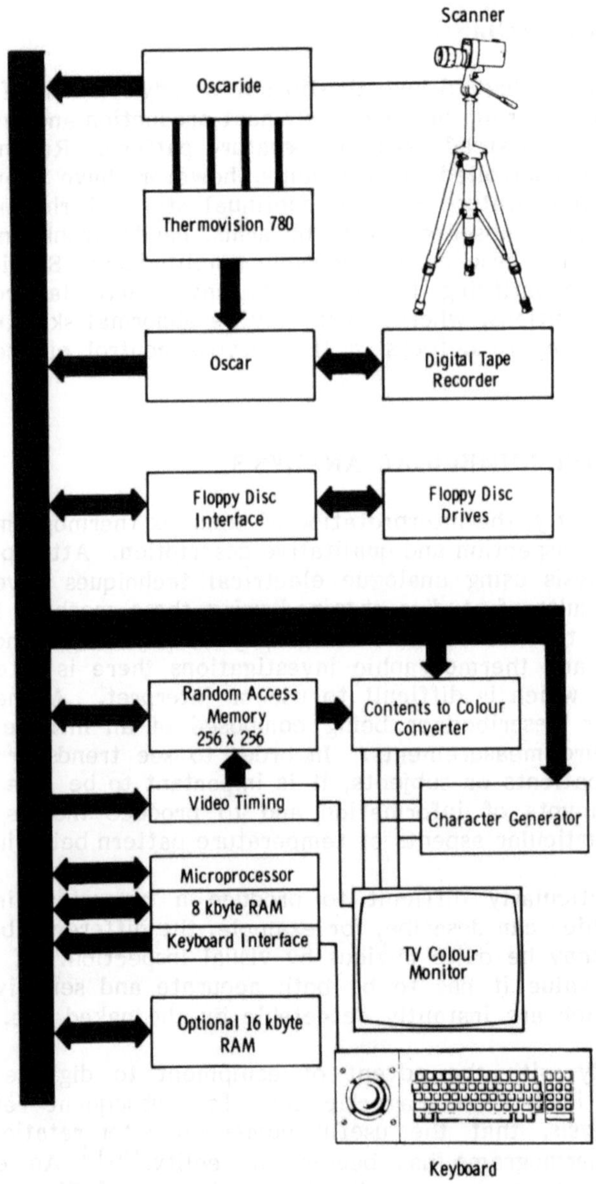

Fig. 4. Block diagram of the AGA Pericolor Computerized thermographic system.

its temperature, down to the smallest picture element of one pixel. Figure 5b shows how this facility may be extended to a particular area of interest over the hand which may be circumscribed (either manually by the operator or by instructing the computer cursor to follow a specific set of co-ordinates). This analysis shows the total circumscribed area in pixels together with the mean temperature of the enclosed area, which in this example is 33.8°C. Such features are particularly valuable in clinical assessment when, for example, the effect of specific drugs on joint inflammation at weekly or monthly intervals may be investigated. Figure 6 shows an example of another type of numerical analysis which may be useful in identifying particular temperature patterns or quantifying their change with time. This example shows a histogram of the relative frequency of the different temperature levels found over the back of a subject.

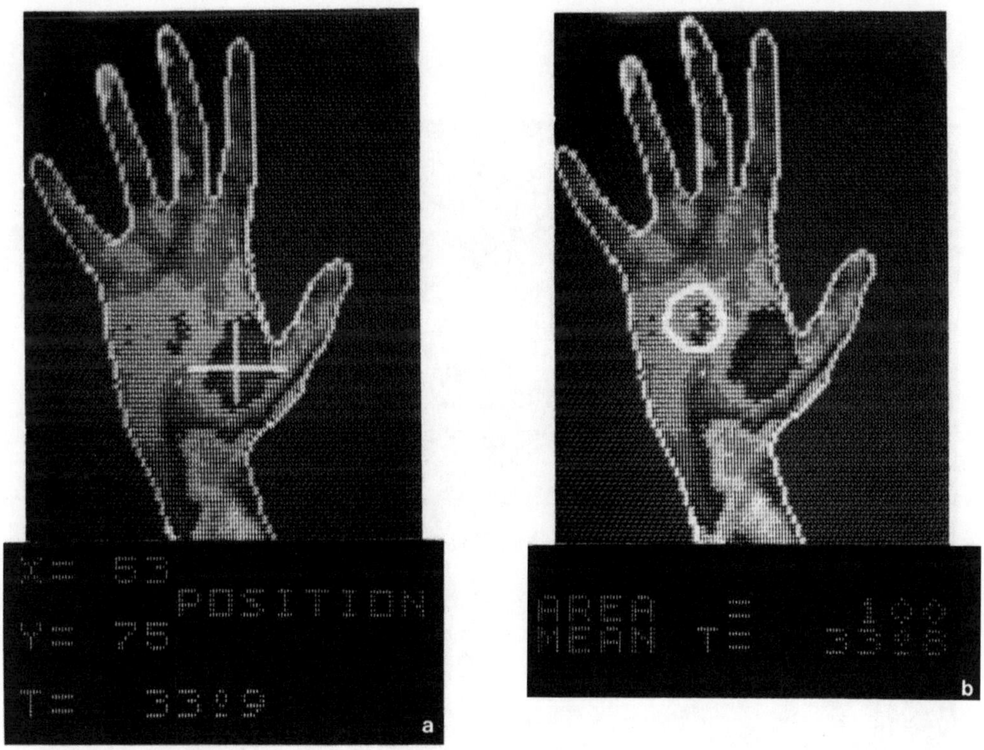

Fig. 5. Computer analysis of thermograms can produce spot temperatures at defined positions for single pixel areas (a) and mean temperatures of circumscribed areas of interest (b).

CONCLUSIONS

Quantitive infrared thermography is likely to find an increasingly important place in human thermal physiology and clinical diagnosis and assessment, and can be expected to provide considerable advances in our understanding of the role of temperature in relation to health and disease. It is most important that pathological studies are complemented by investigations of "normal" healthy controls; this is an area which has often been neglected in the past.

In any thermographic investigation it is vital to understand the physiological and physical heat transfer mechanisms involved, and to appreciate the state of the cutaneous vasculature relevant to the patterns that are being observed. In addition, the influence of environmental conditions must be fully appreciated when preparing investigative protocols and interpreting the results of thermographic analysis.

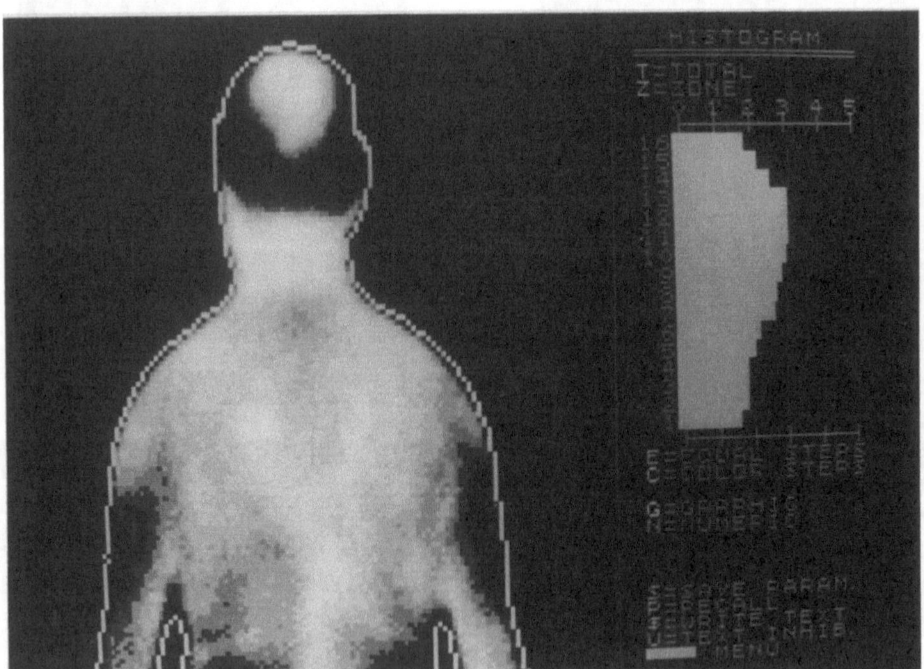

Fig. 6. Temperature/area frequency distribution of thermal images can be presented numerically or graphically using computerized thermographic equipment.

REFERENCES

1. in: "Family Physician," By physicians and surgeons of the principal London hospitals, Cassel, Petter, Galpin and Co. (1882).
2. D. McK. Kerslake, in: "The stress of hot environments," Monograph of the Physiological Society No. 29, Cambridge University Press (1972).
3. Y. Houdas, and E. F. J. Ring, "Human Body Temperature. Its Measurement and Regulation," Plenum Publishing Corporation, (1982).
4. M. Nielsen, Die regulation der Karper temperatur bi Muskelarbeit, Scand. Arch. Physiol. 79:193-230 (1938).
5. R. P. Clark, Human skin temperature and convective heat loss in: "Bioengineering, Thermal Physiology and Comfort," K. Cena and J. A. Clark ed., Elsevier Scientific Publishing Co., (1981).
6. W. M. Park, and B. L. Reece, Fundamental aspects of medical thermography. British Institute of Radiology - Teaching Booklet No. 3. ISSN 0306-8862 (1976).
7. R. P. Clark, and J. K. Stothers, Neonatal skin temperature distribution using infrared color thermography, J. Physiol. 302:323-333 (1980).
8. R. P. Clark, M. R. Goff and J. S. Garrow, Poster presentation at International Conference on the adipocyte and obesity; cellular and molecular mechanisms. June, 1982. Toronto University. In Press.
9. R. P. Clark, B. J. Mullan, and L. G. C. E. Pugh, Skin temperatures during running - a study using infrared color thermography, J. Physiol. 267:1:53-62 (1977).
10. R. P. Clark, M. R. Goff and B. J. Mullan, Skin temperatures during sunbathing and some observations on the effect of hot and cold drinks on these temperatures, J. Physiol. 267:1:8-9 (1977).
11. R. P. Clark, The effect of hot and cold drinks on skin surface temperature distributions, Proceedings of the IXth International Colloquium on Coffee - London. June 16-20. Vol. 11, p.405-411 (1980).
12. R. H. Fox and S. M. Hilton Bradykinin formation in human skin as a factor in heat vasodilatation, J. Physiol. 142:219-232 (1958).
13. R. H. Fox, R. Goldsmith and D. J. Kidd, The cutaneous vasomotor control in the human nose, lip and chin, J. Physiol. 150:22-23 (1960)..
14. R. P. Clark and M. R. Goff, Human skin temperature during rest and sleep visualized with color infrared thermography, J. Physiol. 300:14-15 (1979)..
15. E. F. J. Ring, Computerized thermography for osteo-articular diseases, Acta Thermographica 1:166 (1976).
16. R. P. Clark, M. R. Goff and J. K. Stothers, Microprocessor image analysis of neonatal thermograms, J. Physiol. 352:4 (1982).

THE SPECIAL THERMAL PHYSIOLOGY OF NEWBORN INFANTS

J K Stothers

The London Hospital Medical College
Turner Street, London E1
UNITED KINGDOM

As long ago as 1900 Budin suggested that keeping babies warm decreased infant mortality. However, it was not until some 60 years later that clinicians began to revise their attitudes towards the thermal care of newborn infants. This was possibly because of the recognition of the "neonatal cold injury" syndrome at that time. Prior to this, babies had been regarded as either miniature adults or, as Father Roland Knox described them, "merely a loud noise at one end and absolutely no sense of responsibility at the other", more scientifically as poikilotherms.

Bruck[1] showed that babies, even when born prematurely, did have vasomotor control, that they responded to cold stress by increasing their oxygen consumption and also when hot were able to sweat. Even more detailed work by Hey,[2] using the closed circuit metabolism chamber described by Hill and Rahimtulla,[3] more accurately quantified these responses. He showed that the infant was indeed a homeotherm but differed both qualitatively and quantitatively from the adult.

One reason for the difference is the relative weight to surface area ratio. In the adult the "equivalent" surface area of 1 kg is approximately 0.02 m^2. For the term infant this figure is 0.06 m^2 and for a 30 week pre-term baby is 0.1 m^2. The pre-term infant is therefore thermally disadvantaged by a factor of 5 when compared to the adult.

The tissue insulation of the infant is lower than that of the adult as is the resting metabolic rate. In addition, adults can generate heat by shivering; infants cannot shiver and it is as yet unknown at which age this faculty develops.

It has already been stated that the infant, when placed in a cool environment, can produce the extra heat required to maintain body temperature, albeit within a rather narrower range than the adult. This phenomenon of non-shivering thermogenesis was originally demonstrated in the rabbit by Dawkins and Hull[4] and was found to be due to the presence of multi-locular fat cells at various sites. These cells are popularly called 'brown fat cells' and have been shown to be sympathetically mediated. This has recently been confirmed in the human neonate[5] (Fig 1). Interestingly, the dosage of β-blocking drug required for the human infant appears to be considerably less than that required for a similar effect to be demonstrated in the newborn rabbit.

The study of fat stores in normal infants first brought thermography into the neonatal physiological laboratory. Having previously attempted to demonstrate skin temperature change using cholesteric crystals without success, Bolton, Fox and Kinneard[6] used an AGA gray-tone Thermovision System to visualize the backs of cold-stressed infants. In this study an interscapular warm zone was detected in some infants. In contrast to some more recent reports this was not observed in the adults which they studied. However, there were other warm areas of the back which they suggested were not as easily interpreted as evidence of underlying brown fat. In 1972, Rylander, Pryblova and Lind[7] also examined the dorsal surfaces of human newborn infants using thermography. They reported that most of the 43 infants that they examined showed warm areas in the nape of the neck and interscapular region. In addition, an aspiration biopsy from a 'warm' area was performed in 10 of these infants. Multi-locular fat cells were found in 6 of these biopsies. Such a procedure we would consider to be entirely unethical.

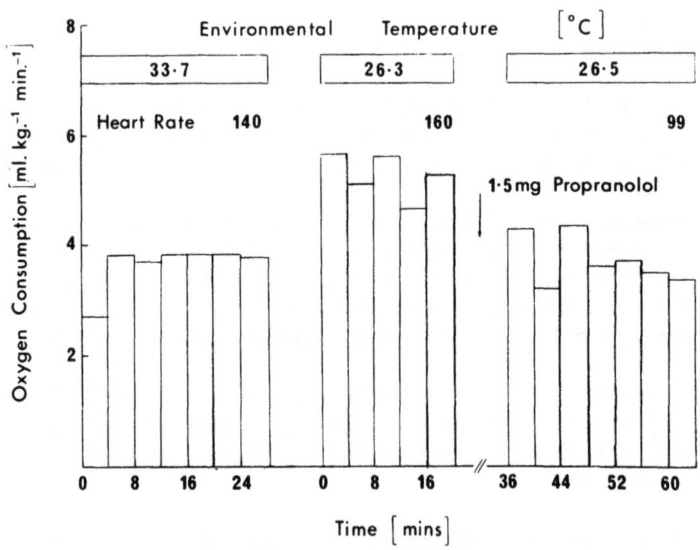

Fig. 1. O_2 consumption at 2 environmental temperatures and the effect of propranolol in reducing O_2 consumption.

These inconclusive results did not deter other research workers from using infrared thermography to investigate neonatal surface temperature. In our own laboratory, Cross et al[8] applied the technique to another basic physiological problem - that of neonatal brain energy metabolism.

Earlier work in the same laboratory had resulted in an estimate being made of the neonatal brain oxygen consumption. This estimate was based on a difference between aural and esophageal temperatures of some 0.15°C. If one assumes the aural temperature to be representative of the "mean" brain temperature and the esophageal temperature as an index of the aortic (carotid) blood temperature, then using the principles of flow calorimetry it was calculated that the infant's brain utilized some 70% of the total oxygen consumed at rest. In the calculations a cerebral blood flow of 100 ml/100 g brain/min[9] was used. In the light of more recent investigations, this value is probably rather high.

It was also noted that the infant's aural temperature as measured by the zero-gradient aural thermometer rose post-prandially. Bearing in mind the already high proportion of oxygen consumption of the brain it was thought unlikely that the brain itself was producing the extra heat merely by metabolism - an increase of some 30% would be required to produce the observed 0.2°C increase in brain temperature. A more likely explanation was that the cerebral blood flow was reduced, thus diminishing the cooling effect of the inflowing arterial blood. Dear,[10] using jugular occlusion plethysmography which is an ethically acceptable method of measuring cranial blood flow (CBF) in the normal neonate, confirmed that the infant's CBF did fall post-prandially. The skull, being thin, offers little insulation and it was felt that this temperature rise might be detectable on the skin surface using infrared thermography.

A total of 7 infants were studied; 6 normal and 1 hydronencephalic. They were nursed in a modified incubator (Vickers Medical Model 79) during the observation. Details of these infants are shown in Table 1. A hole was cut in one end of the incubator canopy to accommodate an open-ended perspex cylinder which was placed between the camera and the infant's head to minimize environmental temperature variations. This procedure did not affect the thermal stability of the incubator. In all cases the vertex temperature rose following feeding; the magnitude of the rise varied from 0.6 - 0.8°C. The time from the end of feeding to the time when the maximum response was observed varied from 25-45 minutes. Pre-feed temperatures were usually re-established after 60-80 minutes. However, this change in vertex temperature could also have been due to an increase in skin blood flow. The opportunity to study the effects of feeding on skin blood flow arose when a hydronencephalic infant was born at the London Hospital (Baby N). Pre-feed vertex temperature of this infant was some 2°C lower than that of the healthy infants at similar environmental temperatures. This difference in temperature was probably related directly to the absence of a heat source, that is the brain, in the abnormal baby. In this infant the response of feeding was also somewhat different. The temperature rose within 15

Table 1. Details of the infants involved in the study, including the hydronencephalic (baby N)

Baby	Wt (kg)	Age (days)	Max rise in temp (°C)	Time of max rise (mins post prand)
Normal				
K	2.4	7	0.6	35
P	1.4	10	0.8	40
Ph	2.1	4	0.8	40
B	2.44	2	0.6	35
R-P	3.3	6	0.8	35
D	1.98	13	0.6	40
Abnormal				
N	2.75	5	0.6	15

minutes from the end of the feed and fell rapidly to the pre-feed level within 45 minutes.

It was therefore suggested that the presence of such a potent heat source in the normal infant masked the transient alteration in skin blood flow observed in the abnormal baby.

Apart from these special studies, little detailed information existed about the skin temperature distribution and its variation in the neonate or the surface response to cool environments until the observations reported by Clark and Stothers.[11] In summary, these showed that surface temperature distributions of infants within the first seven days of life are similar in equivalent environmental conditions, although fluctuations in individual patterns appear to occur and are at times quite marked.

These observations were extended by calculating mean skin temperatures using thermographic techniques. The results were compared with calculated mean temperatures, using the more traditional method of multiple spot measurements. The readings were obtained from several sites using either a thermistor or thermocouple thermometer. A knowledge of the mean skin temperature for the whole body, together with the operative environmental temperature and the air movement, enables heat transfer between an infant and the surroundings to be evaluated. Apart from the physiological aspect of this concept it is also of practical clinical importance. Ill babies are frequently nursed in incubators, where the environmental temperature within the canopy is servo-controlled from a temperature sensor sited at a rather arbitrary point on the skin. Since the work of Agate and Silverman in 1963,[12] this point has been "on the abdomen". However, the measurements made using thermography and skin surface temperatures show that this site is higher in temperature than the overall mean temperature of the skin and also that its response to a cooling environment is less than the mean skin temperature itself (Fig 2).

This is not surprising, for below the skin within the abdominal-thoracic cavity there is another potent heat source, the liver. Although not as active in the neonate as the adult, the liver still utilises a large quantity of oxygen. A further point, possibly even more important than the last, is that nursing practices have changed in recent years. At the time of Agate's work, babies were, in general, nursed supine. It is now common practice to nurse the baby prone. If the infant is nursed in the prone position with an abdominal probe attached it will be totally insulated from the incubator environment, and will not respond to air temperature fluctuations. It was found that the nearest approach of a single area to the mean skin temperature was over the insertion of the deltoid in the upper arm. The difference between the mean skin temperature as measured using thermography and that found using weighted thermocouple measurements is shown in Table 2.

Numerical analysis of whole body thermal images obtained in this study was originally very time consuming and dependent on color transparencies from selected frames of a thermographic cine film which were projected on to a screen. By tracing around different colored areas, cutting the areas from card, and weighing them, a mean skin temperature could be obtained. Obviously, only a limited quantity of data could be fully analysed in this way. Computerised thermography has provided

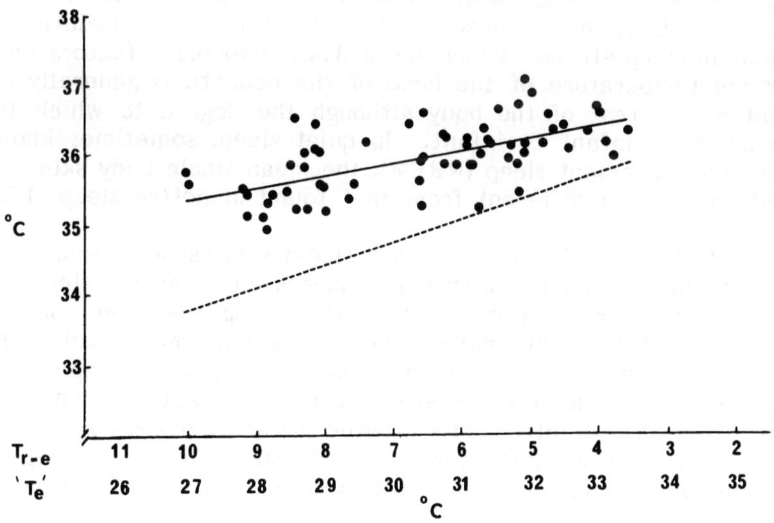

Fig. 2. Linear regression of abdominal temperature. Variation of abdominal skin temperature in a cooling environment.

Table 2. Difference between the mean skin temperature as measured using thermography and that found using weighted thermocouple measurements

T_e (°C)	T_r (°C)	$\bar{T}_s(t)$ (a)	$\bar{T}_s(p)$ (b)	a-b
28.7	36.9	34.2	33.8	0.4
31.7	37.0	34.5	34.8	0.3
29.0	37.3	33.3	33.8	0.5
31.7	36.8	34.0	34.2	0.2
31.2	37.1	33.9	34.2	0.3
32.6	37.2	34.8	34.8	0.0
33.6	37.3	35.7	36.0	0.3

T_e, operative environmental temperature; T_r rectal temperature; $\bar{T}_s(t)$ mean skin temperature obtained from thermograms; $\bar{T}_s(p)$ mean skin temperature obtained from multiple 'spot' measurements.

the storage of digitised data on magnetic tape for subsequent analysis using a microprocessor. This enables dynamic surface temperatures and their changes to be quantified, and related to underlying physiological variables.

Although such studies are in the early stages of development, variations in skin temperature, probably representative of changes in vasomotor control, have already been observed in association with alterations in sleep states. From the analysis, two basic factors emerge. The surface temperature of the head of the neonate is generally higher than that of the rest of the body although the degree to which this is true varies from infant to infant. In quiet sleep, sometimes known as non-rapid eye movement sleep (NREM), the mean whole body skin surface temperature is little different from that found in active sleep (REM).

As yet, the significance of these findings is rather obscure. It is known that the oxygen consumption is increased in active sleep. With the skin surface temperature in both states being the same, one would expect that the total body heat content, or heat storage, would increase in that sleep state. This is not the case. One possible explanation is that the water loss at the surface is greater in active than in quiet sleep[13,14] (Fig 3). The effect of increased surface evaporative loss would tend to obscure more subtle vasomotor changes. In quiet sleep, when the surface evaporative loss is minimal these changes would become more apparent.

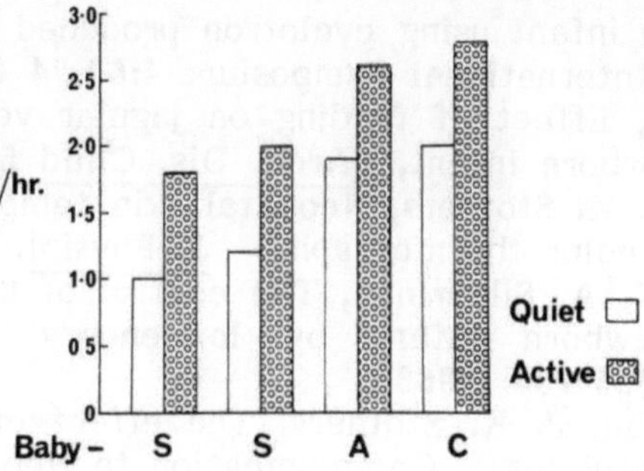

Fig. 3. Weight loss g/hour. Active and quiet sleep. O_2 consumption in quiet and active sleep for 3 neonates.

Thus, already, the application of thermography has demonstrated several phenomena previously undescribed and certainly incompletely understood. This is just one example of the fact that proper application of such thermographic techniques will enable the interaction of the neonate and the microenvironment to be non-invasively investigated. This has not been possible in the past.

REFERENCES

1. K. Bruck, Temperature regulation in the newborn infant, Biol. Neonate 3:65-119 (1961).
2. E. N. Hey, The relation between environmental temperature and oxygen consumption in the newborn baby, J. Physiol. 200:589-603 (1969).
3. J. R. Hill, K. A. Rahimtulla, Heat balance and the metabolic rate of newborn babies in relation to environmental temperature; and the effect of age and of weight on basal metabolic rate, J. Physiol. 180:239-263 (1965).
4. M. J. R. Dawkins, D. Hull, Brown adipose tissue and the response of newborn rabbits to cold, J. Physiol. 172:216-238 (1964).
5. K. W. Cross, R. M. Warner, β-blockage and non-shivering thermogenesis in the newborn infant, J. Physiol. 313:18 (1981).
6. D. D. G. Bolton, A. M. Fox, D. L. Kennaird, Preliminary observations on the application of thermography to the study of brown adipose tissue in the human newborn, J. Physiol. 208:23-24 (1970).
7. E. Rylander, H. Pribylova, J. Lind, A thermographic study of infants exposed to the cold, Acta Pediat. Scand. 61:42-48 (1972).
8. K. W. Cross, J. K. Stothers, R. M. Warner, R. W. Woodruff, The application of thermography to the detection of energy metabolism of the brain in the newborn infant, J. Physiol. 252:44-45 (1975).
9. R. N. Arnot, H. I. Glass, J. C. Clark, J. A. Davis, D. Schiff, C. G. Picton-Warlow, Methods of measurement of cerebral blood flow

in the newborn infant using cyclotron produced isotopes, Vortrage am Gasteiner International Symposium 4:60-74 (1970).
10. P. R. F. Dear, Effect of feeding on jugular venous blood flow in the normal newborn infant, Arch. Dis. Child 55:365-370 (1980).
11. R. P. Clark, J. K. Stothers, Neonatal skin temperature distribution using infrared color thermography, J. Physiol. 302:323-333 (1980).
12. F. J. Agate, W. A. Silverman, The control of body temperature in the small newborn infant by low-energy infrared radiation, Pediatrics 31:725-733 (1963).
13. M. H. Thompson, J. K. Stothers, The effect of forced convection on neonatal water loss, Communication to Neonatal Soc. 3rd. July 1981.
14. L. Curzi-Dascalova, C. Dreyfus-Brisac, Distribution of skin potential responses according to states of sleep during the first months of life in human babies, Electroencephalography and Clinical Neurophysiology 41:399-407 (1976).

SKIN TEMPERATURE CHANGES AND THEIR SIGNIFICANCE IN THE HUMAN NEWBORN

J K Stothers*, R P Clark, M R Goff, S J Linnett*, A J Livingston*

Clinical Research Centre, Harrow, and
*** The London Hospital Medical College**
UNITED KINGDOM

Infrared thermography has, for many years, been used to study skin surface temperature and its distribution in both healthy and sick humans. Until recently, such studies were dependent on "single shot" photographic comparisons taken after varying periods of time had elapsed in order to achieve some degree of equilibration between the subject and the environment. These limitations were partially overcome in a previous investigation of skin temperatures of infants nursed in incubators.[1] The technique used involved continuously recording thermographic pictures on 16 mm cine color film (at a frame rate of 8/sec) with a camera electronically synchronized to make it compatible with the thermographic imaging system which, at that time, was scanning at 16 fps. This was set up to avoid "roll over" and other synchronization difficulties. The infant incubator had to be modified, as infrared radiation transmission is attenuated through perspex panels. Figure 1 shows the modification. An aperture is cut in the incubator roof and a surface silvered mirror angled at 45° to reflect the thermal image of the infant to the infrared scanner. In addition, to obtain adequate magnification and thermal resolution, it was necessary to "pan" the scanner across the image of the infant. Selected frames from the cine film were then identified and photographed to produce a color transparency. By joining three transparencies of the film in this way a thermogram of the whole infant was constructed. This picture was then projected on to a white card screen and the different colored areas, representing different temperatures, were traced around, cut out and weighed (Fig 2). The process was difficult and time-consuming, and numerical results could only be obtained from a limited number of images. Sequential studies over a moderate length of time were also impracticable.

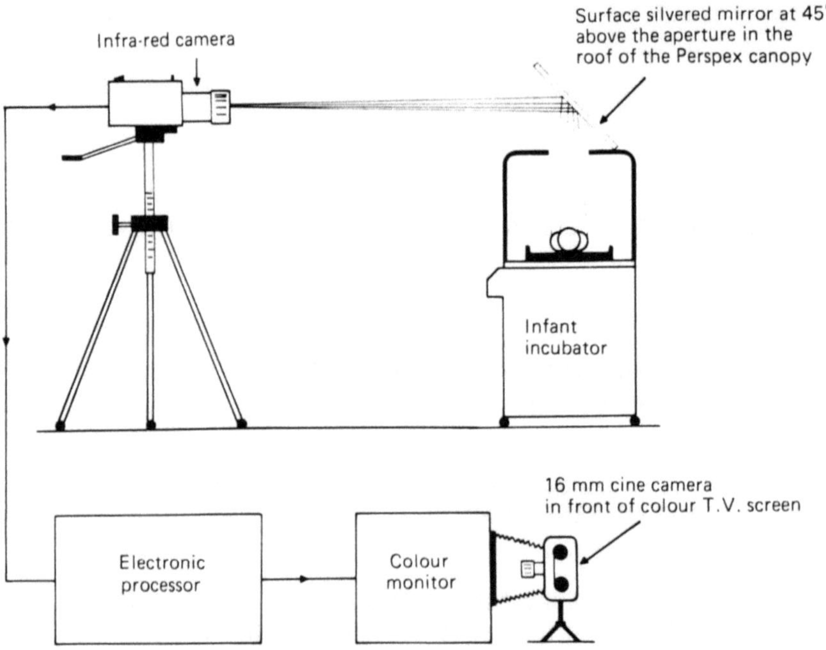

Fig. 1. The arrangement of the thermographic camera and associated equipment and film camera for determining neonatal skin temperature.

With the advent of computerized thermography, it is now possible to digitize and store thermograms, and to perform complex numerical analyses. This system is being applied to the study of dynamic surface temperature changes in infants. Different sleep states in infants are being investigated while they are nursed in thermo-neutrality, in temperature controlled incubators. These investigations can cast some light on the normal variation of neonatal vasomotor control. This has only previously been possible using some form of segmental occlusion plethysmography[2] or hand plethysmography.[3] Both of these techniques are mildly intrusive, and interfere with minor vasomotor changes. However, the changes measured could be representative of those found in other areas of the body, which are unsuitable for either plethysmographic technique.

In the neonate, as in the adult, there are two clearly defined sleep states. Active sleep occurs when movements, respiration and heart rate are irregular; this is sometimes known as rapid eye movement (REM) sleep. Quiet sleep occurs when the child's respiration and heart rate are regular but there are few, if any, body movements. It is known that oxygen consumption is raised in active sleep compared with quiet sleep, and consequently heat production is also increased (Fig 3). The fact that deep body temperature shows little change between the states suggests

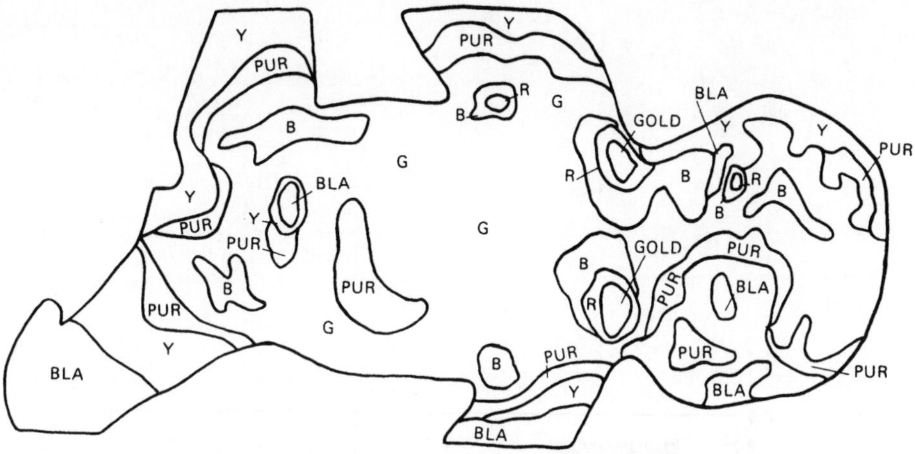

Fig. 2. A tracing from a typical neonatal thermogram showing each isothermal area outlined and labelled. The areas were subsequently cut out from card, and weighed to enable average skin temperatures to be calculated. Y-yellow, G-green, BLA-black, PUR-purple.

that an increased heat loss should be present in active sleep. It has been previously thought that this increased heat loss was via the dry "partition" that is either by radiation, convection or conduction. Sequential studies using the AGA Pericolor Thermovision system have shown that the mean overall surface temperature is very similar in both sleep states. It is therefore unlikely that the radiative component of the dry loss is increased. As the conditions of air temperature and air movement are the same within the controlled environment of the incubator, convective losses are also unlikely to be altered. There is evidence that the increase in heat loss is due to the evaporative route, and is not solely confined to heat loss from the respiratory tract. Why this should be the case is not at all clear; in the adult the reverse is true.[4] In active sleep evaporative loss is much diminished.

Infrared thermography can be expected to cast light on neonatal vasomotor control systems and the factors which alter them. There is also a very practical application of the same technique. It has been postulated[5] that the cerebral blood flow in the full-term neonate falls after feeding. Obviously the cardiac reserve, although much less than that of the adult, allows this to occur without embarrassment to brain oxygenation. In the pre-term infant, however, the cardiac reserve is likely to be minimal. The question to be asked is, to which region or organ is the blood flow diverted? It is probable that the splanchnic flow is increased but there is some variance of opinion on the changes in skin blood flow. Computer analyzed infrared thermography appears to be a most promising tool in this field of investigation.

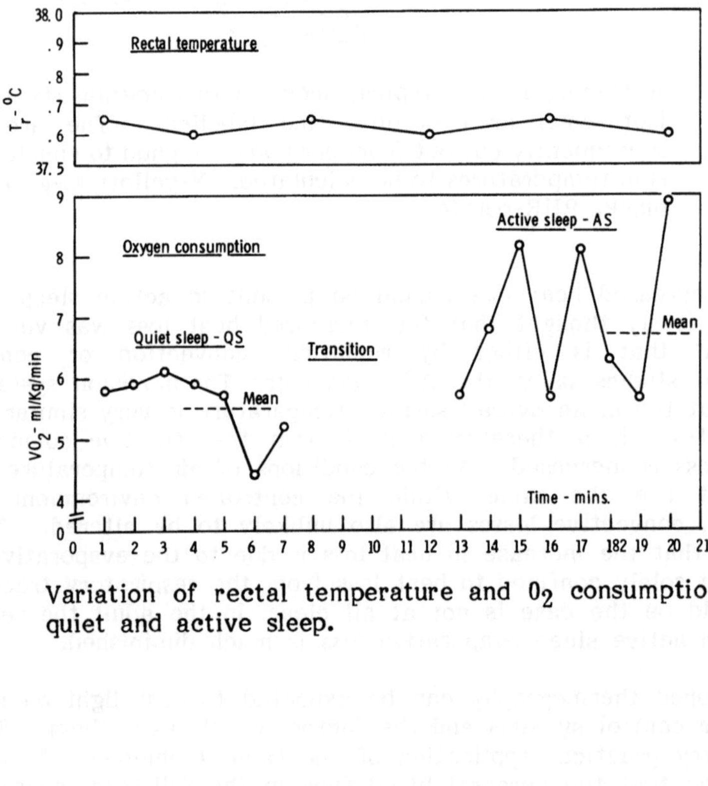

Fig. 3. Variation of rectal temperature and 0_2 consumption during quiet and active sleep.

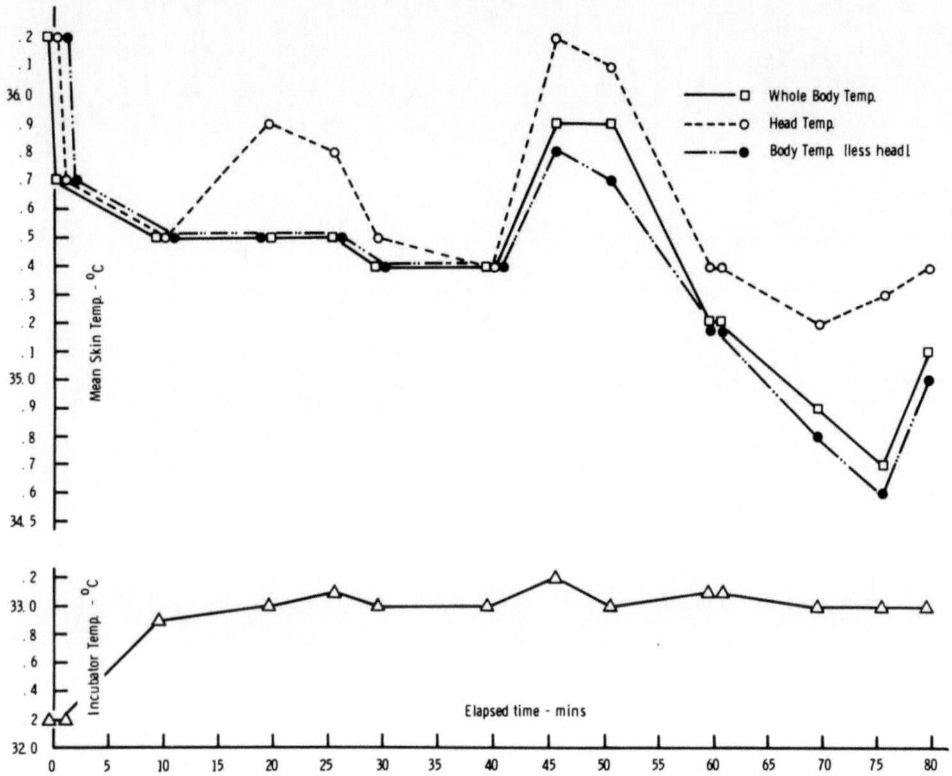

Fig. 4

REFERENCES

1. R. P. Clark, J. K. Stothers, Neonatal skin temperature distribution using infrared color thermography, J. Physiol. 302:323-333 (1980).
2. O. Celander, K. Morild, Reactive hyperemia in the foot and calf of the newborn infant, Acta Pediat. Stockholm 51:544-552 (1962).
3. E. N. Hey, G. Katz, The range of thermal insulation in the tissues of the newborn baby, J. Physiol. 207:667-681 (1970).
4. R. Henane, A. Buguet, B. Roussel, J. Bittel, Variations in evaporation and body temperatures during sleep in man, J. Appl. Physiol. 42:1:50-55 (1977).
5. P. F. R. Dear, Effect of feeding on jugular venous blood flow in the normal newborn infant, Arch. Dis. Child 55:365-370 (1980).

CHANGES IN THERMAL PHYSIOLOGY WITH ADVANCING AGE

K J Collins

Medical Research Council, School of Medicine, University College, London and Department of Geriatric Medicine St. Pancras Hospital, London, UNITED KINGDOM

The functional capacity of the thermoregulatory system can be shown to alter with ageing and this is believed to underly the reduced resistance of elderly people to excessively cold or warm ambient temperatures.[1] The lability of thermal homeostasis appears to be an important factor in the increased incidence of cases of urban hypothermia during the winter in the UK[2] and the summer epidemics of heat stroke in the elderly in cities in the United States.[3] There is, however, little detailed knowledge of thermoregulatory impairment as a concomitant of the ageing process. Investigations have therefore been made of the control and function of different aspects of the thermoregulatory system in healthy elderly as compared with young individuals.

The finely adjusted neural and endocrine control of body temperature, together with behavioral responses and longer-term adaptive changes, normally provide efficient temperature homeostasis in man. Two main types of disturbance, internal and external, are the usual cause of thermal lability. Internal factors arise from disorders of the thermoregulatory system due to primary dysfunction or as a secondary effect resulting from disease or from pharmacological interference. External causes stem from the environment when adverse temperature conditions are so severe or prolonged that physiological defence mechanisms are overwhelmed. Internal and external factors are additive, such that a moderately cold environment, for example one which is normally easily tolerated by a healthy young individual, could be too severe for an elderly patient to contend with and lead eventually to a fall in core temperature. One general feature of the observed changes in thermo-regulatory capacity in old people needs to be emphasized. Standard tests of thermoregulatory function show that there is a considerable inter-subject variability and a markedly different rate of decline of thermoregulatory processes in different individuals with ageing.

SHIVERING THERMOGENESIS IN THE ELDERLY

When challenged with a cold environment, the metabolic response appears to be diminished in the elderly; shivering thermogenesis has even been reported to be absent in old age. Older subjects (age range 52-76 years) sitting unclothed for up to 30 minutes in a 10°C environment with minimal air movement were found not to shiver nor to complain of feeling cold, whereas younger subjects in the same conditions all shivered and felt uncomfortable.[4] In thermoregulatory function tests on 144 old people,[1,2] only 18 shivered during a cold stress lasting 15 minutes. Shivering did not occur in all the young control subjects either, but the proportion who did shiver was greater than in the elderly people. The tests were not intended to demonstrate whether or not shivering could be induced by prolonged exposure to cold. The importance of the shivering response is, however, illustrated in Figure 1, where the effects of the absence of shivering on core temperature are predicted by means of a computed model of thermoregulation.

More detailed investigations of the shivering responses have been made using a Body Cooling Unit[5] which was found to induce shivering within a few minutes by convective cooling in air at a temperature of 20°C moving at a velocity of 0.5 meters sec^{-1}. In 80-year old subjects, shivering, monitored by oxygen consumption, electromyography and accelerometry, was less intense than in young subjects. At the end of a 30-minute cooling phase, mean deep body temperature had fallen by

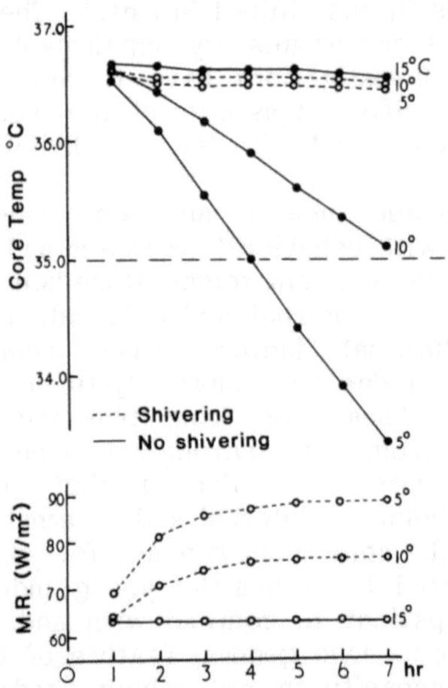

Fig. 1. Predicted changes in core temperature and metabolic rate with and without shivering in air temperatures of 15°C, 10°C and 5°C using the computed model of human thermoregulation described by Nishi and Gagge.[16] The data refer to a sedentary person wearing 1.0 clo of insulation.

0.14°C (± 0.09 SD) in 7 healthy elderly subjects compared to only 0.03°C (± 0.12) in 7 young control subjects. Three of the elderly and two of the young did not shiver at all, but the mean metabolic heat production increased by 14.4 Wm^{-2} (± 4.2) in the elderly and by 25.1 Wm^{-2} (± 22.9) in the young. Mean responses are shown in Table 1. The difference in metabolic response to cooling in these subjects did not appear to be related to body fat insulation; the highest increments in metabolic response per unit body weight in both age groups in fact occurred in those with the highest percentage of body fat.

VASOMOTOR RESPONSES

Hand blood-flow has been studied using venous occlusion volume plethysmography and a photoelectric pulsimeter during thermoregulatory tests in cold and warm ambient conditions. In the vasomotor zone of thermoregulation three types of response were observed. The normal response, that is, with rapid vasoconstriction on cooling and vasodilatation on warming, occurred in young and most elderly subjects. Some elderly people, however, did not constrict significantly during cooling. A third type can be distinguished by a lack of vasomotor response to both cooling and warming.

In longitudinal studies on a group of healthy elderly volunteers, the proportion showing non-constrictor responses to cooling increased during

Table 1. Mean respones (± SD) of elderly and young adult subjects during 30 min convective cooling in dry air at 20°C (air speed 0.5 m sec^{-1})

	Young (n=7)	Elderly (n=7)
Age (years)	26.4 ± 4.5	80.3 ± 4.0
Body fat (%)	15.0 ± 5.4	22.5 ± 4.7***
Aural temperature before cooling (°C)	36.76 ± 0.10	36.56 ± 0.39
Aural temperature change by end of cooling (°C)	-0.03 ± 0.12	-0.14 ± 0.09*
Skin temperature before cooling (°C)	30.65 ± 1.45	29.94 ± 1.78
Skin temperature change by the end of cooling (°C)	-3.92 ± 0.45	-4.17 ± 1.70
VO_2 before cooling (1 min^{-1})	0.21 ± 0.09	0.20 ± 0.05
VO_2 change during cooling (1 min^{-1})	+0.12 ± 0.12	+0.07 ± 0.02

(Unpaired t-test *** $p < 0.001$, * $p < 0.05$)

the course of a 4 year period[1] and this was even more marked after 8 years. The non-constrictor response was characterized by an increase in the mean half-time for maximum vasoconstriction, eg 6.9 ± 8.1 minutes in the elderly and 1.1 ± 0.7 minutes in the young (Fig 2), and a diminished intensity of vasoconstriction. If similar changes in conductance occur on the body surface as a whole, and evidence from earlier studies[1] suggest that this might be so, then heat losses from the skin surface of elderly people with non-constrictor responses would be significantly increased in cold conditions and physiological tolerance reduced.

Pulsimeter records from the index finger of young adults demonstrate transient bursts of vasoconstrictor activity. The vasoconstrictor bursts occur in a rhythmical fashion, unlikely to be attributable to random arousal stimuli. A striking similarity in timing of the vasoconstrictor bursts has been observed in neurograms from the median nerve in man,[6] which is consistent with a rhythmic central nervous generation of vasoconstrictor impulses. In young subjects in a neutral climate of 30°C such a vasoconstrictor rhythm was invariably present, increasing in frequency during cooling. Many of the elderly subjects, however, exhibited a much diminished rhythm and amplitude, though they responded to cooling by an overall increase in vasomotor tone.

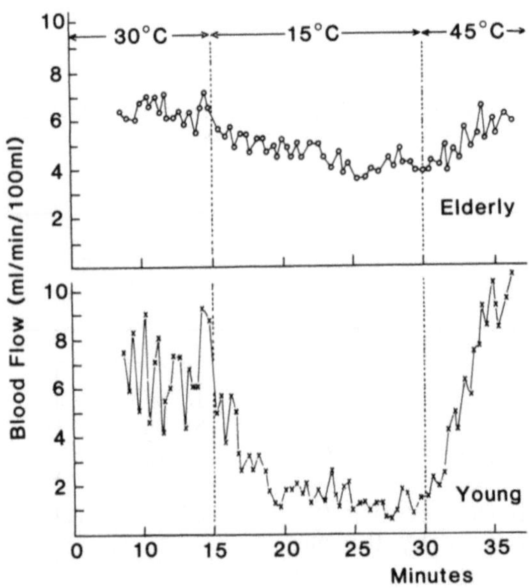

Fig. 2. Hand blood flow (ml/min/100 ml tissue) in a neutral environment (air temperature 30°C), during body cooling (air temperature 15°C) and rewarming (air temperature 45°) in an elderly (0) and young adult (x) subject.

BODY TEMPERATURE AFTER-DROP

Another manifestation of impaired vasomotor control in the elderly can be observed during rewarming after cold exposure.[7] It is well-recognized that core temperature continues to fall for a time after removal from immersion in cold water. The after-drop in core temperature can also be shown to occur during warming after convective surface cooling in air, and this is an important consideration in the rewarming and management of hypothermic patients. The long-held view that the after-drop in body temperature is caused by cold blood held in the peripheral blood vessels being returned to the core after the release of vasoconstrictor tone[8] has been challenged.[9] An alternative explanation, based on cold-water immersion studies, contends that the after-drop arises primarily from a latency in the reversal of conductive cooling in the tissues when rewarming commences. In our experiments investigating the effects of convective cooling in air at 20°C followed by rewarming in air at 40°C, two distinct forms of after-drop were observed. In elderly subjects with poor vasoconstrictor responses to cold the rate of fall of core temperature on rewarming was the same as that at the end of cooling. In those who shivered or possessed an effective peripheral vasomotor response to cold, the rate of fall of core temperature on rewarming was greater than that during cooling. Though the purely physical process of conductive cooling between the shell and core can be demonstrated in elderly subjects with poor vasomotor control, the evidence suggests that the redistribution of cold blood from the skin is also involved in producing the after-drop. Further experiments have demonstrated that peripheral vasoconstriction induced by lowering skin temperature during an established after-drop has the effect of rapidly arresting the fall in core temperature.

THERMAL PERCEPTION IN THE ELDERLY

Sensory systems in general become less efficient with age and there is some evidence that a number of elderly people are more insensitive to warm and cold stimulation as well as to other sensory modalities on the skin surface. Several observers have remarked on the apparent lack of discomfort experienced by old people in the cold in spite of the reduced ability to maintain a constant body temperature. It is possible that the tendency to accept low-temperature environments might be related to a change in threshold for cold discrimination. To study the relationship between thermal comfort, sensation and temperature perception, investigations have been made on healthy elderly people (68 to 87 years of age) and young adults, (a) in a series of static controlled environments (24°, 21°, 18°, 15° and 12°C)[10] and (b) using the technique of remote control of room temperature so that the individual is able to select his own preferred ambient conditions.[11] During these investigations thermal balance, thermal comfort and temperature discrimination thresholds were determined. About a third of the elderly subjects were found to have poor peripheral temperature discrimination and reported that they were more satisfied with the cold environments than the other elderly or young subjects. Signal-detection analysis suggested that although there were significant differences in perception thresholds, there were no differences in the criteria upon which the temperature discrimination decisions were based.

The evidence from tests of digital temperature perception performed under controlled conditions suggests that some elderly people are less sensitive to warm and cold stimulation at the skin surface (Fig 3). A decrease in blood supply to the skin tissues and in the density of functional nerve cells or thermoreceptors may contribute to these changes in peripheral perception in old age.[12]

THERMAL COMFORT AND BEHAVIOR

In laboratory experiments in the United States and Denmark[13,14] the neutral ambient temperature for thermal comfort was found to be the same for both young and elderly. This in some respects is surprising in view of the fact that temperature control mechanisms and thermal perception appear to deteriorate in a proportion of elderly people. It is argued[13] that since the lower level of metabolic heat production in the elderly is balanced by a reduced evaporative heat loss, the elderly do not necessarily prefer higher ambient temperatures than young adults. The tendency to assume that old people prefer warmer temperatures is nevertheless generally true, for it is an assumption based on the fact that most elderly are less active and therefore require more external

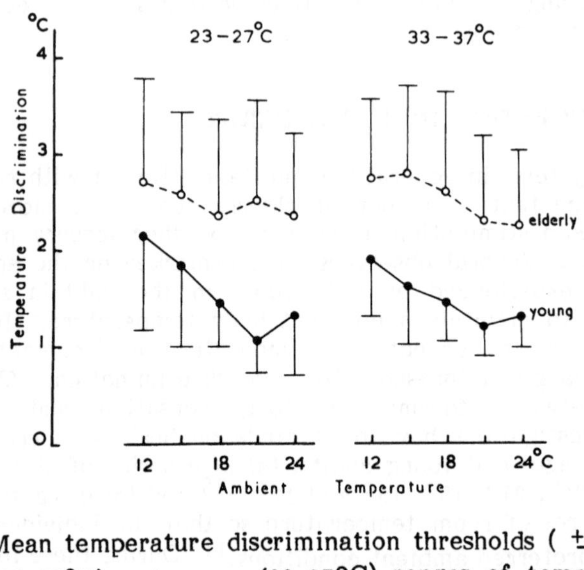

Fig. 3. Mean temperature discrimination thresholds (±S.D.) for cool (23-27°C) and warm (33-37°C) ranges of temperature in 16 young adult and 16 elderly subjects tested in different ambient temperature conditions (from Collins and Hoinville[10]).

heat to maintain thermal balance and comfort. Investigations of thermal comfort in controlled temperature conditions more appropriate to old people's housing in Great Britain[10] confirm that there is no statistical difference in the thermal comfort sensation experienced by the majority of healthy, young adult and elderly groups, given that activity and clothing insulation are the same. As indicated above, however, a small proportion of elderly people display a blunting of peripheral temperature perception together with an abnormal sense of thermal comfort. This observation is consistent with the findings of Watts[15] who concluded that elderly people were quite capable of experiencing cold or cold discomfort but some experienced cold only at a lower environmental temperature than might be expected.

Thermal discomfort may be regarded as providing arousal responses operating to condition thermoregulatory behavior. It may, furthermore, also have significant survival value in that it will stimulate behavioral mechanisms to promote increased heat gain or reduce heat loss in extremely cold conditions. Behaviorally, some old people are at a disadvantage, for it has been shown that, when given control over their environment, the elderly prefer the same mean ambient temperature as young adults but manipulate the temperature much less effectively.[11]

SUMMARY

Research into the biology of ageing has shown that there is a general decline in the efficiency of thermoregulation in old age. The range of individual variability in susceptibility to cold is wide but a proportion of apparently healthy elderly people show impairment of the ability to control body temperature in cold conditions. An appropriate diagnostic test to demonstrate thermal lability in the aged is difficult to devise, but a number of basic physiological observations may be taken to indicate the condition. These include low core temperature (urine or rectal temperature), an inability to increase metabolic heat production in the cold by shivering, an unusually small temperature gradient difference between deep body and skin temperature, poor circulatory responses to ambient temperature change with an after-drop in core temperature on rewarming which is not accelerated, and a blunted sense of temperature perception compared to normal.

ACKNOWLEDGEMENTS

The author wishes to acknowledge the support given by Professor A N Exton-Smith, Barlow Professor of Geriatric Medicine, School of Medicine, University College London, in whose department much of this research has been undertaken, and the scientific assistance of Dr J C Easton who is collaborating in these investigations.

REFERENCES

1. K. J. Collins, C. Dore, A. N. Exton-Smith, R. H. Fox, I. C. MacDonald, P. M. Woodward, Accidental hypothermia and impaired temperature homeostasis in the elderly, Br. Med. J. 1:353 (1977).
2. R. H. Fox, P. M. Woodward, A. N. Exton-Smith, M. F. Green, D. V. Donnison, M. H. Wicks, Body temperatures in the elderly: a national study of physiological, social and environmental conditions, Br. Med. J. 1:200 (1973).
3. F. P. Ellis, Mortality from heat illness and heat-aggravated illness in the United States, Envir. Res. 5:1 (1972).
4. S. M. Horvath, C. E. Radcliffe, B. K. Hutt, G. B. Spurr, Metabolic responses of old people to a cold environment, J. Appl. Physiol. 8:145 (1955).
5. K. J. Collins, J. C. Easton, A. N. Exton-Smith, Shivering thermogenesis and vasomotor responses with convective cooling in the elderly, J. Physiol. 320:76p. (1981).
6. G. Bini, K-E Hagbarth, P. Hynninen, B. G. Wallin, Thermoregulatory and rhythm-generating mechanisms governing the sudomotor and vasoconstrictor outflow in human cutaneous nerves, J. Physiol. 306:553 (1980).
7. K. J. Collins, J. C. Easton, A. N. Exton-Smith, Body temperature after drop; a physical or physiological phenomenon? J. Physiol. 382:72. (1982).
8. A. C. Burton, O. G. Edholm, in: "Man in a Cold Environment," Edward Arnold, London, p. 216 (1955).
9. F. St. C. Golden, G. R. Hervey, The mechanism of the after-drop following immersion hypothermia in pigs, J. Physiol. 272:26p. (1977).
10. K. J. Collins, E. Hoinville, Temperature requirements in old age, Build. Serv. Eng. 1:165 (1980).
11. K. J. Collins, A. N. Exton-Smith, C. Dore, Urban hypothermia: preferred temperature and thermal perception in old age, Br. Med. J. 282:175 (1981).
12. K. J. Collins, Hypothermia and thermal responsiveness in the elderly, in: "Indoor Climate," P. O. Fanger and O. Valbjorn eds., Danish Building Research Institute, Copenhagen (1979).
13. P. O. Fanger, "Thermal Comfort," McGraw-Hill, New York (1972).
14. F. H. Rohles, M. A. Johnson, Thermal comfort in the elderly, Trans ASHRAE 78:131 (1972).
15. A. J. Watts, Hypothermia in the aged: a study of the role of cold-sensitivity, Envir. Res. 5:119 (1972).
16. Y. Nishi, A. P. Gagge, Effective temperature scale for hypo and hyperbaric environments, Aviation Space and Env. Med. 48:97 (1977).

BODY COOLING AND THE TREATMENT OF HEAT STROKE

K J Collins, J C Easton, S D Rowlands and the late J S Weiner

Medical Research Council (Annexe)
242 Pentonville Road, London
UNITED KINGDOM

Uncontrolled hyperpyrexia, resulting from an imbalance between excessive heat gain and the inability of the thermoregulatory system to provide an equivalent rate of heat loss from the body, is a serious heat disorder which carries a high mortality.[1] Rapid cooling of the core is of utmost priority in treating heat stroke cases. Customarily, emergency treatment involves sponging the patient with cold water, sometimes ice-cold, and exposing the body surface to rapid movement of cool air. A useful degree of cooling, often life-saving, can be obtained by simply plunging the patient into a bath of cold water or applying ice packs.

Rapid cooling of the skin, by whatever means, is counter-productive in that induced vasoconstriction greatly diminishes the flow of blood and heat from the core to the periphery, while evaporation from the cold skin is simultaneously reduced. The need to avoid vasoconstriction during cooling treatment and to provide the means of managing violent, delirious and often incontinent hyperpyrexial patients has lead Weiner and Khogali[2] to devize a Body Cooling Unit (BCU) (Fig 1) for use in treating heat stroke among pilgrims to the Hajj at Mecca. The patient is first placed on a net support over a bath unit and strapped down if necessary. Cooling starts at once by spraying atomized water at 20°C and by blowing air at 50°C over the body surface. The skin temperature is thereby maintained above 32°C, vasoconstriction is prevented and the rate of evaporative cooling from the warm skin remains extremely rapid.

Investigations have been made of several different cooling procedures applied after exercise-induced hyperpyrexia when the subject's core temperature has been raised to 39.5°C. Rates of body cooling were compared in the following situations:

1. At rest in still air at 21°C dry bulb, 18°C wet bulb,

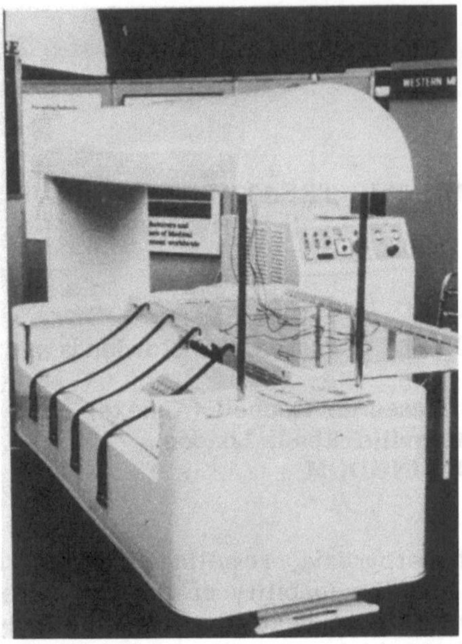

Fig. 1. The Body Cooling Unit consists of a cooling bed of moulded fiberglass with atomized water and air blown from ducts above and below the patient who is placed on netting on a sliding stretcher. A tank for the water supply is incorporated in the compressed air line above the bed. Heaters and fans for the warmed air supply are housed in a separate control console.

2. At rest on a mattress filled with water at 20°C,
3. Immersion in a stirred water bath at 15°C,
4. Spraying atomized water at 15°C in the BCU,
5. Spraying warm atomized water so that it reaches the body surface of the subject at temperatures ranging from 25°C to 33°C.

The three levels of warm air spray used were:

(a) Warm air at outlet 35°C reaching the subject at 25°C with an air movement of 0.4 m sec^{-1},
(b) Warm air at outlet 40°C with 30°C at the body surface and air movement of 0.4 m sec^{-1},
(c) Warm air at outlet 45°C with air spray reaching the body surface at 32-33°C and an air movement of 0.5 m sec^{-1}.

The mean cooling rates using these different procedures are shown in Table 1. The difference in rate of cooling by the warm air spray method with body surface temperature maintained above 30°C and that obtained by other cooling methods was statistically highly significant

Table 1. The rates of body cooling with different cooling procedures applied when core temperature has reached 39.5°C (From Weiner and Khogali[2])

Procedure		No of experiments	Time for 1°C fall in core temp (min)		Time for 2°C fall in core temp (min)		Mean cooling rate
			Mean	Range	Mean	Range	(°C min^{-1})
1. Still air	(21°C)	6	13.9	11.5 17.5	33.7	27.0 39.0	0.06
2. Water mattress	(20°C)	6	7.7	6.5 8.5	19.3	16.0 26.0	0.10
3. Water bath	(15°C)	3	5.2	4.0 6.5	18.4	14.0 22.0	0.11
4. Cold air spray	(15°C)	6	6.1	5.0 7.0	17.2	10.0 26.0	0.12
5. (a) Warm air spray	(25°C)	6	6.0	5.0 7.0	15.5	14.0 17.0	0.13
(b) " " "	(30°C)	5	5.3	3.5 6.5	11.4	7.5 13.0	0.18
(c) " " "	(32–33°C)	6	3.3	2.0 4.3	6.5	4.3 7.5	0.31

($p < 0.001$). Cooling by warm air spray with the body surface at 32–33°C compared with that at 30°C was also significantly different ($p < 0.001$).

These results show that immersion of hyperpyrexial patients in cold water does not result in a faster rate of body cooling than that produced by forced convective and evaporative cooling. The most efficient method of cooling involves warm air spray treatment. In tests in the field where the BCU was installed at treatment centers along the pilgrimage route at Mecca considerable success has been achieved in the treatment of heat stroke. Between October 29 and November 2 1979, 18 cases of severe heat stroke (with mean rectal temperature 42.3°C, range 41.2 to 43.1°C) were treated.[3] Sixteen of these patients successfully recovered and two, who were diabetic as well as hyperpyrexic, did not recover.

The BCU can also be used to induce body cooling by convection alone.[4] Investigations of thermogenesis using the BCU have shown that shivering can be induced within a few minutes in young adults by convective cooling in dry air at 20°C.[1]

REFERENCES

1. J. S. Weiner, K. J. Collins, Heat stress and heat stroke, in: "Hunter's Tropical Medicine," 6th edition. G. T. Strickland ed., Saunders, Philadelphia (1983).
2. J. S. Weiner, M. Khogali, A physiological body-cooling unit for treatment of heat stroke, Lancet 1:507 (1980).
3. M. Khogali, J. S. Weiner, Heat stroke: report on 18 cases, Lancet 2:276 (1980).
4. J. C. Collins, K. J. Collins, J. C. Easton, A. J. Hackett, S. D. Rowlands, J. S. Weiner, A bed unit for controlled rapid cooling, J. Physiol. 320:2 (1981).

1. The BCU heat stroke treatment system is manufactured by the Western Medical Group, 26 New Cavendish Street, London W1, UK

THE PHYSIOLOGY OF HEAT PRODUCTION

Y Houdas, G Carette

Laboratoire de Physiologie
Faculte de Medecine de Lille
FRANCE

THE ORIGIN OF THERMAL ENERGY

Life on earth is possible because a source of energy is available. This energy comes from the sun and is transferred by electro-magnetic radiations, mainly in the visible and infrared ranges.

Solar energy is directly used by vegetable cells for the synthesis of biological molecules, and this energy can be considered as stored in the molecules. When these molecules are eaten by an animal, this provides a supply of energy for that animal. In the body, the major part of this energy is released and used by the animal cells for various tasks, eg synthesis of new molecules, mechanical activity, etc.

The First Law of Thermodynamics states that the various forms of energy can be changed from one form to another one, but cannot be created nor destroyed. However, the First Law cannot predict the direction of a spontaneous energy transformation. This requires the Second Law of Thermodynamics which states that all energy transformations can occur spontaneously only in one direction, that is toward the greatest entropy. The concept of entropy is difficult to define clearly, but there is a similarity between this concept and those of uniformity and randomization. Therefore, energy is spontaneously always tending towards degradation into the ultimate random motion we call heat.

For these reasons, life first appears to contradict the Second Law of Thermodynamics because living organisms are able to organize atoms into new molecules, which is opposite to that of degradation. However, this contradiction is only apparent because a living organism is not, physically speaking, a closed system, since it requires an external supply of energy, obtained from food.

In acccordance with the Second Law of Thermodynamics, only a small part of the energy (at maximum 25%) can be transformed into "superior forms" of energy by the body. The major part, at least 75%, appears as thermal energy.

USE OF ENERGY

Biological molecules can be divided into two groups, according to the amount of heat that can be given up: below 5 Kcal (21 KiloJoules per mole) for the first group, above this figure for the second one.

As shown by Table 1, one group contains mainly the peptide and phosphoric ester bindings, and corresponds to the glucids and proteins.

In the body, the energy supplied by these compounds can be stored in molecules with "high level" energy, such as the molecules of ADP (adenosine-di-phospate) and, especially ATP (adenosine tri-phosphate).

These transfers of energy are made by a series of chemical reactions that can be divided into three stages (Fig 1).

First Stage

The molecules of the three groups of foods, namely, the glucids, lipids and proteins, are hydrolized by specific reactions.

However, the final chemical compound is the same for the three groups. This is the acetyl-coenzyme A (ac-CoA).

Second Stage

The acetyl-coenzyme A is then oxidized in carbon-dioxide in the

Table 1. Free energy of some important compounds (in KiloJoules)

Low Level	Glucose-1-Phosphate	-20.9
	Glucose-6-Phosphate	-13.8
	Glycerol-1-Phosphate	-9.2
High Level	1.3-Diphosphoglycerate	-49.3
	Creatine-Phosphate	-43.0
	ATP — ADP	-30.1
	ADP — AMP	-30.1

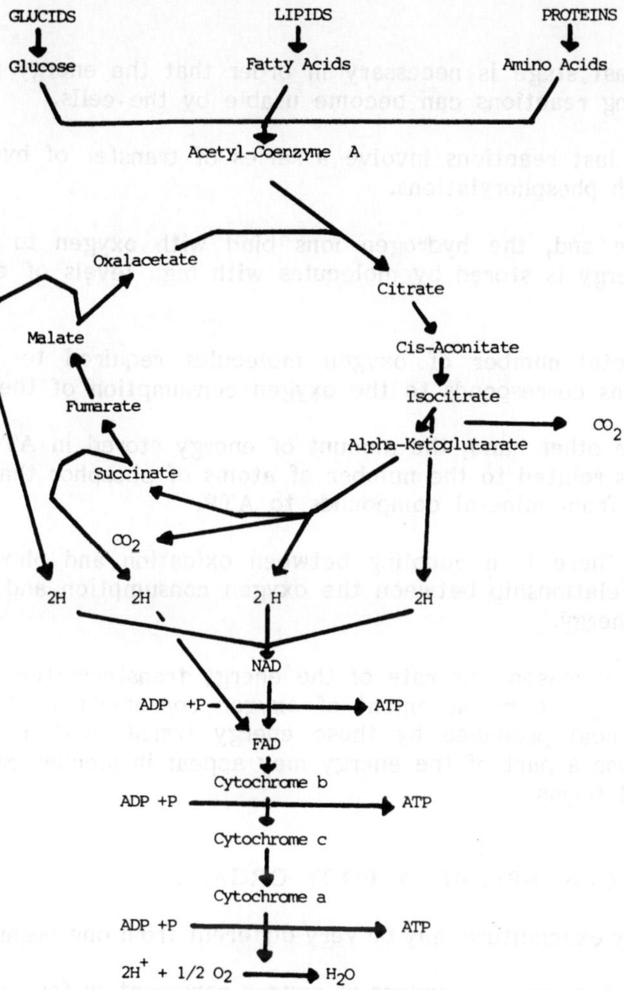

Fig. 1. Metabolic energy transfer pathways.

Kreb's cycle. This oxidation is coupled with the reduction of other coenzymes, mainly the nicotinamide-adenine-dinucleotide (NAD) and the falvine-adenine-dinucleotide (FAD) that become NADH2 and FADH2 respectively. At the end of this stage, the energy of the food is not yet usable.

Third Stage

This last stage is necessary in order that the energy produced by the preceding reactions can become usable by the cells.

These last reactions involve a series of transfer of hydrogen ions, coupled with phosphorylations.

At the end, the hydrogen ions bind with oxygen to give water, whereas energy is stored by molecules with high levels of energy, such as ATP.

The total number of oxygen molecules required to capture the hydrogen ions corresponds to the oxygen consumption of the body.

On the other hand, the amount of energy stored in ATP and other molecules is related to the number of atoms of phosphor that have been transferred from mineral compounds to ATP.

Since there is a coupling between oxidation and phosphorylation, there is a relationship between the oxygen consumption and the amount of stored energy.

For this reason, the rate of the energy transformation is generally determined by the measurement of oxygen consumption. However, the amount of heat produced by these energy transformation is generally lower because a part of the energy may appear in mechanical, electrical or chemical forms.

OXYGEN CONSUMPTION BY BODY ORGANS

Energy expenditure may be very different from one organ to another.

Table 2 shows some values of oxygen consumption for various organs. As can be seen, the greatest energy requirement is that of the heart.

For the other muscles, the oxygen consumption varies according to the level of activity. Therefore, muscle can be an important source of thermogenesis.

PRODUCTION OF HEAT BY THE WHOLE BODY

Heat production of the total body depends mainly on two factors:

1. The level of muscular activity
2. The requirements for thermal homeostasis.

Table 2. Oxygen consumption of some human organs (in ml min^{-1}. 100 g^{-1})

Heart (Left ventricle)	7.8 - 10.3
Brain	3.4 - 3.8
Kidney	about 10
Muscle	
at rest	0.15 - 0.20
exercizing at 180 W	about 11
Liver (in vitro)	about 6

The Level of Muscular Activity

At rest and in neutral thermal environments, heat production is due to the basal functioning of the cells and organs, mainly the respiratory and circulatory functions.

This thermogenesis is close to the total energy expenditure. It averages 50 W/sqm in the adult. This value is higher in children but tends to decrease in the elderly.

Muscular activity markedly increases this energy expenditure and, for this reason, heat production may be enhanced, to a value of 500 W/sqm.

Requirements for Thermal Homeostasis

When the ambient temperature falls, the rate of heat removed from the body by the environment increases.

To prevent hypothermia, an important physiological mechanism is an increase in the heat production of the body called "cold thermogenesis".

Cold thermogenesis. This can result from two different processes: either the special activity of the muscles called "shivering", or an increase in the heat produced by other organs. This second mechanism is called "non-shivering thermogenesis".

Non-shivering thermogenesis. It has been assumed for a long time that all cells could raise their metabolism and therefore produce more heat when the body was cold-stressed but this point of view has now been challenged. Recently, attention has been focused on the role of a special tissue, called brown fat. This tissue presents particular characteristics - its location in some regions of the body, its structure and mainly its important nervous supply by fine adrenergic fibers.

Numerous experiments have shown that adrenergic stimulation of the brown fat induced an increase in its heat production and that this response was normally observed when the body was cooled.

However, although the non-shivering thermogenesis due to the activation of the brown adipose tissue seems to play a great role in some species, its role in the human is of minor importance, because brown fat is almost absent in adult man. Therefore, the increase in heat production in cold-sressed man is almost totally due to the shivering mechanism.

Shivering. Shivering is a particular kind of muscular contraction in which the mechanical efficiency is very low. Therefore, almost all the energy used in this process appears as thermal energy.

All muscles can shiver, but this activity is more marked at the proximal muscles of the limbs.

Although shivering may be partly under the control of the will, its origin is reflex and its intensity closely depends on the importance of the cold stress.

Shivering appears as successive bursts of muscular contraction. However, its total intensity closely depends on the importance of the cold stress.

Measured by the oxygen consumption, the rate of heat production occurs in successive bursts. During a burst, the maximal value has been found to be as high as ten times the basal thermal production. However, the maximal average rate of shivering thermogenesis that the human body can maintain over a period of several hours does not exceed 2 or 3 times the basal level. Therefore, the physiological capacity of man to withstand cold stress is low.

CONCLUSION

The concept of heat is generally associated with that of life. However, for the body as for all other chemical or physical transformations, heat is a waste. The loss of this thermal energy into the environment is the primary pre-requisite of life.

FURTHER READING

1. A. C. Burton, O. G. Edholm, "Man in a cold environment," Arnold, London (1955).
2. J. Bligh, Thermography. What is regulated and how, in: "New Trends in Thermal Physiology." Y. Houdas, J. D. Guien eds. Massan, Paris, p. 1-9 (1978).
3. K. Cena, J. A. Clark, "Bioengineering, thermal physiology and

comfort," Elsevier, Amsterdam (1981).

4. O. G. Edholm, J. S. Weiner, Thermal physiology, in: "Principles and practice of human physiology," Academic Press, London, p. 111-190 (1981).
5. Y. Houdas, E. F. J. Ring, "Human Body Temperature," Plenum Publishing Corporation, New York (1982).
6. D. Mitchell, A. R. Atkins, C. H. Wyndham, "Essays on temperature regulation," J. Bligh, R. E. Moore eds. North Holland Publishing Co. p. 37-54 (1972).
7. C. E. A. Winslow, L. P. Herrington, Temperature and human life, Princeton University Press. Am. J. Physiol. 184:664 (1941).

THERMORECEPTION AND TEMPERATURE REGULATION IN MAN

The late H Hensel and K Schafer

Institute of Physiology, University of Marburg
Deutschhausstrasse 2, D-3550 Marburg
FEDERAL REPUBLIC OF GERMANY

THERMORECEPTION AND TEMPERATURE SENSATION

The study of thermal physiology in man includes the investigation of temperature sensation and thermal comfort, recording of afferent impulses from thermoreceptors as well as the study of autonomic temperature regulation and thermoregulatory behavior. Thermal sensors in man are not only involved in conscious temperature sensations but also play an important role in autonomic and behavioral responses of the organism to its thermal environment. In order to account for this variety of functions, the concept of "Thermoreception" was introduced.[1] In 1974 the Encyclopedia Britannica adopted this term,[2] defining it "as a process in which different levels of heat energy (temperature) are detected by living things".

Temperature sensations can be described as a continuum of a dual or polar character: cold pain - very cold - cold - indifferent - warm - hot - heat pain. The intensity of a temperature sensation depends both on the absolute temperature and on the rate of temperature change. Whereas the influence of the rate of change has long been known from our everyday experience and from numerous investigations, the role of absolute temperature has been studied only recently.

Constant thermal stimuli were applied to various skin areas of 75 cm^2 for at least 30 min. At the end of this period, when the temperature had reached a steady state, the subject was asked to estimate his static temperature sensations, either in ^{o}C or in a magnitude scale between heat and cold.[3] All subjects were able to discern constant temperatures between 25-38°C (Fig 1). The magnitude estimates were approximately proportional to the constant stimuli temperatures, the static discrimination for forehead and abdomen decreasing in the range between 29-25°C.

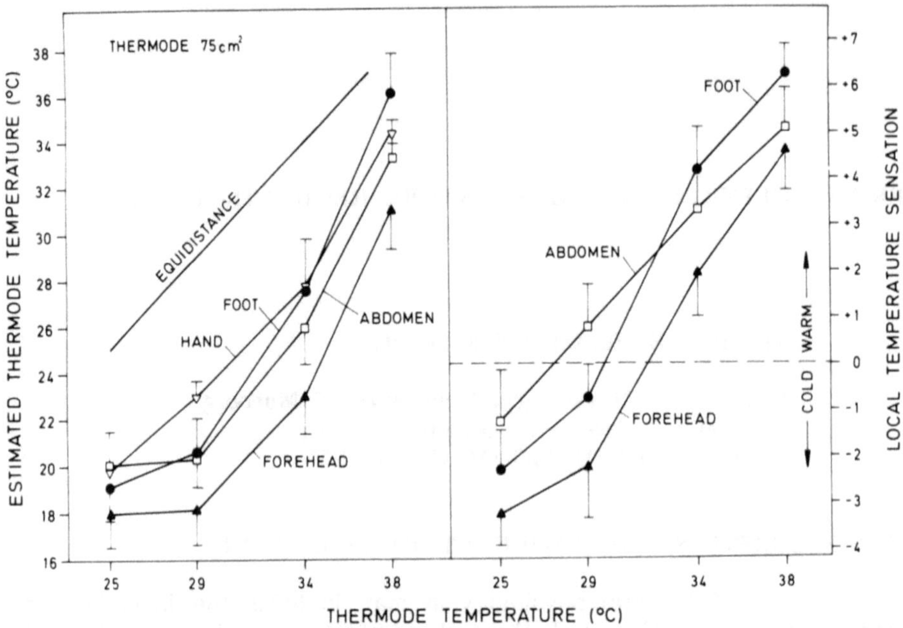

Fig. 1. Static temperature estimates as function of constant temperature of palm, forehead, abdomen and foot. Left: estimates in °C; "Equidistance" line indicates identity between physical and estimated temperature in °C. Right: estimated intensity of thermal sensations. Thermode 75 cm². Average values from 12 subjects; bars indicate standard error of mean. (From Issing and Hensel[3]).

Temperature estimates in °C for abdomen, forehead, hand and foot were lower than the corresponding stimuli temperatures and produced a nonlinear curve. In the range between 29-25°C, the discrimination decresed for the foot and became zero for abdomen and forehead, that is, the subjects could not discern between 29 and 25°C.

HUMAN THERMORECEPTORS

Considerable progress has recently been achieved in recording afferent impulses from single thermosensitive fibers in human subjects by means of microelectrodes (Fig 2). In the hairy skin of the human hand, specific warm receptors[4,5,6,7] and cold receptors[7,8] have been identified. The warm receptors were supplied by unmyelinated fibers with conduction velocities of 0.5 to 1.0 ms^{-1}. It can be assumed that the cold receptors are supplied both by thin myelinated fibers and unmyelinated fibers. All warm and cold fibers had spot-like receptive fields, whereas the receptive fields of fibers supplying polymodal

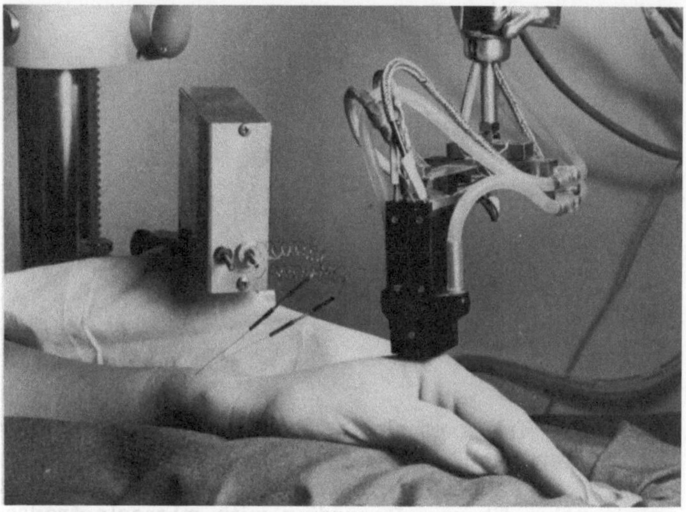

Fig. 2. Experimental setup for recording with microelectrode from single afferent nerve fibers in the superficial branch of the human radial nerve. The subject's arm rests in a mould of plasticine. Microelectrode (foreground) introduced into the nerve; reference electrode (background) introduced into the skin. The electrodes are connected with flexible wires to a preamplifier. A Peltier thermal stimulator is above the receptive field at the back of the hand.

nociceptors often covered a larger skin area (Fig 3).

Figure 4 shows the static discharge frequency as a function of steady temperature for specific cold and warm fibers from the human radial nerve. The receptive fields were on the back of the hand. The temperature-frequency characteristics of both groups of receptors are well separated with only little overlapping. Usually the warm fibers had higher maximum frequencies than had the cold fibers. As can be seen, the maxima of the cold fibers are distributed over a range between 16-28°C and the maxima of the warm fibers over a range of 40-45°C.

It seems likely that the sensation of heat is mediated mainly by the group of high-threshold warm receptors. The majority of polymodal nociceptors is activated in a higher temperature range that corresponds to heat pain, not to the typical sensation of heat.

Human warm receptors responded to warming steps with a dynamic overshoot in frequency and to cooling steps with a transient decrease in frequency, the magnitude of the overshoot being a function of the rate of temperature change (Fig 5) and of the initial temperature.[5] Human

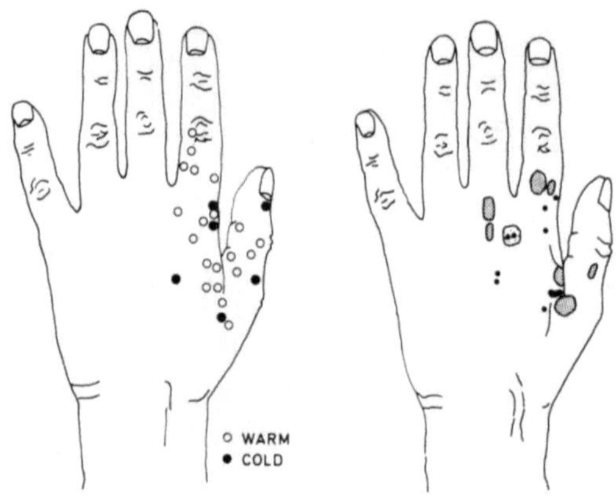

Fig. 3. Receptive fields of single fibers supplying the hairy skin of the human hand. Left: receptive fields of single warm and cold fibers. Right: receptive fields of single polymodal nociceptive fibers. Recording with microelectrodes. (From Konietzny[7]).

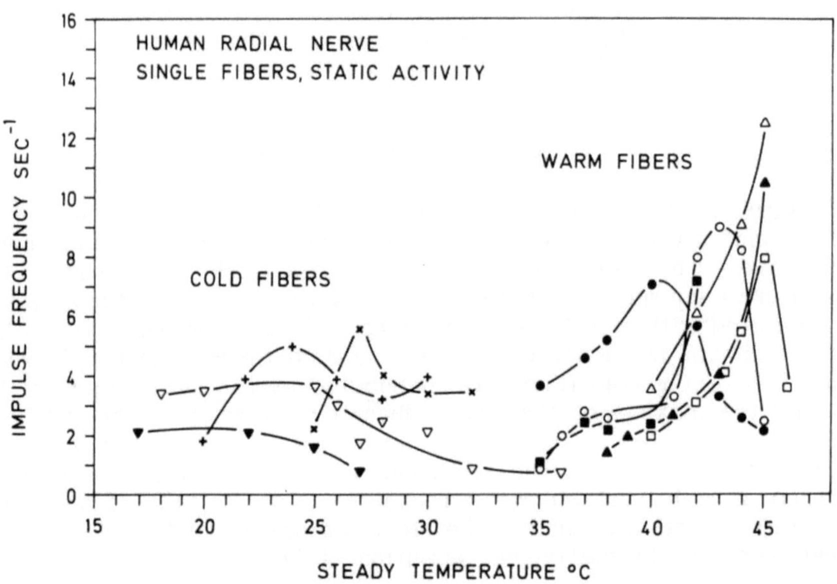

Fig. 4. Static discharge frequency of single warm and cold fibers from the hairy skin of the human hand as a function of constant temperatures. Recording with microelectrodes. (Modified from Konietzny and Hensel[6] and Konietzny[7]).

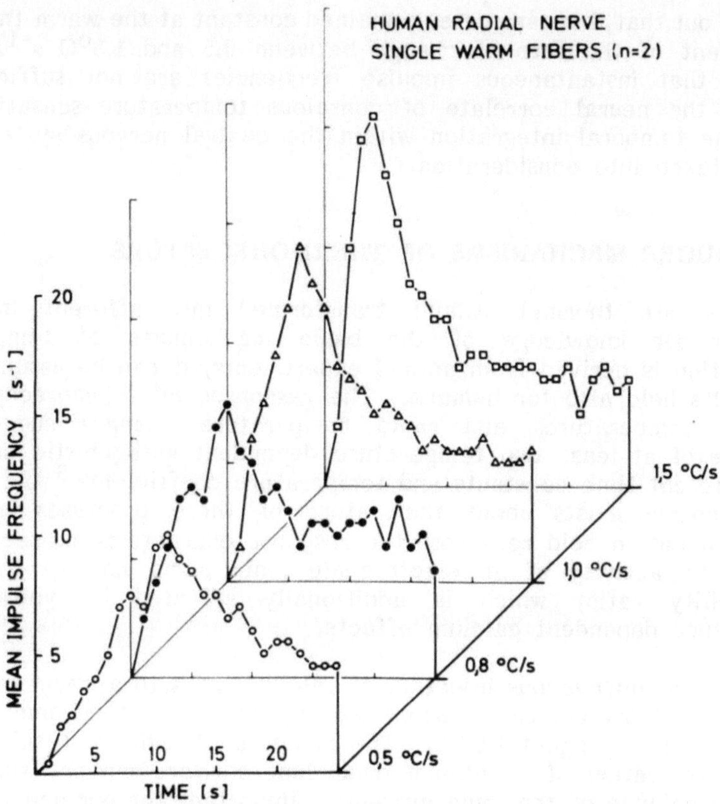

Fig. 5. Average impulse frequencies of two single warm units from human hairy skin at various rates of warming (dT/dt = 1.5, 1.0, 0.8 and 0.5°C s^{-1}) from 32 to 37°C. (From Konietzny and Hensel[5]).

cold receptors showed an opposite response, namely an overshoot on cooling steps and an undershoot on warming steps,[7] with a dependence of the dynamic peak frequency on the initial temperature and the rate of temperature change.

Recording from human nerve fibers offers the possibility of a direct correlation between afferent impulse discharge and sensation. It turned out that warm sensations are mediated by specific warm fibers and cold sensations by specific cold fibers. When stimulating a single warm spot in human subjects with a small thermode, starting from 35°C with a slope of 0.8°C.s^{-1}, the average threshold of warm sensations was 37.8°C. This would correspond to an instantaneous frequency of 9 s^{-1} or to a total number of 28 impulses in the warm fiber supplying the spot. If the average impulse frequency during warming was taken as a correlate, it turned out that this parameter remained constant at the warm threshold, independent of the rate of change between 0.5 and 1.5^{o}C s^{-1}.[7] This suggests that instantaneous impulse frequencies are not sufficient to describe the neural correlate of conscious temperature sensations but that some temporal integration within the central nervous system must also be taken into consideration.

TRANSDUCER MECHANISMS OF THERMORECEPTORS

How are thermal stimuli transformed into afferent impulses? Although our knowledge of the basic mechanisms of temperature transduction is derived from animal experiments, it can be assumed that the results hold also for humans. The responses of thermoreceptors to constant temperatures and rapid temperature changes suggest the existence of at least two temperature dependent antagonistic processes with different time constants and temperature coefficients.[9] At present no conception exists about the nature of these processes in warm receptors, but in cold receptors the response characteristics seem to be due to the activity of an electrogenic ionic pump and to the Na-K permeability ratio, which is additionally affected by voltage and temperature dependent calcium effects.

In man, intravenous injection of calcium leads to a warm sensation. The cold receptor activity is suppressed by calcium and enhanced by the calcium chelating agent EDTA. The effect is thought to result from a changed activation of a calcium dependent outward current which will alter the balance of the ionic currents. Important for our understanding of the underlying membrane process is the so-called burst discharge, that is, a pattern of periodic groups of impulses interrupted by silent intervals (Fig 6). Burst discharges are seen in most species including man.[9] A systematic analysis of bursting cold fibers in cats and monkeys has led to the postulation of an underlying endogenously oscillating receptor potential that initiates impulses in the afferent nerve when exceeding a threshold value.[10,11] Frequency and amplitude of the hypothetical oscillation increases with temperature (Fig 6). We assume that the

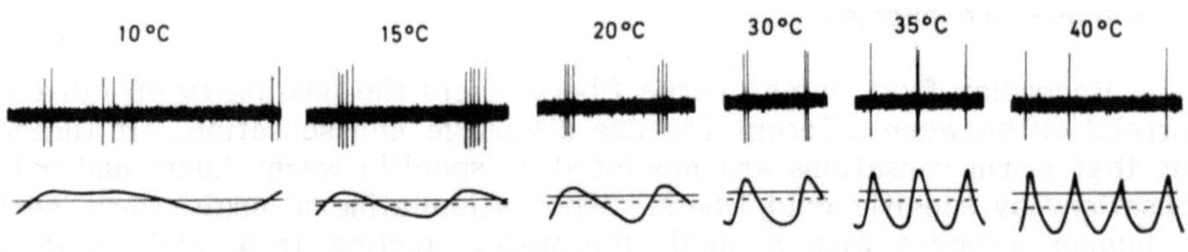

Fig. 6. Burst discharge of a single cold receptor from cat's tongue and hypothetical oscillation of the receptor potential. Solid line: threshold for initiation of impulses. Interrupted line: mean value of oscillating potential. (From Braun et al[11]).

depolarizing phase of the oscillation is induced by an imbalance of ionic currents, whereas the burst is terminated by a voltage dependent and calcium sensitive outward current which is activated during depolarization and temporarily hyperpolarizes the neuron. The assumed ionic imbalance will be increased by calcium reduction and decreased by calcium elevation. Experiments with artificially changed calcium levels support these assumptions: the burst discharge was suppressed by calcium and enhanced by EDTA; in non-bursting cold fibers, EDTA elicited a burst discharge.[12]

The peak frequency of the dynamic response to rapid cooling was only slightly affected by changes in calcium level. Thus, the dynamic response does not seem to originate from the change in passive currents. Application of the inhibiting substance ouabain has led to the assumption that besides the Na-K permeability ratio an electrogenic ionic pump contributes to the membrane potential.[13] In recent experiments in cold receptors using an artificially perfused isolated tongue preparation, we found that 10^{-7} to 10^{-6} mol/l ouabain reduced the dynamic response to cooling and increased the discharge rate at constant temperature. No change in receptor sensitivity was observed. Figures 7 and 8 give a summary of the hypothetical processes underlying the cold receptor potential and the discharge pattern.

AUTONOMIC AND BEHAVIORAL CONTROL OF BODY TEMPERATURE

The diagram in Figure 9 shows some features of autonomic temperature regulation and thermoregulatory behavior in man. Thermal disturbances have two main sources:

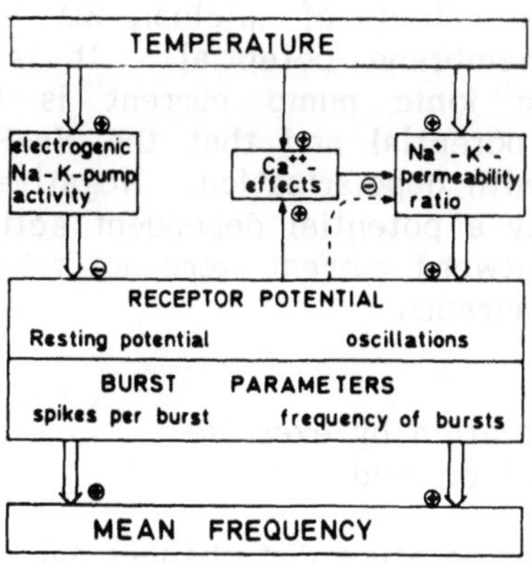

Fig. 7. Model of temperature transduction in bursting cold fibers. Electrogenic ionic pump activity and Na-K permeability ratio are assumed to increase with increasing temperature, the permeability ratio being additionally affected by temperature and potential dependent calcium effects on currents.

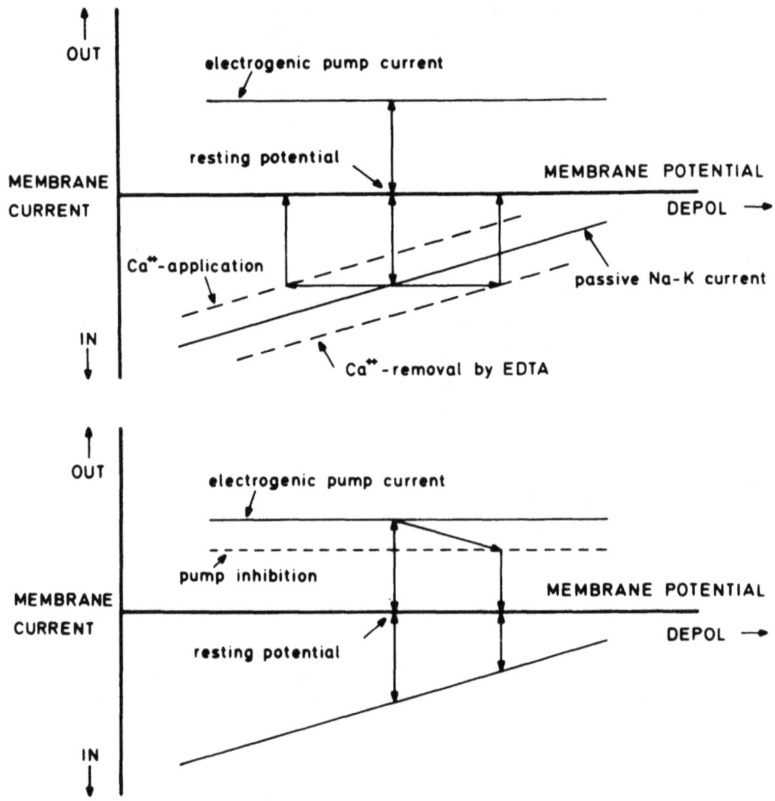

Fig. 8. Hypothetical effects of calcium, EDTA and oubain on cold receptor membrane potential. It is assumed that the electrogenic ionic pump current is independent of the membrane potential and that the Na-K permeability ratio decreases with depolarization. Negative feedback processes mediated by a potential dependent activation of a calcium sensitive outward current were not taken into consideration in these diagrams.

1. Internal heat generation by exercise
2. Environmental heat or cold.

Signals from external temperature disturbances can rapidly be transferred to the central nervous system by means of cutaneous thermoreceptors via afferent nervous pathways before the disturbance has reached the core of the body. Important in this respect is the ability of cutaneous thermoreceptors to respond not only to the temperature but also to the rate of temperature change. The rate control is so effective that rapid

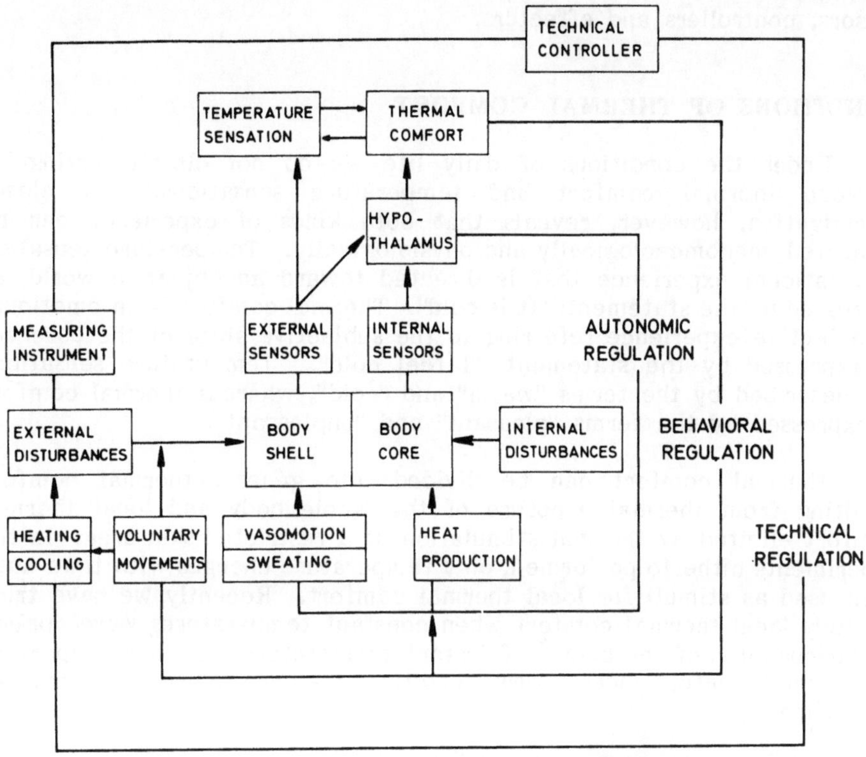

Fig. 9. Schematic diagram of autonomic, behavioral and technical temperature regulation in man. (Modified from Hensel[9]).

external cooling or warming may result in a transient opposite change of internal temperature.

Behavioral responses to heat and cold modify the relations between organism and environment and thereby modify the need for autonomic thermoregulatory responses. Usually we prefer to regulate by behavioral means before our autonomic thermoregulation is activated.

In man, behavioral thermoregulation is associated with conscious temperature sensations as well as with emotional feelings of thermal comfort and discomfort. Whereas temperature sensations depend mainly on the activity of cutaneous thermoreceptors, thermal comfort and discomfort reflect a general state of the thermoregulatory system, or, in neurophysiological terms, an integration of afferent signals from both cutaneous and internal thermoreceptors, preferably from those in the hypothalamus. In a wider sense, the technological inventions of man for artificial control of his thermal environment can be considered as part of his behavioral thermoregulatory ability. In this case the regulation of temperature is shifted from the body to the environment, with artificial sensors, controllers and effectors.

CONDITIONS OF THERMAL COMFORT

Under the conditions of daily life we do not clearly distinguish between thermal comfort and temperature sensations. A closer investigation, however, reveals that both kinds of experience can be separated phenomenologically and physiologically. Temperature sensation is a rational experience that is directed toward an objective world, as expressed by the statement: "It is cold". Thermal comfort is an emotional or affective experience referring to the subjective state of the observer as expressed by the statement: "I feel cold". Temperature sensations are described by the terms "warm" and "cold", whereas thermal comfort is expressed by the terms "pleasant" and "unpleasant".

Thermal comfort can be divided into general thermal comfort resulting from thermal exposure of the whole body and local thermal comfort elicited by thermal stimulation of a restricted skin area. In the experiments hitherto performed, only temperature changes with time have been used as stimuli for local thermal comfort. Recently we have tried to study local thermal comfort when constant temperatures were applied to various sites of the body.[3] Constant temperatures are more important for thermal comfort under natural conditions than are temperature changes with time.

In human subjects in a climate chamber, a water-circulated thermode of 75 cm^2 was placed either on the palm of hand, sole of foot, abdomen or forehead. The temperature of the thermode was adjusted between 25 and 38°C and kept at a constant level for 30 min. The subjects estimated static thermal sensations as described before. In addition, they estimated local and general comfort in a magnitude scale between pleasant and unpleasant. The temperature of the climatic chamber was linearly increased between 12-45°C within 45 minutes; this corresponded to a change in mean skin temperature from 29-36°C.

Figure 10 shows the ratings of local thermal comfort elicited from constant temperatures at forehead, abdomen and foot for various mean skin temperatures. In each region, low local temperatures are rated as pleasant when mean skin temperature is high and as unpleasant when this temperature is low; the opposite holds for high local temperatures. At thermally neutral mean skin temperatures around 33°C, the comfort rates are near the neutral value for all local temperatures, although there is still a certain dependence on mean skin temperature. The crossing point of the curves is not at zero but slightly in the pleasant range.

In contrast to local thermal comfort, local static temperature sensations show a monotonic change with local temperature, the ambient or mean skin temperature, respectively, having only little influence.

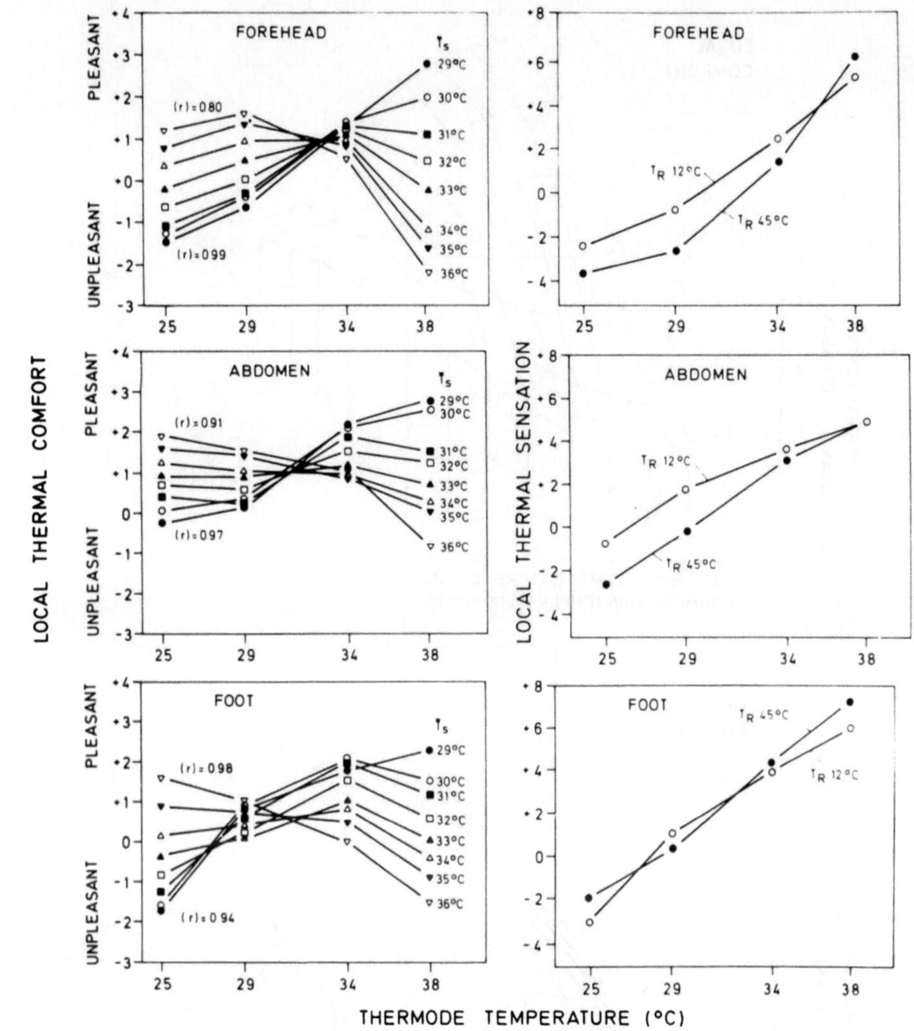

Fig. 10. Left diagrams: Local thermal comfort elicited from forehead, abdomen and foot as function of constant local temperatures (75 cm^2) for various mean skin temperatures ($\bar{T}_S$). Right diagrams: Estimated local temperatures of forehead, abdomen and foot as a function of constant local temperatures (75 cm^2) for ambient temperatures of 12 and 45°C. (From Issing and Hensel[3]).

These results suggest that local temperature sensations are mediated through separate sensory channels, whereas local thermal comfort reflects the neural integration of thermal afferents from various sites of the body.

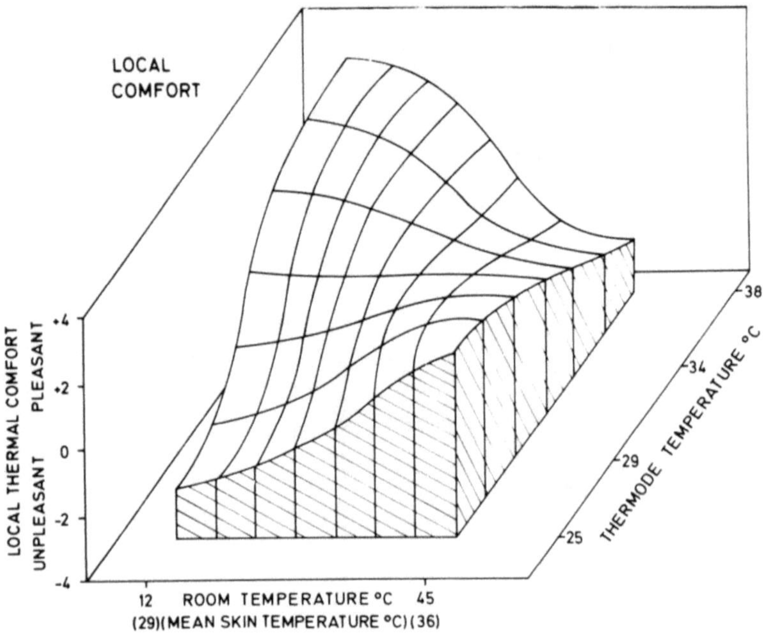

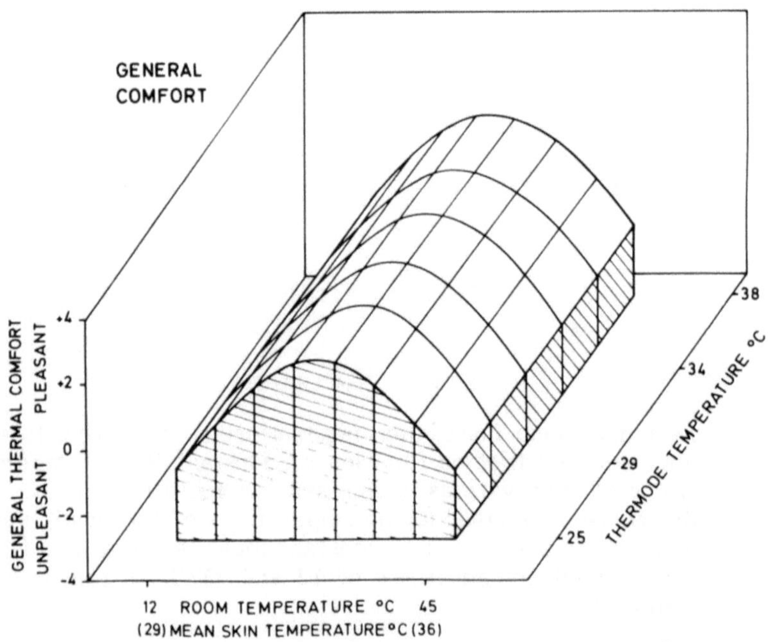

Fig. 11. Upper: Local thermal comfort as a function of constant local temperature (forehead, abdomen, foot) and room temperature (T_r) or T_s, respectively. Lower: General thermal comfort as a function of constant local temperature (forehead, abdomen, foot) and T_r or T_s, respectively. (From Issing and Hensel[3]).

The schematic three-dimensional diagram in Figure 11 shows local and general thermal comfort as a function of local skin temperature and room temperature or mean skin temperature, respectively. Whereas local static thermal comfort changes drastically with local skin temperature, general thermal comfort is virtually independent of this parameter. Instead, there is a marked dependence of general thermal comfort on mean skin temperature, the judgements changing from slightly unpleasant to pleasant and to unpleasant again when the mean skin temperature is increased from 29-36°C.

It was only recently that particular attention was paid to local thermal comfort. This factor plays an important role under the conditions of asymmetric temperature fields, cold floors, heat loss through windows, and draught. Avoiding local thermal discomfort may save considerable amounts of heating energy.[14]

The biological significance of thermal comfort can be seen in the fact that it motivates behavioral responses according to the load error of temperature regulation. On the other hand, the temperature sensations are mediated through separate pathways, and therefore allow us to judge the thermal environment more objectively and thus to predict what might happen to the organism if it were exposed to that particular environment.

REFERENCES

1. H. Hensel, Physiologie der thermoreception, Ergebn. Physiol. 47:166 (1952).
2. H. Hensel, Thermoreception, in: Encyclopedia Britannica. Volume 18, Encyclopedia Britannica Inc., Chicago (1974).
3. K. Issing, H. Hensel, Static temperature sensations and static thermal comfort, Deutsche Forschungsgemeinschaft, Symposium on Mechanisms of Temperature Regulation and Adaptation, Bad Nauheim (1981).
4. F. Konietzny, H. Hensel, Warm fiber activity in human skin nerves, Pflugers Arch. 359:265 (1975).
5. F. Konietzny, H. Hensel, The dynamic response of warm units in human skin nerves, Pflugers Arch. 370:111 (1977).
6. F. Konietzny, H. Hensel, The neural basis of the sensory quality of warmth, in: "Sensory Functions of the Skin of Humans," D. R. Kenshalo ed. Plenum Press, New York and London (1979).
7. F. Konietzny, "Neurophysiologische untersuchungen an mechano und thermosensiblen sowie nozizeptiven cutanen afferenzen beim menschen," Inaug Diss, Marburg (1981).
8. T. Jarvilehto, H. Hamalainen Touch and thermal sensations: Psychophysical observations and unit activity in human skin nerves, in: "Sensory Functions of the Skin of Humans," D. R. Kenshalo ed. Plenum Press, New York (1979).
9. H. Hensel, "Thermoreception and Temperature Regulation," Academic Press, London, New York, (1981).

10. A. Iggo, D. W. Young, Cutaneous thermoreceptors and thermal nociceptors, in: "The Somatosensory System," H. H. Kornhuber ed. Thieme, Stuttgart (1975).
11. H. A. Braun, H. Bade, H. Hensel, Static and dynamic discharge patterns of bursting cold fibers related to hypothetical receptor mechanisms, Pflugers Arch. 386:1 (1980).
12. K. Schafer, H. A. Braun, H. Hensel, Static and dynamic activity of cold receptors at various calcium levels, J. Neurophysiol. 47:1017 (1982).
13. F. K. Pierau, R. D. Wurster, T. Neya, T. Yamasato, J. Ulrich, Generation and processing of peripheral temperature signals in mammals, Int. J. Biometeorol. 24:243 (1980).
14. P. O. Fanger, O. Valbjorn, "Indoor Climate," Danish Building Res. Inst., Copenhagen (1979).

COLD FIBER DISCHARGE PATTERN UNDER IN VITRO CONDITIONS: OUABAIN AND K^+-EFFECTS

K Schafer, H A Braun and the late H Hensel

Physiologisches Institut der Philipps-Universitat
Deutschhausstrasse 2, D-3550 Marburg
WEST GERMANY

INTRODUCTION

Until recently it has only been possible to give broad descriptions of the mechanisms involved in cold receptor temperature transduction. Cold receptor responses to constant and to rapidly changing temperatures were ascribed to two temperature dependent antagonistic processes with different time constants and temperature coefficients.[1] Evidence as to the nature of these processes came from studies of molluscan neurons. It was assumed that the response characteristics of cold receptors were due to the activity of an electrogenic ionic pump and to the Na-K permeability ratio, which is additionally affected by voltage and temperature dependent calcium effects.[2,3,4,5]

Important for our understanding of the underlying mechanisms is the so-called burst discharge, that is, a pattern of periodic groups of impulses interrupted by silent intervals. A systematic analysis of bursting cold fibers in cats has led to the postulation of an underlying endogenously oscillating receptor potential that initiates impulses in the afferent nerve when exceeding a threshold value.[4] From experiments with reduction and elevation of the extra-cellular calcium activity in cats, we assume that the depolarizing phase of the oscillation is induced by an imbalance of ionic currents, whereas the burst is terminated by a voltage dependent and calcium sensitive outward current, which is activated during depolarization and temporarily hyperpolarizes the neuron.[6]

Manipulations of either the passive ionic permeabilities (by changing extracellular ionic activity) or the electrogenic ionic pump (by application of the inhibiting substance ouabain) are possible only with the use of isolated preparations. Experiments of this kind have so far been based on nerve-skin preparations and because of experimental difficulties no systematic analysis of the effect of ouabain and K^+-free solutions on response characteristics of cold receptors is available. Figure 1 shows a hypothetical model of temperature transduction in cold receptors and the methods so far applied (or to be applied) to gain insight into the underlying mechanisms.

METHODS

To study the effects of ouabain and changed extracellular K^+-activity on cold receptor response characteristics, we developed an isolated cat tongue preparation. This contains numerous cold receptors, the properties of which are well examined[7] and it can be easily perfused by the two lingual arteries. The isolated tongue is placed upside down in a perspex chamber filled with mineral oil. The tip of the tongue is placed on a metal thermode which allows the application of heat and permits rapid temperature changes in the range 10° - 40^oC. The tongue is perfused with a modified Krebs solution. Extra-cellular calcium has a strong dependence of cold fiber response and because of this the calcium activity in the perfusing solution was adjusted to 1.35 mM using a calcium sensitive electrode. This value corresponds to that measured by us in blood plasma. Figure 2 shows a diagram of the recording chamber.

RESULTS

In the experiments so far performed, a total of 25 cold fibers could be recorded. Data from 12 cold fibers were sufficient for complete analysis.

The response characteristics of the analysed cold fibers were

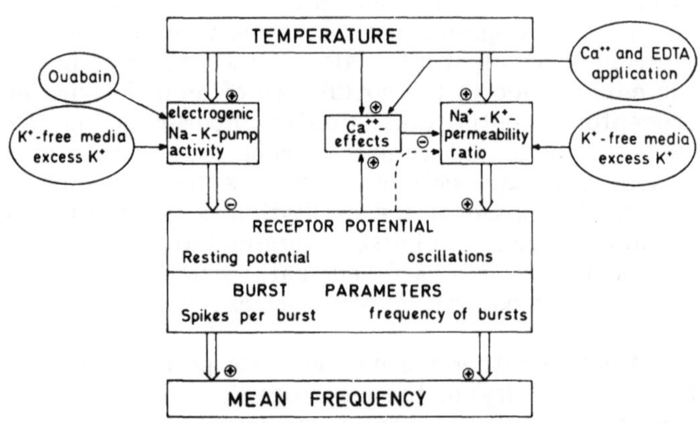

Fig. 1. Hypothetical model of temperature transduction in cold receptors, modified from Braun et al.[4] Encircled factors influence underlying membrane processes.

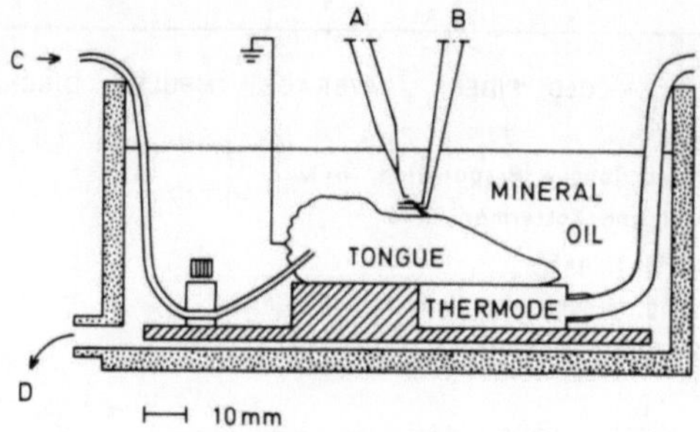

Fig. 2. Diagram of recording chamber. A: recording electrode; B: dissection plate; C: inflow of saline into lingual arteries; D: outflow of fluid; E: thermode water supply from thermostats.

comparable to those recorded in the intact animal. A comparison of the data from the isolated tongue and data from different studies performed in intact animals can be seen in Figure 3.

At constant temperatures, ouabain in high doses ($5x10^{-4}$ mol/l) immediately leads to a transient high frequency discharge, after which the fiber in most cases remains silent. With lower doses (10^{-6} mol/l) such transient high frequency discharges occur occasionally, in some fibers even repeatedly. In every case impulse activity after such transient discharge is low, but dynamic responses to cooling are still obtainable.

The effects of 10^{-6} mol/l ouabain and K^+-free perfusion on mean discharge rate at constant temperatures are different. In general, ouabain leads to an increse in discharge rate (Fig 5, A1), whereas K^+-free perfusion results in a decrease of cold fiber activity (Fig 5, B1). In some cases a change in discharge pattern after application of ouabain or K^+-free perfusion can be observed (Fig 4); ouabain induces a burst discharge in a non-bursting fiber whereas bursting activity is suppressed by K^+-free perfusion. As can be seen in Figure 5 A2 and B2, the appropriate changes in average discharge rate follow either from the appearance of shorter intervals during ouabain treatment or from the suppression of the shorter intervals by K^+-free perfusion. The changes in mean discharge rate therefore reflect the changes of discharge pattern.

The effects of ouabain and K^+-free perfusion on the dynamic response to cooling are similar. Both decrease the peak frequency, mainly at lower temperatures.

DISCUSSION

The results show that the isolated cat tongue preparation is well suited to investigate the effects of ouabain and changed ionic

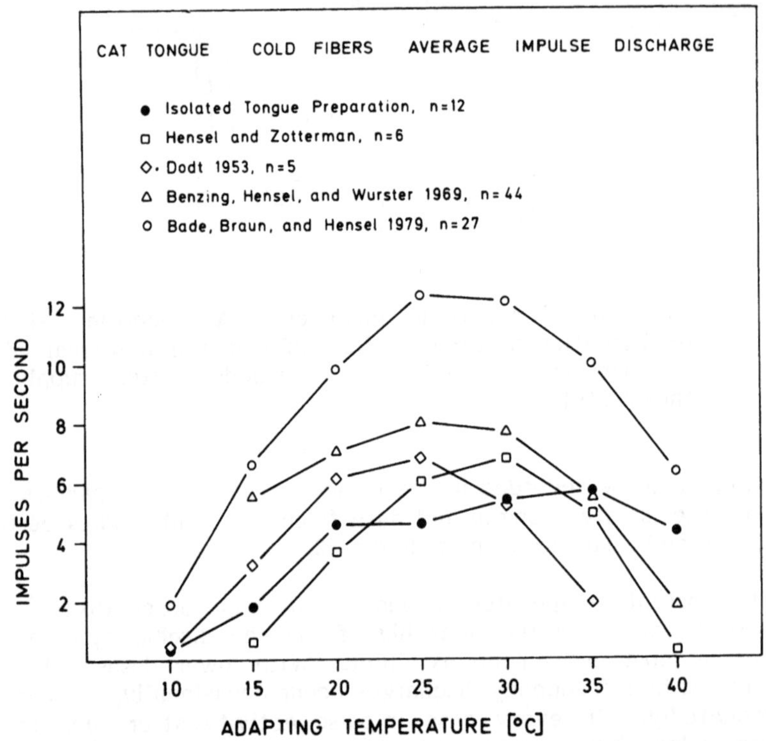

Fig. 3. Comparison of mean impulse discharges of cold fibers from the cat tongue. Data of Benzinger[8] cited after Hensel.[9]

concentration on cold receptor activity. From the data, it seems reasonable to assume a role for an electrogenic ionic pump in the process of temperature transduction in cold receptors as already suggested by Pierau et al.[2] The transient high frequency discharges at high doses and the increase in mean discharge rate at lower doses of ouabain indicate a depolarizing shift of receptor potential due to an inhibition of the hypothetical electrogenic ionic pump. The observed decrease of the dynamic response to cooling can likewise be considered to result from an electrogenic pump inhibition.

Our data do not support the hypothesis that application of ouabain and K^+-free solutions equally affect the process of temperature transduction in cold receptors by a comparable inhibition of the hypothetical electrogenic pump. The observed changes in mean discharge rate and in discharge pattern following perfusion with K^+-free solutions indicate a modification of the passive currents by changed extracellular K^+-activities. Likewise the absence of transient high frequency discharges during K^+-free perfusion corroborates the assumption of different mechanisms. The results support the hypothesis, that K^+-free solutions

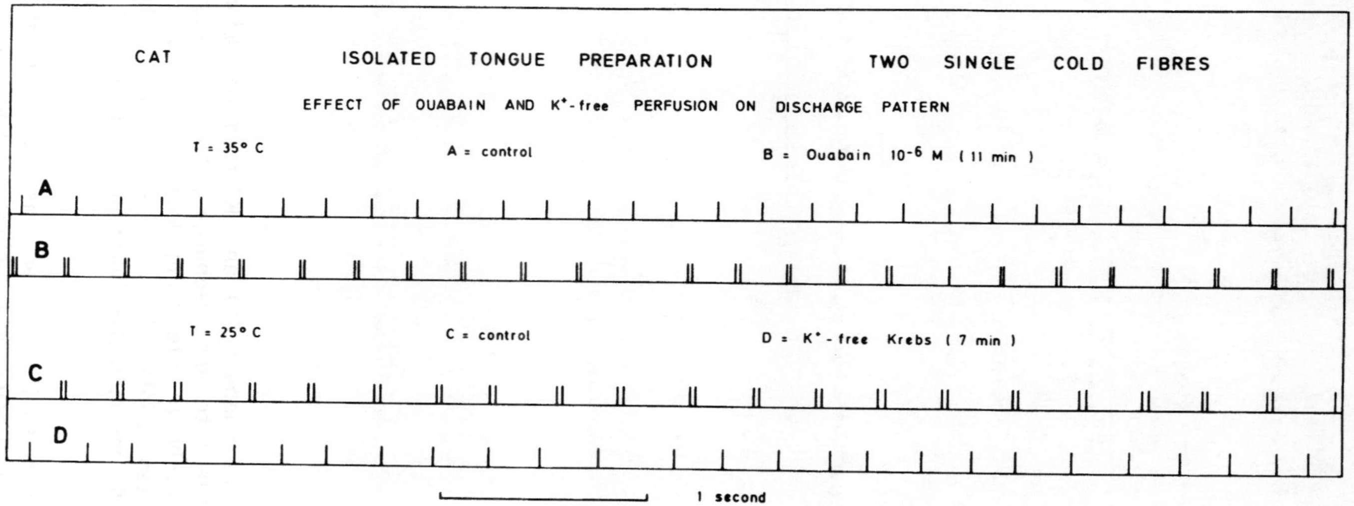

Fig. 4. Changes in impulse discharge pattern following administration of ouabain or K^+-free solutions.

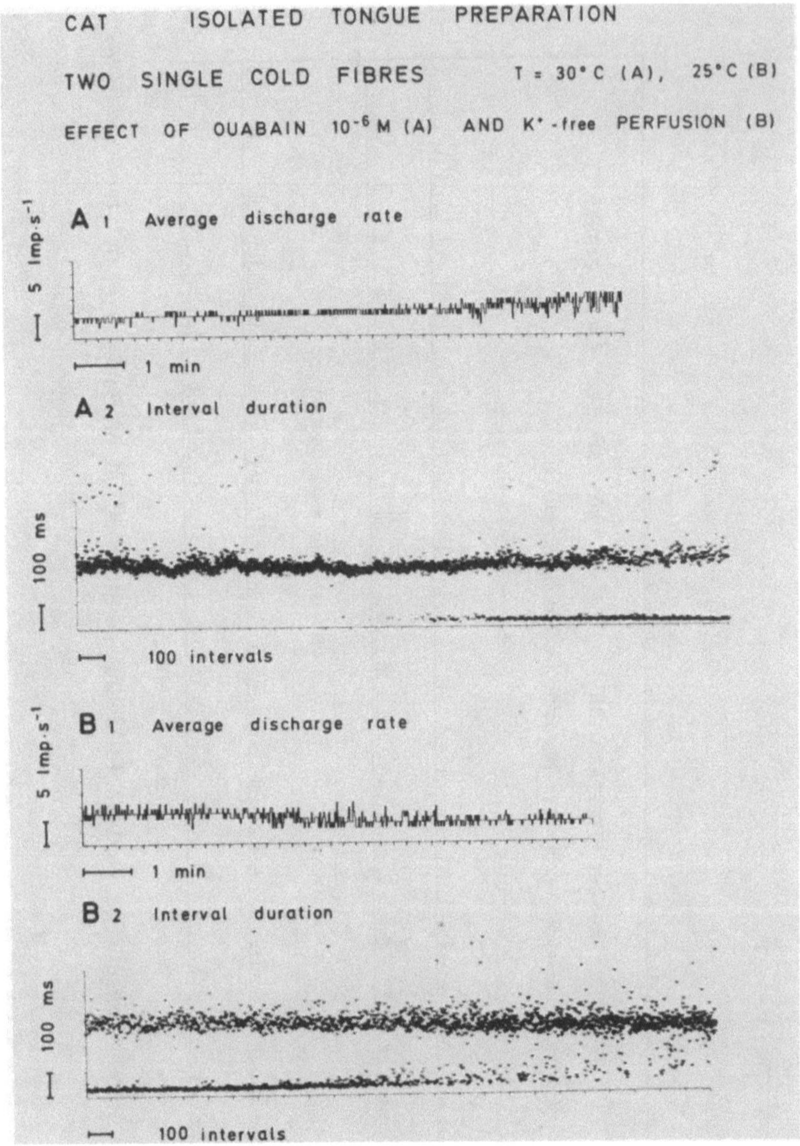

Fig. 5. Effect of ouabain (A) and K^+-free perfusion (B) on mean discharge rate and discharge pattern of single cold fibers. Mean discharge rate in impulses per second. Interval duration of successive intervals in ms. Appearance of intervals shorter than 100 ms indicates burst or grouped discharge.

affect the passive permeabilities as well as the hypothetical electrogenic pump. This can be confirmed by the similar changes of dynamic response during ouabain and K^+-free perfusion.

REFERENCES

1. H. Hensel, "Thermoreception and temperature regulation," Academic Press, London (1981).
2. F. K. Pierau, P. Torrey, D. Carpenter, Effect of ouabain and potassium-free solution on mammalian thermosensitive afferents in vitro, Pflugers Arch. 359:349 (1975).
3. R. Eckert, H. D. Lux, A voltage sensitive persistent calcium conductance in neuronal somata of Helix, J. Physiol. (London) 254:129 (1976).
4. H. A. Braun, H. Bade, H. Hensel, Static and dynamic discharge patterns of bursting cold fibers related to hypothetical receptor mechanisms, Pflugers Arch. 386:1 (1980).
5. D. O. Carpenter, Ionic and metabolic bases of neuronal thermosensitivity, Fed. Proc. 40:2808 (1981).
6. K. Schafer, H. A. Braun, H. Hensel, Static and dynamic activity of cold receptors at various calcium levels, J. Neurophysiol. 47:1017 (1982).
7. H. Bade, H. A. Braun, H. Hensel, Parameters of the static burst discharge of lingual cold receptors in the cat, Pflugers Arch. 382:1 (1979).
8. T. H. Benzinger, Heat regulation: homeostasis of central temperature in man, Physiol. Rev. 49:671-759 (1969).
9. H. Hensel, Neural processes in thermoregulation, Physiol. Rev. 53:948-1017 (1973).

PHYSICAL PRINCIPLES GOVERNING BODY HEAT EXCHANGE

D McK Kerslake

Yateley
UNITED KINGDOM

Physiologists are taught to partition heat exchanges according to the various physical processes involved - convection, radiation etc. This approach is useful in many situations, but it has limitations which are seldom discussed and may be overlooked. This paper draws attention to some of the difficulties which may arise in the practical application of these principles.

The heat balance equation for a human subject is presented in forms which are simple variations of equation (1).

$$M - W = H_k + H_c + H_r + H_e + L + S \qquad (1)$$

In this form the rate of heat production (metabolic energy production minus rate of external work) is on the left, and on the right are the rates of skin heat loss by conduction, convection, radiation and evaporation, followed by L, the rate of heat loss from the respiratory tract and S, the rate of storage of heat within the body.

Conventionally all rates of heat exchange are expressed per unit skin area, A_D. This is the total skin area, a difficult quantity to measure, and normally estimated from height and weight by a formula due to DuBois and DuBois.[1] This does not give a particularly good estimate of total skin area, but so long as everyone uses the same formula it matters very little. The subscript, D, is for DuBois.

Each of the skin heat loss terms can be represented by an equation of the same general form. Taking convection as an example,

$$H_c = \frac{A_c}{A_D} . h_c \ (T_s - T_a) \qquad (2)$$

On the right hand side there is first an area term, A_c/A_D, which is the ratio of the skin area available for heat exchange by the channel in question (in this case convection) to the total skin area estimated by the DuBois formula. For convection the available area may be considerably affected by posture, since skin surfaces touching one another are not available for convective heat exchange. The next term is a coefficient, dimensionally a conductance, which in this case depends on the physical properties of the air and the way it is moving over the skin. This in turn depends on the shape of the subject (posture) and the external air movement. Since posture affects both h_c and the area term there is little point in separating them, and they are usually lumped together as a single coefficient. The final term in brackets is the driving force for heat transfer, here the difference between mean skin temperature and air temperature.

The environmental factor in the bracket is different for each channel and it is assumed that the heat exchange has settled to its steady state level under the prevailing values with which the equation is entered. This is plausible for heat transfer across the boundary air layer or by radiation, but not necessarily for conduction.

CONDUCTION

This means heat transfer from the skin or clothing surface to a solid surface with which it is in contact. Conduction makes a regular appearance in general statements about heat balance, but is rarely referred to outside that context. In the laboratory it is usually made negligible by keeping the area of contact small. The subject stands or walks. If he sits it is on an ergometer saddle. Exceptionally he may recline, but in a net, thus keeping the contact area small. In real life people spend a good deal of time sitting in chairs or lying down. On flat, hard surfaces the areas of contact may remain fairly small, but if the hard surface of the chair is contoured to suit the body, which in turn contours itself to fit the chair, the situation is more comfortable, but the consequent conductive heat exchange can no longer be neglected. With upholstered chairs and soft mattresses conduction is sometimes dismissed on the ground that the material is of high insulation so that heat transfer will be negligible. One might suppose that this should be the case for a subject in a bed having a thick quilt and blankets. In a cold room the addition of a further blanket may convert discomfort into comfort, showing that heat loss through the bedclothes cannot be neglected.

The principle difficulties about dealing with conduction in real situations, or indeed in the laboratory, are the time relations and the mathematical problems of conduction in three dimensions. Consider the case of a subject who sits down on a flat slab of wood, previously at air temperature. Sideways conduction in the slab, from the not very simple geometrical shape of the contact area, is hard to compute, even for the steady state. The transient changes are complicated by this, and by the response of the body tissues to heat exchange. It cannot be assumed that the upper surface of the wood rises suddenly to a fixed temperature when the subject sits down. The time relations can be very roughly inferred from the case of a thick wooden slab with no sideways conduction, the upper surface suddenly rising to skin temperature and remaining steady thereafter. At first the subject's heat loss is infinite for this rather unrealistic model, but it soon diminishes so that after perhaps five minutes it has fallen to about the level it was at before he sat down (say 10 W/m^2 per ^{o}C difference between T_s and T_a), continuing to fall thereafter (as $t^{-\frac{1}{2}}$). For a concrete block it takes about thirty times as long to reach the same rate of heat transfer. In the early stages the rate of heat loss to the concrete is many times what it was in the air before contact, and much greater than it will be in the steady state. (In a real case there will probably be a finite steady state rate of heat transfer). Clearly the steady state equation of the form of equation (2) will rarely be applicable and should only be used with great caution.

CONVECTION

This means sensible heat transfer between the skin and the air. In forced convection the movement of air near the skin is dominated by the general air movement, whereas in natural convection the air movement is generated by changes in density of the air as it exchanges heat with the skin. Equation (2) works well for forced convection, but is somewhat misleading when used for natural convection, because the convection coefficient then depends on the temperature difference, usually increasing as $(T_s - T_a)^{0.25}$. Using the subscript, nc, for natural convection, this process is better represented by

$$H_{nc} = h_{nc} (T_s - T_a)^{1.25} \qquad (3)$$

The curvilinear relation between heat exchange and temperature difference presents some minor problems when solving for T_s in the presence of other skin heat exchanges. A more serious difficulty arises if evaporation is occurring at the same time as natural convective heat transfer. Moist air is less dense than dry air, so convection currents can be generated or altered by evaporation from the skin. An increase of 0.35 kPa (3.5 mb) in water vapor pressure produces about the same change in air density as a rise in temperature of $1^{o}C$. If h_{nc} in equation (3) is thus affected by the rate of evaporation, the independence of the different heat loss channels is vitiated, diminishing the utility of the partitioning process. When the skin is covered with a film of water a combined equation for natural convection and evaporation could be used. More commonly evaporation is less than it would be from fully wet skin, being determined by physiologically controlled sweat gland activity. It seems probable that the effect of such a fixed rate of water vapor transfer on convective heat exchange will depend on the magnitude of the latter. Natural convection has been shown by Schlieren techniques to be important in many common situations for both adults and infants, but there have not been many direct measurements of h_{nc} in living subjects.[2]

RADIATION

The correct form of the equation for radiant heat transfer is

$$H_r = \frac{A_r}{A_D} F\varepsilon \cdot \sigma (T_s^4 - T_r^4) \tag{4}$$

$F\varepsilon$ is an emittance factor, usually so close to unity for infrared radiation that it can be omitted, and σ is the Stefan-Boltzmann constant, 5.67. 10^{-8} W/m².K^4. The temperatures are in Kelvin. If the range of temperature to be considered is not great, a first power approximation can be used, which is more compatible with the equations for the other heat exchange channels,

$$H_r = \frac{A_r}{A_D} h_r (T_s - T_r) \tag{5}$$

where h_r is the linear radiation coefficient, equal to $\frac{1}{2}\sigma (T_s + T_r)^3$, so the advantage of a physically rather than an empirically based coefficient independent of posture is not lost. Posture affects A_r, which is reduced if some parts of the body partially screen others from the radiant environment. T_r is the mean radiant temperature at the place occupied by the subject. It is defined as the temperature of a uniform black enclosure in which the radiant heat exchange of the subject would be the same as it is in the real environment. This makes equation (5) a rather useless tautology unless it is possible to measure or calculate T_r in some way independent of this equation. This can often be done since T_r is an expression of the mean radiant heat flux coming from the surrounding surfaces, weighted by their importance to the subject. For a spherical subject all directions are of equal importance and the components of the radiant environment are weighted by the solid angles subtended at the center of the sphere. Observations could be made by radiometry or a globe thermometer could be used to determine the weighted mean.

For the case of a human subject, one should strictly substitute a person-shaped object for the globe thermometer. A standing person is more like a cylinder than a sphere (hopefully) and conditions above his head or below his feet are less important than those at the front, back and sides, because the projected area of the end is less than that of the side of the cylinder. A suitably shaped 'globe' thermometer will provide the correctly weighted mean. The instrument can be much smaller than the subject provided that the subject is small compared with the radiant enclosure, or more correctly with the distances from the nearer parts of the enclosure. As Henry Ford might have said, the subject can be any size you like as long as he is small. If he is close enough to the environment to fall into the category 'not small', it becomes necessary to specify his size and it may be difficult to impossible to measure T_r.

If the subject, let us say a cylinder, is small compared with its distance from the nearest part of the environment, the radiant temperature in any given direction is the same at all parts of the surface and would be the same for a smaller cylinder situated at the subject's center. On the other hand, if the subject is close to the environment, the direction of a given point in the radiant environment varies from place to place on the subject's surface, and the weighting of T_r may be altered. In our schooldays some of us were fortunate enough to win the desk by the radiator. This provided warmth on one side from the waist down, but, apart from convected heat, was of little benefit above the waist. It remains possible in such circumstances to calculate T_r for the whole body so as to satisfy the definition, but the only globe thermometer type of instrument which would give the right weighting would be the same size and shape as the subject.

Another important effect of environmental proximity is that the presence of the subject may screen some parts of the environment from others. A measurement of T_r made in the absence of the subject will not be appropriate for the calculation of his radiant heat transfer. An example is the shadow cast on the ground by a subject standing in the sun. The shadowed area will cool, and as some of it at least is near the subject, T_r will be reduced. The reduction will take time, and problems of conduction in the shadowed area will arise. The classical infantryman in the desert is conveniently assumed to be marching, so that these effects can be neglected.

If the subject is lying down on flat ground the shadow area is all close to him, occupying a large proportion of the lower hemisphere of the radiant environment. In this case it is difficult to measure the temperature of the ground under the subject and it may be impossible to make a reliable estimate of T_r.

The temperature of the environmental surfaces may be significantly affected by heat exchange with the subject. An example is the case of a baby in an incubator. Here there is a finite, though not usually large area of contact between the baby and the mattress. Elsewhere below the baby the skin is near the mattress and it may be difficult to measure the temperature of either. If we neglect heat transfer through the mattress, the mattress temperature at any point will be somewhere between the air temperature and the baby's skin temperature. The mattress will gain heat from the baby by radiation and lose heat to the air by convection and perhaps to the canopy by radiation. These exchanges are hard to estimate even if the air flow in the incubator is large and forced convection prevails. The canopy is not very far from the baby and may be warmed like the mattress. The regions most strongly heated by radiation from the baby are also those most important in his radiant environment.

If we can neither measure nor calculate T_r, equation (5) will not provide a value for H_r. However it is possible to measure the mean radiant temperature in the absence of the baby. Call this T_{ro}. With the baby present, T_r will exceed T_{ro} by an amount proportional to the baby's radiant heat loss, say by $a.H_r$. Thus

$$H_r = \frac{A_r}{A_D} . h_r \ (T_s - (T_{ro} + a.H_r)\ ) \qquad (6)$$

which can also be expressed

$$H_r = \frac{A_r}{A_D} . h_r' \ (T_s - T_{ro}) \qquad (7)$$

where h_r' is a modified radiation coefficient, not calculable from physical principles as is h_r. If the value of h_r' could be measured experimentally this would provide a solution to the problem, but direct measurement using a heated model as a subject will not work because T_{ro} cannot be varied without affecting the convective heat exchange of the model.

There seems to be little use in an equation which defines the unknown in terms of the unknowable. If we suspect that the presence of the subject is making it impossible to measure T_r (and it would be useful to have a formal way of recognizing this) we must modify the application of the classical partitioning system. That system broke the heat exchange into channels, each of which described the exchange of heat between the subject and one discrete and measurable environmental heat sink. The case of the baby can be approached in the same way. The air in the incubator is one sink, and its temperature can be conveniently measured at the inlet. As we have seen, T_r in the incubator cannot be measured satisfactorily and is influenced by the internal air temperature, the outside conditions (ward temperature) and the baby's skin temperature. Internal air temperature is already one sink, and ward temperature (actually a suitably weighted mean of radiant and air temperatures in the ward) is another. Sensible heat exchange between the baby (T_s), the incubator air (T_i) and the ward (T_w) takes place by a network of radiant, convective and conductive processes, but as with electrical conductances these can be reduced to three conductances each linking one pair of temperatures. Thus the heat exchanges can be expressed

$$H_{si} = h_{si} \ (T_s - T_i) \qquad (8)$$

$$H_{sw} = h_{sw} \ (T_s - T_w) \qquad (9)$$

$$H_{iw} = h_{iw} \ (T_i - T_w) \qquad (10)$$

The total sensible heat exchange of the baby is the sum of H_{si} and H_{sw}.

$$H_s = (h_{si} + h_{sw})\ T_s - h_{si}.T_i - h_{sw}.T_w \qquad (11)$$

The coefficients can be measured most readily by means of a heated model. They do not represent physically distinct processes. Both refer to combinations of convective and radiative exchange. Equation (11) shows the relative importance of T_i and T_w. If T_w falls by 1°C, T_i should be increased by (h_{sw}/h_{si})°C to keep the baby's heat loss constant.

It is possible to use other temperatures in the system and to arrive at a modified form of equation (11). Thus, suppose that instead of T_w we use the temperature of a point on the canopy, T_c. This is affected by all three sink temperatures, and can be represented as a weighted mean of them.

$$T_c = p.T_i + q.T_w + r.T_s \tag{12}$$

By substitution in equation (11),

$$H_s = (h_{si} + h_{sw} + \frac{r}{q} h_{sw})\ T_s - (h_{si} - \frac{p}{q} h_{sw})\ T_i - \frac{h_{sw}}{q}.\ T_c \tag{13}$$

The coefficients can be obtained experimentally as for equation (11) but equation (13) is somewhat misleading. It disguises the effect of T_s on T_c and suggests that the effect of a change in T_s will be greater than it really is (cf equation [11]). If T_s increases, so does T_c. Similarly since T_c depends in part on T_i, the effect of a change in T_i appears smaller than it really is. These considerations do not preclude the use of T_c as an input to the incubator temperature control.

REFERENCES

1. D. DuBois, E. F. DuBois, The measurement of the surface area of man, Archs. Intern. Med. 15:868 (1915).
2. D. A. McIntyre, Indoor Climate, Applied Science, London (1980).

ENVIRONMENTAL HEAT LOSS

K Cena

Technical University, 50-370 Wroclaw, POLAND, now at Occupational Health Program, McMaster University Hamilton, Ontario, CANADA

INTRODUCTION

The thermal environment affects humans and the temperature patterns on their skin surface through the transfer processes between their bodies and surroundings. The physical processes involved are basically the same both outdoors and indoors but their relative magnitude is dependent on the conditions.

The main modes of heat transfer - conduction, radiation and evaporation - play different roles in the heat balance of homeotherms and are dependent on the state of the physical environment and also on a number of physiological responses. The environment of a thermographic examining room is less complex than that in the open where wind, sunshine or thermal radiation from surroundings of complicated geometry make accurate analysis even more difficult. The optimal conditions for comparable thermographic examinations of human subjects would, of course, be provided by a climatic chamber where all the main micrometeorological variables (ie air temperature and speed, radiant conditions and humidity) can be easily set and maintained. In practice such chambers are not generally available and thermographic measurements are routinely performed in ordinary rooms, although often with some control of air temperature. This paper will describe the interactions between the most relevant physical characteristics of standard rooms and the subject's physiological responses as measured by infrared thermography. A particular emphasis will be made on demonstrating that for accurate measurements uniform radiant conditions are necessary and relevant examples will be discussed.

METABOLISM AND ENVIRONMENTAL TEMPERATURE

Application of the First Law of Thermodynamics to steady heat transfer at the surface, which can be either from skin or clothing, requires that the sum of heat inputs, outputs and storage must be zero. This assumes that the conditions do not vary with time or that their changes can be neglected. The heat produced by a subject during metabolic processes under given conditions is mainly a function of activity level. A simple algebraic expression for the heat balance is:

$$M + R + H + G + J = 0 \qquad (1)$$

M denotes the metabolic heat production. When the energy conservation equation is presented in this form, the balance is between terms which are heat gains by the body and those which involve heat dissipation. Therefore M is always positive and the other terms are usually negative.

Changes in the body heat mapping are the usual object of thermographic examinations but are often overshadowed by the other factors representing heat losses and specified in equation (1). R is the net radiative exchange between the skin and surroundings. H is the total heat transfer with the air and because the environment of, for example, an examining room is specified in terms of air temperature and humidity, H is the sum of convection (C) and evaporation (E), so that

$$H = C + E \qquad (2)$$

G is the heat transfer by conduction to substrates like beds, chairs and floors, and J is the heat stored in the body. In the state of thermal equilibrium, when $J = 0$, and assuming for simplicity that also $G = 0$, equation (1) takes the form

$$M + R + C + E = 0 \qquad (3)$$

R + C is the total sensible heat exchange. E represents heat losses from the skin and the respiratory tract and the latter does not change independently of metabolism in steady conditions. Excluding behavioral thermoregulation, the relation between metabolic heat on one hand and the sensible and insensible heat losses on the other hand, defines the physical and physiological processes maintaining the human homeothermy.

A classical graphic representation of the interrelations involved is the metabolic diagram[1] presented in Figure 1. The environment is defined, in the simplest way, by a single environmental temperature T which is related to the metabolic heat production. The distance between the lines C and D in figure 1 represents a limited range of temperatures in which the metabolism is minimal and no effort is required for thermoregulation. The width of this zone may vary from tens of degrees in large animals to a few degrees in small neonate animals. It depends on both the size of the subject and the insulation available to protect the body, as well

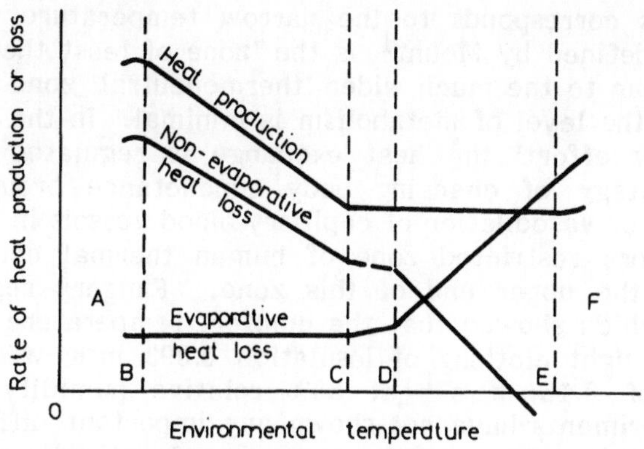

Fig. 1. Diagrammatic representation of relationships between heat production, evaporative and non-evaporative heat losses. A: zone of hypothermia; B: temperature of summit metabolism and incipient hypothermia; C: critical temperature; D: temperature of marked increase in evaporative loss; E: temperature of incipient hyperthermia; F: zone of hyperthermia; CD: zone of least thermoregulatory effort; CE: zone of minimal metabolism; BE: thermoregulatory range. From Mount.[1]

as on other variables, such as food intake. The 'lower' critical temperature (C in Figure 1) for a 70 kg naked man is about 26 - 27°C, whereas for a 5 kg arctic fox it is below -40°C. Similarly, B indicates a temperature of summit metabolism (and incipient hypothermia) and ranges from about 30°C for small neonate mammals to temperatures theoretically below those at which air liquefies for some arctic mammals or humans in well constructed sleeping bags. In general, the larger the subject the greater the available metabolic energy per unit of skin surface area, and also the greater the thickness of insulation it can carry. Above the upper temperatures (D in Figure 1) of the zone CD, increased evaporative heat loss is necessary for heat balance. In man and other sweating animals the limit is when the whole skin is wet. In this condition the rate of heat loss by evaporation is limited mostly by the physical resistance to the transfer of water vapor to the surrounding air, rather than by the physiological control of sweat secretion. The minimum value of the evaporative heat loss in man is about 6% only of the rate from a fully wet surface at the same temperature. The minimum level of the evaporative heat loss (left of C in Figure 1) is also governed by the total amount of heat loss during evaporation.

Thermographic examinations in medicine are usually performed in air temperatures of 20-23°C that is a few degrees less than the expected thermal comfort range. Houdas and Guieu[2] pointed out that this condition may significantly change the vasomotor tone of the cutaneous vessels and they recommend air temperatures much closer to thermal neutrality

(26-27°C). This corresponds to the narrow temperature zone between lines C and D, defined by Mount[1] as the 'zone of least thermoregulatory effort' rather than to the much wider 'thermoneutral' zone between lines C and E where the level of metabolism is minimal. In the 'zone of least thermoregulatory effort' the heat exchange is regulated by the most economical strategy of changing body conductance brought about by vasoconstriction or vasodilation of capillary blood vessels in the peripheral tissue. The more restricted zone of human thermal comfort may be identified with the upper end of this zone. Fanger[3] reported classic measurements which showed that the comfort temperature for sedentary people, wearing light clothing of insulation 0.093 $m^2K\ W^{-1}$ (0.6 clo), in an air speed of 0.15 $m\ s^{-1}$ at 50% relative humidity was 25.6°C. Laboratory experiments have not shown any important difference in the comfort temperatures between the sexes, people of different age groups or between times of year. A detailed analysis of all aspects of thermal comfort is given by Fanger,[3] McIntyre,[4] and Cena and Clark[5] who discuss the methods of determination of the thermal comfort conditions for humans in enclosed environments. Relevant information on the total and partial heat losses is also given in texts by Kerslake,[6] Monteith and Mount[7] and Mount.[8] Useful examples of calculations of man's physiological responses to changes in the physical environment can be found in a monograph by Campbell[9] and for practical calculations of the thermal comfort conditions the classic text by Fanger[3] is recommended.

As most thermographic measurements in human medicine are performed in conditions close to thermal comfort, the further discussion of equation (1) may be limited to a few major aspects. The problem is also simplified by the absence of clothing and by limited sweating. In fact, the concept of an 'environmental temperature' (Fig 1) as a useful, but vague, index describing the environment of an examining room seems to be justified. An important assumption is also made that thermographic measurements are always performed in the state of thermal equilibrium and a waiting period of about 30 minutes, postulated by Houdas and Guieu[2] is allowed before measurements under a different set of conditions are taken. This does not apply to specialized cases such as studies of human exercise[10,11] but to the standard indoor conditions of an examining room.

EFFECTS OF RADIATION

Electromagnetic radiation in two discrete wavebands plays a dominant role in the heat balance of animals and man.[13] Between 3 and 100 µm, 'long-wave', or thermal, radiation in the environment is absorbed by the surface of all organisms, and radiation in the same waveband is emitted at a rate which depends on the surface temperature. 'Short-wave' radiation in a waveband between 0.3 µm and 3 µm has its origin from the sun, or artificial lighting in indoor conditions. Integrating over both wavebands, the difference between the total radiation emitted and reflected is known as the net radiation exchange.

The rate of emission of energy from a surface by radiation is proportional to the fourth power of the surface (skin) temperature T_s (in Kelvins) and emissivity ε. The emitted radiation flux L (in Wm^{-2}) is given by

$$L = \varepsilon \, \sigma T_s^4 \tag{4}$$

where T_s is the Stefan-Boltzmann constant equal 56.7 x 10^{-9} $Wm^{-2}K^{-4}$. The emissivity of skin and almost all clothing materials is very close to unity and equation (4) can be simplified to a form

$$L = \sigma T_s^4 \tag{5}$$

Unless strong lamps are used, the net radiation exchange indoors is limited to the thermal range. Figure 2 shows a typical radiant environment of an examining room and the net radiation R at a point of interest on the skin at temperature T_s is given by

$$R = \sigma T_w^4 - \sigma T_s^4 \tag{6}$$

where T_w is the temperature of the surrounding walls or screens. A more general concept of a 'radiant temperature' T_r, instead of T_w, allows the inclusion of all sources of radiation present. In general, T_r is considered to be the temperature of a black body source which would give the same value of some measured quantity of radiation as exists in reality. Provided that T_r and the local skin temperature T_s do not differ by more than about 20K, which is the normal case, equation (6) may be linearlized to a simple form[13]

$$R = h_r (T_r - T_s) \tag{7}$$

where h_r is a radiation transfer coefficient with a value of about 6 $Wm^{-2}K^{-1}$ at temperatures close to 300 K (27°C). Equation (7) does not, however, include any factors for the usually complicated shape of objects exchanging radiation.

In addition to the air temperature T_a and the mean radiant temperature $\bar{T}_r$, the simplest derived environmental temperature index for sensible heat exchange is the 'operative temperature' T_o. This is defined as the temperature of a uniform black enclosure in which a human occupant would exchange the same amount of heat by radiation (R) and convection (C), as in the actual non-uniform environment. By this definition

$$R + C = h (\bar{T}_s - T_o) \tag{8}$$

$$R + C = h_r (\bar{T}_s - T_r) + h_c (\bar{T}_s - T_a) \tag{9}$$

where C and R are in Wm^{-2} and $\bar{T}_s$ is the mean skin (or clothing)

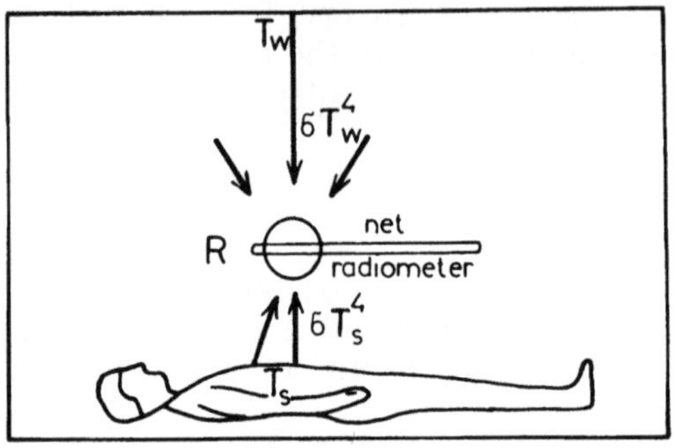

Fig. 2. Net radiant energy exchange between a subject and surroundings. R is the net radiation which can be measured directly with a net radiometer. T_w is the mean radiative temperature of the surroundings and T_s is the local skin temperature.

temperature of the subject. Factor h_r is the linear radiation coefficient in $Wm^{-2}K^{-1}$ given by

$$h_r = 0.72\ (56.7 \times 10^{-9})\ 4\ (0.5\ (T_r + T_a) + 273)^3$$

$$\text{when } T_r > \bar{T}_s > T_a \text{ or by}$$

$$h_r = 0.72\ (56.7 \times 10^{-9})\ 4\ (0.25\ (T_r + T_a + 2\bar{T}_s) + 273)^3$$

when $\bar{T}_s > T_r + T_a/2$. The factor 0.72 is the fraction of the total human body surface contributing to radiation exchange and the second bracket is the Sefan-Boltzmann constant. h_c is the convective heat transfer coefficient which is a function of air velocity but also of the subject's activity and its value can vary widely over the surface of the body. Nishi and Gagge[14] used a naphthalene sublimation method and found that in normal air movement (0.15 - 0.2 ms^{-1}) the mean value of h_c was 3.1 $Wm^{-2}K^{-1}$. The local values of h_c were for head (3.2), chest (2.5), back (2.4), upper arms (4.0), forearms (3.9), hands (4.6), thighs (2.8) and legs (3.7). Clark and Mullan[15] studied the effects of unidirectional downflowing air streams, such as can be encountered in air-conditioned rooms, on the convective heat losses from a model heated cylinder representing a human body. Their results gave higher values in still air for the cylinder positioned vertically (h_c = 6.7 $Wm^{-2}K^{-1}$) and horizontally (6.6 $Wm^{-2}K^{-1}$). The corresponding figures for a downflowing air stream of 0.3 ms^{-1} were 5.9 and 5.7 $Wm^{-2}K^{-1}$, respectively. As completely still air conditions rarely exist in standard rooms a classic formula by Kerslake,[6] based on empirical results, can be used for human subjects in unidirectional airflows

$$h_c = 8.3 \sqrt{v}$$

where v is the mean air speed in m s^{-1}. For v = 0.15 m s^{-1}, h_c = 3.2 $Wm^{-2}K^{-1}$.

From equations (8) and (9)

$$T_o = (h_r\ T_r + h_c\ T_a)\ /\ h \qquad (10)$$

where the combined coefficient $h = h_r + h_c$. The operative temperature can therefore be defined as an average of T_a and weighted by the respective heat transfer coefficients. Equation (10) can be rewritten as

$$T_o = T_a + h_r\ (T_r - T_a)\ /\ h \qquad (11)$$

or

$$T_o = T_a + R_E\ /\ h \qquad (12)$$

where R_E, the effective radiant field, is defined as the radiant energy exchanged by a human subject, whose surface temperature is hypothetically equal to the ambient air temperature.[16] The effective radiant field thus represents the radiant heat received by the subject only from those radiative surfaces in the surrounding enclosure whose temperatures differ from T_a.

The simplest direct measure of the effective radiant field can be obtained by using a Bedford type globe thermometer (15 cm diameter) but conversion from the 'globe temperature' to radiation fluxes is inaccurate as it requires a simultaneous knowledge of the air temperature and the mean air velocity which is difficult to determine. The classic globe thermometer also presents problems associated with its long time constant and as a result it is much easier to measure the radiation field directly.[5]

The effect of small radiant sources on the mean radiant temperature can also be dealt with by a different method, since a source such as an infrared lamp or an electric heater presents neither a simple shape nor uniform temperature. If the source contributes an irradiance I(W m^{-2}) at a test point this increases the mean radiant temperature at that point by

$$T_r = I\ /\ (16\ \sigma T_r{}^3) \qquad (13)$$

At normal temperatures an approximation can be used

$$\Delta T_r = 0.0043\ I \qquad (14)$$

This is accurate for small ΔT_r and the error increases to about 5% for T = 20K. The value of the irradiance may be obtained from the manufacturer's specification for heating panels or measured directly. McIntyre[4] introduced a concept of the 'vector radiant temperature' which can be visualized as the average surface temperature on one half of a room minus the average of the other half. The equivalent of vector radiant temperature T_v, in units of power, is the net radiation, specified in equations (1) and (6). The net radiation R is therefore the difference in irradiance on opposite sides of a plane element

$$T_V = R / 4 \sigma T_r^3 \tag{15}$$

and at room temperature,

$$T_V = 0.17 R$$

The net radiation represents the final effect of exchange between all radiation sources present in a room. It is therefore an ideal measure of radiant asymmetry which impairs the accuracy of thermographic measurements. Despite the fact that a simple and inexpensive instrument has been developed for physiological application by Funk,[17] it has not been much used in practice, although it allows a direct measurement of the radiant heat exchanges.

The net radiometer consists of a thermopile with the sensitive elements exposed on the two opposite faces of a blackened disc (about 1 cm in diameter). If the disc is held above and parallel to a surface which is receiving radiation, such as a part of the human body, the two faces will be differentially heated by the incident thermal (and also short-wave, if present) radiation falling on the lower face. The thermopile output is therefore directly proportional to the net radiation flux, provided the element is shielded from draughts. Air-inflated domes of thin polyethylene, which is virtually transparent at all the wavelengths of interest, provide appropriate draught protection. A typical output of a miniature net radiometer is about 20 $\mu V/Wm^{-2}$. Substitution of a black metal cavity of known temperature for one of the domes allows direct measurement of the effective radiant field. In this form it can also measure the irradiance, and indirectly the mean radiant temperature (equations (13) and (14)), provided it is held close to the object so that the exposed face of the thermopile 'sees' the whole object and nothing else. The instrument can be used for tests for possible radiant assymetry in an examining room in the absence of a subject and again during measurement with the subject. The net radiometer can be recommended as an accessory in a thermographic laboratory, particularly since the cost is comparable with that of a thermistor thermometer. It can be used with little expertise in most cases and the wide applications of this instrument have recently been recognized by an international standard for the radiant temperature assymetry in inhabited rooms.[18]

APPLICATIONS AND CONCLUSIONS

The above discussion emphasizes a well-known fact that small differences in surface temperatures are often associated with local variations in heat exchange with the surroundings and not with the process studied. These make detectability of weak temperature signals much more difficult and the first to suffer is usually the accuracy of absolute temperature measurements. Strong radiant sources such as lamps or heaters produce asymmetry of the radiant field and this should be tested before a measurement. It is not only strong lamps, however, which can affect the measurement but even objects of low temperature can produce a non-uniform radiant field provided that a gradient of a few degrees Celsius exist between the subject's skin surface and the radiating object. Close proximity of surgeons can lower the comfort temperature of a patient undergoing an operation by as much as 4°C.[19]

Figure 3 presents a pair of gray-tone thermograms of a shorn sheep with about 0.15 cm of fleece left on the body. First, the animal is exposed for 15 seconds to radiation from a human hand kept in a distance of 20 cm from the sheep. The second thermogram shows a thermal imprint of the hand after it was removed from the picture. Similar effects could be met in those cases, both in human and animal applications, where similar insulative layers exist. This will particularly be evident in the veterinary applications.[20,21,22]

Draper and Boag[23,24] discussed the relationship of the surface temperature distribution detected by thermography to the energy balance of human subjects. Estimates have also been made of the size and depth at which local subcutaneous heat sources would produce a resolvable signal by perturbing the overall heat balance. They found that the curvature of smooth anatomical surfaces may also cause errors in temperatures estimated by infrared thermography. Where the emissivity of a smooth surface differs only fractionally from unity, the directional emissivity at oblique angles will be significantly reduced and must become zero at glancing angles. More recently, Clark[25] and Martin and Watmough[26] have also considered the contribution of radiation reflected from anatomical surfaces which has originated from the surroundings. This may mask changes in the emitted radiation by contributing to the total radiation reaching the detector. Signals of interest can therefore be obscured without the reduction in apparent temperature predicted by considering the emitted radiation alone, a factor which is particularly important where much of the environment of a surface is composed of other anatomical surfaces of similar temperature. Clark[25] demonstrated that the thermal radiation environment of smooth anatomical surfaces will have a major influence on the error in temperatures observed for angles of view greater than 45°, and that such errors are likely to be proportional to the difference between the temperature of the skin of the subject and his environment, and to the deviation of the skin emissivity from unity. This phenomenon is known to those users of infrared thermography who are familiar with an apparently reduced surface temperature close to the 'horizon' of a curved object.

The effects of anatomical curvature will be different when measurements of the surface temperature distribution of animal coats or clothed human subjects are performed. Clothing and animal coats will be black body emitters whatever the angle at which they are viewed. However, the surface temperature sensed by a directional radiometer will vary with the angle of view to the surface. Cena and Clark[27] showed

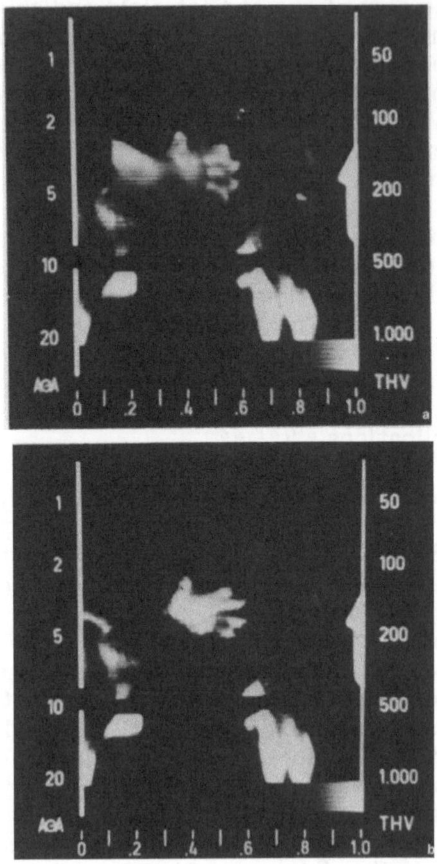

Fig. 3. Thermograms of a shorn sheep exposed to thermal radiation from a hand kept at a 20 cm distance from the sheep (a); after the hand was removed a thermal imprint remains (b). Air temperature 18°C, indoors.

that the change of radiative temperature with angle for hair coats will be related quantitatively to the coat structure and the temperature gradient across the coat. In the simplest terms, as a camera looks closer to the horizon it 'sees' a greater proportion of the hair tips and therefore a surface which is closer to air temperature than when the coat is viewed along the normal.

The radiant environment of animal or human subjects with coats or clothing may also produce variations in the local energy balance with consequent differences in surface temperature related only indirectly to the skin temperature. These can be due to three causes, and are particularly important when thermographic measurements are prmed outdoors. When the surface is exposed to direct solar radiation, the angle of incidence of the radiation on the body causes variations in the incident energy flux density which depends on the subject's posture, the coat or clothing roughness and the solar elevation. Usually, it is the difference between diffuse solar radiation and the total radiation which produce a marked temperature difference between the sun and shade sides of a subject. Coats or clothing with variable pigmentation or color will absorb different proportions of incident sunlight or short-wave radiation from electric lamps. For example, Figure 4 shows a giraffe in sunshine which exhibits a temperature pattern due to differences in radiation absorption associated with the coat pattern. Shadows may produce similar effects though perhaps more obviously. Solar radiation will therefore make thermography difficult in all but dull daytime conditions. This also applies to some specialized measurements performed indoors, and in particular, to all conditions where a radiant asymmetry can be detected. More care and research must therefore be devoted to the radiant environment in which thermographic examinations are made, especially when high degrees of accuracy are required.

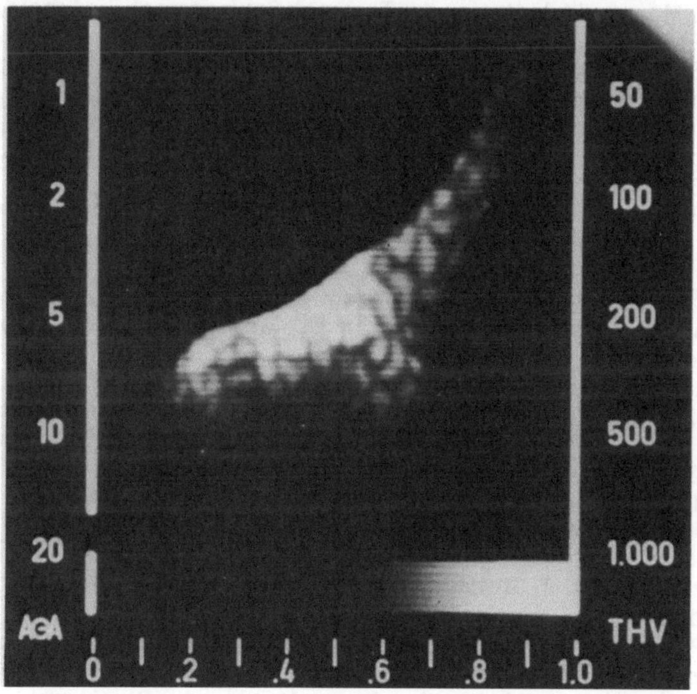

Fig. 4. Thermogram of a giraffe in sunshine. The temperature distribution on the coat is due to differential absorption of solar radiation associated with the familiar visible pigmentation. Solar irradiance on the horizontal plane was 500 to 700 Wm^{-2}. In these conditions the maximum measured temperature difference between adjacent dark and light areas was 6°C. From Cena and Clark.[22]

REFERENCES

1. L. E. Mount, The concept of thermal neutrality, in: "Heat Loss from Animals and Man," J. L. Monteith, L. E. Mount eds., Butterworths, London (1974).
2. Y. Houdas, J. D. Guieu, Environmental factors affecting skin temperatures, in: "Thermography," N. J. M. Aarts, M. Gautherie, E. F. J. Ring eds., Karger, Basel (1975).
3. P. O. Fanger, "Thermal Comfort," Danish Technical Press, Copenhagen, (1970), also: McGraw-Hill Book Co., New York 1973.
4. D. A. McIntyre, "Indoor Climate," Applied Science Publishers, London (1980).
5. K. Cena, J. A. Clark, "Bioengineering, Thermal Physiology and Comfort," Elsevier, Amsterdam (1981).
6. D. McK. Kerslake, "The Stress of Hot Environments," Cambridge University Press (1972).
7. J. L. Monteith, L. E. Mount, "Heat Loss from Animals and Man," Butterworths, London (1974).
8. L. E. Mount, "Adaptation to Thermal Environment," Arnold, London (1979).
9. G. S. Campbell, "An Introduction to Environmental Biophysics," Springer-Verlag, New York (1977).
10. K. Cena, J. A. Clark, Thermographic observations of skin temperatures of trained and untrained runners, J. Physiol. 257:8p. (1976).
11. R. P. Clark, B. J. Mullan, L. G. C. E. Pugh, Skin temperatures during running - a study using infrared color thermography, J. Physiol. 267:53 (1977).
12. K. Cena, Radiative heat loss from animals and man, in: "Heat Loss from Animals and Man," J. L. Monteith, L. E. Mount eds., Butterworths, London (1974).
13. J. A. Clark, A. J. McArthur, J. L. Monteith, A. E. Wheldon, The physics of the microclimate, in: "Bioengineering, Thermal Physiology and Comfort," K. Cena, J. A. Clark eds., Elsevier, Amsterdam (1981).
14. Y. Nishi, A. P. Gagge, Direct evaluation of convective heat transfer coefficient by naphthalene sublimation, J. Appl. Physiol. 29:830 (1970).
15. R. P. Clark, B. J. Mullan, Convective body cooling in air-conditioned buildings, J. Physiol. 267:8p. (1977).
16. A. P. Gagge, Effective radiant field, an independent variable that describes thermal radiation, in: "Physiological and Behavioral Temperature Regulation," J. D. Hardy, A. P. Gagge, J. A. J. Stokwijk eds., Thomas, Springfield Ill. (1970).
17. J. P. Funk, Direct measurement of radiative heat exchange of the human body, Nature 201:904 (1964).
18. American Society of Heating, Refrigerating, and Air-Conditioning Engineers (ASHRAE) (1981) Standard ANSI/ASHRAE 55-1981
19. E. Sliwinska, J. Laszczynska, K. Cena, Heat loss from patients during anesthesia, in: "Proceedings of the Third International

Congress of Thermology," Plenum, New York (1982).
20. J. A. Clark, K. Cena, The application of thermal imaging techniques to animals, Dtsch. tierarztl. Wschr. 79:289 (1972).
21. J. A. Clark, K. Cena, The potential of infrared thermography in veterinary diagnosis, Vet. Rec. 100:402 (1977).
22. K. Cena, J. A. Clark, Veterinary applications of thermography, Acta Thermographica 3:135 (1978).
23. J. W. Draper, J. W. Boag, The calculation of skin temperature distributions in thermography, Phys. Med. Biol. 16:201 (1971a).
24. J. W. Draper, J. W. Boag Skin temperature distributions over veins and tumors, Phys. Med. Biol. 16:645 (1971b).
25. J. A. Clark, Effects of surface emissivity and viewing angle on errors in thermography, Acta Thermographica 1:138 (1976).
26. C. J. Martin, D. J. Watmough, Thermal scanning of curved surfaces, Acta Thermographica 2:18 (1977).
27. K. Cena, J. A. Clark, Thermal radiation from animal coats: coat structure and measurements of radiative temperature, Phys. Med. Biol. 18:432 (1973).

THERMOGRAPHIC ESTIMATION OF THE EFFECTIVE HEAT EXCHANGE AREA BETWEEN A SUBJECT AND SUBSTRATE

A Solon, W Stepniewski, K Cena*

Technical University of Wroclaw, Wroclaw, POLAND
*** Occupational Health Program, McMaster University Hamilton, Ontario, L8N 3Z5, CANADA**

INTRODUCTION

Two classic methods have been used for evaluating the area of heat exchange between a subject and substrate. Kelly et al[1] proposed to measure the animal-floor contact area by simply drawing the animal's contour. Grommers et al[2] photographed, from the underside, the subject reclining on a transparent 'perspex' panel. The contact area was then determined by planimetry. Neither of these methods allows direct estimation of the area of heat exchange because it is impossible to define the proportion of contour area which exchanges heat. Furthermore, Grommers' method assumes that heat is exchanged by conduction only.

This paper describes a method for measuring the effective contact surface area between a subject and substrate and is suitable for use in evaluating heat exchange by conduction and radiation.

METHOD

Thermovision apparatus (AGA type 680) was used in these experiments. A subject was placed on a wooden panel covered with 2 mm felt sheet; some 30 minutes were allowed for thermal equilibrium in a room at 16°C. After steady state conditions were reached the subject was removed from the panel and a thermogram was taken of the remaining thermal imprint. Decrease of the apparent radiant temperature of the imprint was recorded with a thermoprofile adapter. The thermal imprint area was determined by planimetry from a millimetric scale grid superimposed on the thermograms. Alternatively, an electronic system was used to measure the areas of interest. This was based on digital analysis 'off-line' from a static thermogram and produced direct readings

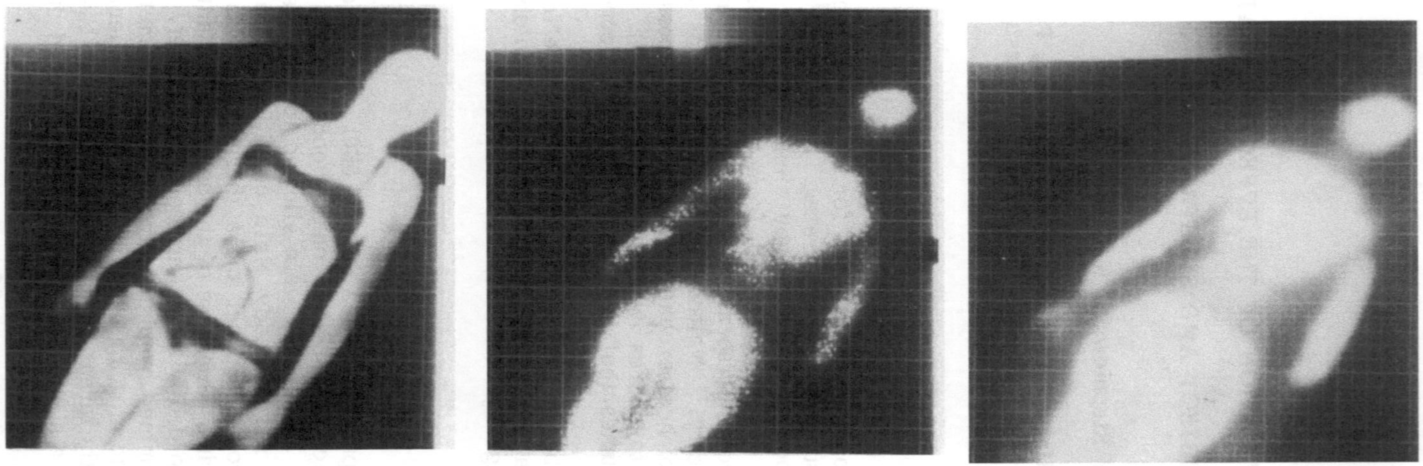

Fig. 1. Thermograms of a human subject. (a) Contour area - equivalent to Kelly's method. (b) Contact area - equivalent to Grommers' method. (c) Effective area of the heat exchange (conduction + radiation) - present thermographic method.

of temperature and the area of thermal contact.[3]

RESULTS AND DISCUSSION

Typical thermograms obtained by this method are shown in Figure 1. In Figure 1(a) an ordinary thermogram of a subject with a DuBois area of 1.73 m^2 is presented. A contour of the subject is here equivalent to the contour area determined by Kelly's method. In Figure 1(b) the warmest parts of the thermal imprint are shown. The area marked on this thermogram corresponds to the contact area estimated by Grommers' method.[2] The corresponding areas calculated by Kelly's and Grommers' methods were 0.75 and 0.37 m^2, respectively. The effective contact surface area between the subject and substrate determined with the present method (Fig 1(c)) was 0.55 m^2. This represents both the conduction and radiation heat exchange areas. In cold discomfort conditions at an air temperature of 15°C the area of conductive heat transfer (as in Fig 1(b)) was less than the area estimated by Grommers' method. The substrate thermal resistance and capacity were calculated from the thermographic measurements taken every 30 seconds from the moment when the subject left the panel.

The area recorded on a thermogram depends both on the distance between the camera and the subject and also on the camera's angle of view. For example, a measurement area of 1.24 m x 1.24 m was possible from a distance of 4.9 m with an angle of view of 15°, the distortion error being less than 0.6%. For a 45° angle of view the required distance was only 1.6 m, but the error of reproduction increased to 12.7%. Much of this error was attributed to the camera mounted mirror.

CONCLUSIONS

Thermography allows accurate measurement of the effective conductive and radiative heat exchange area between a subject and substrate. The effective area determined is usually smaller than the geometric contour area (Kelly's method) and larger than the contact area (Grommers' method). The difference is due to the extra radiative heat exchange area. The method presented may be applied to physiological, medical and technical applications.

REFERENCES

1. C. F. Kelly, H. Heitman Jr., J. R. Morris, Effect of environment on the heat loss from swine, Agr. Eng. 29:525-528 (1948).
2. F. J. Grommers, G. I. Christison, S. E. Curtis, Estimating animal-floor contact area, J. Animal Sci. 30:552-555 (1970).
3. W. Stepniewski, Ph.D Thesis, Technical University of Wroclaw, (1982).

HEAT LOSS FROM PATIENTS DURING ANESTHESIA

E Sliwinska*, J Laszczynska, K Cena*+**

*** Technical University of Wroclaw, POLAND**
**** Institute of Experimental and Clinical Medicine, Warsaw POLAND**
+ Occupational Health Program, McMaster University Hamilton, Ontario, L8N 3Z5, CANADA

INTRODUCTION

Constant body temperature in an unanesthetized human is due to thermoregulatory system activity. In thermoneutral and warm environments, sweating and cutaneous vaso-activity both influence heat loss from the body surface. In a cold environment, shivering increases metabolic heat production. Shivering does not often occur during anesthesia because of the use of muscle relaxants. The basal metabolic rate during surgery can also be lowered by the anesthetic agents which suppress oxidative metabolism.[1] Most anesthetics (halothane, nitrous oxide) produce vasodilatation.[2] This is followed by a rise in the skin temperature and increased heat loss from the body. The latter effect can be associated with a considerable decrease in the central body temperature. As the compensatory mechanisms are impaired during anesthesia, physical environmental conditions are most important in governing the heat exchange between patients and their surroundings.

The aim of this paper is to verify the influence of both the thermal environment and the type and site of surgery on the heat loss from patients. The optimal environmental temperature for patients undergoing intra-abdominal surgery is estimated.

METHODS

Measurements were taken in an operating theater during surgical operations. 16 patients of similar body structure and age (30-40 years old) were studied. The general anesthesia used was a mixture of 1% halothane, 30% nitrogen and 69% air.

Thermocouples, rectal probe and heat flow meters were attached to the patient at the beginning of surgery. At the same time the preparatory activities usual before operations were performed. Anesthetization was begun about 20-30 minutes after the preparations had started. At this time the first temperature and heat loss measurements were performed. The next set of readings was taken after 5 minutes, immediately following intubation. Measurements were then taken every 10 minutes during surgery and finally when the operation was ended. Dry and wet bulb temperatures, globe temperature and air velocity were monitored in the middle of the operating theater and at the operating table. Dry and wet bulb temperatures of air inspired and expired by the patient were measured. Heat production was estimated by means of oxygen consumption measurements.

Thermal insulation of the standard draping used through this study was measured on a thermal manikin. The insulation value was found to be 1.1 clo (Madsen, private information).

RESULTS

The air temperature and the mean radiant temperature in the operating theater ranged from 23.5 to 24.5°C, with a mean of 24.0°C. Air velocity was usually less than 0.1 $m.s^{-1}$. Relative humidity ranged from 50% to 60%. The mean value of the operative temperature at patient level (27.8 - 28.4°C) was found to be about 4°C higher than in the rest of the room. This was probably caused by the presence of surgical staff and by radiation from the lamps.

At the beginning of surgery high values of dry heat loss were observed (50-60 Wm^{-2}). The dry heat loss (H_D in Figure 1) decreased with the surgery progression to:

20 Wm^{-2} for intra-abdominal operations
33 Wm^{-2} for lower abdomen operations
50 Wm^{-2} for extra-abdominal operations.

The basal metabolic rate changed during intra-abdominal surgery. The mean value between intubation and completion of an operation was 30 Wm^{-2}. This is lower by about 10 Wm^{-2} than the metabolic heat production of sleeping man and is probably an effect of the anesthetic.[3]

Respiratory heat loss was estimated using classic formulae for the convective and evaporative heat exchange, and gave a mean value of about 3 Wm^{-2}. This is much lower than in unanesthetized humans and is probably due to the lowered basal metabolic heat production (hence the decrease in oxygen uptake) and also to the humidification of the air supplied to the patients.

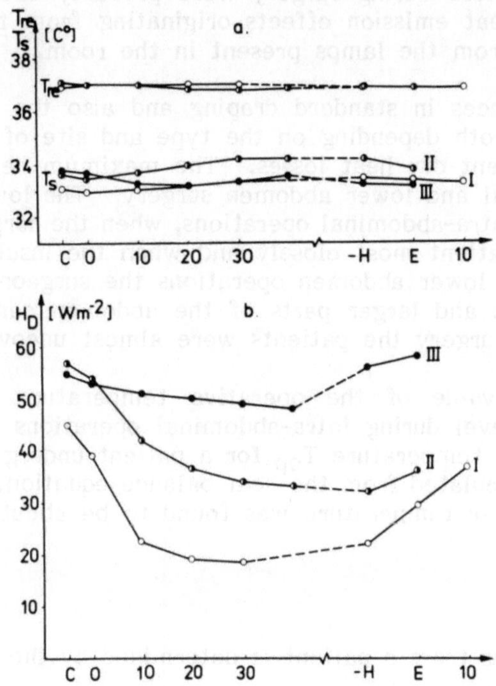

Fig. 1. (a). Changes of rectal T_{re} and skin T_S temperatures during anesthesia against time in minutes. (b) Changes of dry heat loss H_D during anesthesia against time in minutes.
C: control, H: removal of halothane, E: end of surgery.
I: intra-abdominal operations; II: lower abdomen operations;
III: extra-abdominal operations.

Rectal temperature (T_{re} in Fig 1) did not change greatly during typical operations. The small fall was due to a decrease in the metabolic heat production and was in good agreement with the results of Morris,[4] Shanks,[5] Holdcroft and Hall.[6]

Individual values of the mean skin temperatures T_S varied from 32°C to 34°C. Small changes of T_S of the order of 0.5°C were associated with changes in draping, increased radiation from lamps and also the application of the anesthetic agents. The fall of T_S at the end of surgery was mainly caused by removal of the draping sheets.

DISCUSSION

The reason for high values of dry heat loss at the beginning of surgery was the comparatively large body surface area exposed prior to draping and then covered with sheets at room temperature. Low values of the dry heat loss during surgery were probably caused by radiative and convective heat emission effects originating from the surgeons and nurses and also from the lamps present in the room.

The differences in standard draping and also the presence of the operating staff (both depending on the type and site of surgery (Fig 2)) caused the different dry heat losses. The maximum heat loss occurred in extra-abdominal and lower abdomen surgery. The lowest values were recorded during intra-abdominal operations, when the surgeons and nurses surrounded the patient most closely and when the insulation of draping was maximal. In lower abdomen operations the surgeons stand close to the patient's legs and larger parts of the abdomen can emit heat. In extra-abdominal surgery the patients were almost uncovered.

The mean value of the operative temperature T_o observed at operating table level during intra-abdominal operations was 28°C. The optimal operative temperature T_{op} for a patient undergoing this type of surgery, when calculated from the heat balance equation, was 27.3°C and the optimal theater temperature was found to be about 23.3°C.

CONCLUSIONS

1. The heat loss from a patient is determined by the type and site of surgery.

2. The air temperature recommended for thermal balance for patients in this study was 23°C. This relatively low value was caused by the presence of surgeons and other heat sources in the operating theater.

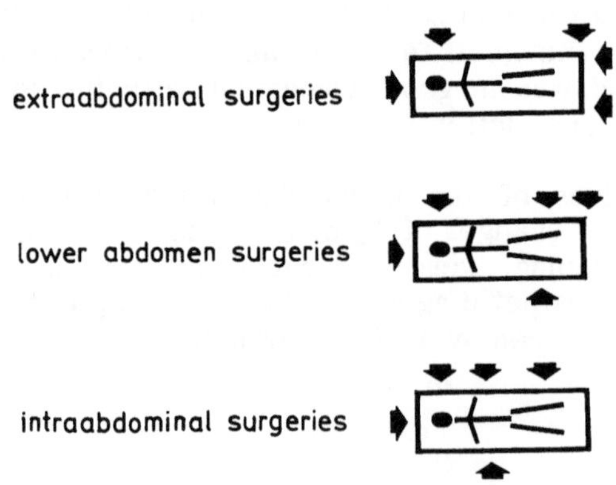

Fig. 2. Arrangement of surgeons and nurses (indicated by arrows) around patients for 3 different types of surgical operations.

REFERENCES

1. M. J. Goldberg, F. C. Roe, Temperature changes during anesthesia and operations, Arch. Surg. 93:365 (1966).
2. R. J. Vale, Normothermia: its place in operative and postoperative care, Anesthesia 28:241 (1973).
3. C. F. Roe, The influence of body temperature on early postoperative oxygen consumption, Surgery 60:85 (1966).
4. R. H. Morris, Operating room temperature and the anesthetized paralysed patient, Arch. Surg. 102:95 (1971).
5. C. A. Shanks, Mean skin temperature during anesthesia: an assessment of formulae in the supine surgical patients, Br. J. Anesth. 47:871 (1975).
6. A. Holdcroft, G. M. Hall, Heat loss during anesthesia, Br. J. Anesth. 50:157 (1978).

ENVIRONMENTAL INFLUENCE ON THE VASCULAR RESPONSE OF THE HANDS TO COLD CHALLENGE: AMBIENT TEMPERATURE AND TIME

E D Cooke, S A Bowcock

Department of Medical Electronics
St Bartholomew's Hospital, London
UNITED KINGDOM

Despite the often-reported recommendations that thermographic studies should be performed at an ambient temperature of 20°C ± 1.°C[1,2,3,4], our empirical observations and studies of blood flow in a number of ambient temperatures (Fig 1) suggested that this advice may not hold good for thermographic studies of the peripheral vasculature.

During these investigations patients may remain in the prevailing ambient temperature for periods of longer than one hour and are subject to some form of vascular challenge, usually mild cold stress. Figure 1 shows the behavior of blood flow at various ambient temperatures. Since laboratory conditions under 20°C soon become intolerable for staff and patients alike and since temperatures greater than 26°C produce invisible perspiration in lightly clad subjects only the temperature range of 20°C >25°C need be considered (the continuous line in Fig 1); the arrow indicates 24°C. It will be seen that blood flow at this point is just about to increase and that vasoconstriction is not evident, ie a neutral position. Clearly vasodilatation or vasoconstriction from this point may occur readily. However, this graph takes no account of time in the response to a particular ambient temperature nor the effect of cold challenge. An experiment, therefore, was constructed to test the vascular response at two ambient temperatures and to include both the elements of time and mild cold challenge.

PATIENTS AND METHOD

Ten asymptomatic normal subjects, 8 hypertensive patients, BP 160/105, and 8 patients with Raynaud's phenomenon were examined using thermography (AGA 680 Medical) of the hands and photoplethysmography (PPG-Medisonics PA13) with the transducer placed on the pulp of the thumb.

Cold stress, 20°C for one minute, was applied to the right hand of each patient. The subject remained in the same ambient temperature seated at rest for 30 minutes after which the same cold challenge was repeated.

This procedure was used on each subject on two separate occasions, separated by a period of one week, in ambient temperatures of 20°C ± 1°C and 24°C ± 1°C, the first ambient temperature used with any particular subject being allocated at random.

RESULTS

Raynaud's phenomenon: cold sensitivity occurred in all subjects at both ambient temperatures and tended to be increased with time.

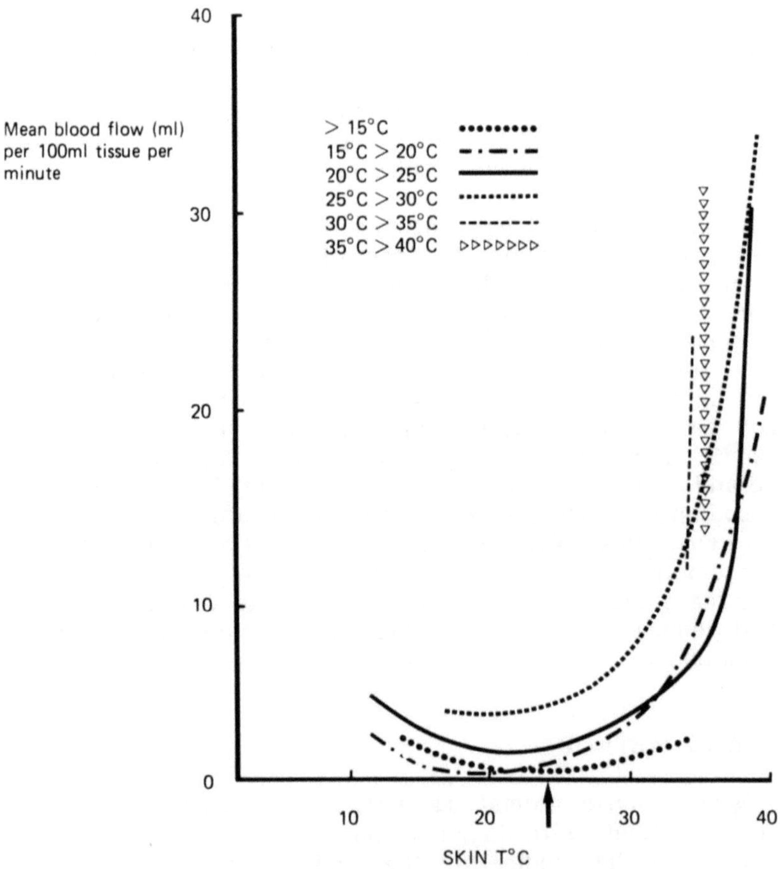

Fig. 1. Blood flow at a number of ambient temperatures (see text) after Montgomery, NASA publications.

At 20°C ± 1°C 6 of the 8 hypertensive subjects showed immediate cold sensitivity and all subjects showed delayed cold sensitivity. At 24°C ± 1°C 4 hypertensives showed immediate and 6 delayed cold sensitivity.

Of the 10 normal subjects 2 showed immediate and 8 showed delayed cold sensitivity at 20°C ± 1°C. At 24°C ± 1°C neither immediate or delayed cold sensitivity was observed. These observations are summarized in Table 1. Figures 2 and 3 illustrate the typical thermographic and PPG findings in a normal subject at 20°C ± 1°C and 24°C ± 1°C.

DISCUSSION

These findings confirm that an ambient temperature of 20°C ± 1°C may be sufficiently vasoconstrictive to influence the response to mild cold stress even in some normal subjects, an effect that is more severe with increasing time of exposure. The marked reduction in vasomotor activity and the vasoconstriction with time which occurs at 20°C as compared with the retention of normal vasomotor activity, which occurs at 24°C, even following a second cold challenge, is well seen on the PPG tracing (Fig 2). Figure 3 shows the attentuation of the response to mild cold stress which occurs in a normal subject at 20°C, compared with the normal response found at 24°C.

It is evident, therefore, that the lower temperature may lead to anomalous results when testing the vascular function of the hands and, by extrapolation, this comment would be expected to be true for other parts of the peripheral circulation.

Table 1. Raynaud's phenomenon: cold sensitivity was further increased with time at both ambient temperatures in all subjects

Hypertensive subjects	Ambient temperature	
Cold sensitivity	24.0°C	20.0°C
Immediate	4	6
Delayed	6	8
Normal subjects	**Ambient temperature**	
Cold sensitivity	24.0°C	20.0°C
Immediate	0	2
Delayed	0	8

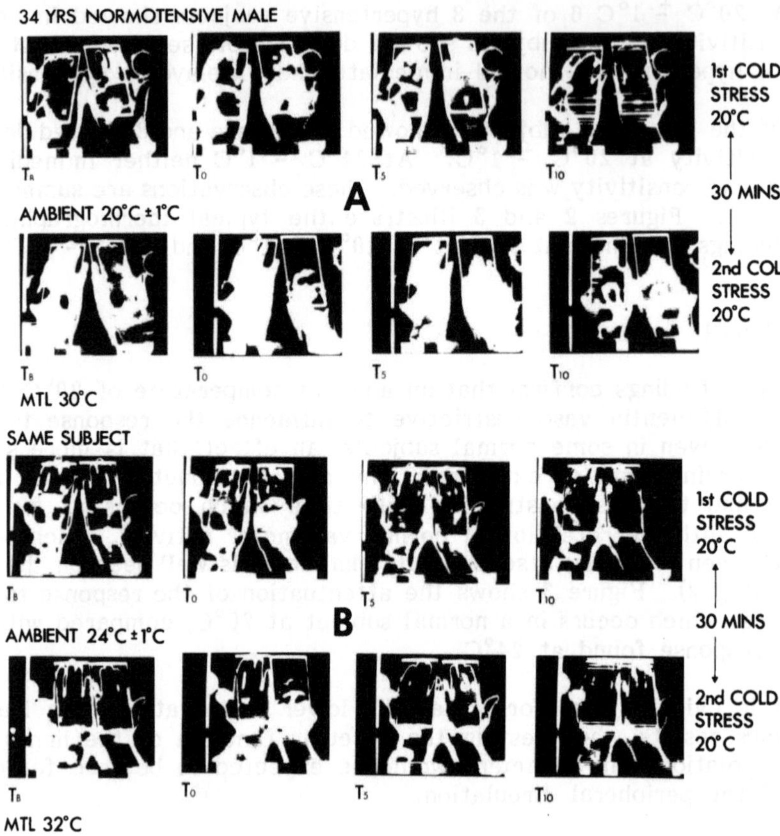

Fig. 2. Thermographic record of mild cold stress: normal subject. At 20°C the normal response (upper row) is attenuated with time (second row). At 24°C (third and fourth rows) no attenuation of the response to mild cold stress occurs. Tβ, T_0, T_5 and T_{10} are the thermograms at baseline, immediately following and at 5 and 10 minutes after cold stress.

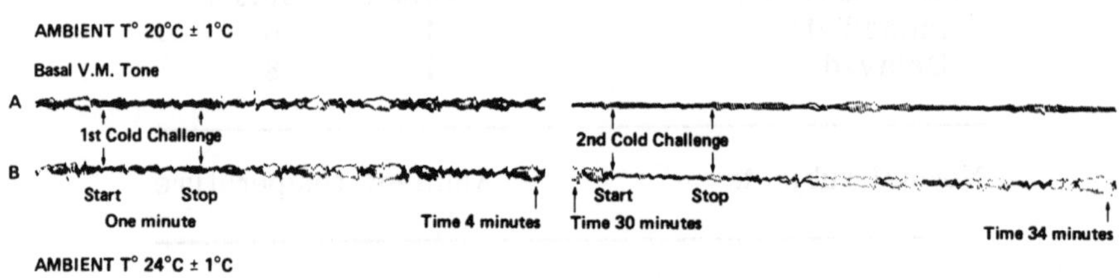

Fig. 3. Photoplethysmography trace obtained from the same patient as illustrated in Fig. 2. The upper trace taken at 20°C shows the almost complete loss of resting vasomotor tone with time. Vasomotor tone is unaffected with time at 24°C (lower trace).

It follows that tests of vascular performance using thermography should be carried out in an ambient laboratory temperature of 24°C. Lower temperatures, which reduce vasomotor activity, particularly when exposure time is prolonged, should be used only to enhance the contrast between hot and cold areas of the body, eg when examining the female breast or examining the legs for deep vein thrombosis.

REFERENCES

1. Locomotor Diseases Group Report of the Anglo-Dutch Thermographic Society. Thermography in locomotor diseases - Recommended Procedure (1978).
2. L. Langer, S. E. Fagerberg, C. Johnsen, Peripheral circulation in diabetes mellitus - a study with infrared thermography, Acta Med. Scand. 191:17 (1972).
3. C. H. Jones, W. P. Greening, J. B. Davey, S. A. Mescinna, V. J. Weaver, Thermography of the female breast. A five year study in relation to the detection and prognosis of cases, Br. J. Radiol. 48:532 (1975).
4. J. Nyirjesy, M. R. Abernathy, G. S. Billingsley, P. Bruns, Thermography and detection of breast carcinoma: a review and comments, J. Reprod. Med. 18:4:165 (1977).

TEMPERATURE CHANGES CAUSED BY DRUGS

J Frens

Institute of Veterinary Pharmacology, Pharmacy and Toxicology, Utrecht University, Biltstraat 172, Utrecht THE NETHERLANDS

The basis of thermology is the temperature and conductance of the skin. This parameter may be affected in many different ways: by the processes that are under study, or by processes that have no relation whatsoever to those that the thermologist pursues.

One of the physiological processes that does affect skin temperature is the regulation of body temperature. This thermoregulation may be affected by drugs that are taken by the patient. In addition, drugs may act directly on skin temperature or conductance.

Body temperature is a regulated entity. In normal subjects, the thermoregulatory system tries to maintain the balance between heat production and heat loss, so that a stable core temperature is achieved. Other temperatures in the body (eg skin temperature at the extremities) are subordinate to core temperature and are used in the regulatory process to keep core temperature constant.

Thermoregulatory mechanisms have the characteristics of a servo system. The components of this system are temperature sensors, an integrating mechanism in the CNS and thermal effectors like shivering, vaso-activity, and metabolism. The nervous system plays a predominant role in the regulation, and different neurotransmitters have a function.

Obviously, there is a relation between core and skin temperature. The higher the core temperature, the steeper the gradient to the environment will become, with generally higher skin temperature. Some things, however, obscure this relationship considerably, for example vaso-activity in the skin and evaporation by sweating. Therefore it does not necessarily follow that a rise in core temperature will always be accompanied by a rise in skin temperature. Nor will parts of the skin react in a uniform way. In man, peripheral blood vessels involved in thermoregulation are essentially limited to the extremities, while sweating may occur over practically the whole skin surface. This means that the relation between core and skin temperature is complex. A cold stimulus to the skin produces vasoconstriction at the extremities, a warm stimulus leads to vasodilatation at the extremities accompanied by sweating over the total body surface. The magnitude of these responses is modulated by the core temperature.

Interference, either with the thermoregulatory system or with the thermal effectors, may lead to changes in core and skin temperature that are not necessarily similar. Many drugs have such an effect on different levels in the system.

Several substances change the firing rate of thermosensors. These thermosensors are neurones that change their activity when the temperature changes. They are of two types, the so called warm sensors that increase their activity when temperature rises and the cold sensors that increase their activity when temperature falls. Notorious for its effect on warm sensors is capsaicin, the pungent principle in the red pepper. By stimulating the warm sensors, capsaicin induces the well known "hot" feeling, and so gives information to the thermoregulatory system that is not in accord with reality. The thermoregulatory system reacts to this by inducing vasodilation and thus elevating skin temperature. Other substances have similar or opposite effects and may have been taken by patients undergoing tests with thermography.

The central regulating system also can be affected by drugs. On experimental evidence it was formerly assumed that the hypothalamus was the center for the control of body temperature. Although the hypothalamus is important, it is no longer tenable to give an exclusive role to this part of the brain. Neuronal circuits involved in thermoregulation range throughout the central nervous system and participate not only in thermoregulation, but in a variety of other functions as well. A neuronal network to describe the thermoregulation in a model, therefore, is bound to be incomplete. Several models exist and most have the following characteristics: signals from warm sensors tend to excite pathways involved in heat loss while simultaneously inhibiting pathways involved in heat production. Signals from cold sensors operate in the reverse manner. Built into several such models is the concept of a "set-point" which results either from comparison of activity in the warm and cold sensor pathways, or by assuming a comparison to a reference signal that is temperature insensitive. Many neurotransmitters are involved in the central thermoregulatory processes. Strong evidence has been found for the involvement of acetylcholine, 5-hydroxytryptamine (5-HT), dopamine, noradrenaline and histamine. These same neurotransmitters also have a function in many other cerebral processes.

Drugs that affect the regulating mechanism have to pass the blood brain barrier and interfere with central neurotransmission. Many drugs are able to do this and are even designed to do so, such as tranquillizing drugs, MAO inhibitors, 5-HT or dopamine antagonists etc. The effect they have on thermoregulation may vary, depending on the type of neurotransmitter affected, the thermal status of the patient on administering the drug, and the ambient temperature. Most neurotransmitters have more than one function and most drugs are not specific as to the neurotransmitter involved, so the effect a drug has may vary considerably.

An example of this may be found in the effect of methysergide (Deseril R), a 5-HT antagonist used in - among other things - the treatment of migraine. This drug was studied in relation to thermoregulation in the goat and rabbit. Both species have a thermoregulation similar to the human, and use 5-HT as a neurotransmitter in the central neuronal pathways that induce heat loss.

When methysergide is given, it would be expected that this 5-HT antagonist would inhibit heat loss and produce a rise in body temperature. This actually happens in a warm environment, when heat loss is active. At such an ambient temperature, administration of methysergide results in a rise of core temperature accompanied by a lowering of skin temperature in the extremities, because vasodilatation is inhibited. In thermography such an effect obviously will influence thermograms, but interpretation will be difficult, because a rise in corė temperature in this case will show as a lowering of skin temperature in the extremities.

When methysergide is administered at a low ambient temperature, the effect is completely different. Because there is no initial vasodilatation, inhibition of this phenomenon will not take place and there is no change in skin temperature, while core temperature sometimes even falls, because methysergide mimics the effect of 5-HT in a low temperature environment. In a semi-thermoneutral environment, such as that normally used in thermography, the effect of methysergide will be a lowering of skin temperature without a clearcut effect on core temperature.

These effects at different ambient temperatures can be explained theoretically, but it is difficult to define a consequent effect of methysergide on thermoregulation, and even more difficult to define its effect on skin temperature. This is also true for other drugs affecting central thermoregulation, making standardization of experimetnal conditions essential in studies involving thermoregulatory drugs.

Other drugs that may extensively influence the central regulating mechanism are tranquillizers of the phenothiazine type, amphetamines, reserpine and the like. All these drugs may have been taken by a patient undergoing thermography, either with knowledge of the thermologist or not.

The effects they have are just as complicated as those of methysergide. Some generalizations can be made however. It appears that drugs of the type that lower body temperture (eg chlorpromazine, reserpine etc) do so by inducing vasodilatation in the extremities. In a thermoneutral environment this means that a rise in skin temperature can be observed even though core temperature is below normal. Core temperature may be measured directly or may be derived from the skin temperature of the trunk, which is correlated to core temperature in this case. Drugs of this class, that elevate body temperature in a thermoneutral environment (eg amphetamines), generally have no such primary action on the peripheral blood vessels. Amphetamine toxicity is generally accompanied by flushing and peripheral vasodilatation.

Drugs that interfere directly with the thermal effectors like vaso-activity, metabolism, sweating etc, generally also have an effect on body temperature. Their effects are more clearcut than those of centrally active drugs. Stimulation of metabolism produces a rise in core temperature and vasodilatation lowers it. Nevertheless there are some complications. Thermoregulation per se tries to counter any changes that are induced.

A good example of this are the cannabinoids, which are mainly used abusively. Some - but not all - cannabinoids have hypothermic activity. It is not clear, however, whether a significant decrease in body temperature accompanies the doses commonly employed, because tolerances to the hypothermic effect may develop quite easily and there is a considerable individual variation in response. The cause of the hypothermia appears to be a lowering of metabolism. It is accompanied by a peripheral vasoconstriction, because the normal physiological mechanism of heat conservation is still intact and becomes activated by the drop in core temperature.

In conclusion, it appears that any drug-induced change in body temperature will necessarily be complicated in its effects on the human body. Thermology has to reckon with these effects, however, because thermograms only show part of the complex events that take place to keep core temperature constant, while on the other hand a change in skin temperature or some other thermological parameter may very well be a thermoregulatory event, not connected to the process under study.

MONITORING THE FACIAL FLUSH: A METHODOLOGICAL STUDY

S A Bowcock, S Medbak*, E D Cooke

Departments of Medical Electronics and
*** Chemical Endrocrinology**
St Bartholomew's Hospital, London, UNITED KINGDOM

Facial flushing following the oral administration of a hypoglycemic agent (chlorpropamide) and alcohol may be genetically determined and is associated with an elevation of met-encephalin levels in plasma.[1,2,3,4]

It is desirable, therefore, to have an accurate, objective and reproducible method of measuring the facial flush.

This investigation was constructed to compare the accuracy of thermography, a thermistor and subjective observations in determining the presence or absence of facial flushing.

PATIENTS AND METHOD

The investigation was performed at an ambient temperature of 21.0°C $\pm$ 0.5°C, humidity 50 $\pm$ 10 per cent. The instrumentation and position of the patient is shown in Figure 1.

The thermographic system used was the AGA 680 Medical with an integrator which measured the mean infrared emission of the forehead and right cheek using a digital analog scale (one unit = 0.1°C). The thermistor, which had an output of 10 mV per 1°C temperature change and was accurate to $\pm$ 0.2°C within the range of 26°C to 35°C, was placed on the bony prominence of the cheek just below the outer canthus.

The subjects were 25 type II diabetic patients aged 51-73, and 25 healthy individuals (control group) aged 32-60.

Chlorpropamide 250 g was given orally to all subjects approximately 18 hours before the commencement of the investigation. Alcohol was

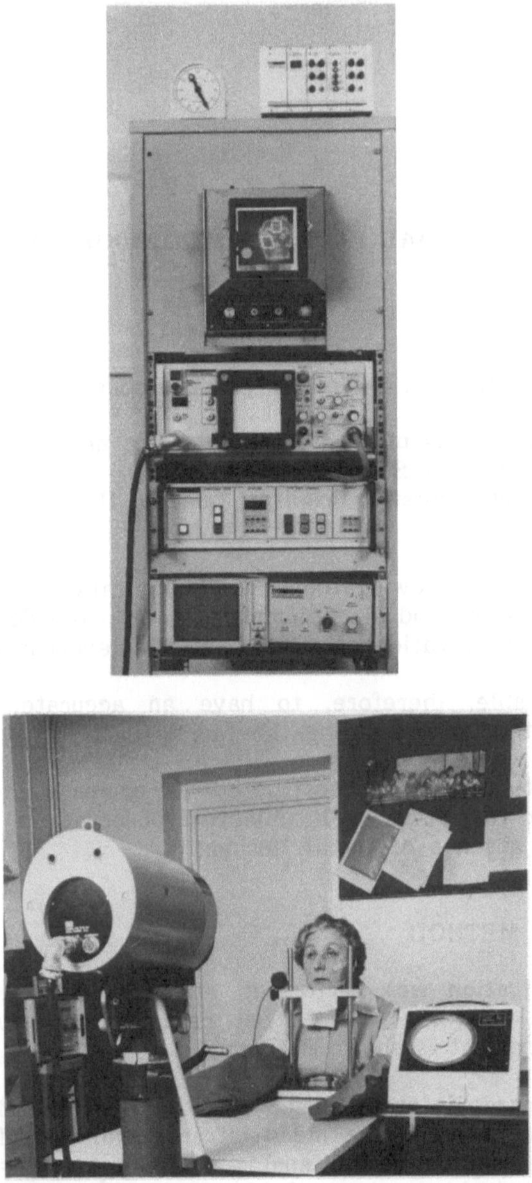

Fig. 1. Showing the position of the subject and thermographic system and site of thermistor during monitoring the effect of chlorpropamide-alcohol challenge.

given orally as 40 ml dry sherry (or ethanol) following stability of the facial temperature. The period of monitoring was 70 minutes during which patients were asked to assess the occurrence of flushing by comparing their facial appearance with that at the beginning of the experiment. A feeling of flushing was also recorded as a positive result.

Each subject was studied twice with and without the administration of chlorpropamide. A separate study was performed in a few subjects with the thermistor placed on the cheek without alcohol and chlorpropamide challenge but with the ambient temperature at 24°C ± 0.5°C, ie when vasodilation is likely to be present.

RESULTS

In patients who flushed, the flush commenced 10-15 minutes after the ingestion of alcohol and was maximal at about 20 minutes. Characteristically it was sustained for at least 30 minutes and in most patients until the investigation was complete, ie one hour after the ingestion of alcohol. The presence of a flush was most easily determined by thermography, ie the mean infrared emission of the forehead or cheek rather than by measuring point temperature by thermistor. The reading from the latter lacked stability and was subject to wide, unsustained fluctuation so that in the great majority of cases it was not possible to

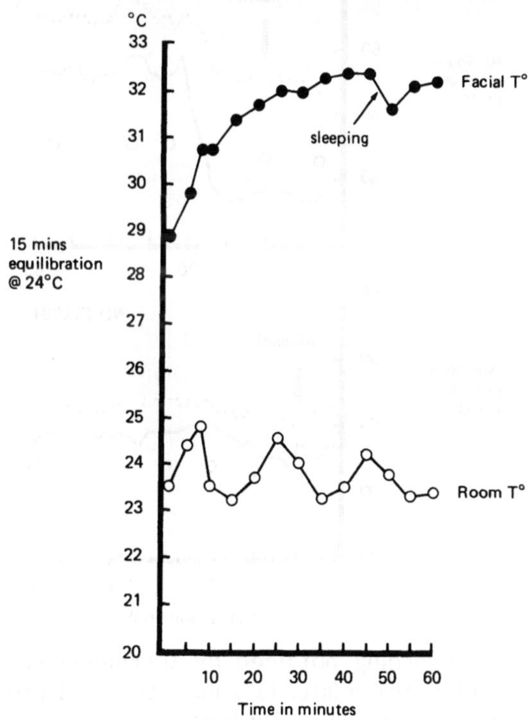

Fig. 2. The temperature recorded by a thermistor placed on the cheek just below the outer canthus varies spontaneously and may rise in the absence of a chlorpropamide-alcohol challenge.

determine whether a flush had occurred or not. Figure 2 shows an example of a rise in thermistor (cheek point temperature) in the absence of challenge. In contrast, when no flush occurred the mean infrared emission of the cheek and forehead was stable. This was remarkably so in the case of the forehead where the reading is virtually a straight line with time (Fig 3).

In the 31 subjects who were asked to make a subjective assessment of the presence of a flush, 15 flushed and 13 recognized a flush correctly. However, of the 16 that did not flush, seven subjects believed that they had (Fig 4).

DISCUSSION

From the results it is evident that the mean temperature (infrared emission) of the cheek or forehead, which is remarkably stable under standard ambient conditions, should be used to determine the presence or absence of facial flushing rather than the temperature of a single point

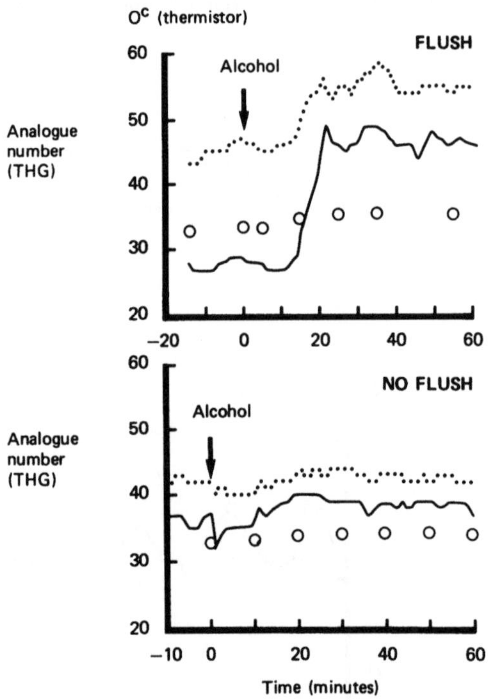

Fig. 3. Typical recording obtained by thermography using an analog scale and by thermistor (absolute temperature) in the presence and absence of facial flushing.

.....	forehead	: THG
———	cheek	: THG
ooooo	cheek	: thermistor

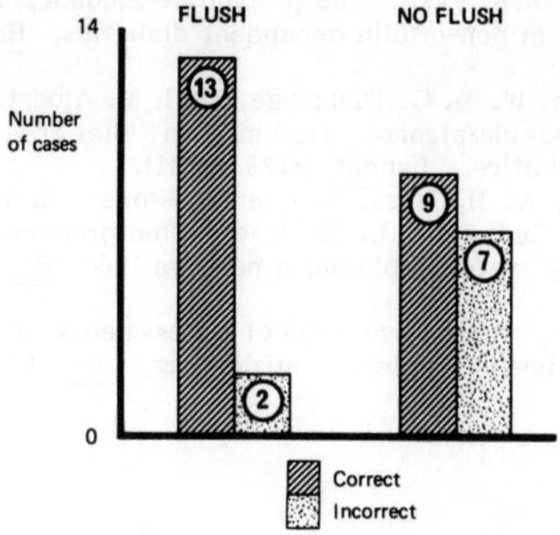

Fig. 4. Subjective assessment of flushing may be grossly misleading.

in the cheek which is chosen arbitrarily and which is subject to wide spontaneous fluctuation. This is scarcely surprising when the rhythmic change of vasodilation and vasoconstriction due to resting vasomotor tone are considered. The phenomenon associated with fixing the thermistor must also be considered. By its nature it is not possible to be certain that the approximation of the thermistor to the skin is the same every time, and that the position will remain unchanged by the facial movements of the patient, which inevitably occur. The inaccuracy of determining a facial flush by subjective assessment was not unexpected, considering previous experience with this method of measurement.[5]

CONCLUSION

Subjective assessment or measurement of point temperature by a thermistor placed on the cheek may be grossly misleading and are inaccurate determinants of facial flushing with chlorpropamide-alcohol challenge. In contrast the mean infrared emission of the cheek or forehead (the latter is in our experience the most stable part of the body) determines the presence or absence of a facial flush with accuracy.

REFERENCES

1. R. D. Leslie, D. A. Pyke, Chlorpropamide-alcohol flushing:a dominantly inherited trait associated with diabetes, Br. Med. J. 2:1519 (1978).

2. A. H. Barnet, D. A. Pyke, Chlorpropamide-alcohol flushing and large-vessel disease in non-insulin dependent diabetics, Br. Med. J. 2:261 (1980).
3. N. E. De Silva, W. M. G. Tunbridge, K. G. M. Alberti, Low incidence of chlorpropamide-alcohol flushing in diet-treated non-insulin dependent diabetics, Lancet 1:128 (1981).
4. S. Medbak, J. A. H. Wass, V. Clement-Jones, E. D. Cooke, S. A. Bowcock, A. Cudworth, L. M. Rees, Chlorpropamide-alcohol flush and circulating met-encephalin; a positive link, Br. Med. J. 283:937 (1981).
5. E. C. Poulton, Qualitative subject assessments are almost always biased: sometimes completely misleading, Br. J. Psychol. 60:408 (1977).

TEMPERATURE CHANGES IN SKIN DISEASE

G Stuttgen

Hautklinic und -Poliklinik der Freien Universitat
FB 03 - WE 15 Rudolf-Virchow-Krankenhaus
D-1000 Berlin 65 WEST GERMANY

INTRODUCTION

The clinical features of skin disease are well known, but many are difficult to quantify, particularly with minor changes which are very much observer-dependent. For monitoring the course of disease a verbal description may be insufficient, and too crude for analysis. Infrared thermography can be helpful in analyzing surface thermal patterns in dermatoses and in supplementing morphological and histological findings. A most important application in dermatology is the evaluation of various pharmacological effects of drugs, especially those applied topically.

For dermatological diagnostics, thermography must be complemented by other clinical and technical examinations, in order to facilitate the interpretation of skin temperature patterns.

Dermatological thermograms may be difficult to interpret because of the relative cooling effect of intradermal edema, vesicles or blisters in upper skin layers, and evaporative water loss in response to sweating. Anatomical and physiological factors must be considered.

Apart from the well known clinical signs of redness, swelling and surface damage, dermatoses are frequently characterized by changes in blood flow, and thus by alterations in heat transport from the body core to the skin microcirculation. Skin blood flow is one of the most important factors influencing surface temperature, but heat production by metabolic processes in the different skin layers is less significant. Capillary blood flow and sweat gland secretion are most important in skin temperature regulation, whereas the thickness of different skin layers influences heat conduction from the level of the microcirculation to the surface.

Reviewing the literature, there are few papers dealing with thermography in dermatology, and consequently the following clinical descriptions are mostly based on this author's own experience. The use of thermography in dermatology is at an early stage with more details about the clinical value, limitations and perspectives needed for the future. In this paper, only first impressions can be given about the clinical possibilities for the different pathologies.

ERYTHEMA

Skin erythemas are characterized by an increase in the number of erythrocytes within the subpapillary capillary plexus. In cases of increased blood flow through the capillary network, erythema may be associated with high skin surface temperatures owing to greater heat transport. Skin temperature is dependent more on blood flow than on the amount of blood in the skin. For example, hemangiomas are normally "cool" lesions with a low surface temperature because they are not integrated into the normal blood circulation but are by-passed by shunts. Therefore, the greater volume of blood in a hemangioma is not associated with high temperature.

In contrast, erythema of the skin after alcohol ingestion may well be associated with higher skin temperatures - not because of a direct vasodilator effect of alcohol but because of the vasodilator neural reflex after irritation of the gastric mucosa by alcohol contact and uptake. This erythematous reflex is pronounced in the head and neck region and is distinct from the alcohol-induced vasodilation that occurs 10-15 minutes after consumption. The temperature may rise by 6°C in the first minutes after alcohol consumption and gastric contact. The level of temperature increase is not directly correlated to the intensity of erythema, because the development of erythema is dependent on the number of erythrocytes passing through open shunts and dilated vessels, whereas temperature increase is dependent on convective heat transport by blood plasma.

To further evaluate this difference we measured the temperature of different erythematous skin regions induced experimentally by either ultraviolet B irradiation or by topical application of dithranol. Both provocations were performed simultaneously on comparable areas on the patient. After UVB-irradiation the peak temperature was reached within 24 hours, but the dithranol treated area only showed the maximum temperature after a further 24 hours. The physiopathological cause of this time difference is the liberation of mediators with different vasodilator activity. The difference can only be assessed with thermography as the erythema formation is similar after both treatments.

URTICARIAL ERUPTIONS

Urticaria Factitia

This can be provoked by pressure on the patient's skin. The combined symptoms (redness, swelling, warmth) may be suppressed by pre-treating the skin with topical drugs, eg anti-histamines. Thermography can show the suppression of urticarial eruptions and the temperature rise after treating the skin topically with anti-histamines four hours before urticarial provocation. On the other hand, pre-treatment with topical preparations of corticosteroids prevents or suppresses wheal eruption, but the temperature rise remains unchanged, as clearly demonstrated with thermography. Corticosteroids have a greater influence on the permeability of capillary walls than on blood flow, whereas anti-histamines have a direct influence on vasodilation.

Cold-induced Urticaria

This is also a mediator related process. Thermographic results are similar to those mentioned above. In patients pre-treated on one side of their back with a topical corticosteroid preparation 5 hours prior to the application of ice pack to provoke urticaria, we could find no significant changes in skin temperature between steroid pre-treated and control sides. Wheal eruption, however, was significantly less pronounced on the steroid pre-treated side.

Thus, thermographic methods are well suited to clinical trials of anti-histamine, anti-urticarial, topical or systemic drugs as well as to differentiating the role of various mediators in the inflammatory process.

Cholinergic Urticaria

Cholinergic urticaria is stimulated by sweat secretion. After physical work or exercise the clinically visible urticarial eruptions are mostly not very impressive. However, the temperature pattern shows around the sweat glands. After topical applications of anti-histamine these temperature peaks disappear. This is another example of thermography as a complementary, objective and quantitative index in clinical dermatology.

Urticaria Pigmentosa

This is characterized by a local increase in mast cells, continuously liberating histamine and other irritating mediators which stimulate melanocytes. This leads to the typical brown skin coloration owing to higher melanin content. At the same time, histamine and other mediators have a marked vasodilator effect, especially on arterioles. This vasodilation causes a rise in skin surface temperature, which can be detected by thermography. Thermography also reveals the effect of histamine on lymphatic and venous vessels, a phenomenon that has to be studied in further detail.

EPIDERMAL LESIONS

After epidermal damage, the evaporative cooling effect of intercellular and lymphatic fluid can disguise a rise in local temperature. Without vesiculation and epidermal damage, skin inflammation - as in allergic contact eczema - will be characterized by local hyperthermia owing to the liberation of vasodilator mediators of inflammation. The extent of inflammation can be quantified by thermography.

Burns

Burns can be monitored by thermography to evaluate the degree of epidermal damage.

Frost bite

In severe burns and frost bite local capillary circulation ceases, thus lowering skin temperature. Necrotic tissue is thermographically a hypothermal, cold area.

In deciding on therapy and assessing progress, thermography can be a very valuable tool for the clinician. In addition, whenever the center of visually necrotic or burned tissue is warm, this may be taken as evidence that the capillary network in deeper layers of the skin is not completely destroyed (with a consequently bad prognosis), but dilated by local liberation of mediators. These facts can be demonstrated experimentally in toxic skin reactions after short time chemical contact.

BULLOUS SKIN LESIONS

Intradermal blisters inhibit heat conductivity and extend the distance between the skin capillary network and surface, thus masking inflammatory vasodilation in the skin microcirculation.

Porphyria cutanea tarda and pemphigoid lesions

In these conditions the discrepancy in surface temperature between blistered and non-blistered inflamed skin regions can be demonstrated. This heat transport inhibiting effect of blisters is dependent on the area location within the skin.

Zoster

In zoster patients the small vesicles only show small cold spots within a hyperthermal area at the level of the dermatomes affected. After clinical improvement and also after the erythematous lesion has disappeared, a more or less marked rise in surface temperature remains for some time, indicating the persisting lack of vasoconstrictor sympathetic activity after zoster infection.

ERYTHEMA

In this dermatosis there is always a hyperthermal thermographic pattern which extends beyond the visible skin lesion. Visually grouped skin lesions thus show a more or less uniform hyperthermal thermogram and cannot be separated in the thermal image because of the extensive vasodilation in the skin.

ERYTHEMA NODOSUM

The thermographic pattern can be correlated to the location of local vasculitis. Again, the area of higher temperature is larger than the visible lesion owing to the increased blood flow in the skin microcirculation around the inflammatory process.

SCLERODERMA CIRCUMSCRIPTA

Increased surface temperature is a common thermographic feature of scleroderma. In this disease the subcutaneous fatty tissue becomes more and more atrophic thus increasing heat conductivity from deeper layers, underlying sub- and epifascial microcirculation networks. In contrast to the temperature distribution at general locations, finger tips in scleroderma are mostly cold when the generalized process also affects the fingers.

The cause of this is a peripheral vasculitis with diminished blood flow and a marked vasospastic effect of cold ambient temperature when the sclerotic process has diminished the skin thickness and insulation.

RAYNAUD'S PHENOMENON

In contrast to scleroderma lesions at the fingers where there is a continuous temperature gradient from hands to finger tips, Raynaud's disease (or syndrome) exhibits a sharp decrease in temperature at the

MCP level depending on the vasospastic reaction of finger blood vessels. Differential diagnosis is only possible by dynamic thermographic tests.

MALIGNANT TUMORS

Increased surface temperature is one of the most important features of malignant skin tumors, such as lymphangiosarcoma, carcinoma or malignant melanoma. Temperatures are well correlated to the degree of malignancy. The more malignant the neoplastic process the greater the blood flow to and through the tumor. If the center of the tumor becomes necrotic there is a significant decrease in temperature showing a cold spot within a hot region.

BENIGN TUMORS

In contrast to malignant tumors, benign tumors mostly show a lower local temperature owing to the insulating effect of tumorous tissue mass, compression of local blood vessels and lower blood flow through the tumor. Basaliomas, fibromas, neurofibromatosis Recklinghausen and sometimes hemangiomas show this thermal effect.

CONCLUSIONS

There are a number of factors influencing surface temperature, one of the most important being the skin microcirculation. Skin lesions may influence one or more of these factors. Thermography is a valuable tool for complementary quantification of these thermal effects, which often extend beyond the visible lesion. Dermatologists now have the opportunity to use thermographic procedures in daily clinical routine, but they must learn to interpret information accurately and to explore and evaluate new thermographic procedures.

FOR FURTHER READING

1. C. Gros, P. Bourjat, M. Gautherie, L'image thermographique normale et pathologique de la face, J. Radiol. Electrol. Med. Nucl. 51:333-348 (1970).
2. J. D. Hardy, P. A. Gagge, J. A. J. Stolwijk eds., "Physiological and behavioral temperature regulation," Charles C. Thomas, Springfield (1970).
3. J. H. W. Mah, Some physiological aspects of the temperature of the body surface. in: "Medical thermography," S. F. C. Heerma van Voss, P. Thomas, eds., S. Kerget, Basel, New York (1969).
4. A. Parodi, M. Guarrera, A. Rebora, Flushing in rosacea: an experimental approach. Arch. derm. Res. 269:269-274 (1980).

5. D. Sinclair, Motor nerves and reflexes, in: "The physiology and pathophysiology of the skin," A. Jarrett ed., Vol. 2., Academic Press, London, New York, pp. 475-493 (1973).
6. G. Stuttgen, U. Flesch, H. Witt, H. Wendt, Thermographic analysis of skin test reaction using AGA Thermovision, Arch. derm. Res. 268:113-128 (1980).
7. G. Stuttgen, W. G. Forssmann, Pharmacology of the microvascularization of the skin, in: "Normale und Pathologische Physiologie der Haut. III," G. Stuttgen, H. W. Spier, E. Schwarz eds. Springer, Berlin, Heidelberg, New York (Handbuch der Haut- und Geschlechtskrankheiten Bd 1, Teil 4B, pp 379-540) (1981).

INFLUENCE OF TISSUE INJURY ON SKIN TEMPERATURE

L D Reed*, R C Black**

* Environmental Protection Section, Defence Research Establishment, Ottawa, Ontario, CANADA
** Medical Engineering Section, National Research Council of Canada, Ottawa, Ontario, CANADA

ABSTRACT

Thermography has application in the investigation of the nature and extent of tissue trauma. However, the effects of tissue injury on skin temperatures may also interfere with other diagnostic uses of thermography. Specific examples relating to the main pathological agents, (pressure, heat, cold, nuclear radiation, etc) are briefly reviewed. The effects of trauma are discussed with reference to the pathophysiology of tissues, the resulting thermal responses, and the diagnostic process. Particular emphasis is given to the example of pressure sores. In this application it has been possible to extend the subjective interpretation of thermograms (obtained after removal of pressure) to detailed studies of the time-dependence of thermographic patterns and the quantitative analyses of the areas involved. It was found in this example that the surface temperatures, after the application and relief of sufficient tissue loading, increased to a maximum and then decreased. The temperature decrease (observed after peak temperature was attained) was approximately two degrees centigrade within the first 15 minutes in the case of normal subjects. The corresponding decrease was markedly less in the case of regions involving compromized tissue.

INTRODUCTION

Following many forms of tissue injury there are changes in one or more of: the vascular systems, the microcirculatory system, the local metabolic rate, and the thermal conductivity of tissue. These changes cause alterations in the rate at which heat is generated at the site of injury, or in the rate at which heat is transferred either to or from the site. Since any of these changes can cause alterations in the patterns of skin temperature,[1] thermography can sometimes be employed to investigate the nature and extent of injuries or the course of healing. Thermography can even be used, in situations involving pressure sores, to prevent injury. It is the purpose of this brief paper to review such applications of thermography.

DIRECT APPLICATIONS OF THERMOGRAPHY RELATING TO TISSUE TRAUMA

Many thermographic studies exist concerning the trauma which results when soft tissues are cut, burned, frozen, irradiated, or subjected to damaging applications of pressure.

Wounds

Kliot and Birnbaum[2] studied the evolution of skin temperature patterns near surgical wounds and found warm areas extending to as far as 5 cm on either side of the cold cut line. Temperature increases of up to 2°C reached maxima within 6 days following surgery, and variations in temperature were closely linked to those of biochemical assays. When infection was present the temperature increases were larger and tended to subside more slowly. Viitanen and Viljanto[3] extended the previous work to include children. A different, age-dependent variation of temperature with time was observed in young patients. Acciari et al[4] have described thermographic observations which were made following the re-attachment of severed digits. They noted the usefulness of thermography in early detection of stenosis or thrombosis. It was found that four phases of thermal patterns of the hand frequently occurred in sequence while nerves regenerated: these were the hyperthermic, hypothermic, thermal 'amputation', and stabilization phases.

Burns

In an early application of thermography to the study of burns, Lawson et al[5] were able to achieve 90% reliability in differentiating between deep and superficial burns in experimental animals. Others[6-12] have extended this application of thermography. Hackett,[10] in particular, has reported high accuracy in a study of over 300 burn sites in which full thickness burns were usually at least 2.5°C cooler than surrounding tissue. Areas from 1.5 to 2.5°C cooler were diagnosed as deep dermal burns, likely to benefit from excision. Burn areas which were warmer than surrounding tissue were diagnosed as superficial burns.

Frostbite

Frostbite cases have been studied in much the same way as those of burns. In Lawson's[5] animal experiments, for example, it was possible to predict accurately which frostbite regions would become gangrenous. More recently, Hamlet et al[13] have studied the sharp thermal gradient between (cold) areas of deep frostbite and (warm) areas of viable tissue in the feet of rabbits. The sharp gradient demarcated those tissue areas which eventually sloughed.

Radiation

It has long been known that radiation treatment produces an erythema and associated thermal effects. The 'thermal tide'[14] of erythematous hyperthermia has been observed by many thermographers, particularly those involved in breast cancer clinics. Gautherie et al,[15] for example, reported increases of 1.5 to 2.0°C in mean breast temperatures, which disappeared within a period of one year. Amalric et al[16] studied the course of interference between the thermal effects produced by radiation and those produced by breast carcinomas, and were able to distinguish between therapy which was successful (linear cooling), temporarily successful (cooling followed by rewarming), unsuccessful (no cooling) and successful therapy which produced a 'radiotherapic' breast (ie with some cooling but only to a stable temperature which remained higher than normal). Silberstein and Kattan[17] have proposed the use of thermographic techniques for dosimetry in case of accidental exposure to radiation. However, they were unable to detect skin temperature effects from doses smaller than 960 rads and it has long been known[18] that the dose required to produce a threshold erythema depends upon the individual and on the location of the irradiated areas.

Sports Injuries

Tissue trauma arising from sporting activities has also been studied thermographically.[19] Bergtholdt,[20,21] for example, has observed skin temperature changes in the case of sprains, dislocations, and fractures, but not in the case of contusions and muscle strains.

Sores

The thermal patterns of open sores have been studied by many workers. In the case of pressure sores, Barton and Barton[22] have distinguished three main types:

1. Indolent sores occurring in relatively avascular sites (especially in geriatric patients) in which the temperature elevation is less than 1.0°C.
2. Sores in relatively healthy persons in which the temperature elevation approaches 2.5°C.

3. Hyperactive sores having a temperature elevation of 5°C, which require careful management.

Importance of Pathophysiological Bases

In each of the preceding successful applications of thermography, increasing use has been and will continue to be made, of knowledge of the basic pathophysiological processes involved in traumatic injury and subsequent recovery. With an awareness of such processes, the thermographer can best complement the information derived from other clinical methods. Thermographic research methods themselves will continue to be useful in studying such processes. One example is the work of Branemark and Nilsson[23] who have combined vital microscopic examinations and thermographic observations in a study of the microcirculatory damage caused by burns. A second example of the use of thermography to complement basic clinical studies is in demonstrating what temperature changes, if any, are associated with local changes in skin color. Killian,[24] for example, has recently reviewed the complex details of processes which affect skin color following frostbite.

USE OF HYPERTHERMIC RESPONSES IN THE PREVENTION OF TISSUE TRAUMA DUE TO PRESSURE SORES

Provoked Thermal Responses

In some situations it is possible to apply provocative thermal techniques to determine important properties in the tissues of interest. The "thermal recovery" technique of Nagasawa and Okada,[25] for instance, has been used to study the relative contributions of circulatory and metabolic influences on areas having unusual skin temperatures. To apply the technique, one merely heats or cools the skin, and then monitors the rate at which the skin temperature returns to its original value. The remainder of this paper will deal with another provoked response (the hyperthermic response of tissue after the application of pressure) and uses of this response in preventing tissue trauma caused by pressure sores.

Pathogenesis of Pressure Sores

The complex mechanisms responsible for pressure sores have been studied intensively. Kosiak[26] studied the combinations of magnitude and duration of pressure required to induce pressure sores in dogs. An inverse relationship was found, ie sores were produced both by high pressures applied for short durations, and by low pressures applied for long durations. Kosiak was also able to detect microscopic pathological damage in tissues subjected to pressures as low as 4.0 kPa for only 1 hour. Brand et al,[27] in their study of insensitive limbs, have noted that tissues could also proceed to an inflamed condition as a result of the application of many small impulses, none of which alone was large enough to cause damage. Brand[28] has reviewed his experiments in which such repetitive stresses were applied to the feet of rats and to human fingers. He has called the damaging process "autolysis", on the assumption that the tissue damage was the result of the action of enzymes released as the inflammatory process progressed. Brand has observed a marked increase in surface temperatures as the process continues and a slow recovery after stress repetition ends.

Other aspects of the pathophysiology of pressure sores have been reviewed by Barton[29] and by Branemark[30] (and by other authors participating in the 1975 Seminar on Tissue Viability and Clinical Applications (reported in the same proceedings).

Hyperthermic Response

Many authors have reported a hyperthermic response of tissue following the removal of applied pressure.[31-36] For high enough pressures, it was found that the skin temperature rose to a maximum and then decreased, usually to a level which was comparable to the original temperature observed before the application of pressure. It was found that the temperature returned to its original value more slowly when the duration of the pressure was increased. The maximum temperature attained was higher for higher values of applied pressure. For low values of applied pressure little or no hyperthermic reaction was detected. Indeed, Mahanty[36] and Goller et al[32] observed a slight decrease in temperature after the application of low pressure. Mahanty and Roemer[37] have formulated a mathematical model of the hyperthermic response based on known tissue properties and on assumptions concerning an increase and a subsequent decrease in blood flow which are presumed to follow the release of the applied pressure.

Application to Insensitive Limbs

In attempting to find a proper custom fit of shoes (or prostheses) for insensitive feet (or stumps), Bergtholdt and Brand[38] have used, as a 'pain substitute', the hyperthermic responses (if any) which have followed the use of the device, ie the response was viewed as a signal of localized stress. Their aim was to detect potential tissue problems before breakdown. Brand[39] has acknowledged that interpretation of the thermograms involved in this method is not an easy matter. However, he has also stated that the combination of an approximate estimate of pressure distribution, with a progressive watch on hot spots, has yielded the most successful and practical method he has ever known for monitoring the patient.

Application to Wheelchair Seats

Black and co-workers[40,41] have investigated the use of the hyperthermic response in the assessment of the seats used by wheelchair patients. In a series of experiments on normal subjects, it was found that a large hyperthermic response could be produced by seats made of stiff materials, especially seats of asymmetrical design. It was possible to detect some hyperthermic responses more sensitively when quantitative methods were used. An electronic areameter[42] was employed which allowed simultaneous measurement of the areas associated with each of ten adjacent isotherms and which also allowed rapid left-right comparisons (of such area sets) within the field of view. To plot these results, as in Figure 1, the sums of the areas hotter than each integral temperature, from warmest to coolest isotherm (eg 35, 34, ... 28, 27°C), were plotted versus time.

Plotting the data in this way produced families of curves which tended to be concave downward for the times at which hyperthermic responses were occurring in the previously stressed tissue. Black and his co-workers have extended these measurements to the case of wheelchair patients who were examined before and after episodes of sitting in seats

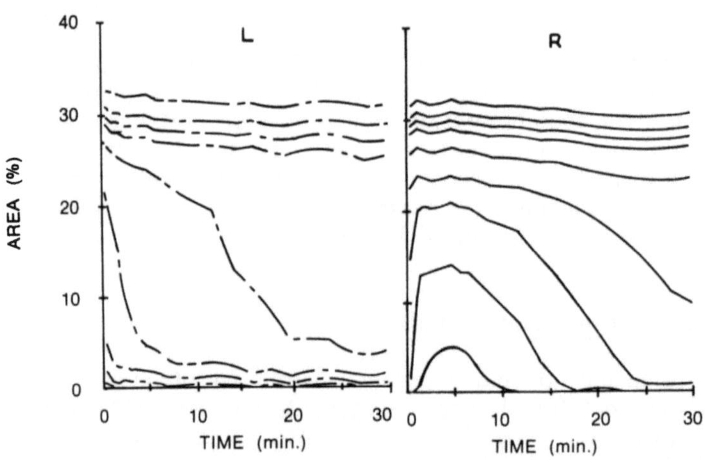

Fig. 1. The dependence of left (L) and right (R) isotherm areas (in %) on the time (in minutes) following the removal of asymmetrical stress of the buttocks (greater on the right). The curve nearest the time axis is the cumulative area warmer than 35°C, etc. These curves were obtained with a normal subject who had been sitting on a plywood seat having a cutout, on the left side, which markedly increased the pressure applied in the region of the right ischial tuberosity.

in which it was hoped that seating stress would be minimal. For normal subjects the decline in temperature, after the peak occurred, was approximately 2°C in the first 15 minutes. In the case of badly compromized tissues, or of poorly designed seats, it was found that hyperthermic responses were of greater magnitude and took longer to decrease to 'baseline' temperatures.

CONCLUDING COMMENTS

The hyperthermic response is one form of provoked response which requires further basic studies which could provide the basis for extending its application. A very different application of hyperthermic responses was recently reported by Moss et al.[43] Intestinal ischemia was experimentally induced in dogs for periods of from 2 to 8 hours. At the end of the ischemic period, a thermographic examination was made and the nature of any hyperthermic response was noted. These results were later compared with observations made on the viability of the intestines. It was found that all dogs which demonstrated uniform hyperthermic reactions survived with bowels that were histologically normal. Dogs which exhibited non-uniform hyperthermic responses all showed some bowel damage and some deaths resulted. The dogs which demonstrated no hyperthermic responses all died as a result of bowel infarction.

In a sense, the uses of hyperthermic response which have been described above may represent quite different 'domains' of applied pressure level and of tissue condition. In pressure sore applications hyperthermic signals from tissue are viewed as danger signs of localized stress, while in the study of intestinal ischemia, the tissues best able to send hyperthermic signals were those most likely to survive the long periods of extreme stress.

It is strongly felt that most of the future noteworthy advances in the application of thermography to any of the above forms of tissue trauma will involve research personnel who are abreast of recent pathophysiological advances, have access to instrumentation systems which complement thermographic observations, and who are actively involved in careful detailed analyses of the relationships between the tissue processes of interest and the measurements made.

ACKNOWLEDGEMENTS

It is a pleasure to acknowledge past collaborative work related to tissue trauma in seating conducted by Dr G Hahn and Mr A F Filippone of the Ontario Crippled Children's Center, and by Dr S D Livingstone and Mr R E Limmer of the Defence and Civil Institute of Environmental Medicine. The above project was supported by the federal department of Health and Welfare Canada (Reference number 6606-1510).

REFERENCES

1. K. Lloyd Williams, Pictorial heat scanning, Phys. Med. Biol. 9:433 (1964).
2. D. A. Kliot, S. J. Birnbaum, Thermographic studies of wound healing, Am. J. Obst. and Gynec. 93:515 (1965).
3. S. M. Viitanen, J. Viljanto, Wound healing - a thermographic study, Annles Chirurgial et Gynecologiae Fenniae 60:101 (1972).
4. L. Acciarri, L. Cugola, R. Maso, L. Nogarin, The thermographic hand, Acta Thermographica 3:65 (1978).
5. R. N. Lawson, G. D. Modek, D. R. Webster, Thermographic assessment of burns and frostbite, Can. Med. Ass. J. 84:1129 (1961).
6. R. Mladick, N. Georgiade, F. Thorne, A clinical evaluation of the use of thermography in determining degree of burn injury, Plastic and Reconstructive Surgery 38:512 (1966).
7. G. Buwalda, Thermographic Assessment of Burns and Frostbite, in: "Medical Thermography, Bibliol. Radiol. No. 5," Karger, Basel (1969).
8. M. Hackett, Color thermography in the diagnosis of burn depth, in: "Transactions of the Fifth International Congress of Plastic and Reconstructive Surgery, Melbourne," J. T. Hueston ed. Butterworth, Sydney, London (1971).
9. M. Hackett and H. P. Henderson, Infrared thermography in medicine, in: Industrial and Civil Applications of Infrared Technology, SPIE Vol. 110, p. 74-77 (1977).
10. M. Hackett, The use of thermography in the assessment of depth of burn and blood supply of flaps, with preliminary reports on its use in Dupuytren's contracture and treatment of varicose ulcers, Brit. J. of Plastic Surgery 27:311 (1974).
11. M. Hackett, The place of thermography in medicine, Acta Thermographica 1:176 (1976).
12. A. Sowa, Thermographic estimation of burn depth, Acta Thermographica 4:30 (1971).
13. M. P. Hamlet, J. Veghte, W. D. Bowers, J. Boyce, Thermographic evaluation of experimentally produced frostbite of rabbit feet, Cryobiology 14:197 (1977).
14. M. Tonegutti, L. Acciarri, A. Racavelli, Fundamentals of contact thermography in female breast diseases, Acta Thermographica supplement, Bertoncello Artigrafiche, Padova (1980).
15. M. Gautherie, D. Gros, C. H. Gros, Contribution of infrared thermography to the surveillance of irradiated breast carcinomas, Acta Thermographica 2:23 (1977).
16. F. Amalric, D. Giraud, C. Altschuler, R. Amalric, J. M. Spitalier, Infrared thermographic follow-up after breast cancer curative radiotherapy, Acta Thermographica 4:54 (1979).
17. E. B. Silberstein, J. Kattan, The use of thermography in radiobiological dosimetry, Acta Thermographica 2:91 (1977).
18. H. E. Johns, J. R. Cunningham, "The Physics of Radiology," Third Edition, Charles C. Thomas, Springfield, Ill. (1969).
19. F. Lelik, G. Kezy, Contact thermography in sports medicine, Acta Thermographica 4:24 (1979).

20. H. T. Bergtholdt, Thermography and athletic injuries, in: "Medical Thermography, Theory and Clinical Applications," S. Uematsu ed., Brentwood Publishing Corp., Los Angeles (1976).
21. H. T. Bergtholdt, Thermography on insensitive limbs, in: "Medical Thermography, Theory and Clinical Applications," S. Uematsu ed., Brentwood Publishing Corp., Los Angeles (1976).
22. A. A. Barton, M. Barton, The clinical and thermographical evaluation of pressure sores, Age and Ageing 2:55 (1973).
23. P. I. Branemark, K. Nilsson, Thermographic and microvascular studies of the peripheral circulation, in: "Medical Thermography. Biol. Radiol. No. 5," Karger, Basel (1969).
24. H. Killian, "Cold and Frost Injuries - Rewarming Damages Biological, Angiological and Clinical Aspects," Springer-Verlag, Berlin (1981).
25. A. Nagasawa, H. Okada, Thermal recovery, in: "Medical Thermography," K. Atsumi ed., University of Tokyo Press, Tokyo (1973).
26. M. Kosiak, Etiology and pathology of ischemic ulcers, Arch. Phys. Med. Rehabil. 40:62 (1959).
27. P. W. Brand, T. D. Sabin, J. F. Burke, Sensory denervation, study of its cause and its prevention in leprosy and of management of insenstive limbs, SRS Project No RC-40-M Final Report, Carville USPS Hospital, Carville, LA (1971).
28. P. W. Brand, Pressure sores - the problem, in: "Bedsore Biomechanics," R. M. Kenedi, J. M. Cowden, eds., MacMillan, London (1976a).
29. A. A. Barton, The pathogenesis of skin wounds due to pressure, in: "Bedsore Biomechanics," R. M. Kenedi, J. M. Cowden, eds., MacMillan, London (1976).
30. P. I. Branemark, Microvascular function at reduced flow rates, in: "Bedsore Biomechanics," R. M. Kenedi, J. M. Cowden, eds., MacMillan, London (1976).
31. P. J. Verhonick, D. W. Lewis, H. O. Goller, Thermography in the study of decubitus ulcers, Nursing Research 21:233 (1972).
32. H. Goller, D. W. Lewis, R. E. McLaughlin, Thermographic studies of human skin subjected to localized pressure, Am. J. Roent. Rad. Ther. Nucl. Med. 113:749 (1974).
33. R. S. Trandel, D. W. Lewis, P. J. Verhonick, Thermographic investigation of decubitus ulcers, Bull. Prosthetics Research, Fall 1975;137 (1975).
34. J. Rogers, L. F. Wilson, Preventing recurrent tissue breakdown after "pressure sore" closures, Plastic and Reconstructive Surgery, 56:419 (1975).
35. G. Pye, P. Bowker, Skin temperature as an indicator of stress in soft tissue, Engineering in Medicine 5:58 (1976).
36. S. D. Mahanty, Thermal response of the skin to the application of localized pressure, Arch. of Phys. Med. and Rehabil. 60:584 (1979).
37. S. D. Mahanty, R. B. Roemer, Thermal and circulatory response of tissue to localized pressure applications: a mathematical model, Arch. of Phys. Med. Rehabil. 61:335 (1980).

38. M. S. Bergtholdt, P. W. Brand, Thermography: an aid in the management of insensitive feet and stumps, Arch. of Phys. Med. and Rehabil. 56:205 (1975).
39. P. W. Brand, Patient monitoring, in: "Bedsore Biomechanics," R. M. Kenedi, J. M. Cowden eds., MacMillan, London (1976b).
40. R. Black, A. F. Filippone, Thermography in a seating program, in: Frontiers in Biomedical Engineering, Proceedings of 8th CMBE Conference. Canadian Medical and Biological Engineereing Society, Ottawa (1980).
41. R. Black, G. Hahn, Color thermography in a pediatric tissue trauma program, in: "Medical Thermography," R. Ghys ed., European Press, Ghent (1978).
42. L. D. Reed, M. R. Howat, J. E. Ulrichsen, Thermographic Areameter, US Patent 4,218,707 (1980).
43. A. A. Moss, H. Y. Kressel, A. C. Brito, Use of thermography to predict intestinal viability and survival after ischemic injury: blind experimental study, Investigative Radiology 16:24 (1981).

DEVELOPMENT AND APPLICATION OF AN IMPROVED THERMAL DOLORIMETER

I Fukumoto*, M Saito*, K Ono**

*** Institute of Medical Electronics, Faculty of Medicine University of Tokyo, 7-3-1 Hongo, Bunkyo-ku, Tokyo**
**** Seiwa ME Laboratory, 5-19-2 Honkomagome, Bunkyo-ku Tokyo, JAPAN**

The thermal dolorimeter of W Hardy[1] has been used to measure thermal pain threshold (Fig 1). The original model is accurate but has the following drawbacks:

1. Measurements must be repeated to determine the thermal pain threshold
2. It is too large and requires elaborate preparation of the subjects being investigated.

A new thermal dolorimeter has been constructed overcoming these drawbacks. In order to avoid repetition of measurements, we have used the period of heating to tolerable temperature as the thermal pain threshold, while keeping the stimulating energy fixed. Figure 2 shows the time-temperature relationship of the new dolorimeter.[2] In order to reduce the size and to make it more convenient, a small lamp is used for heating by direct thermal conduction. Figure 3 shows the block diagram of the measuring system. The time for measurement is small, enabling the instrument to follow fast changing phenomena.

The dolorimeter has been used to measure thermal pain thresholds under the following conditions.

EFFECT OF ELECTRICAL STIMULATION

Figure 4 shows the effect in a patient undergoing treatment with an electrical pain reducer. As the output of the reducer increases, the pain threshold also increases.

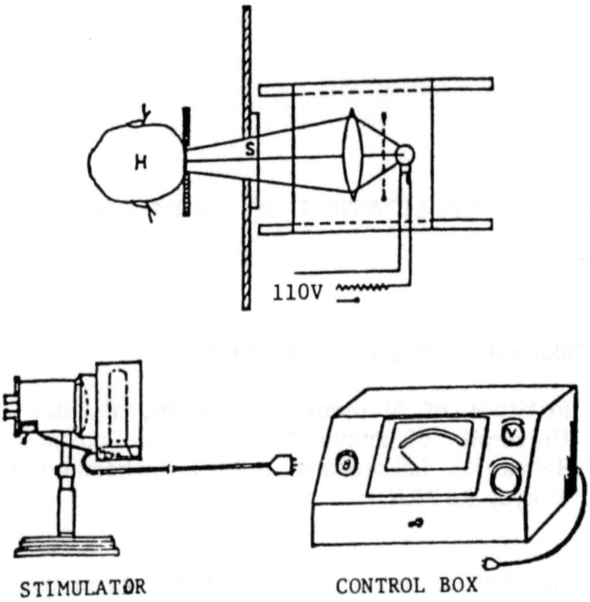

Fig. 1. Hardy's Thermal Dolorimeter.

EFFECT OF AMBIENT TEMPERATURE

Pain thresholds of a normal subject were measured for a period of one month (Fig 5). The room temperature was also plotted on the same graph as a half-dashed line. A negative correlation was found indicating

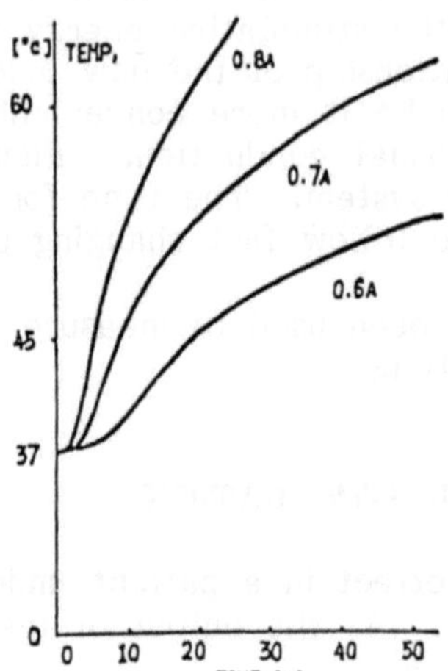

Fig. 2. The transient property of the thermal dolorimeter.

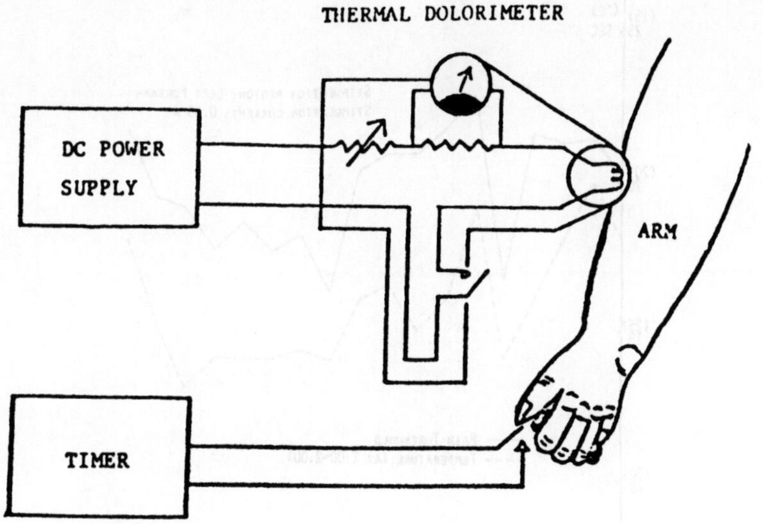

Fig. 3. The block diagram of the method of measurement of pain threshold.

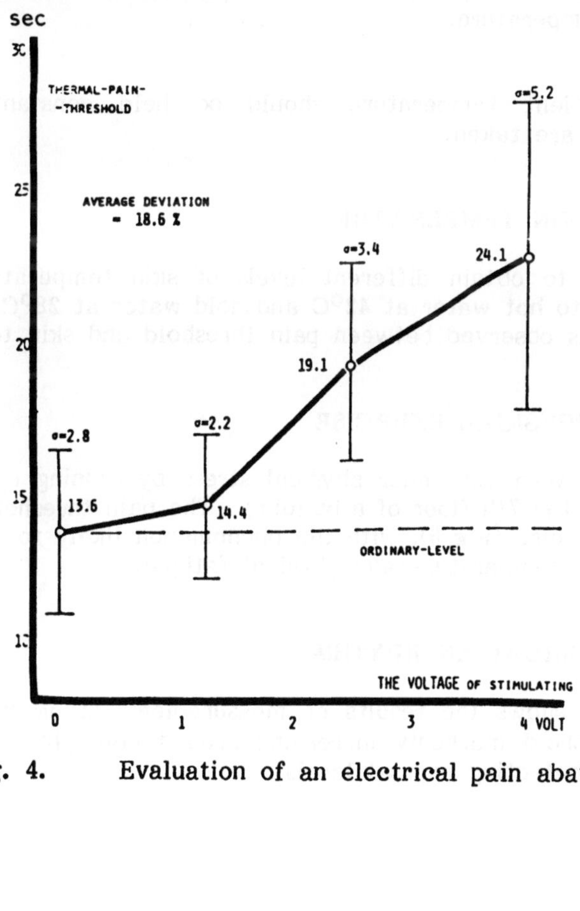

Fig. 4. Evaluation of an electrical pain abater.

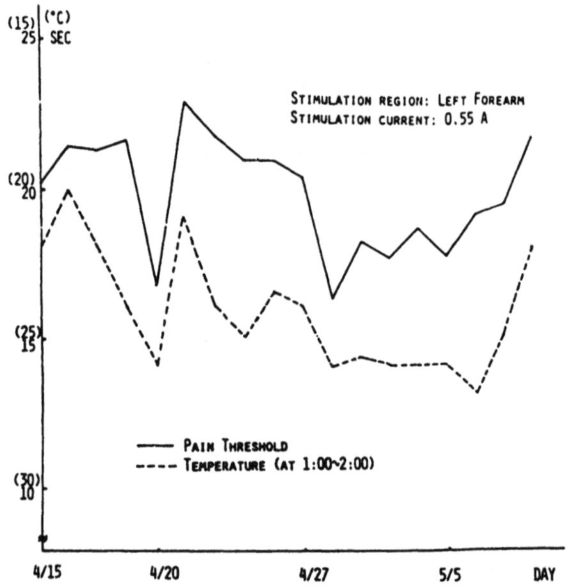

Fig. 5. Relationship between the pain threshold and ambient temperature.

that the ambient temperature should be held constant when the measurements are taken.

EFFECT OF SKIN TEMPERATURE

In order to obtain different levels of skin temperature, subjects were exposed to hot water at 42°C and cold water at 28°C. A positive correlation was observed between pain threshold and skin temperature.

EFFECT OF PHYSICAL EXERCISE

Subjects were put under physical stress by running up stairs from a basement to the 7th floor of a building. The pain threshold was raised during the exercise (Fig 6), with the phenomenon likely to be influenced by raised body temperature and physical fatigue.

EFFECT OF CIRCADIAN RHYTHM

Figure 7 shows the results of measurement during 24 hours with the pain threshold markedly increasing around midnight suggesting the possibility of detecting arousal levels.

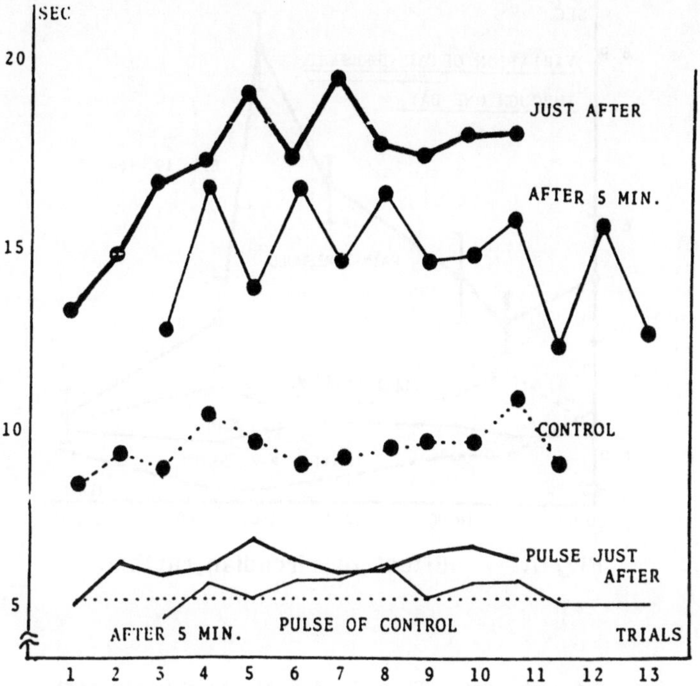

Fig. 6. The variation of pain threshold after physical exercises.

EFFECT OF VISUAL BIO-FEEDBACK

When a subject is permitted to look continuously at his pain threshold it was found that he becomes more tolerant.

ESTIMATION OF THE DEPTH OF HEAT SENSATION BY COMPUTER SIMULATION

Computer simulation was explored in order to estimate the depth at which heat is sensed in human skin. Thermal conduction calculations were based on Fourier's law and heat radiation and convection at the skin surface were estimated by Stefan-Boltzman's law and Newton's cooling law respectively. From this study we have estimated the depth of the heat sensor as about 0.5 mm.

RELATIONSHIP BETWEEN PAIN THRESHOLD AND SKIN ELECTRIC RESISTANCE

In oriental medicine, the location of acute points is important to

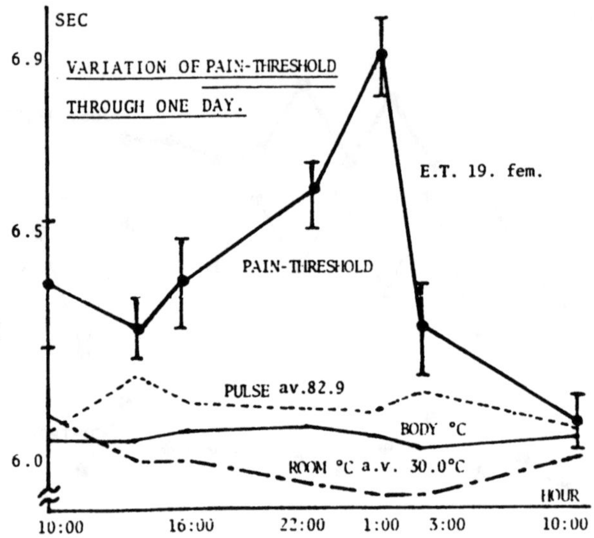

Fig. 7. Effect of circadian rhythm.

both diagnosis and treatment. The electrical resistance method is usually adopted but it would be advantageous if it were possible to find the acute points by heat stimulation.

Figure 8 shows the time variation of pain threshold at 136 acute points in oriental medicine (Fig 9), with the electrical resistance of the same points also plotted as a half-dashed line. This study indicates that it is possible to use the thermal dolorimeter instead of electrical resistance measurement to locate the acute points.

In these investigations it is important to accurately recognize the thermal pain threshold, as this strongly affects the variation of measured data. Using the stinging sensation after the short painless period we have obtained similar results to Hardy.

We may summarize the study as follows:

1. The thermal dolorimeter is useful in evaluating the effectiveness of electrical pain relief.
2. Thermal pain threshold has a negative correlation with ambient temperature.
3. Skin temperature and physical exercises have strong effects on the thermal pain threshold.
4. Circadian rhythm of the pain threshold is observed, possibly related to the level of arousal.
5. Visual biofeedback appears to raise the pain threshold.

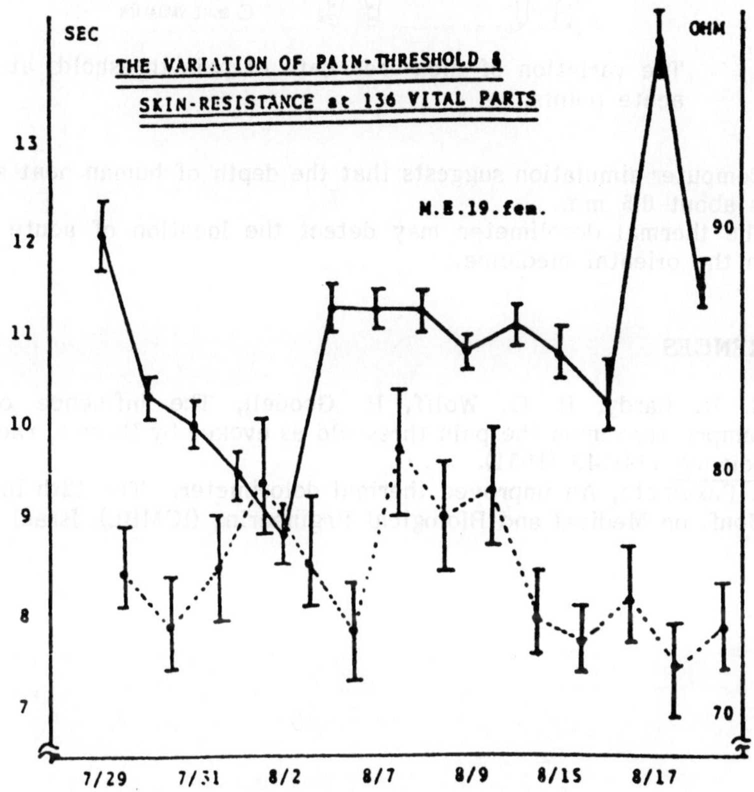

Fig. 8. Pain threshold at acute points. Solid line: average of the pain threshold at 136 acute points. Dashed line: average of the electrical resistance at the same points.

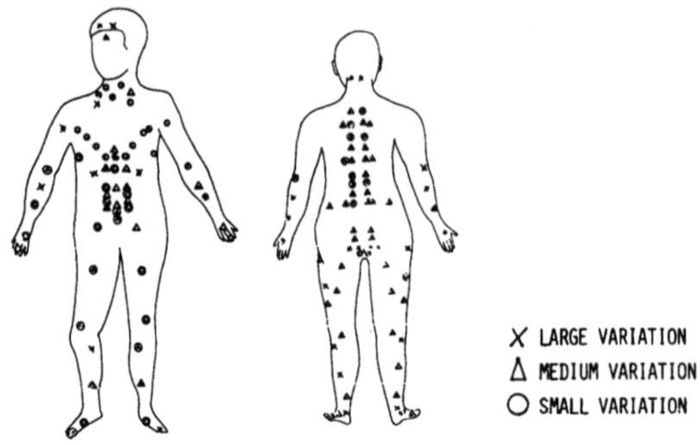

Fig. 9. The variation of measured data of pain thresholds at human acute points.

6. Computer simulation suggests that the depth of human heat sensing is about 0.5 mm.
7. The thermal dorolimeter may detect the location of acute points in the oriental medicine.

REFERENCES

1. H. D. Hardy, H. G. Wolff, H. Goodell, The influence of skin temperature upon the pain threshold as evoked by thermal radiation, Science 114:149 (1951).
2. I. Fukumoto, An improved thermal dolorimeter. The 12th Int. Nat. Conf. on Medical and Biological Engineering (ICMBE), Israel (1979).

TECHNICAL DEVELOPMENTS IN THERMOLOGY - INTRODUCTION

C H Jones

Institute of Cancer Research and the Royal Marsden Hospital
Fulham Road, London SW3
UNITED KINGDOM

In recent years the technical developments in imaging have been considerable. Thermology, in particular, has benefited through the improvements that have been made in detector systems and the application of image and data processing techniques. However, in spite of technical improvements, careful consideration must always be given to the fundamental physical basis of IR emission. In his review of this subject, Prof J Steketee has emphasized the important aspects of emissivity, cooling, surface curvature and spatial resolution. The importance of emissivity in clinical investigations is also discussed by Dr J M Engel, who demonstrated unequivocally that ointment can alter the skin temperature by affecting the microcirculation and the emission characteristics of the skin surface.

Setting the scene for the 1990's, Dr E H Putley of the Royal Signals and Radar Establishment, Malvern (UK) describes in his review of infrared systems the sequence of thermal detector systems using single element transducers, linear arrays, matrix detectors and the strip type SPRITE detector. Commercial development of the SPRITE detector has already led to an 8 element linear array which is relatively easy to fabricate. Dr Putley suggests that future detectors will probably be hybrid devices constructed of detector arrays (possibly consisting of 10^4, or even 10^6 elements) and a matching Si CCD read-out chip, the detector array being connected point by point to the Si chip. Such detectors would be capable of high resolution imaging at room temperatures. A Pyroelectric-CCD hybrid with a sensitivity about 10 times that of the present Pyroelectric Vidicon is also a possibility for the future. The current state of the art for Pyroelectric IR sensors is covered comprehensively by Dr D E Burgess. Considerable interest was shown in the Pyroelectric Vidicon cameras demonstrated at the Technical Exhibition of the Conference. These lightweight, portable cameras do not require cooling and are relatively inexpensive; being TV compatible, accessories can be added according to the need and resources of the user. Further development of these cameras will take place: current devices are able to resolve temperature differences of about 0.3 ^{o}C with a limiting resolution of 250 lines.

Thermal imaging using liquid crystals is reviewed by Dr U Flesch. The reader's attention is drawn to the physical factors influencing the resultant temperature distribution displayed by this contact method. Hitherto, most clinical work has ben carried out with mixtures of cholesteric liquid crystals, but more recently chiral nematic thermochromic liquid crystals have been prepared for thermography. In the Poster Workshop Mr K G Archer and Dr I Sage demonstrated the advantages of these materials. Improvements in resolution, a slower photochemical degradation, no selective crystallization and a fast response time make these liquid crystals specially attractive for clinical thermography.

In his review lecture, Dr M Gautherie describes the principles of microwave thermography and the role that $9GH_z$ and $3GH_z$ detector systems have in the investigation of breast and brain lesions.

Although microwave detectors are capable of measuring temperature differences of $0.1^{o}C$, in clinical use it is not possible to achieve such precision. The depth of measurement cannot be determined accurately and to obviate this difficulty, multifrequency systems are being developed. This problem was also discussed by Prof Leroy and Dr M Robillard, and from a mathematical standpoint by Prof S Caorsi.

Image processing and the application of computer techniques is addressed by several authors. The subject is reviewed by Prof M Negin who considered the options available for different kinds of investigation, and his paper emphasizes aspects such as image enhancement, computer assisted measurement and automated interpretation. Dr P Bosiger demonstrates how a low cost microprocessor can be used to provide an objective analysis of breast thermograms. By recording in digital form, and with the use of a gray scale or color TV and on-line processing, the microprocessor may be used to assist evaluation with respect to abnormal thermographic features.

Throughout the Technical Development and Instrumentation session, it was clear that microprocessors are being used frequently for computer processing. Contributions by Mr E F J Ring describe basic test pattern imaging and temperature standard measurements and Mr G R Parr demonstrates how a microprocessor could be used as a means of standardizing the use of the AGA 680M Thermovision System.

Other papers and posters included work on the measurement of permittivity and temperature of tissues, skin blood flow systems, digital thermography and liquid crystal sheets.

In summary, it is clear that the quality of thermal images in terms of spatial and thermal resolution has improved progressively over the last 25 years. Unfortunately, this gain has been associated with a relative increase in equipment cost. Unlike industrial thermography, where commercial benefits and advantages of thermography may be assessed specifically in financial terms, clinical thermography has to rely on the statistical improvements accrued by using thermography in place of, or supplementary to, other techniques. For this reason it is clear that cheaper systems of imaging with add-on TV compatible accessories are likely to be of continuing interest for many clinical applications. For medical purposes stability and reliability are probably of greater importance and value than ultra high spatial and thermal resolutions - providing temperature differences can be measured to within better than 0.3^{o}C. In the next few years technical development is likely to progress in two directions: firstly, hybrid type multielement detectors which will provide superb real time images but will be relatively expensive, and secondly, the development of simpler, robust, portable systems designed for specific tasks. Will the time come when every large hospital will have its own thermographic camera? Probably this question will be confidently answered at the next Congress in four years' time.

THE DEVELOPMENT OF THERMAL IMAGING SYSTEMS

E H Putley

Royal Signals and Radar Establishment
Malvern
UNITED KINGDOM

EARLY HISTORY OF THERMAL RADIATION

The first scientific study of thermal radiation began with the optical experiments of della Porta at the end of the 16th century. His experiments were repeated and extended up to the time of Rumford and Davy in the early 19th century, and these experiments led to a considerable knowledge of the characteristics of thermal radiation. This culminated with the work of Pictet and his colleagues de Saussure and Prevost who in Geneva by 1790 had established the concept that all bodies radiate heat, the higher the temperature the greater the radiation such that when two bodies at different temperatures are within sight of each other the hotter will supply heat to the colder until they reach the same temperature. This result, known as Prevost's Theory of Exchanges was the starting point from which the modern physical theory of thermal radiation was developed in the 19th century, eventually leading to Planck's radiation formula at the beginning of the present century.

Pictet and his colleagues in the course of their work made another discovery particularly relevant to medical thermology. They had to take cognisance of the fact that the human body emitted thermal radiation and that this could have a significant effect upon their experiments.[1] De Saussure describes one method of eliminating this, by the use of a pair of liquid-in-glass thermometers, whilst Pictet used a sensitive air thermometer viewed at a distance with a telescope. In her account of their experiment, Mrs Marcet mentions a third method, the use of Leslie's differential thermometer. Although she came from Geneva and was personally acquainted with the experimenters, this is in fact anachronistic since the differential thermometer was not introduced by Leslie until 1797, several years after the first publication of Pictet's results. She was writing in 1809 and must be describing a later version of the experiment.

Sir William Herschel's discovery of infrared spectroscopy occurred some years after Pictet's work. It was the start of a new branch of science and it stimulated the debate on the nature of thermal radiation. The experimental problems faced by the early workers were formidable, so that the true picture of the nature of thermal radiation and its relation to light did not become clear until the middle of the 19th century.

One of those who contributed to the mid-19th century work was Sir John Herschel, Sir William's son. Sir John Herschel was a pioneer of photography and he saw the application of photographic techniques to visible spectroscopy where it enabled permanent records to be obtained and greater sensitivity to be achieved. Sir John was therefore anxious to apply the technique to the infrared to extend his father's work on the solar spectrum. To do this he devised the first technique for making infrared images visible. This used a paper moistened with alcohol. Heating by the absorbed radiation produced a staining pattern as the alcohol evaporated which enabled the infrared image to be visualized. His first experiment used pure alcohol which dried out completely in about half an hour but later, by using dye from the plant viola tricolor, he was able to produce a fixed image.

Herschel applied his method to solar spectra, but his contemporary John Tyndall was concerned with the relationship between visible and infrared radiation and wished to demonstrate that the infrared was just as capable at forming images as the visible. To do this he set up an experiment in which he imaged the electric arc upon a blackened platinum foil. The lamp housing was fitted with a rock salt window (which had been provided by Sir John Herschel) to transmit the infrared rays and a filter consisting of iodine dissolved in carbon bisulphide was inserted to cut off the visible. Tyndall found that after removing the visible the intensity of thermal radiation was sufficient to heat the blackened platinum receiving foil to white heat and to reduce the black to the metallic form leaving a permanent image of the arc stamped upon the foil. Figure 1 shows an engraving of the results, which must be the first thermal image of a terrestrial object.[2]

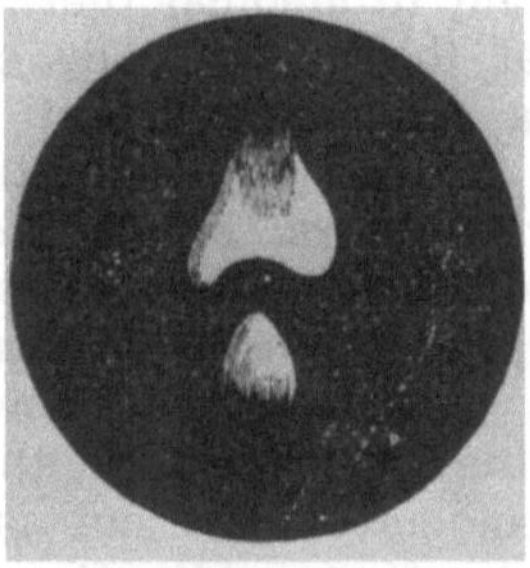

Fig. 1. Tyndall's infrared image of a carbon arc - the first infrared image of a terrestrial object.

There the matter rested until the early 1930's when the spectroscopist Czerny continued the development of the evaporographic technique. His interest like that of Sir John Herschel's, was in producing a recording spectroscopic detector for the infrared so that techniques similar to those used in the visible with photographic plates could be employed. Although Czerny did succeed in recording some spectra, the sensitivity of his method was not good enough to compete with the best detectors (such as the thermopile) then available. It was capable, however, of visualizing objects at temperatures down to about 90°C.

By this time military preparations for World War II were in full swing. Military applications for infrared were pursued by all the major participants, and in particular Germany. Czerny's evaporograph was improved until it could detect a person, whilst an even more important advance was the development of the lead sulphide photoconductive detector. The advances in infrared techniques continued in the immediate post war years. One of these developments was the perfection by Baird in the USA, and also by workers in the Soviet Union, of the evaporograph. Figure 2 is an example of the results obtained with the Baird evaporograph. The exposure time for this picture was about half a minute and the sensitivity must have been a few tenths of a degree. The first medical application of thermography by Dr R N Lawson in Canada was made using an equipment of this type. In more recent years the evaporographic principle has been used to produce simple light weight equipments such as the Gretag Pannicon developed in Switzerland.

An important advantage of the evaporographic technique is that it is a two-dimensional sensor, like the eye, and does not use any form of mechanical scanning as in most thermal imagers now in use. However, it suffers from three major disadvantages. These are:

1. Unsuitable for continuous operation because of need to re-form the liquid film periodically.
2. The long response time. At first sight this might not seem a disadvantage for looking at static objects but associated with the long time is wide thermal diffusion of the hot spots. Hence the spatial resolution is severely limited.
3. The optical read-out makes preservation of the results difficult and is not easily compatible with modern television based picture processing techniques.

For these reasons the evaporographic instruments have been replaced by ones using photon or pyroelectric detectors.

THE DEVELOPMENT OF THE MODERN THERMAL IMAGER

Infrared Detectors

The birth of modern thermal imaging took place in World War II.

Fig. 2. Picture taken with the Baird Evaporograph. With acknowledgement to Applied Optics, where it was first published in 1962.[4]

Infrared was seen as a possible means of detecting hostile ships and aircraft and also by identifying ground features a means for aircraft to locate their position. In World War II most of these tasks were more satisfactorily performed by radar but the first basic developments of infrared techniques were accomplished.

The evaporograph was discussed in the last section. It was too slow for either detecting aircraft or for use aboard them. But to do better, fast long wave infrared detectors were needed. The first step here was the development of the PbS detector by the Germans in the early 1940's. This in itself still did not have a long enough spectral response (cut-off $\sim 3\ \mu m$) but it led the way to the related PbSe and PbTe detectors which can be used in the 3 - 5 μm band. The development of these longer wavelength detectors began in Germany and was continued at RSRE from 1945 onwards. The PbTe detector cooled with liquid air was used in some of our first thermal imaging experiments but it was soon realized that despite its good performance it would be very difficult to manufacture. At this point, the first germanium transistors were being studied. Germanium, it was thought, would not be satisfactory for long term transistor development, but the problems of preparing high purity silicon were yet to be solved. Hence there was an intensive search for new semiconducting materials. The II-V compounds received intensive study. It was soon found that one of them, InSb, would not be suitable for transistor use but it would be a better material than PbTe for a 3 - 5 μm detector. An intensive programme of work at RSRE in collaboration with Mullard led to the development of the InSb photoconductive detector. This was used in the Smith's Pyroscan,[5] one of the first thermal imagers designed for medical use and InSb is still used in a number of systems

It was realized at a very early stage that a photoconductive detector for the 8-13 μm window would be very valuable and a search for a suitable material was launched at RSRE. This led to the discovery of the cadmium mercury telluride family of semiconductor mixed compounds[6] (and also of lead tin telluride[7]) which has proved to be the most widely used photon detector material for the 8-13 μm band. Recently an elegant method of using it was developed by Dr C T Elliott[8] at RSRE. This greatly simplifies the problem of making multi-element detector arrays and has enabled several British manufacturers to produce thermal imagers that are amongst the best in the world.

Alongside the development of the high performance systems we have also given attention to the development of cheap systems. This has led to the design of the pyroelectric vidicon system[9] which is described elsewhere in this volume. RSRE has thus developed not only some of the best but also some of the cheapest thermal imaging systems now available.

Thermal Imagers

We first became interested in the possibilities of thermal imaging before the necessary detectors and other components had been developed. It was realized that the effort to develop these was likely to be considerable and therefore some early imaging systems were constructed with the best components available to allow us to assess the future potentialities, to check our understanding of the design problem and to enable us to justify a major research programme.

Figure 3 shows a system built in 1948. The detector was a thermistor bolometer (8-13 μm photon detectors were not yet available) produced by B T H Rugby. The optical system was a modified searchlight and the detector was scanned by moving plane mirrors across the scene. Figure 4 is an example of the results. The girl had to sit still for about 30 minutes. The results obtained with this system did a lot to establish the potential usefulness of thermal imaging.

For aircraft navigation and reconnaisance the type of system now called linescan was first suggested in the 1940's. In this a detector is scanned along a line perpendicular to the axis of the aircraft. As the aircraft moves in flight the line progresses across the ground. By combining successive lines a picture of the terrain is built up. Such a system will require a fast detector to keep pace with the speed of even a slow aircraft. Despite this, both the Germans and the Americans during

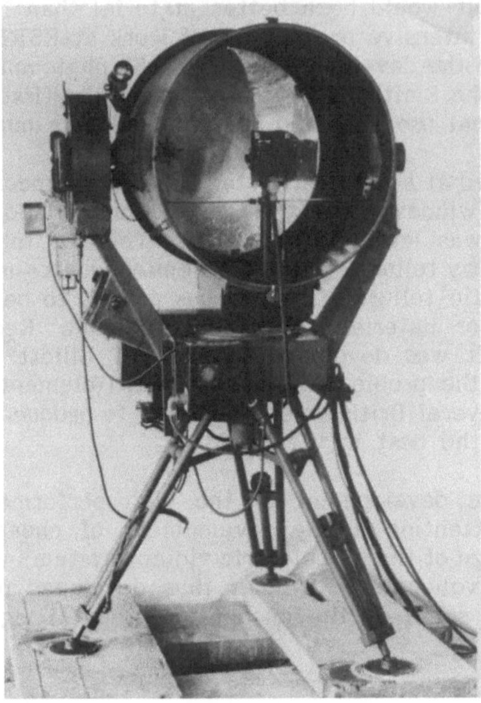

Fig. 3. Thermal imaging system constructed by Salford Electrical Instruments in 1948 for the Admiralty Research Laboratory, Teddington. The detector was a thermistor bolometer manufactured by B T H Rugby. Figure 4 was obtained with this apparatus in 1949.

World War II considered this system. They both hoped that a fast thermal detector such as a metal film bolometer would be adequate. In the USA, the first superconducting bolometers (using niobium nitride elements cooled with liquid hydrogen to $\sim$ 15 K) were considered. We became interested in this type of system towards the end of the war with the problem of detecting submarines at sea particularly in mind. We began to construct superconducting bolometers but we also investigated the use of our new PbTe cells for this task. As a test bed for checking the use of these detectors in thermal imaging a small laboratory imaging system[10] was built which used a Nipkow disc scanner (as in the first television) to generate a two-dimensional scan. The PbTe cells were also used in some of the early airborne experiments. Before the superconducting bolometers could be flown, extrinsic photoconductivity had been discovered in Si and Ge. This being much more convenient, the early work used first Cu doped and later Hg doped Ge. Even this was a passing phase since the rapid development of the InSb 3-5 μm detector enabled good results to be obtained with a liquid air cooled detector.

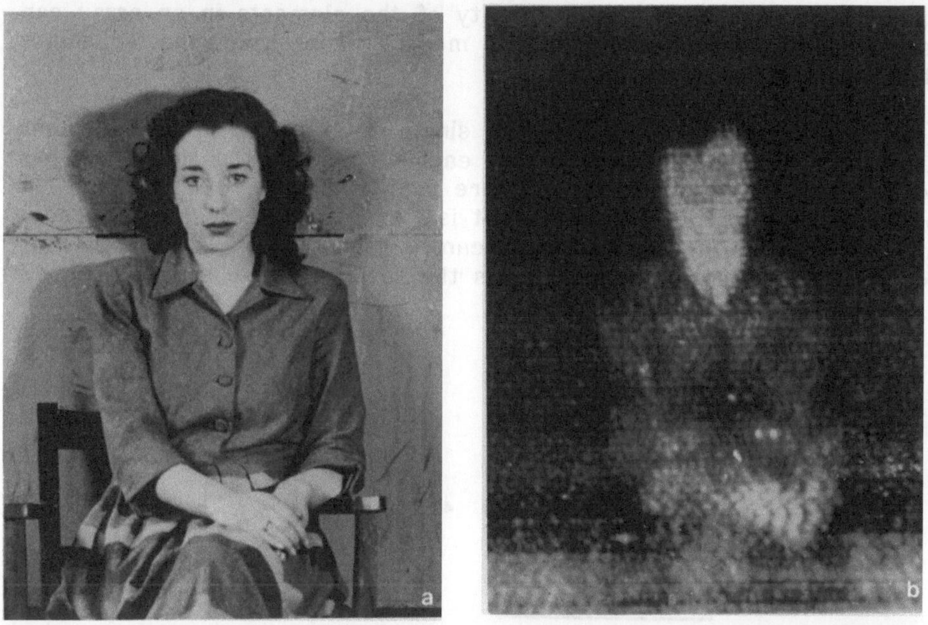

Fig. 4. (a) visible, (b) thermal picture of human target. The thermal picture was taken with the apparatus shown in Figure 3. The ambient temperature was 25°C. The recording time was 40 minutes.

Our first terrestrial results had been obtained with a slow thermistor bolometer system. It is of interest to point out that Bowling Barnes in the USA developed a much more practical system[11] utilizing a thermistor bolometer which still did not operate in "real time" but could get a picture with an exposure time of $\sim$ 20 s. Dr R N Lawson continued his work in Canada with the Barnes system. In the meantime our InSb detectors were used in the Smith's Pyroscan instrument[5] which was developed by Maxwell Cade in collaboration with D W Goodwin of RSRE and K Lloyd-Williams of Bath. The Pyroscan was one of the first instruments developed specifically for medical applications. Although rather slow it gave a good quality image but the display system used was inconvenient for copying or image processing functions.

Many types of system in which a two dimensional picture is built up by mechanically scanning a single element detector have now been designed, of which the most successful are those made by AGA.[12] Initially these systems all used InSb 3-5 μm detectors, but now they can equally well use CMT (or LTT) 8-13 μm detectors. For military applications, however, these systems were still too slow. Since further improvements in the detectors are unlikely, the use of multiple element detectors has become necessary. Before passing to this, it should be said of the single element systems that they lend themselves better to the production of quantitative data since non-uniformity of the elements in an array can lead to serious errors, although with modern signal processing techniques this is less serious than it was.

The advantage of replacing a single detector by a multi-element array of n similar detectors is that each element in the array only has to deal with 1/nth of the picture and when the improvement in performance is properly calculated it is found to increase as $n^{\frac{1}{2}}$. There are several ways in which an array can be used (Fig 5). The simplest is to use a linear array scanned across the scene giving a strip of the

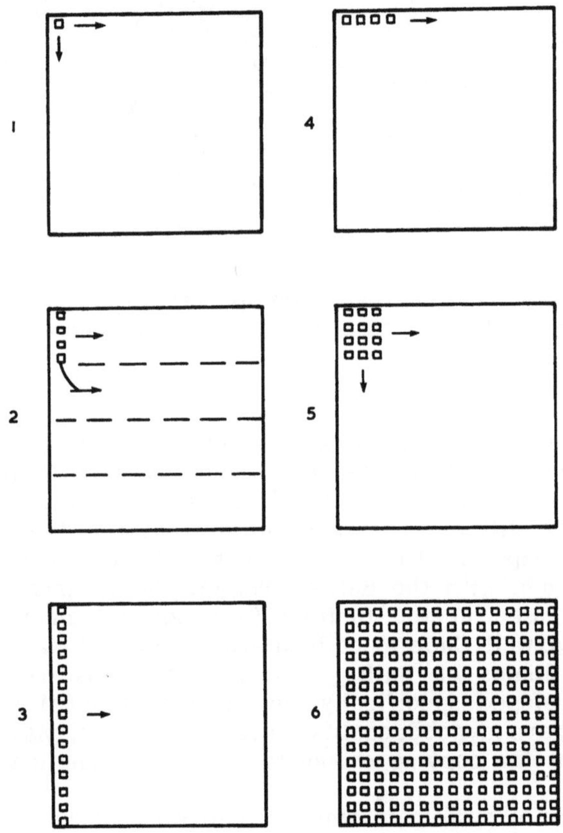

Fig. 5. Scanning methods with detector arrays. 1. single element detector. 2-D scan. 2. small linear array. Banded parallel scan. 3. large linear array. Push broom. 4. small linear array. Serial scan - fast horizontal slow vertical. 5. small 2-D array. Combined parallel - serial scan. A better arrangement is to replace the 2-D array by a TED or SPRITE. 6. large 2-D array. Staring imager.

image. Using a sequence of inclined mirrors the strip is scanned across the image until all the scene has been covered. If N inclined mirrors are used then a total of nN lines will be produced across the image. This system was used in the first multi element imagers designed at RSRE. Figure 6 is an example of the results obtained.

An alternative method of using a linear array which produces a more uniform picture is to mount the array so that each element of the scene is scanned across each detector element in series. At first sight

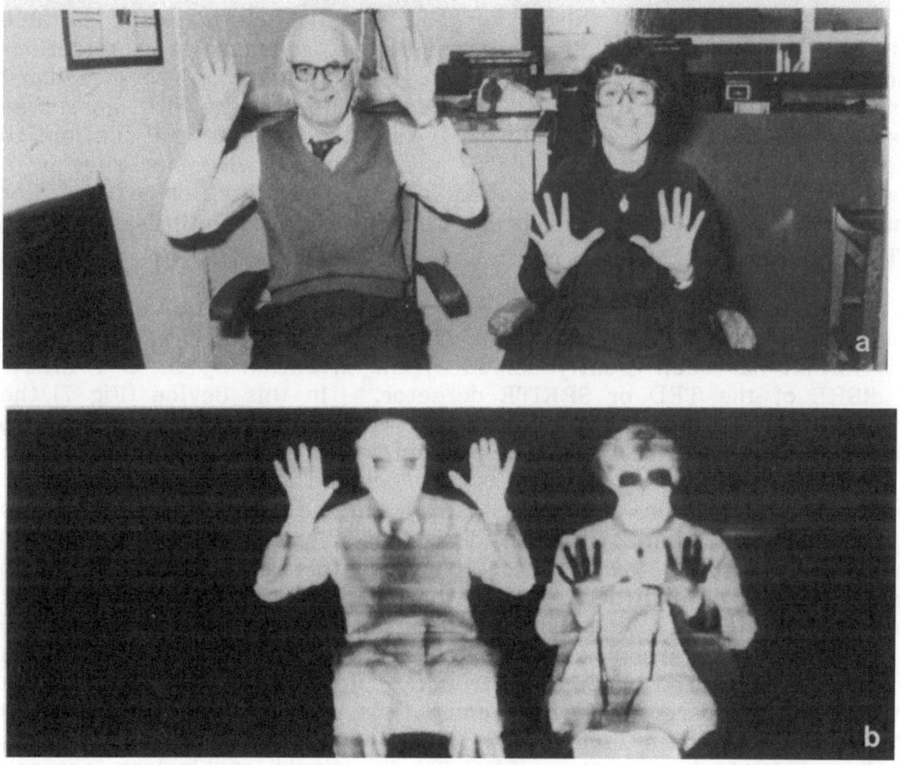

Fig. 6. (a) visible, (b) thermal picture of Mr W D Lawson and his secretary. The thermal picture taken with an imager using a parallel scanned linear array as in Figure 5 diagram 2. White hot. The lady has just moistened her hands with cold water to demonstrate the cooling effect of evaporation. The thermal imager used a linear array of 30 CMT detectors scanned across with a mirror drum with 8 facets to give 8 bands (or a total of 240 lines) across the picture. The angular resolution was 1mR.

the advantage of this serial scanning arrangement may not be apparent. Each detector element will see the same part of the scene but at a different instant in time according to its position in the array. If the outputs from the detectors can be processed by an electronic unit which stores each signal for a time determined by the element's position and then adds together all signals correctly phased the final signal will be equivalent to that obtained by one detector element which had been allowed to observe each point for a time $t = n\tau$ where τ is the dwell time for each actual detector element at a point in the scene. Thus, this system is equivalent to a single element imager operating with a frame time n times longer than that actually used. Again a detailed calculation shows that the performance is improved by a factor $n^{\frac{1}{2}}$. There are several disadvantages to this method. The first is that it requires more complex signal processing electronics. With the rapid development of electronic techniques this is not now a major problem and suitable "time delay integrate" or TDI units have been produced. A more serious shortcoming is that it requires very fast detectors and high speed scanning optics with the result that it is difficult to use arrays with more than about 10 elements. Since, however, it is both desirable and feasible to use much larger arrays (n 100 or more) a practical compromize is to combine the two methods described in a matrix detector array. For example the Rank Prize Fund imager that is in use at Bath uses an 8 x 6 array.

The construction of a matrix detector array is obviously more complicated than that of the same size linear array. The adoption of this format has been greatly helped by the invention by Dr C T Elliott of RSRE of the TED or SPRITE detector.[8] In this device (Fig 7) the integration function of the serial array is performed in a single strip shaped element. Thus an 8 x 6 parallel/series array is replaced by an 8 element linear array (known as an "octoted"). This is simpler to fabricate than the two-dimensional array and the number of output leads is reduced, thus simplifying the design of the detector cryostat and reducing the cooling load. These new detectors are used in the latest British designs of high performance thermal imagers, such as the RSRE sponsored Thermal Imaging Common Modules Systems (TICMS) manufactured by Marconi Avionics and Rank Tayler Hobson and the IR18 imager manufactured by Barr & Stroud.[13] Both systems conform to the TV 625 line standard so are fully compatible with normal TV equipments (Professor R V Jones in his 1982 Royal Institution Christmas Lectures used an IR18 to broadcast on BBC TV the finale of his last lecture). Figure 8 is an example taken with one of these equipments.

The performance obtained with these imagers has reached the point where the factors limiting the medical applications of infrared techniques are no longer instrumental. Unfortunately, these equipments are still very expensive and therefore a simpler equipment with a more modest performance may be more cost effective. I will conclude by considering how we may in future be able to bring these two conflicting requirements closer together.

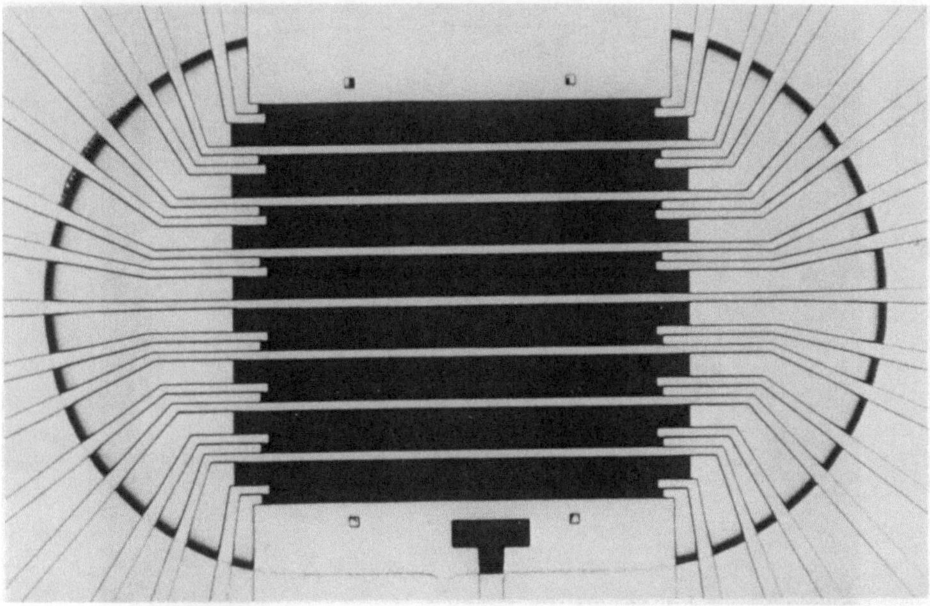

Fig. 7. An eight element SPRITE or TED detector.[8] This provides the most effective way of realizing the scanning mode of Figure 5 diagram 5. It is used in the Barr and Stroud IR18 and in the Marconi Avionics and Rank Taylor Hobson TICMS imagers.

FUTURE DEVELOPMENTS

In the last section I mentioned that the work at RSRE had led to the development of some of the best and some of the cheapest of the currently available thermal imaging systems. Unfortunately, we have not yet succeeded in combining the two. In this last section we will speculate on the possibility of ultimately achieving this.

At the beginning of this paper I mentioned how Sir John Herschel developed his evaporographic technique in an attempt to extend photographic techniques into the infrared. Looking at current developments in the visible may give us some clue as what to expect in the infrared. The most interesting development is that the introduction of very large integrated circuits, initially for data handling systems, has led to a new family of visible imaging systems. This has come about because the basic silicon FET is a very sensitive detector of light. Hence it is easy to make large two-dimensional arrays for use as solid state television cameras. It will be some time before these replace camera

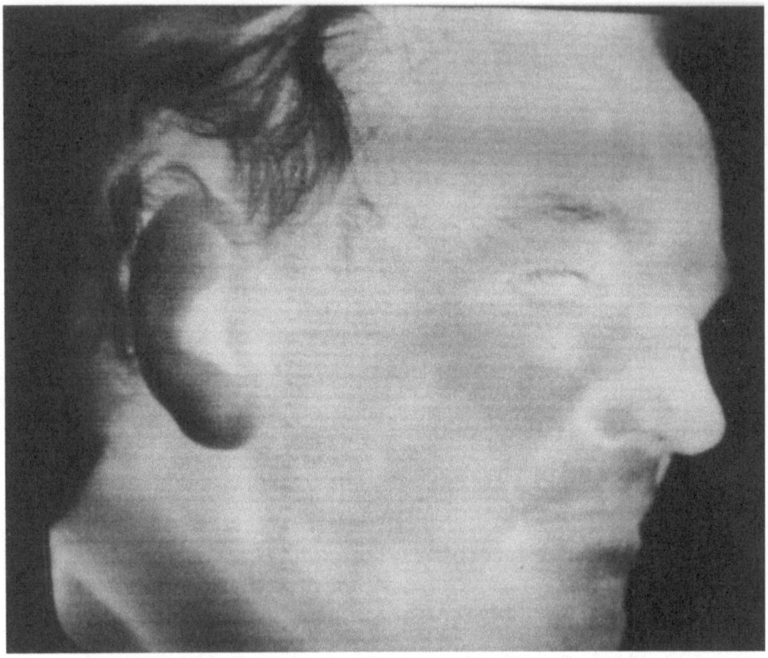

Fig. 8. Thermal picture of a person obtained with an IR18 (By courtesy of Barr and Stroud, Glasgow).

tubes like the vidicon but for scientific work the new silicon detectors have a distinct advantage. For many years (since the time of Herschel, in fact) astronomers and spectroscopists have used photographic plates to record their observations. Now the best photographic materials are very inefficient, utilizing only a percent or so of the available light. On the other hand the silicon detectors have an efficiency of 80% or more. Thus, replacing the photographic plate by an array of Si detectors increases the sensitivity of an astronomical telescope or a spectrograph by nearly two orders of magnitude. To the astronomers this is a valuable advance and they are looking forward to using in the future very large arrays of 10-1,000 cm^2 area containing as many as 10^7-10^9 individual detector elements.[14]

How will this help us in the infrared? If we could make arrays of 10^8 elements then we should be able to improve the sensitivity of a system by a factor of 10^4 over that of a single element system. This means of course that for a level of performance similar to that of present systems we could use detectors of sensitivity 10^4 times less than that of the detector in, say, an AGA imager. Such a low sensitivity is less than that already obtained with an uncooled pyroelectric[15] or CMT detector.[16] Hence we would not require cryogenics. This scheme depends entirely on the feasibility of producing these very large arrays. Figure 9 is a projection of infrared developments compiled from a number of sources which suggests that such things may not be too far in the future. However, it must be remembered that this projection is for the most advanced state-of-the-art systems. These large arrays are expected to be used first in space applications where cooling will be employed to obtain the highest possible sensitivity. The question we have to consider is whether these large units will become cheap enough for general use.

Three types of IR CCD arrays are being considered. First, it would be nice if the whole structure could be made of Si. Unfortunately the long wave cut-off of the normal Si p-n junction is about 1 μm. It is possible to make Si-platinum silicide Schottky barriers with an extended response to 2-3 μm but they do not go to long enough wavelengths to be really useful. The response can be extended to the 8-13 μm band by preparing specially doped silicon and employing extrinsic photoconductivity. Arrays of this type have been constructed successfully. Unfortunately, they require cooling to temperatures of 40 K or below which puts them out

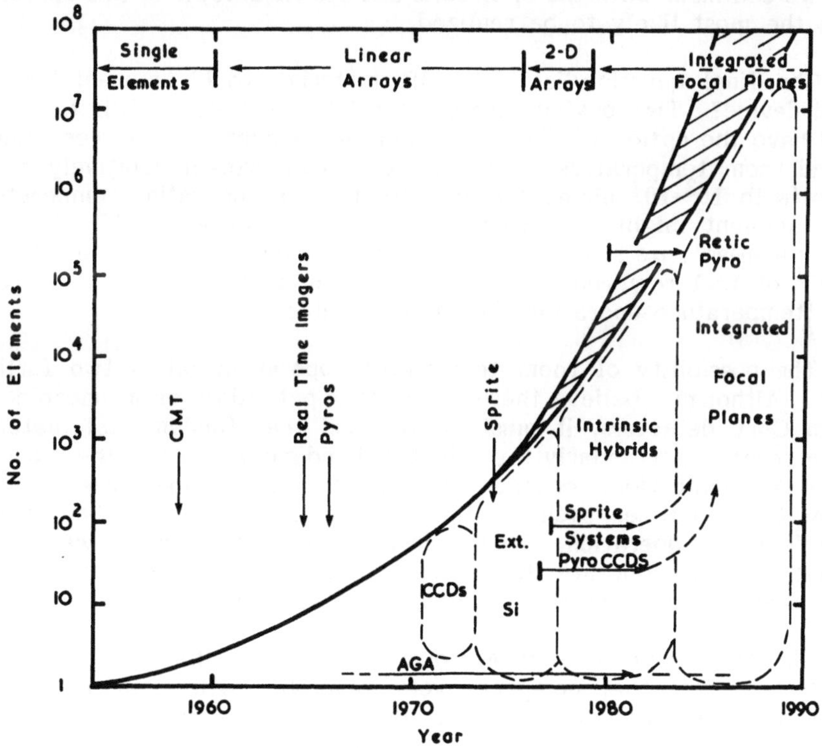

Fig. 9. Projection of the future development of infrared devices predicting the growth of applications for staring array systems.

for most applications. Attempts to employ special doping configurations that would operate at higher tempereatures have not proved successful. Hence an all-silicon device is not likely to meet our requirements.

The second possibility is to fabricate the read out electronics from a narrow gap semiconducter such as InSb or CMT. This would then be a direct IR analog of the visible Si array. Although some experiments along these lines have given encouraging results, the technology of other materials is so far behind that of Si (which was developed in the first place because it was the most suitable material known for integrated circuits) that it is unrealistic to expect these devices to be developed. The effort required would be much greater than that needed for Si, but the requirement for IR devices would be much smaller than that for Si computer circuits. Thus this possibility is not likely to be realized.

The third possibility is the hybrid device. In this two separate units are constructed: a detector array and a matching Si CCD read-out chip. The detector array is then connected point by point to the Si chip. Alternatively, it may be possible to develop a thin film technique for depositing the infrared material onto the Si chip and forming the IR detectors in situ, eliminating the complex bounding operation. Since we know we can make both the Si circuits and the IR detectors, this approach seems the most likely to be realized.

The final question is: Which IR material would be used for the hybrid device? Two obvious choices are InSb or CMT. With these we would have the option of choosing a working temperature between liquid air and room temperature. At first we should expect relatively small arrays (with 10^3-10^4 elements) and with them an operating temperature below ambient might be necessary. When arrays with 10^6 elements become available we should not only need an individual element sensitivity of 10^{-3} of that required for the detector in a single element system. Room temperature operation should be feasible.

The possibility of room temperature operation raises two further points. Although I believe there are further potentialities in uncooled 8-13 μm CMT detectors, it must be realized that fundamental material characteristics, particularly the effective band gap as well as the intrinsic carrier concentration, change so rapidly with temperature that a temperature stabilizer, such as a thermoelectric controller will be required for satisfactory operation. Having installed it, one might as well make full use of it and operate the device at a somewhat lower temperature, say 250 K, and have the maximum available sensitivity.

The final point is that at room temperature the performance of pyroelectric detectors compares very favorably with that of photon detectors at the same temperature. Pyroelectric materials are likely to be considerably cheaper to produce than semiconductors like CMT because their behavior is much less critically dependent upon composition and purity. There is a good chance that techniques for depositing the pyroelectrics direct on the silicon will become available.[17] If, therefore, the cost of making integrated circuits continues to fall, we hope to produce a pyroelectric-CCD hybrid which, in the not too distant future, will have a sensitivity about a factor of 10 better than the present pyroelectric vidicon and ultimately be even cheaper.[18]

I cannot say whether this device will answer all tomorrow's requirements, but I do believe that one of the hybrid systems is likely to be at the heart of the thermal imagers of the future.

REFERENCES

1. E. H. Putley, History of infrared detection - Part I. Infrared Physics 22:125-131 (1982).
2. E. H. Putley, History of infrared detection - Part II. Infrared Physics 22:189-191 (1982).
3. H. L. Gibson, "Photography by infrared," 3rd. Edition. Wily-Interscience, New York (1978).
4. G. W. McDaniel, D. Z. Robinson, Thermal imaging by means of the evaporograph, Appl. Opt. 1:311 (1962).
5. K. Lloyd-Williams, C. Maxwell Cade, D. W. Goodwin, The electronic heat-camera in medical research, J. Brit. I. R. E. 25:241 (1963).
6. W. D. Lawson, S. Nielsen, E. H. Putley, A. S. Young, Preparation and properties of HgTe and Mixed Crystals of HgTe - CdTe, J. Phys. Chem. Solids 9:325 (1959).
7. W. D. Lawson, Private communication.
8. C. T. Elliott, New detector for thermal imaging systems, Electron Lett. 17:312 (1981).
9. R. Watton, D. Burgess, B. Harper, The pyroelectric vidicon: a new technique in thermography and thermal imaging, J. Appl. Sci. and Eng. A2:47 (1977).
10. D. G. Avery, Some experiments with an infrared image forming device using a photoconductive cell, T. R. E. Memo 817 (1953).
11. R. B. Barnes, Diagnostic thermography, Appl. Opt. 7:1673 (1968).
12. S-B. Borg, Thermal imaging with real time picture presentation, Appl. Opt. 7:1697 (1968).
13. D. B. Webb, Thermal imaging via cooled detectors, Rad. and Elect. Eng. 52:17 (1982).
14. J. R. P. Angel, Very large ground-based telescopes for optical and IR astronomy, Nature 295:651 (1982).
15. E. H. Putley, The applications of pyroelectric devices, Ferroelectrics 33:207 (1981).
16. Labimex data sheet Model R006 uncooled 8-12 μm photoresistor, Warsaw 1981.
17. Y. Matsui, M. Okuyama, N. Fujita, Y. Hamakawa, Laser annealing to produce ferroelectric-phase $PbTiO_3$ thin films, J. Appl. Phys. 52:5107 (1981).

18. R. Watton, P. Manning, D. Burgess, J. Gooding, Direct charge injection for the pyroelectric/CCD focal plane hybrid, International conference on Advanced Infrared Detectors and Systems. IEE Conference Publication 204:82-87 (1981).

PHYSICAL ASPECTS OF INFRARED THERMOGRAPHY

J Steketee

Department of Biological and Medical Physics
Erasmus University, PO Box 1738, 3000 DR Rotterdam
THE NETHERLANDS

INTRODUCTION

Thermography in the widest sense of the term can be defined as the recording of temperature. In a similar way the well known rotating drum which records the air-temperature can be thought of as a thermograph. Indeed, such an instrument is often combined with a barograph and/or a hygrograph. However, nowadays the term thermography is mostly used in a narrower sense for a method of forming an image (the thermogram) of the temperature distribution over the surface of a body[1] and this is achieved by using the radiation emitted from the surface of the body. Some authors call this tele-thermography to distinguish the method from contact thermography in which the thermal sensor is in direct contact with the body (eg liquid crystal thermography).[2] In (tele-) thermography we may distinguish:

1. Infrared thermography using the infrared radiation emitted by the surface of the body at wavelengths between 0.8 μm and 1 mm.
2. Microwave Thermography using the microwave energy emitted by the body at wavelengths between 1 mm and 1 m.[3]

A thermographic camera consists basically of an optical scanning system for collecting the infrared radiation and focussing it on to a detector which transduces the radiation into an electrical signal. This signal (after amplification) drives a display unit. Such systems use only that porportion of the radiation spectrum to which the detector is sensitive (eg InSb-detector: 1-5.5 μm; HgCdTe-detector: 2-14 μm).[4]

Spectral Emissivity $\epsilon\lambda$

The radiation emitted from a surface is described by Planck's law

$$L(\lambda,T) = \frac{2\,\Pi hc^2}{\lambda^5}\left[\exp\frac{hc}{\lambda KT} - 1\right]^{-1} \quad Wsr^{-1}m^{-2}$$

and depends on the temperature of the surface and also on the emissive properties of the surface. This means that over the wavelength range to which the detectors are sensitive, knowledge is required of the spectral emissivity, $\epsilon\lambda$, of the surface being examined:

$$\epsilon\lambda = \frac{L(\lambda,T)}{L_b(\lambda.T)}$$

where $L_b(\lambda,T)$ is the spectral radiance of a thermal 'black body' according to Planck's law $L(\lambda,T)$ is the spectral radiance of the surface at the same temperature T.

Figure 1 shows that the spectral emissivity of human skin is close to 1 (0.98 ± 0.01),[5] but this value can change when the skin is covered with cosmetics, eg talcum powder (Fig 2). This decrease in emitted radiation is erroneously interpreted as a lower temperature by an infrared radiometer or thermography (Fig 3).

We calculated that the apparent temperature decrease for InSb detectors, with a wavelength range of 2-5.5 µm is about 0.3 K for talcum venetum which was in good agreement with measurements, whereas users of thermographic scanners equipped with a HgCdTe detector can expect temperature differences more than 1.3 K, owing to the dip at 8.5 µm in the curve of Figure 2.[6]

Emissivity of the Walls

When the surface of a body is not black ($\epsilon\lambda < 1$), part of the radiation from the environment (walls) is reflected by the surface. The question is whether this contribution from the walls depends on the nature of the wall surface or not. Gros[7] published nomigrams in which the temperature deviation ΔT, of a radiation thermometer or thermograph could be read as a function of the emissivity $\epsilon\lambda$, of the environment. On the other hand, Kirchoff's law says: "The rate at which radiant energy is incident on a surface, per unit area ... depends only on the temperature and not on the nature of the surface of the walls of the enclosure." Experimental studies[8] showed that there is no influence of ϵ_e on the total radiation of a non-black surface, even when the walls of the measuring compartment are covered with aluminium foil (ϵ_e = 0.02). Figure 4 shows the results of one of these experiments where the temperature of a lead sample (ϵ = 0.5; T = 45°C) was measured with a Heineman KT 41 radiation thermometer in a compartment enclosed

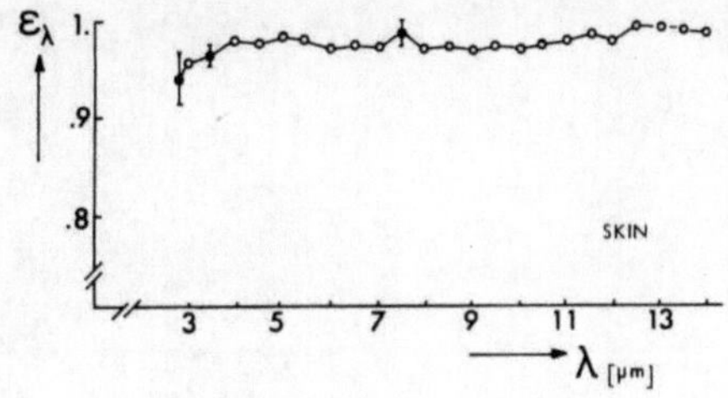

Fig. 1. Spectral emissivity of skin.

with venetian blinds. One side of these blinds was painted black and the other was covered with aluminium foil. The arrow 1 gives the moment at which the venetian blinds were turned, clearly demonstrating that no effect could be measured. From this point of view it is unnecessary to use black-painted walls in a thermography room. Only when low emissivity objects have to be measured in a room with reflecting walls (Al) radiation sources such as incandescant lamps can cause unwanted effects as is shown in the upper part of Figure 4 (arrow 2).

Influence of the curvature of a surface

Watmough[9] examined the extent to which the curvature of a surface influences the apparent temperature as measured by an IR-thermograph by considering the theoretical dependence of the spectral emissivity on

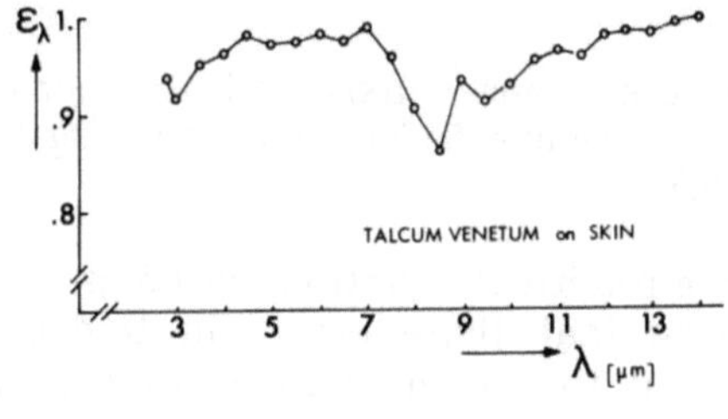

Fig. 2. Spectral emissivity of talcum venetum on skin.

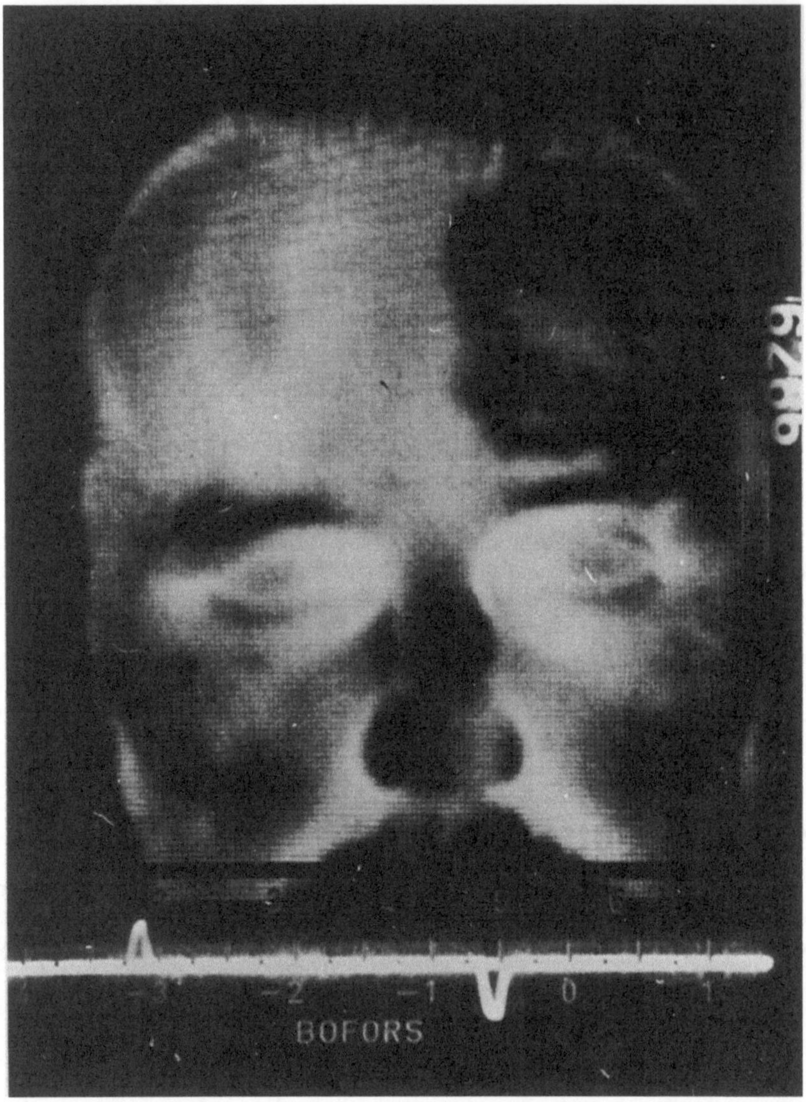

Fig. 3. Thermogram of the forehead partly covered with talcum.

the angle ϕ between the scanner axis and the normal to the surface. The results are given in figure 5 for two wavelengths, λ = 4.4 μm (o) and λ = 3.0 μm (▲).

Clark[10] derived a practical equation, based on the theoretical results of Watmough,[9] showing that these errors in temperature readings are likely to be proportional to the difference between the temperature of the skin and its environment.

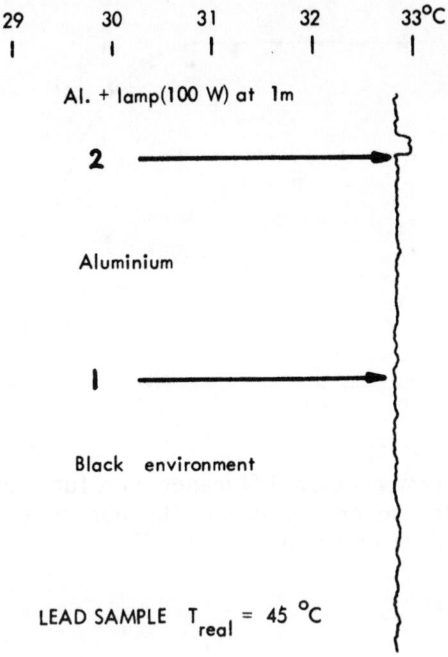

Fig. 4. Recording of the temperature of a lead sample (ϵ = 0.5) before and after changing ϵ_e (arrow 1) and after switching on a radiation source (arrow 2).

Comparing these results with the measured values (Fig 5) on a diffuse radiating body with T = 308 K, $T_{ambient}$ = 293 K, and ϵ = 0.98 (o) and on the forehead of a subject with T = 308 K(x) we can conclude that the apparent temperature differences observed on human skin viewed obliquely are smaller than those predicted theoretically. The temperature differences can generally be neglected for $\phi < 60^o$.

Influence of the Air Velocity

As the temperature of the skin also depends on the air convection in the examination room, the skin temperature of the forehead has been studied as a function of the air velocity, based on a (simple) heat-conduction model of the skin.[11] With this model it is possible to find a theoretical relation between the skin temperature and the heat transfer coefficient h, which is in turn a function of the air velocity v, in the

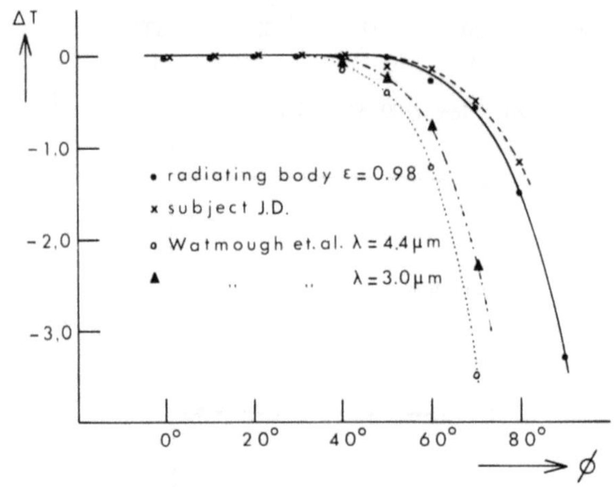

Fig. 5. Apparent temperature difference as a function of the angle ϕ between the scanner axis and the normal to the surface, for an ambient temperature of 293 K.

room. The heat transfer coefficient can be measured with special probes WS21HT (Technisch Physische Dienst, Delft, The Netherlands).

Figure 6 shows the results of measurements of the skin temperature-decrease ΔT as a function of the heat transfer coefficient h for the forehead of one of the subjects (RvP).

In the same figure is also drawn the theoretical curve according to the model with the parameter m, being the blood perfusion and the depth d, under the skin surface at which the skin temperature is supposed to be 37°C. The values of m and d can be measured for each subject after cooling the skin and analyzing the recovery curve. The results of experiments on four other subjects appeared to be closely similar to that shown in Figure 6. This means that the skin temperature of the forehead decreases about 2K when the air velocity is increased from a value below 0.05 ms^{-1} to 0.4 ms^{-1}. Therefore special attention must be paid to the construction and the use of the air condition system in a thermography room.

Spatial Resolution

The accuracy of temperature measurement of small objects with a thermograph strongly depends on the spatial resolution of the camera.

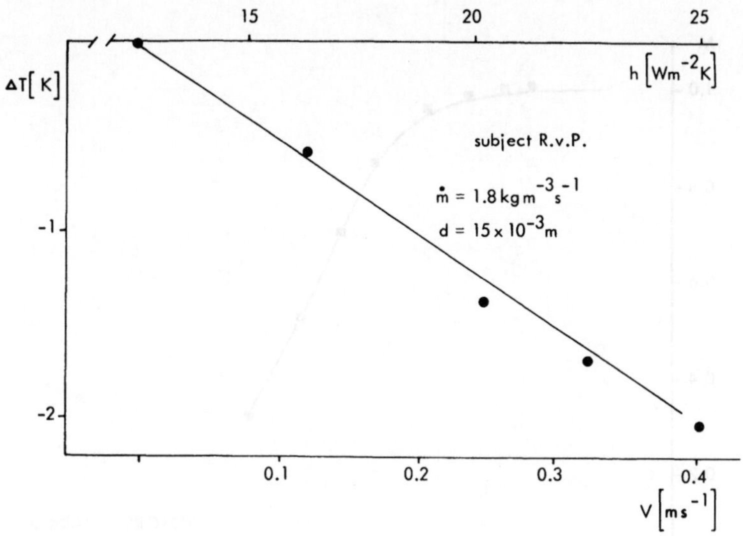

Fig. 6. Skin temperature decrease ΔT as a function of the heat transfer coefficient h, measured on the forehead of subject RvP (o) and compared with a theoretical curve according to the model. On the lower axis the corresponding values of the air velocity are shown.

Spatial resolution may be described by means of the Modulation Transfer Function (MTF).[12] To obtain an impression of the MTF of a number of thermographic cameras, Steketee[13] plotted the relative amplitude M of the videosignal, after scanning a test pattern with a series of gratings, as a function of the spatial frequency (Fig 7). The disadvantages of this method are:

1. The modulation function M is not exactly equal to the MTF.
2. A test pattern is not always available.

Therefore Vermeij[14] introduced a simpler method to check the MTF of an infrared thermograph which is based on thermographic measurement of the line spread function (LSF) with an oscilloscope and a heated (black) wire. The LSF points are Fourier transformed by a programable pocket calculator after which the MTF can be calculated. The authors recommend the use of the whole MTF as the standardized way of specifying the spatial resolution. This is made clear in Figure 8 where two MTF's are shown. Thermograms recorded by an instrument with MTF 1 are very sharp looking (high frequency long tail) but the measuring error at

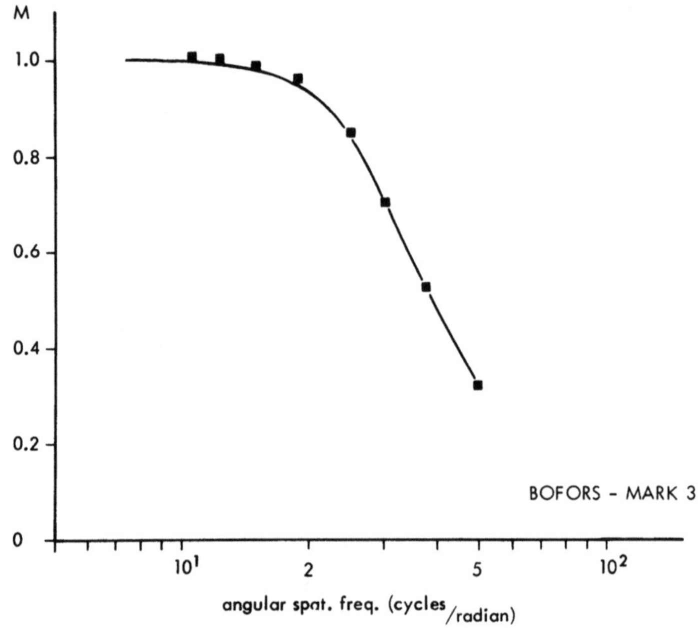

Fig. 7. Modulation function (M) for a Bofors Mark III Thermograph.

frequency f_c is more than 30% whereas the system with MTF 2 produces thermograms that lack the photographic sharpness of 1, but the measuring error at the same frequency f_c is only 10% and from a quantitative point of view the second system is therefore better in performance. However, both systems are identical when only a "50% point" in the MTF is specified.

Object Distance and Background

Attention must also be paid to errors in thermographic measurements due to differences in object distance. For example, two objects with a real temperature difference 5.90 K were measured with a Bofors Mark III system at 0.5m and 1.8m. At the first position a temperature difference of 6.05 K was measured, whereas the second reading was 5.35 K. A correction factor for different object distances is needed but not always produced by manufacturers. It is better to use two reference sources with a known temperature difference. Another phenomenon, especially in older systems, is the enormous shift in temperature readings when (large) objects with a temperature different to the background are introduced into the field of view. It is caused by AC-coupling of the

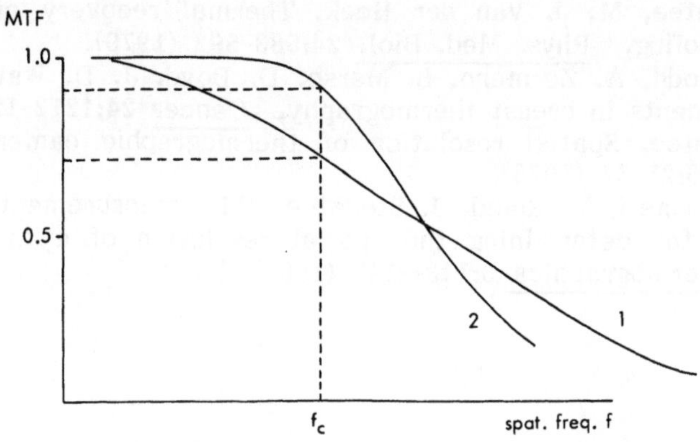

Fig. 8. Modulation transfer functions of two different systems.

amplifiers in the electronic system and when present can only be avoided by continuously calibrating the thermograph with a temperature reference source.

REFERENCES

1. Terminology Commission, Thermographic Terminology, Acta Thermographica (suppl. 2) (1978).
2. U. Flesch, Thermographic techniques with liquid crystals, in: "Recent Advances in Medical Thermology," E. F. J. Ring and B. Phillips eds., Plenum, New York (1983).
3. M. Gautherie, J-L. Guerquin-Kern, A. Kotewicz, Microwave thermal imaging: technical advanced and clinical findings, in: "Recent Advances in Medical Thermology," E. F. J. Ring and B. Phillips eds., Plenum, New York (1983).
4. C. H. Jones, Physical aspects of thermography in relation to clinical techniques, Bibl. Radiol. 6:1-8 (1975).
5. J. Steketee, Spectral emissivity of skin and pericardium, Phys. Med. Biol. 18:686-694 (1973).
6. J. Steketee, Influence of cosmetics and ointments on the spectral emissivity of skin, Phys. Med. Biol. 21:920-930 (1976).
7. C. H. Gros, M. Gautherie, P. Bourjat, C. Vrousos, Les applications medicales de la thermographie infrarouge, Acta Electron. 12:63-119 (1969).
8. J. Steketee, The influence of the environment on infrared thermography, Acta Thermographica 4:43-47 (1979).
9. D. J. Watmough, P. W. Fowler, R. Oliver, The thermal scanning of a curved isothermal surface, Phys. Med. Biol. 15:1-8 (1970).
10. J. A. Clark, Effects of surface emissivity and viewing angle errors in thermography, Acta Thermographica 3:138-141 (1976).

11. J. Steketee, M. J. Van der Hoek, Thermal recovery of the skin after cooling, Phys. Med. Biol. 24:583-592 (1979).
12. G. D. Dodd, A. Zermeno, L. Marsh, D. Boyd, J. D. Wallace, New developments in breast thermography, Cancer 24:1212-1221 (1969).
13. J. Steketee, Spatial resolution of thermographic cameras, Bibl. Radiol. 6:25-32 (1975).
14. G. F. Vermeij, M. Eland, J. Steketee, MTF measurement, a simple method for determining the spatial resolution of a thermograph, Acta Thermographica 3:143--147 (1980).

PHYSICAL AND PHYSIOLOGICAL INFLUENCE OF MEDICAL OINTMENTS ON INFRARED THERMOGRAPHY

J M Engel

Staatliches Rheumakrankenhaus
Baden-Baden
WEST GERMANY

INTRODUCTION

Emission of electromagnetic energy at infrared wavelengths from the human body surface is the physical basis of infrared thermography (IRT). This implies that infrared emission of skin is well known and constant. If this is so then real surface temperatures can be calculated from detected radiance values. Local or temporary differences in emissivity of skin, however, can lead to misinterpretation of IRT measurements taken either for diagnosis or disease follow up.

These problems become even more important if topical ointments are to be studied by IRT. In today's medicine there is a trend towards topical therapy for many different diseases, not only diseases of the skin but also rheumatic and other internal diseases. According to Steketee[1,2] cosmetics and ointments may have an important influence on IRT readings, eg if they contain talcum venetum. But change in emissivity is only one side of the problem; evaporation of water or alcohol from the ointment or jelly can cause a real cooling of the skin surface which will affect the regulation factors in skin microcirculation. Also, heating effects are possible if the ointment contains wax that spreads over the surface and prevents normal evaporation. Apart from these physiological factors, ointments can change the structure of the corny layer and thus alter normal emissivity and heat conduction from deeper layers of the skin. On the other hand, the active pharmaceutical preparation can induce changes in skin blood flow (microcirculation) either as vasodilation or as vasoconstriction (Fig 1). These effects might be long lasting owing to slow resorption of drugs from the ointment applied at the surface. All these different factors should be studied in detail to estimate the errors in IRT temperature measurements after application of topical ointments (Fig 2).

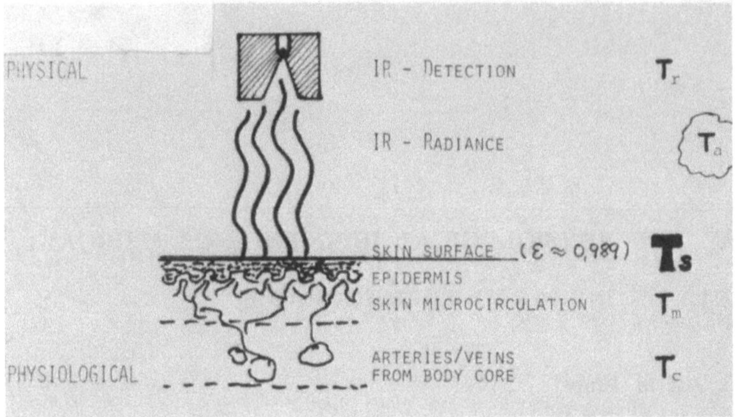

Fig. 1. Physical and physiological principles of medical IRT measurements. IRT detects body surface radiance within the IR wavelength band, the spectral radiance detected being dependent on the camera/detector system used. Under certain standard conditions the surface temperature (T_S) can be estimated/calculated from the radiance detected. Nevertheless, surface temperature is mainly dependent on conductive heat transport by skin microcirculation. This may be affected by cybernetic cycles eg body temperature and blood pressure regulation.

MATERIALS AND METHODS

In a first series of measurements we studied 24 topical ointments currently in use against rheumatic, venous, skin and cardiac disease. As far as possible we tried to compare a placebo preparation (pharmaceutical base of the active ointment).

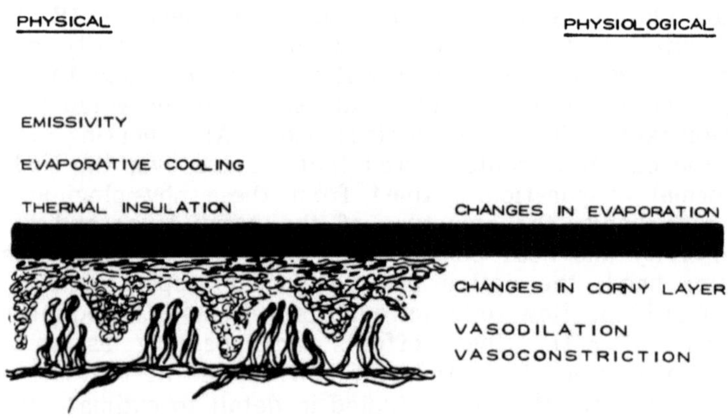

Fig. 2. Factors influencing surface temperatures in IR temperature measurements in man.

All ointments were first measured for changes in spectral infrared emissivity of a Nextel^R covered black body according to the technique described by Steketee.[2]

In a second study we looked for real temperature changes at a flat black body surface (temperature 33°C) to measure the effect of evaporative cooling. After these in vitro experiments we started measurements on patients, where we first compared two antirheumatic jellies and an antirheumatic ointment together with the corresponding placebo (ointment base without drug).

RESULTS

The emissivity measurements showed for none of the ointments studied showed a significant emissivity change in the spectral range of 3 - 14 µm. There were some slight differences for some wavelengths compared with the Nextel^R-base, but the overall error measured and calculated did not exceed 0.3 K. This is well within the error range of the IRT camerea system itself. The emissivity variation with wavelength was within the range of 0.1 - 1.9% but was not big enough to introduce significant corrections for thermographic readings (see Table 1).

The in vitro experiments designed to study the real temperature changes at the surface after application of ointments were carried out first with temperatures measured by Heimann KT41 bolometer.

A second series of measurements was done with the new Ikotherm-Camera system (Carl Zeiss, Oberkochen, FRG). The results obtained were similar to those with the bolometer (Fig 3).

All ointments with a polyglycol base showed a marked decrease of surface temperature for a variable period of 10 to 20 min, caused by evaporative cooling. In other ointments (oil/water, water/oil-emulsions) the temperature decrease corresponded to the water content of the ointment. In some cases the active pharmaceutical preparation showed differences in cooling when compared with the placebo. This was especially true for those active drugs that are volatile like etheric oils.

After these in vitro studies we are continuing with measurements on patients. These results are preliminary as we have not yet completed the whole series. The first results clearly show that the physical effect of evaporative cooling is present on patients. The degree of cooling seems to be little less than in the in vitro experiments. But the time of thermal recovery is much longer in man. In some cases we could show a lowering of surface temperature up to 60 minutes caused by evaporation of alcohol or water from the ointment. After about 15 minutes there was additional interference from pharmacological effects of the active

Table 1. Results of emissivity measurements in 24 medical ointments

Spectral Emissivity	Wavelength (µm)						
Ointment	3	5	7	9	11	13	14
NEXTEL	0.955	0.935	0.895	0.925	0.94	0.935	0.935
Malinert ^{R} - N	0.935	0.935	0.935	0.945	0.93	0.935	0.93
Rheumon ^{R}	0.94	0.94	0.92	0.93	0.93	0.93	0.93
Gelgrundlage zu Rheumon ^{R} Gel	0.955	0.945	0.945	0.93	0.945	0.935	0.93
Topical Gel: Indomethacin	0.955	0.935	0.92	0.92	0.94	0.935	0.935
Topical Gel: Placebo	0.935	0.93	0.92	0.935	0.95	0.94	0.925
Asche ^{R} Cream base	0.93	0.92	0.915	0.93	0.93	0.93	0.92
Neribas ^{R} Ointment	0.95	0.935	0.92	0.945	0.965	0.945	0.92
Neribas ^{R} Lanolin cream	0.93	0.93	0.925	0.935	0.94	0.93	0.92
Ambene ^{R}	0.955	0.935	0.895	0.915	0.955	0.95	0.925
Dolo-Arthrosenex ^{R}	0.95	0.95	0.94	0.94	0.965	0.955	0.94

	6.5µm	11µm
Bayolin ^{R} ointment	0.953	0.939
Bayolin ^{R} Liniment	0.946	0.940
Salirheuman ^{R}	0.947	0.949
Reparil ^{R} Gel	0.934	0.940
Betaisodona ^{R}	0.942	0.940
Hirudoid ^{R} 40 000	0.925	0.930
Isoket ^{R} ointment	0.932	0.931
Fragivix ^{R} ointment	0.930	0.925
Syracerin ^{R} Emulsion	0.935	0.945
Venalitan ^{R} 150 000	0.937	0.942
Venoflexil ^{R}	0.930	0.934
Hydroexan ^{R}	0.953	0.955
Baycuten ^{R} SD Creme	0.949	0.948
Medigel ^{R}	0.949	0.948

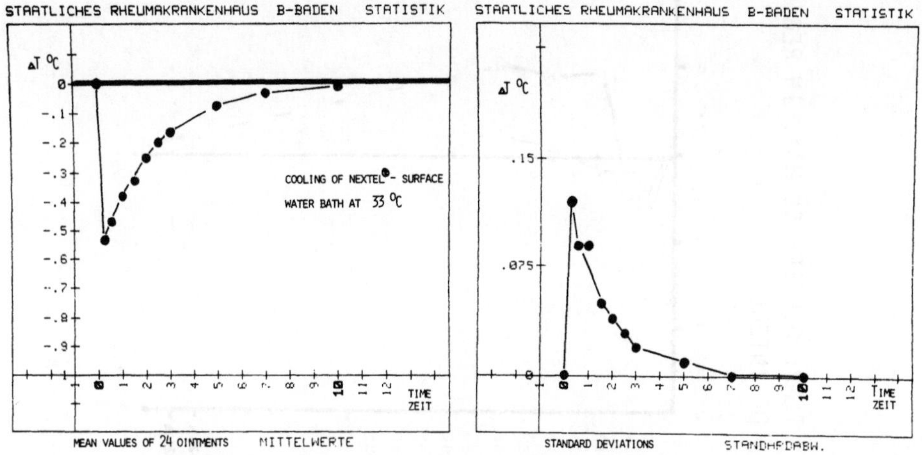

Fig. 3. Changes in surface temperatures over time after ointment application. Immediately after application of an ointment there is a mean surface cooling by evaporation of water, alcohol or other volatile ingredients (left) with the standard deviation maximal within the first few minutes. These immediate effects are dependent on physical properties of the ointment. Therefore, in vivo measurements should not be taken before 10 minutes after application to avoid errors by evaporative surface cooling.

compound in the ointment with the physical effect. If vasoactive substances like nicotinic acid are in the ointment there is a marked vasodilation and rise in surface temperature that may last for more than 24 h, even if the amount of ointment was very small (Fig 4).

DISCUSSION

IRT surface (radiation) temperature measurements may be altered after application of medical ointments or cosmetics. There are two effects: the physical influence and the physiological changes that both affect IRT. The practical consequence must be to avoid any contact with cosmetics and ointments prior to IRT. Pharmaceutical preparations containing vasoactive drugs penetrate very slowly and may thus influence skin microcirculation for hours. Earlier studies[3] showed this vasodilative effect parallels the drug concentration in deeper layers of the skin after one single application of an antirheumatic ointment containing nicotinic acid-ester.

In clinical studies of the therapeutic effect of topical therapy careful experiments must be carried out in vitro to avoid errors in IRT readings

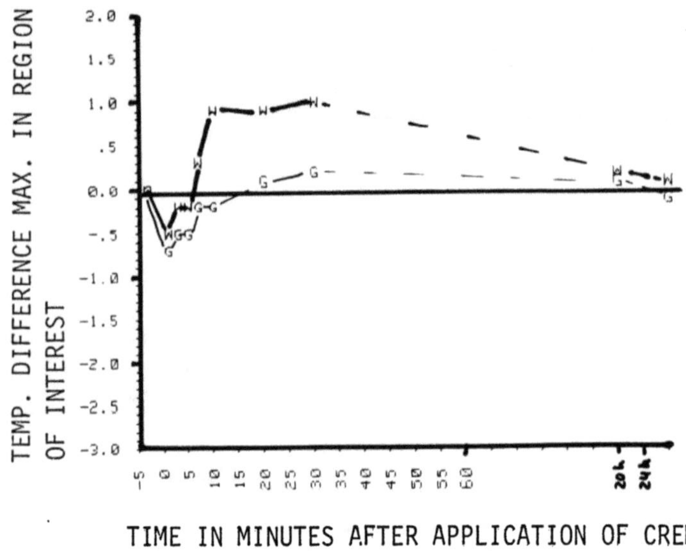

Fig. 4. Surface temperatures in vivo after MALINERT-N application. Both M-N ointment and placebo (G) decrease surface temperature as a result of evaporative cooling. Thereafter the temperature over MN rises continuously due to the vasodilative effect of nicotinic acid. The slight and transient temperature rise over the placebo may be caused by a reflex mechanism, or by the protective effect of the base preparation.

by changes in emissivity and evaporative surface cooling. After our studies we now think that IRT alone possibly might not be the method of choice to study the local effect of topical ointments because of so many possible errors. In any case a very careful design of such a drug trial is essential, considering all physical, physiological and pharmaceutical factors influencing surface temperature.

ACKNOWLEDGEMENTS

All experiments on emissivity have been carried out at Erasmus University, Rotterdam. I wish to thank Prof Dr Johan Steketee for his kind support of this work and his scientific advice as well as H van den Breggen for the careful measurements.

REFERENCES

1. J. Steketee, Spectral emissivity of skin and pericardium, Phys. Med. Biol. 18:686-694 (1973).
2. J. Steketee, The influence of cosmetics and ointments on the spectral emissivity of skin, Phys. Med. Biol. 21:920-930 (1976).
3. W. J. Zerweck, J-M. Engel, Pharmakologische und thermographische objektivierung der wirksamkeit von Malinert-N[R]-Salbe bei nichtentzundlichen rheumatischen erkrankungen, Therapiewoche 28:3131-3142 (1978).

QUALITY CONTROL IN INFRARED THERMOGRAPHY

E F J Ring

Royal National Hospital for Rheumatic Diseases
Upper Borough Walls, Bath
UNITED KINGDOM

INTRODUCTION

The image produced by infrared (IR) thermography is used in a number of different forms. Direct viewing of the monochrome video screen is sometimes sufficient, and is described as thermoscopy. More often the image is processed in some way, with single isotherms, multiple color isotherms or digitization through a computer. Photographic or magnetic storage are frequently used.

The basic optics and electronics of the imaging system can be subject to deterioration over a period of time. The procedures used to display or record the image as described may themselves be subject to drift or to variation in quality. In addition, the whole process is manipulated by a human operator. Subjective variations are therefore introduced, which may not always be obvious. It is important to know these variables between individual users, and on one operator over a period of time. The latter is important not only in a training situation, but also a check on the skilled operator, who may develop procedural habits, which themselves lower the quality of performance of the system. These requirements are highlighted by the increasing trend to quantitative thermography.

TEST PROCEDURES

Three procedures are described which allow the operator to carry out basic quality control for IR thermography.

Spatial Resolution

A test card is prepared which consists of a series of parallel bars of differing widths and spacing. Vertical and horizontal bars are distributed over the surface. The card measuring 20 x 20 cm is a 5 mm aluminium plate, heated from behind by an array of heating elements, as used in a domestic iron. The pattern is applied by the photo-resist technique. The temperature of the plate may be varied between 25 and 35°C by a variable transformer (Fig 1).

Temperature Resolution

A linear array of 5 constant temperature blocks are used. These may be placed in the field of view in vertical or horizontal format. Using different temperature settings for each block, temperature differences across the scene may be checked (Fig 2). Two reference sources could be used as a simpler form for this test.

Phantom

A simple phantom is used, which is designed to be fitted over the front of a large thermal reference source such as the AGA 1010. This is a constant temperature source of 10 cm diameter. A pattern is produced by constructing a sandwich layer of two sheets of polyvinyl acetate of differing thickness, ie 0.06 mm x 0.1 mm. The center has an open hole, to allow the actual temperature of the source to be located in the image (Fig 3a). Using an indium antimonide detector, the acetate material

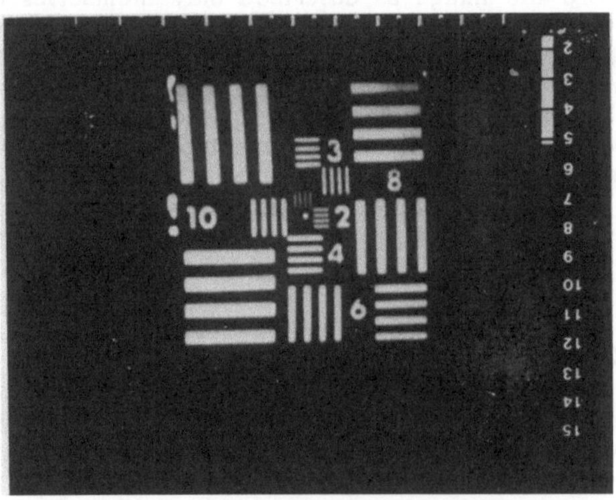

Fig. 1. Thermogram of the test pattern imaged in a 30 x 30 cm field (white = hot).

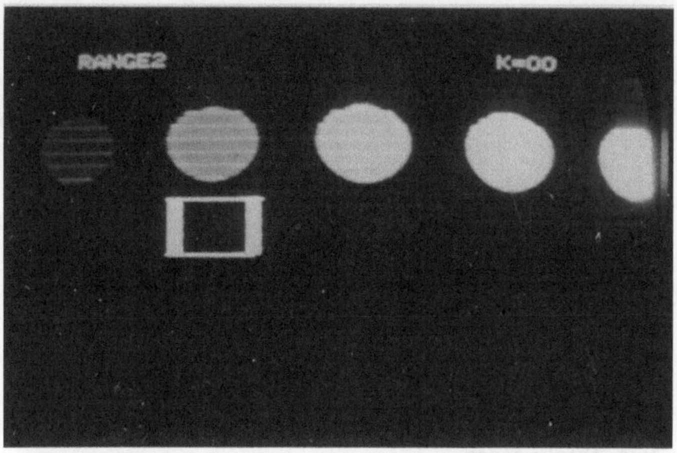

Fig. 2. Thermogram of the five reference sources in horizontal format.

attenuates infrared emission. The system described will provide a pattern ranging from 26 - 36°C, when the source is at the latter temperature (Fig 3b). Cadmium Mercury Telluride detector systems, however, require different materials and thicknesses to achieve a similar effect.

PROCEDURE

Spatial Resolution

The card is erected before the scanner when temperature stability has been achieved. An optimum temperature for operation is 30°C. Three distances are used to compare the performance of the system approximating to the target sizes used in medical thermography. These have been taken at 20 sq cm, 30 sq cm and 50 sq cm. Visual and photographic and/or magnetic stored images can be compared.[1]

Temperature Resolution

A standardized procedure is adopted to compare the temperature differences seen across the scene. This may be recorded across the top and bottom edges and the left and right vertical edges. The isotherm integrator or linescsan facility is used to record the temperature differences.

Thermal Phantom

The phantom plate attached to the front of the reference source

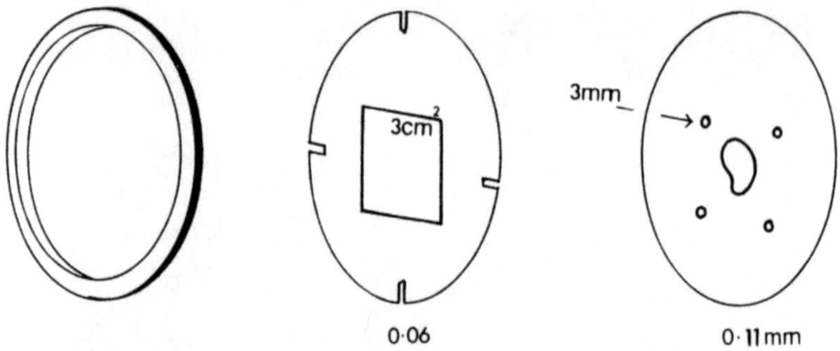

Fig. 3a. Construction of a thermal phantom with acetate sheets at two thicknesses.

is set up to fill the field of view. Temperature measurements, and regions of interest, if appropriate, may be selected from different parts of the image. If a calculated index is used, this may be applied to the whole or to a selection of smaller "windows". As above, the visual, photographic and replayed stored image can be compared for reproducibility.

RESULTS

Spatial Resolution

The ability of thermography to resolve the smaller lines on the test card is reduced at a larger target size of 50 sq cm (Fig 4). Problems

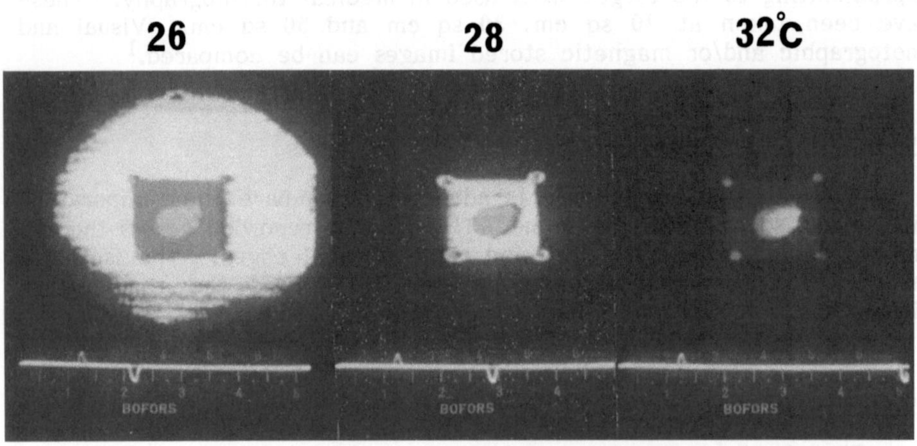

Fig. 3b. Isothermogram of the phantom at 26, 28 and 32°C.

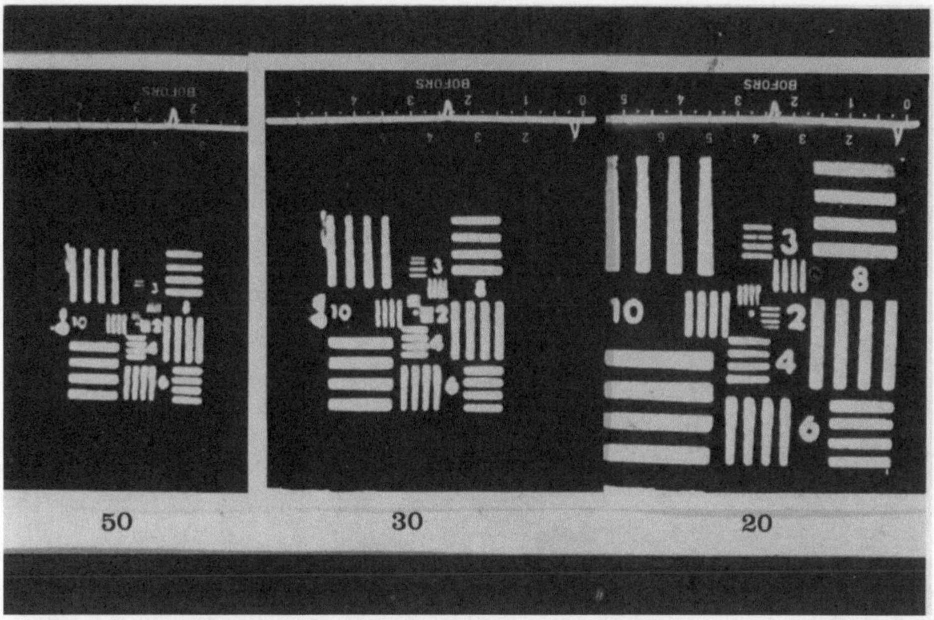

Fig. 4. Comparison of target scanned in 20, 30 and 50 cm sq fields.

of alignment with images produced from a multi-element detector are clearly seen (Fig 5). In photographic records of the image, vertical elements of the test card may prove more easily defined than horizontal (Fig 6). Direct visual appreciation of the image is often much better, due to merging of the flickering image.

Distortion of the image may occur from a lens system (Fig 7). Older thermographic systems may show a loss of linearity across the image, which increases with age. Most of these defects may have little serious effect on the interpretation of medical thermograms. However, it is important to note the decreased resolution at greater scanning distances.

Temperature Resolution

Temperature differences from the highest of the group of five sources were recorded using the isotherm system of the Bofors IR camera. Figure 8 shows that the temperature differences on all four edges of the field were within 0.2°C. The greatest discrepancy was recorded in the bottom left corner (A-D) at 0.3°C. The values recorded from a digitized scan were also within the same limits. The test was also applied to a prototype system. This system is not equipped with isotherms, but with

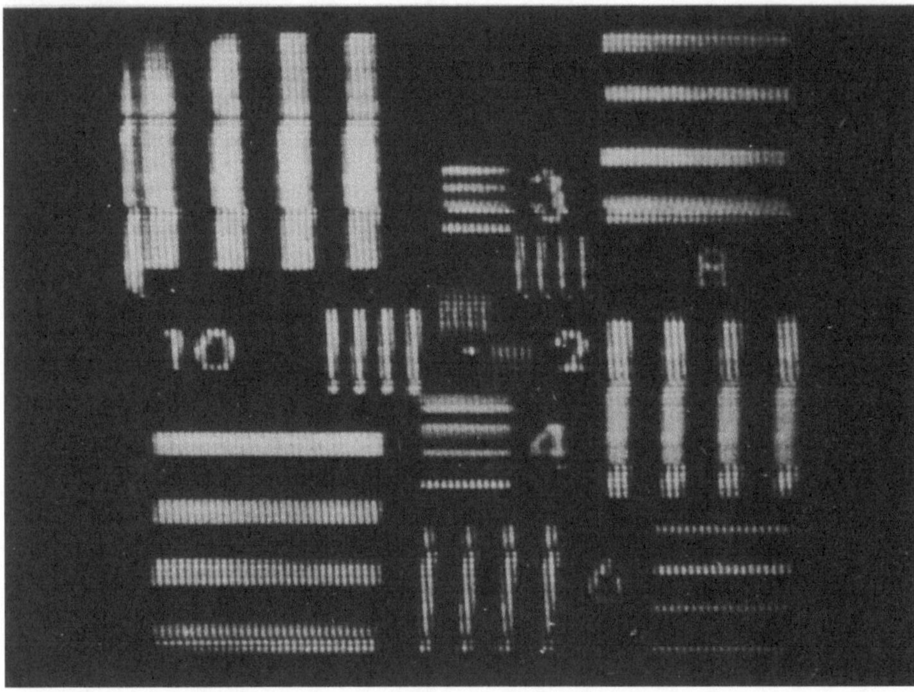

Fig. 5. Resolution loss due to a misaligned multi-element detector system.

an integrator, measuring mean temperature from an adjustable window. As the window is restricted in movement, measurement at the extreme edges of the field was not possible. The sources were used across the center field in the vertical and horizontal (Table 1).

Thermal Phantom

Figure 3b illustrates the temperature difference across the phantom shown by isotherms at 26, 28 and 32°C. On a color scale, this will be represented in four colors ranged over degrees. Three regions of interest were selected from the thermogram by a single operator over a period of 10 weeks. A thermographic index was calculated on each occasion using a small computer by the formula $TI = \frac{\Sigma(\Delta t_{26} \times a)}{A}$

Table 1. Integrator values

Reference 1 = max 5 = min	Horizontal L (max) R (min) °C	Vertical Up (max) Down (min) °C	Δ
1-2	3.0	3.1	0.1
1-3	4.5	4.8	0.3
1-4	6.0	6.2	0.2
1-5	7.6	7.3	0.3

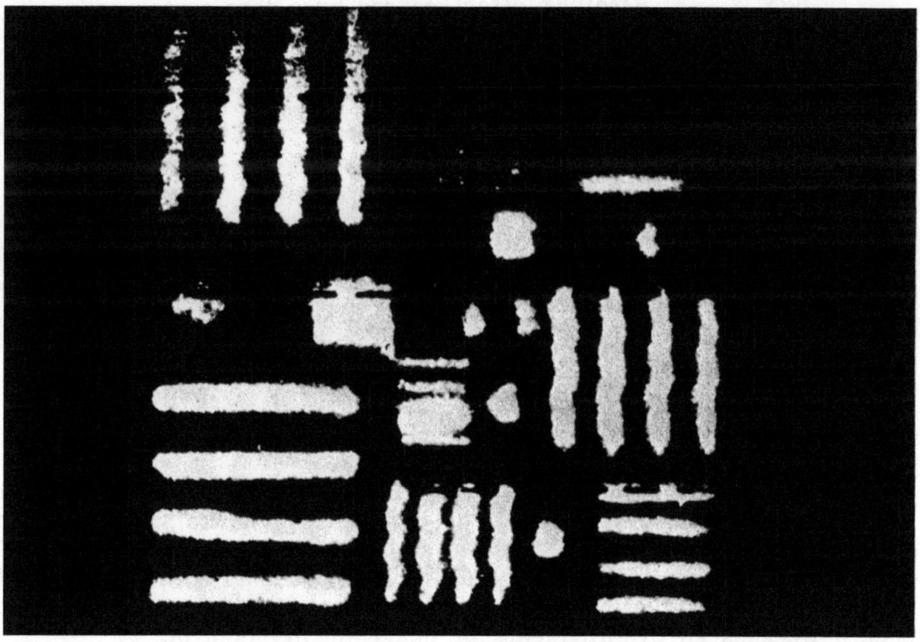

Fig. 6. Poor resolution thermogram made worse by over exposure of photograph.

Fig. 7. Barelling effect caused by a lens system.

where a = isotherm areas and A = total area of the region of interest.[2] The results (Fig 9) show that the index was less reproducible for the central area measured over a wider temperature range. The other two areas, being more homogenous, gave less variable results.

CONCLUSION

The three test procedures are useful for the routine checking of a clinical thermographic system. They can be used in training operators to appreciate the performance of the system and the repdroducibility of technique. On a routine basis, the first two procedures are recommended at 3-4 monthly intervals. The thermal phantom should be used at least once a week as an overall means of image quality control.

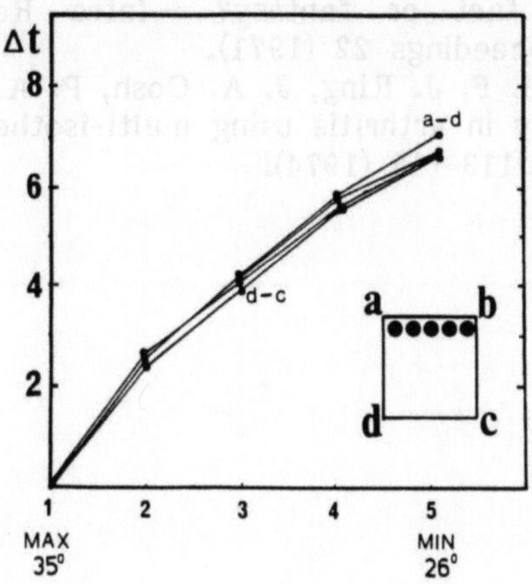

Fig. 8. Temperature differences measured from isotherms, using 5 sources placed at the four edges of the field a-b, b-c, d-c, a-d.

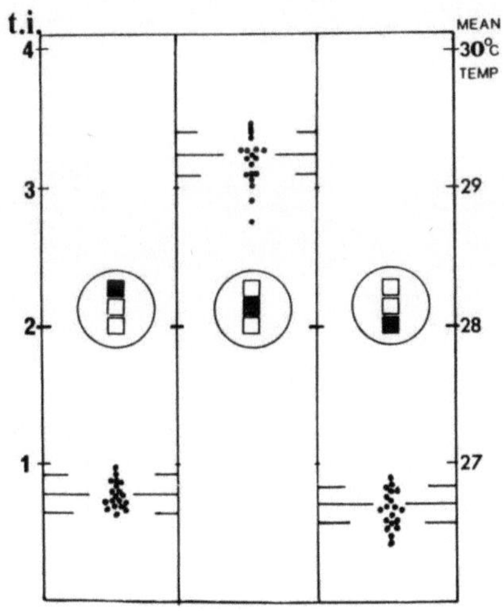

Fig. 9. Variation in computed values from three regions of interest from the phantom.

REFERENCES

1. K. Lloyd-Williams. K. D. Patil, B. Phillips, E. F. J. Ring, Quantitative thermography, fact or fantasy? Infra Red Techniques IERE Conference Proceedings 22 (1971).
2. A. J. Collins, E. F. J. Ring, J. A. Cosh, P. A. Bacon, Quantitation of thermography in arthritis using multi-isothermal analysis, Ann. Rheum. Dis. 33:113-115 (1974).

COMPUTER PROCESSING OF THERMOGRAPHIC IMAGES

M Negin

Computer and Information, Sciences Department
Temple University, Philadelphia
Pennsylvania 19122, USA

INTRODUCTION

The organization of system components for acquiring the thermographic images for processing by digital techniques is not unique. Numerous system architectures may be constructed and investigators will develop their own varied approaches to the issue. One general architecture is given in Figure 1, and will serve to orient the reader to several important topics.

Each block in the system diagram will be briefly reviewed with emphasis on some of the more commonly encountered problems and their associated techniques for resolution.

The choice of input format must be addressed early in the systems consideration. Is input going to be in the form of photographs, digital archival sources (magnetic tape or disk or other computers), or direct from an on-line thermographic scanner? Computer techniques must be defined. What types of algorithms are to be applied to the pictorial imagery? Hardware must be selected. Will redisplay of the data via computer enhancement programs be required? Is user interaction required? Will the data need to be sorted in a data base for future analysis comparisons? These are but a few of the questions that will be spawned by the more detailed considerations to follow.

IMAGING INSTRUMENT CONSIDERATIONS

There are numerous considerations required to be made when preparing to perform digitization of thermographic data from a sensor. For the most part this paper will be restricted to general sensor types

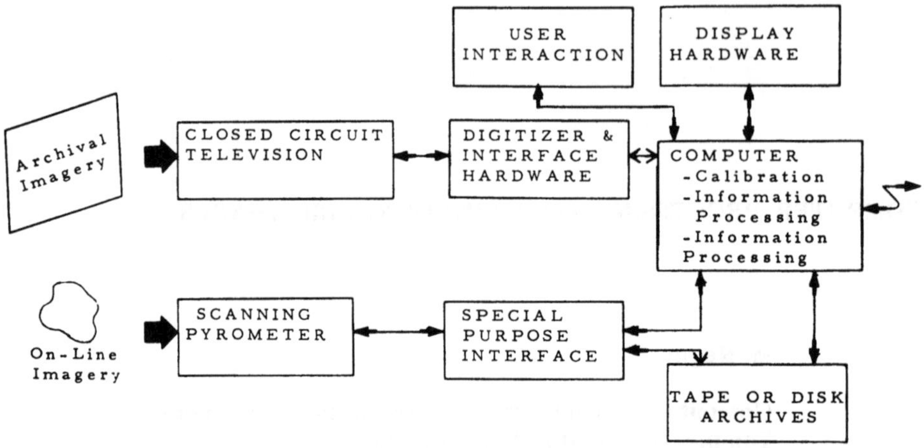

Fig. 1. System architecture for computer processing of thermograms.

that naturally produce images. Point sensors, thermocouples, hand-held radiometer instruments and the like that produce an electrical signal may of course be digitized, but this paper will not be concerned with those devices. Imaging devices that inherently produce three or more dimensional data (scanning pyrometers, pyroelectric vidicons, liquid crystal sheets) will be of prime consideration.

Interfaces and Acquisition

There are several key considerations when interfacing a thermal imaging system to a computer. If one is using archival photographic media, a good quality closed circuit television system (American or European Standard) may be used to view the images and off-the-shelf digitizers may be purchased to accomplish the optical-to-digital conversion. For color photographs from liquid crystal or pseudo-color image inputs, tri-stimulus decomposition must be accomplished, generally using three appropriately chosen color filters. The digitizers that are generally available may also require a separate interface board so the computer may communicate and control the digitizer.

When on-line real-time digitization is required, one may choose a system that uses standard television timing and signal levels, in which case interfacing to the computer may be accomplished by the use of standard digitizers and interface boards. Many thermographic units have non-standard scanning and frame rates and will require special purpose hardware to implement the optical-to-digital conversion. This may require special purpose intermediate data buffering, as well as special purpose timing and synchronization. In some cases, this special purpose hardware may be available from the thermographic manufacturer and will probably only function with the company-specific imaging system. At least one manufacturer offers a digital tape output from its unit. This tape may be later input to the computer via industry standard tape formats.

Calibration

The calibration requirements of the system may fall into several categories. One of the most important is the temperature calibration of the thermographic sensor. Normally, one would place a known thermal source with at least two thermal calibration intensities, and image the known standards. Response curves may then be fitted using curve fitting procedures or piecewise linearization and interpolation. Obviously, the computer may perform the required calculations. Other calibration for spatial distortion due to non-linear sensor scanning may be included. This might require digitization of a known geometric grid pattern, and then applying a spatial correction (eg one based on a bilinear spatial interpolation). After sensor calibration, one may also need to account for perspective and parallax distortions based on the geometric relation between the test object and the plane of the sensor.

Storage

Data storage may be performed on tape, disk, or high speed memory. In all cases the amount of raw data stored per image depends greatly on the application. For most medical applications, representation of the image data in a two-dimensional array is useful, where the dimensions of the array are of the order of 128 x 128 with each element having a resolution of 8 bits. The whole image then would require 16,384 8-bit memory cells (16,384 bytes, one byte = 8 bits). In some applications image digitization of 512 x 512 picture elements may be desired. Currently in our laboratory at Temple University, digitization of four liquid crystal plate thermograms is being performed during one television frame scan. Three balanced color filters are used for color separation. In this case, the raw data requires 512 x 512 x 3 bytes (786,432 bytes = 6,291,456 bits) for storage.

Information Processing and Display

The representation of images as two dimensional arrays of data points is an important conceptual framework from which virtually all image or picture processing algorithms arise. The arrays may be thought of as matrices, and consequently many computer procedures may be easily and formally structured. Each data point represents a "picture element", referred to hereafter as a "pixel", and may be considered to be a "gray level" at a specific spatial location. Stored images may then be thought of in terms of their "gray levels" or their "gray scale".

Information Processing and Information Display

After the thermographic image has been acquired, a variety of different information processing techniques may be applied in an attempt to enhance or extract pertinent information. One categorization of the general types of image processing is given below and in Figure 2.

1. Image enhancement
2. Computer assisted measurement
3. Automated interpretation

This list is ordered according to the amount of information that is required to fully implement the process. Simple image enhancement techniques require very little knowledge about the detailed structure of the image itself. For example, enhancing the contrast of an image may be as simple as linearly expanding the gray scale range of a display device (eg turning up the "contrast knob" on a television monitor). Other enhancement techniques (eg edge enhancement or enhancement of venous patterns) may require more information about the image structure in order to optimize the computer software or hardware algorithm. Image enhancement techniques are generally used when a human observer is going to interpret the image. The enhancement methods may make the thermogram easier for interpretation. Gray scale modification alters the brightness of each pixel so as to more effectively utilize the human eye-brain system. Gray scale modification may be contrasted to artifical

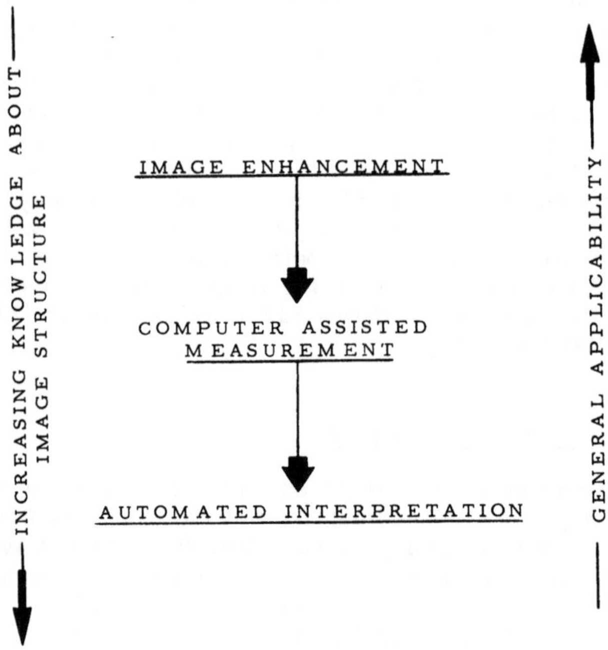

Fig. 2. Classes of computer methods for analysis of thermograms.

coloring (so called "pseudocolor") of the image. Many commercially available thermographic units provide for the pseudocolor display of the thermal pattern. The operator usually must select placement and width of thermal bands for mapping into colors. Given that the thermal images are stored in a computer, one can use computer generated color-look-up tables to code the images for viewing.

Simple methods like gray scale expansion alter the pixel values via point-to-point transformation. Point-to-point type transforms are of the form showin in Figure 3, where a pixel at (x,y) in the input image is transformed into a pixel at (x,y) in the output image. Transforms may also be constructed whereby groups of points in one image are transformed into single points in another. Figure 3 shows this via a local averaging technique, an edge detector, and a line sum. Groups of points in one image may also be transformed into groups of points in another image and certain special purpose transforms (eg Fourier Transform) may be useful.

Beyond general image enhancement techniques, one can structure a series of computer assisted measurement methods. Figure 2 illustrates an important point about quantitation. As an investigator understands the pertinent image categories more completely, more quantitative methods will be naturally developed. For example, in thermography of the diabetic, one may choose to define very specific thermal measurements on the plantar surface of the foot in order to follow a patient's progress. In breast thermography, one may measure the average breast temperature, or localized sector or quadrant temperatures to assess the breast thermal patterns. Computer assisted measurements may take the form of gray

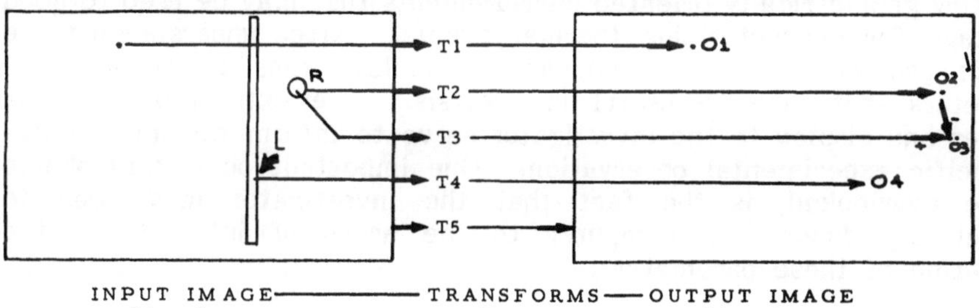

T1 -- Point-to-point Transform -- O1 = I•k + b (Linear Contrast Increase)

T2 -- Region-to-point Transform - O2 = Average of Region R (Blurring)

T3 -- Region-to-point Transform - O3 = I - O2 (Edge Enhancement)

T4 -- Line - to -point Transform - O4 = Sum of Vertical Region "L" (Vertical Projection)

T5 -- Region-to-Region Transform- O5 = Frequency Decomposition of I(x, y) into O(x, y), (Fourier Transform)

Fig. 3. Image transformations - general structure.

level statistics on a global or local basis. For example, one may measure the average gray level (average temperature) of the entire region of interest. Statistical-type measurements are quite straight-forward to compute, are experimentally reproducible, and are quite insensitive to minor changes in pattern and thermal distribution. Commonly used measurements include:

1. Average value
2. Variance
3. Higher order moments
4. Minimum value
5. Maximum value
6. Range of gray level
7. Gray level distribution or histogram

Obviously, this statistical analysis and summary may be displayed on a CRT terminal or printed out for study or archival purposes. Trending of the gray level statistics often provides valuable insight into the underlying thermal behavior of the subject material. These statistical summaries may be computed globally or locally.

Statistical parameters computed from a gray level histogram contain no direct information regarding the geometry of the thermal patterns. Some of this information may be preserved by using perimeter or edge information as well as absolute location information. For example, an isolated hot spot in a breast thermogram may be characterized by temperature, size, shape (eg perimeter vs area factor), location (eg upper-outer-quadrant), or symmetry (eg bilateral focal thermal elevation). These are a few geometrically oriented measurements that may be used to good advantage for characterizing thermal events. After understanding the specific problem, one can frequently formulate tens to hundreds of parameters that may be useful in analysis. The extraction of these features and display to the investigator serve to inform the investigator of specific experimental observation. One important point that should not be overlooked, is the fact that the investigator may need to interactively direct the computer to the areas of interest for the calculation of these parameters.

Parameterization of the images in this fashion also is useful in significantly reducing the data storage requirements. A raw, unprocessed thermogram may require 65,536 bytes or more of storage, while a parameterized thermogram (eg 1,024 parameters) may require less than 10% of the raw, un-parameterized storage requirements.

The last stage of information processing is at the level of automated interpretation, where the parameters that have been intelligently extracted may be combined in some fashion to develop a pattern classifier. This requires the collection of a data base, the extraction of features, and a truth index or measure. For example, one may have archival records of breast thermograms of patients with known clinical outcomes.

One can compose programs to extract the parameters, and to associate the features with classes (eg normal breast, benign disease, malignant disease). The pattern classifier is trained according to the input data base, and is then able to classify new, previously unseen data. If the parameters selected are indeed characteristic of the thermal process presentation, and the data base and classes are appropriately known, then a classifier has a reasonably good chance of succeeding. Obviously, bad data may cause the classifier to be trained poorly, and therefore to perform poorly during testing.

Classifiers commonly used are linear discriminant techniques (parameters are combined linearly), and nearest neighbor (measures the closeness of the test parameters to previously known classes). The linear discriminant classifier (see Figure 4) combines multiple parameters in order to develop a new parameter that discriminates among the classes better than individual parameters. Figure 4 illustrates this point, since the individual (one-dimensional) parameters cannot separate the classes, while a two-dimensional scatter plot can be used to accomplish separation.

Nearest neighbor approaches use the "distance" of a test point to either a representative point or a class centroid, and are especially useful

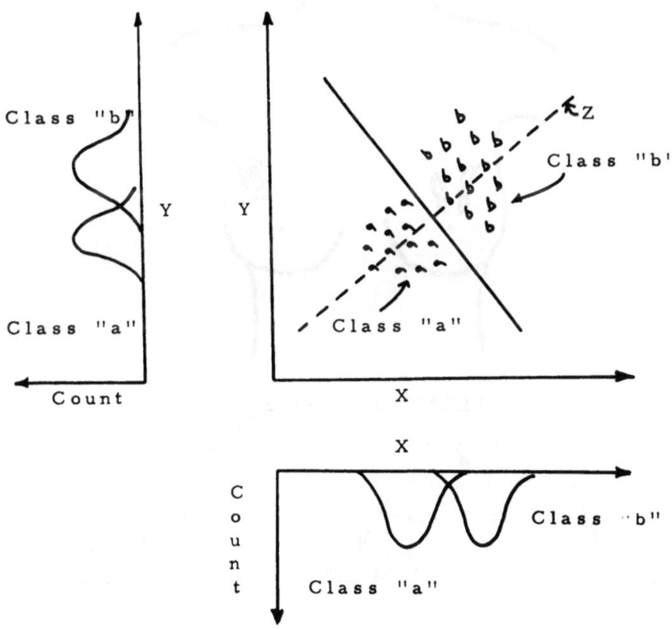

$$Z = \lambda_1 X + \lambda_2 Y$$

Fig. 4. Linear discriminant classifier.

for non-linearly separable data structures. Nearest neighbour techniques suffer from a dimensionality problem however. The linear discriminant parameter (Z) is a one-dimensional parameter produced from a linear combination of other parameters and will tend to be Gaussian distributed because of the central limit theorem from statistics. The density of points in a multi-dimensional space may be very low, and clusters may not be clustered compactly. In this event, spurious outlying points may cause spurious shifts in centroids and therefore precipitate and propagate mis-classification. This is the so-called "curse-of-dimensionality".

Beyond the statistical pattern recognition approaches, one can look to artificial intelligence techniques for assistance in automated interpretation. Techniques such as linguistic structuring, hierarachical trees, and relational graphs may be useful (see Figure 5 for example), but at the present time, little work in thermography is being addressed by these techniques.

CONCLUSIONS

Numerous techniques have been briefly reviewed and illustrated.

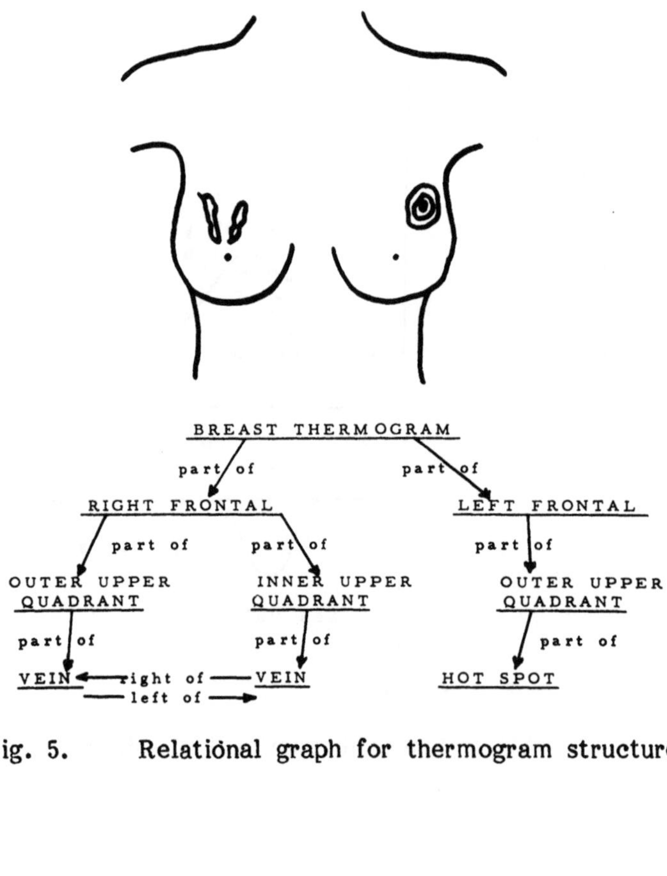

Fig. 5. Relational graph for thermogram structure.

The references should permit the interested investigators to obtain all the detail necessary to initiate their own computer processing studies. The investigator is strongly urged by the author to use as much a priori knowledge as possible in establishing parameters for analysis.

BIBLIOGRAPHY

1. AGA Thermovision Softa Manual, AGA Infrared AB, Stockholm (1980).
2. H. C. Andrews, Digital Image Processing IEEE Computer Society, IEEE Catalog EHO-133-9, New York (1978).
3. M. Anliker, P. Friedli, Evaluation of high resolution thermograms by on-line digital mapping and colour coding, App. Radiol. and Nucl. Med. 5;3:114-118 (1976).
4. I. M. Barash, B. S. Pasternak, L. Venet, W. I. Wolff, Quantitative thermography as a predictor of breast cancer, Cancer 21:769-776 (1976).
5. L. M. Biberman, S. Nudelman, "Photoelectronic Imaging Devices," Volumes 1 and 2, Plenum Press, New York (1971).
6. M. P. Cronin, Thermography in the diabetic clinic, Appl. Radiol. July/August (1975).
7. J. W. Draper, C. H. Jones, Thermal patterns of the female breast, Brit. J. Radiol. 42:401-410 (1969).
8. R. O. Duda, P. E. Hart, Pattern classification and scene analysis, Wiley Interscience, John Wiley and Sons, New York (1978).
9. J. D. Foley, A. Van Dam, "Fundamentals of interactive computer graphics," Addison-Wesley, Reading, Mass. (1982).
10. M. Gautherie, The logic and perspective of thermography, Bibl. Radiol. 6:255-262 (1975).
11. M. Gautherie, C. H. Gros, Contribution of infrared thermography to early diagnosis, pre-therapeutic prognosis and post irradiation follow up of breast carcinomas, Medicamundi 21;3:135-149 (1976).
12. R. C. Gonzalez, P. Wintz, "Digital image processing," Addison-Wesley, Reading, Mass. (1977).
13. P. G. Hoel, "Introduction to mathematical statistics," John Wiley and Sons Inc., New York (1971).
14. C. H. Jones, Thermal patterns of the healthy female breast, Bibl. Radiol. 6:57-64 (1975).
15. M. S. Lapayowker, H. L. Kundel, M. C. Ziskin, Thermographic patterns of the female breast and their relationship to carcinoma, Cancer 27:819-822 (1971).
16. M. S. Lapayowker, I. Barash, R. Byrne, C. H. J. Chang, G. Dodd, C. Farrell, J. D. Haberman, H. J. Isard, B. Threatt, Criteria for obtaining and interpreting breast thermograms, Cancer 38;5:1931-1935 (1976).
17. M. Negin, M. C. Ziskin, Feature specification for an automatic breast thermographic interpreter, Milwaukee symposium on automatic control and computation. (1976).
18. M. Negin, M. C. Ziskin, C. Piner, M. S. Lapayowker, A computerized

breast thermographic interpreter, IEEE transactions on biomedical engineering, B. M. E. 24;4:347-352 (1977).
19. P. Newman, A. L. Evans, M. Davison, A system for the automated diagnosis of abnormality in breast thermograms, Brit. J. Radiol. 50:231-232 (1977).
20. H. Ooi, Y. Miki, Computer analysis of breast cancer thermograms, Jap. J. Med. Electr. 14;2:111-117 (1976).
21. W. K. Pratt, "Digital image processing," John Wiley and Sons, New York (1978).
22. J. Winter, M. A. Stein, Computer image processing techniques for automated breast thermogram interpretation, Comp. and Biomed. Res. 6:522-529 (1973).
23. M. C. Ziskin, M. Negin, C. Piner, M. S. Lapayowker, Computer diagnosis of breast thermograms, Radiology 115;2:341-347 (1975).

INFORMATION THEORY AND THERMOGRAPHIC QUANTIFICATION

D P Page Thomas

Rheumatism Research Unit E6
Addenbrookes Hospital, Hills Road, Cambridge
UNITED KINGDOM

The electronic display of a thermal image from an infrared camera is comprized of a number of time scans which a computer converts into an array of a number of points (pixels) of different thermal values. For quantification purposes each pixel may be considered to possess three positional characteristics:

1. Location along the horizontal axis (x_i)
2. Location vertically (y_i)
3. A thermal location (z_i) in a plane directed towards the camera at right angles to the x,y plane.

From this standpoint, therefore, the thermal image may be considered as being three dimensional, with each pixel in the image possessing three coordinates (x_i, y_i, z_i). Frequently, we also add the fourth dimension, that of time, when we wish to make serial observations of the same patient, eg during follow-up visits, to study the effects of therapy, or to study dynamic changes such as in response to cold stress.

The interfacing of microcomputers with thermographic systems obviously enables more sophisticated analysis of the thermal image to be performed than has previously been possible, and the time has now come for a reappraisal of what is the best method of thermal image analysis for a particular task, ie what form of analysis describes most efficiently the thermal changes in question.

Quantification of the thermal image is thus possible in any one, or in various combinations of the dimensions previously mentioned. We may utilize only thermal information within a fixed area of interest - one dimensional analysis. Alternatively, thermal changes may be related to distance, along either 'box' axes - two dimensional analysis, eg 'line scans'. Finally, all three components may be considered simultaneously when changes of the thermal 'surface' are being examined.

In this communication consideration is only given to one dimensional analysis, partly because it has been most used, and partly because its possibilities have not as yet been fully exploited, or explored. Most commonly in this type of analysis measurements of maximum, minimum, or average temperature are made over an 'area of interest' of known dimensions. Such measurements are technically simple and quick to make. They do, however, have certain disadvantages in that their values can be quite markedly altered by diurnal variation. It is possible, however, to use the same thermal values in a different way, that is, to examine the relative frequency distributions of the pixels within the various 'gray levels'. This method of analysis has been relatively little used to date because it requires the interfacing of a microcomputer to the thermographic equipment, but it is being put to increasing use in both tumor[1] and arthritic studies.[2,3]

In order to find the relative frequency distribution (RFD) of pixels from a selected area of interest it is first necessary to digitize the analog signal from the IR camera for each point scanned. Because computers function in binary code or simple multiples thereof, it is convenient to convert the temperature range, eg a 10°C range into 64 or 2^6 gray levels, so that each gray level corresponds to 10/64 = 0.15625°C.

As each point is scanned the analog-to-digital convertor then assigns a particular gray level, and a count is then made of the number of pixels at each gray level. If the number of pixels at each level is then divided by the total number of pixels processed, the relative proportion of pixels at each gray level is found. An example of this procedure is shown in Figure 1, where two curves are shown. The thermograms were taken from the lateral aspects of both knees of a patient with unilateral inflammatory arthritis. The distribution in the normal knee (dashed line) is shown to be very narrowly distributed, whereas in the arthritic knee (solid line) the curve is much more irregularly distributed over many more gray levels. As the arthritis heals the RFD gradually assumes a progressively normal type of distribution.

By neglecting the contribution of pixels, which comprize less than 15% of the modal frequency of a particular curve, it is possible to fit a normal distribution curve to the data and the width of the major frequency components can be calculated at ± 1 Standard Deviation from the mean. This method has been successfully applied to the assessment of disease activity in arthritis, as is discussed elsewhere in these proceedings,[3] and has been shown to have definite advances over purely thermal measurements. It is obvious, however, from inspection of Figure 1 that whilst a calculated normal curve will approximate extremely closely to the leptokurtic shape of the major frequency components of the normal knee, no way can it represent the irregular frequency plateau of the arthritic knee although it will fit very closely to the width of the curve's

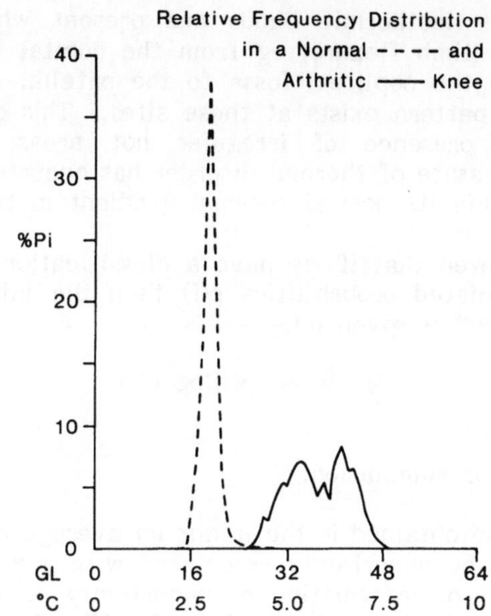

Fig. 1. Relative frequency distribution in a normal and arthritic knee.

'shoulders'. I believe that information theory can be usefully employed to describe both varieties of curve and also the transitional forms in between.

Quastler,[4] in an extremely lucid introductory article about information theory, followed the historical development of the idea in physics that there is a measurable relation between information and the degree of orderliness (entropy) which dated from Boltzmann's classic definition of entropy in 1872,

$$S = k \int p(x)\log p(x)dx$$

through the major contribution made over the years by Fisher,[5] Szilard,[6] Hartley,[7] Wiener,[8] and culminating in the work of Shannon,[9] which went in several respects beyond previous work. The relationship of Boltzmann's concept of orderliness (entropy) and information has been discussed in depth by Brillouin.[10]

What justification is there for using such concepts in relation to thermal images? Firstly, it is well known that the naked normal human body at equilibration with a cool environment radiates its thermal energy in a particular pattern, being hottest at the flexures of the limbs, over the spine, and in various regions of the head and neck. Closer inspection of isothermic distribution in the limbs shows that a succession of progressively cooler isothermic bands are present when one proceeds outwards from the limb flexure, eg from the cubital fossa towards the olecranon, or from the popliteal fossa to the patella. Thus an ordered state of radiation pattern exists at these sites. This ordered pattern is disrupted by the presence of irregular hot areas as is found in inflammation, ie a state of thermal disorder has occurred, since the body is unable to maintain its normal thermal gradient in these areas.

Shannon[9] showed that if we have a classification of events with n categories of associated probabilities p(i) then the information content of the classification[4] is given by:-

$$H = \sum_i p(i)\log_2 p(i)$$

where Σ is sign for summation.

The number so obtained is the minimum average of biniary symbols needed to abolish the uncertainty associated with a given situation, and is thus the amount of information or uncertainty of the situation. In our application read thermograph for classification, gray level for category, and proportion per gray level as probability. Making allowances for scale, and it not being a continuous function, there is a strong resemblance of the information function 'H' to Boltzmann's entropy S. For practical calculation purposes the expression:-

$$H = -K \sum_i p(i)\log p(i)$$

where $K = 1/\log_{10}2 = 3.3219$ or $K = 1/\ln 2 = 1.4427$ is used.

In our application, if all pixels were to be found in 1 gray level only, then there is no uncertainty and H = 0. Alternatively, at the other extreme, if all the pixels were equally distributed in all the $64 = 2^6$ gray levels (ie equiprobable) then there would be complete uncertainty and H = 6.0. In practice, however, both these extreme conditions are highly unlikely to occur, unless, of course, we make the 'width' of a gray level too great in thermal terms.

"H" can best be visualized perhaps as a probability gradient function. In Figure 2 the same data of the relative frequency proportions (probabilities) shown in Figure 1 have now been rearranged in order of increasing frequency, ie a relative frequency gradient has been made from the data. If an imaginary straight line were drawn approximately to represent these relative frequency gradients then the information function 'H' would tend to behave as the cotangent of the angle subtended with the x axis, tending towards zero as the angle increases, and to a maximal values when parallel with the x axis.

In the poster session, thermographs from a patient, who was observed over eighteen months, were used to illustrate changes in the information

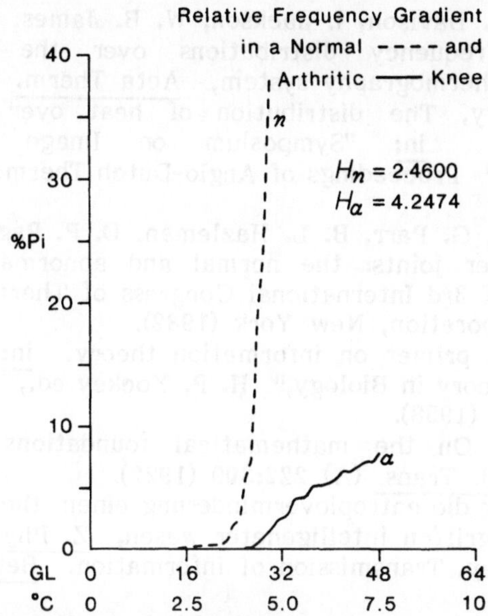

Fig. 2. Relative frequency gradient in a normal (H_n) and arthritic (H_a) knee.

function during exacerbation and remission of the disease. The abnormal thermographs (4 values) had a range of 'H' values of 3.8472 - 4.671, 2 thermographs of borderline appearance had values of 3.2322 and 3.3100, the remaining 4 normal thermographs had a range of 2.2345 - 2.7885.

On one occasion three thermographs were taken from the same patient at two hourly intervals during the day for both the normal and arthritic knees. The 'H' values in either case showed little variation. Left (arthritic) knee 'H' = 4.0773 $\pm$ 0.1701, right (normal) knee 'H' = 2.4088 $\pm$ 0.0512.

In conclusion, I believe that the information function 'H' can be usefully applied to accurately represent the extremely varied distribution of pixel frequencies which occur in health and disease. It provides a sensibly small numerical value of known limits and meaning. It is perhaps not entirely serendipity that has led from its derivation in statistical thermodynamics to an application in the field of biological infrared emissivity.

REFERENCES

1. P. Newman, M. Davison, I. Jackson, W. B. James, The analysis of temperature frequency distributions over the breasts by a computerized thermography system, Acta Therm. 4:3 (1979).
2. R. S. Salisbury, The distribution of heat over joints and its quantification, in: "Symposium on Image Processing in Thermography," Proceedings of Anglo-Dutch Thermographic Society 6 (1981).
3. R. S. Salisbury, G. Parr, B. L. Hazleman, D. P. Page Thomas, Heat distribution over joints: the normal and abnormal pattern, in: "Proceedings of 3rd International Congress of Thermology," Plenum Publishing Corporation, New York (1982).
4. H. Quastler, A primer on information theory, in: "Symposium on Information Theory in Biology," H. P. Yockey ed., Pergamon Press Inc. New York (1958).
5. R. A. Fisher, On the mathematical foundations of theoretical statistics, Phil. Trans. (A) 222:309 (1922).
6. L. Szilard, Über die entropieverminderung einem thermodynamischen system bei eingriffen intelligeneter wesen, Z. Phys. 53:840 (1929).
7. R. V. L. Hartley, Transmission of information, Bell Syst. Tech. J. 7:535 (1928).
8. N. Wiener, "Cybernetics," J. Wiley & Sons, New York (1948).
9. C. E. Shannon, A mathematical theory of communication, Bell Syst. Tech. J. 27:379 and 623 (1948).
10. L. Brillouin, "Science and Information Theory," Academic Press, New York (1956).

MICROCOMPUTER STANDARDIZATION OF THE AGA 680M SYSTEM

G Parr, M Prees, R Salisbury, P Page Thomas, B L Hazleman

Rheumatology Research Unit
Addenbrooke's Hospital
Cambridge UNITED KINGDOM

INTRODUCTION

Although standard techniques for subject preparation and environmental conditions have been established, the increasing use of computers in thermography and the resultant numerical data which is produced emphasizes the importance of adequate calibration and standardization of the equipment.

We describe some minimum test procedures that should be carried out to attempt to establish guidelines for the check of camera and image processing equipment and illustrate how a microcomputer can simplify standardization by creating synthetic thermal test signals.

Our system comprises the AGA 680M camera, color monitor, OSCAR (Offline System for Computer Access and Retrieval), a digital tape recorder, and an Apple II microcomputer with Thermoteknix software. The computer is interfaced with OSCAR to read the OSCAR memory (Fig 1).

METHOD

In this system an incoming signal has to be digitized in order to be stored in the digital buffer memory and on the digital magnetic tape. This is done in the analog-to-digital converter by sampling the continuous video signal at discrete time intervals, quantizing the sample values into a finite number of values and coding that number in binary form. After digitization the incoming thermal image is held in the 128 x 128 x 8-bit memory in the OSCAR unit. Similarly, when a tape recorded digitized image is retrieved, it is first written into the OSCAR memory and then

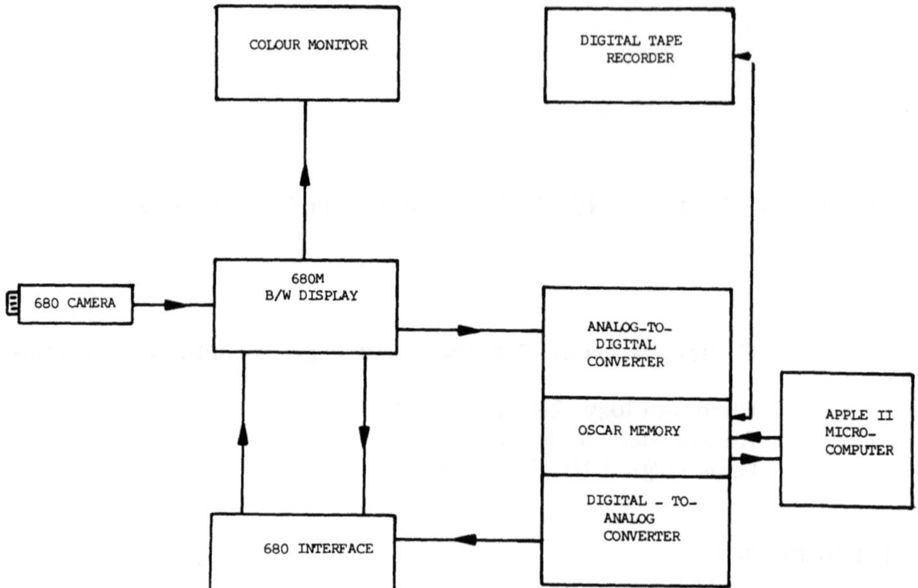

Fig. 1. Simplified block diagram.

undergoes digital-to-analog conversion for visualization on the monitors. The exact balance of the analog-to-digital and digital-to-analog converters is essential for the system to be properly standardized. Once these have been balanced and adjustment made as necessary, it is then possible to generate theoretically known 'synthetic' thermal values in the microcomputer which can be transmitted to OSCAR via its computer interface to undergo digital-to-analog conversion for visualization.

The 'synthetic' thermal values of 256 gray levels are written into OSCAR memory as a series of binary coded values by the Apple II microcomputer and when visualized appear on the color monitor as a scale of eleven colored blocks of equal size and as a scale of increasing intensity on the black and white monitor.

Using such a method it has been possible to produce two thermal scales (Fig 2) one of which is an inversion of the other and these are written to appear directly above the temperature scale which is produced by the 680M unit.

In addition to these horizontal temperature scales the Apple II microcomputer also writes a lantern pattern of known thermal values into OSCAR memory and this appears after visualization on the color and 680M monitor above the horizontal 'synthetic' temperature scales as shown (Fig 2).

The lantern shape and the thermal scales make up a 'test-card' which can be repeatedly reproduced in a few seconds to appear on the monitors. When the OSCAR unit is in the 'HOLD' mode, and after generation of the 'test-card', it becomes possible using the 680M to overlay a single isotherm marker which appears as a horizontal line over the lantern shape and as a vertical line over all three horizontal temperature scales at the same time.

When the isotherm one 'level' control knob is turned in a clockwise direction, the vertical lines over the lower and middle temperature scales, which are aligned, are seen to move to the right, ie in a direction of increasing thermal values. At the same time the vertical isotherm marker line is seen to move in the opposite direction on the upper scale which is an inverted scale. The isotherm one marker is also seen to move down over the lantern shape when the 'level' control knob on the 680M is turned in a clockwise direction.

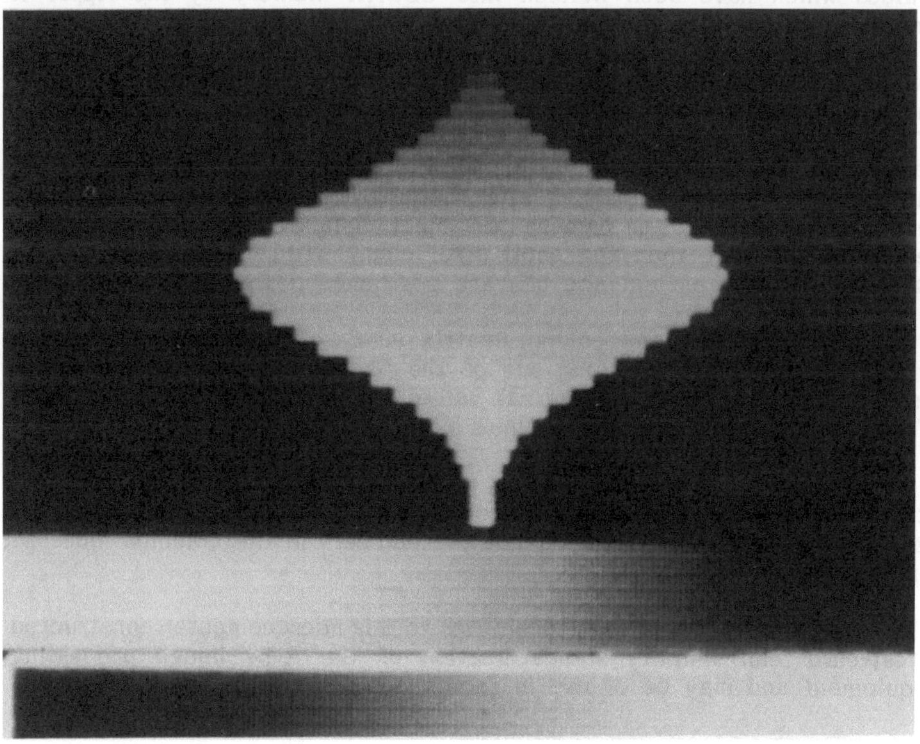

Fig. 2. Thermal 'Test-Card'.

It is apparent that the markers move towards the direction of increasing thermal values over all scales on the 'test-card' when the isotherm one 'level' control knob on the 680M is turned in a clockwise direction and towards a direction of decreasing thermal values on all scales when the same knob is turned counter-clockwise.

The middle temperature level may be set simply by carefully adjusting the isotherm one 'level' control knob of the 680M to a position where all the superimposed vertical isotherm marker lines are in exact alignment. There is only one scale position where all three vertical isotherm marker lines can be lined up in the vertical plane and this point is at the middle temperature level. Once this position has been found then all scales, monochrome and color, are a perfect linear representation of the theoretical thermal values (0 to 256), which have been generated by the Apple II microcomputer. It is then possible, knowing that the visual scales are correctly displayed, to calibrate the black body radiator control box across a selected temperature range by matching each isotherm color to its appropriate temperature setting. A readout of thermal values for the black body radiator may be examined to confirm that the calibration is accurate.

It is possible using this system to examine the theoretical thermal values which have been written into OSCAR memory by the Apple II microcomputer, as a matrix type array, but before this can be done a region of interest needs to be defined. The program allows for boxes which are adjustable in terms of position and size to be displayed over a thermograph on both color and monochrome monitors simultaneously. These boxes may be sorted under patient file numbers on a disk for convenient future retrieval and this facility improves reproducibility when an operator chooses to analyze serial thermographs of the same anatomical area. Provided that the camera-to-subject distance remains constant at each patient visit then the total pixel count within the box should be unchanged if the patient and box are positioned correctly.

When the row and column matrix array of thermal values for a region of interest sited along one of the 'test-card' temperature scales is examined, the rows of thermal values are seen to increase in value from left to right, whereas columns down are the same.

This is not true for a region of interest placed over the length of the lantern shape when the rows of thermal values are seen to be of equal value from left to right, but the numbers in the columns increase in value when reading down.

The application of such techniques to this microcomputer constructed 'test-card' can simplify standardization of the AGA image processing equipment and may be of use in fault finding as well.

MICROPROCESSOR-ASSISTED BREAST THERMOGRAPHY

P Bosiger, F Scaroni, M Geser

Institute of Biomedical Engineering of the University and ETH Zurich, CH-8044 Zurich SWITZERLAND

INTRODUCTION

Microprocessor-assisted breast thermography allows for a quantitative and therefore a more objective analysis of thermograms.[1-4] A low-cost system has been developed which permits:

1. The recording of temperature distributions and their temporal variations in digital form;
2. Gray-scaled or color-coded visualization of the temperature distribution on a TV-monitor and
3. On-line processing and microprocessor-assisted evaluation with respect to abnormal features.

A new method for the analysis of static breast thermograms has been worked out which reduces the copious thermographic information to a set of parameters characterizing abnormalities of the temperature patterns and their gradient distributions. The diagnostic relevance of these parameters has been tested on a group of 94 patients of the University Hospital of Zurich.

TECHNICAL DESCRIPTION AND BREAST THERMOGRAM ANALYSIS

Thermographic signals from a modified thermograph Spectrotherm 2000 are digitized on-line with a resolution of 256 x 256 points and an accuracy of 8 bits and are stored in a RAM memory (Fig 1). The temperature resolution is about 0.1°C; the spatial resolution may be much better than 1 mm^2 depending on scanning distance. The scanning time for a digital thermogram is 1 sec. A Winchester type hard disk unit (Micropolis) stores a thermogram within less than 0.5 seconds. Thus

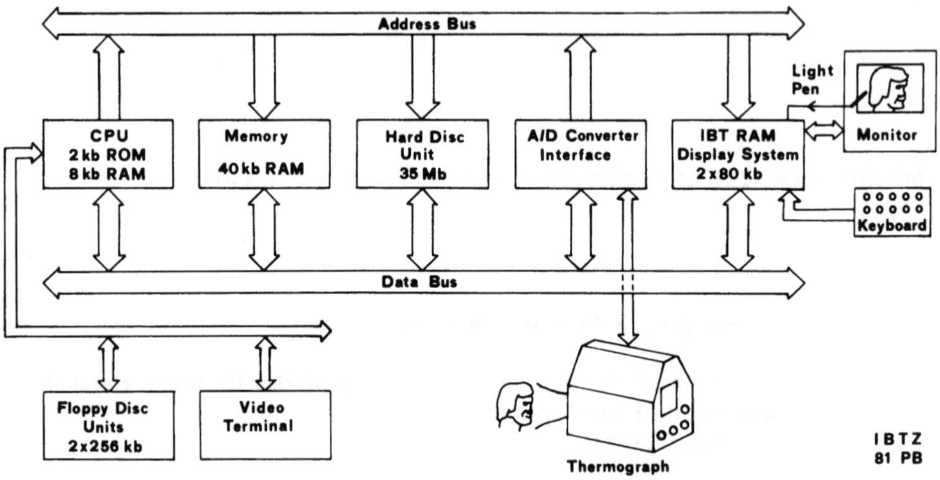

Fig. 1. Block diagram of the microprocessor-assisted thermography system.

dynamic sequences of thermograms can be recorded and stored in intervals of about 1.5 seconds. The analysis of the temperature distributions is performed by a Texas Instruments FS990/4 microprocessor system.

The evaluation of standard breast thermograms is initiated by defining regions of interest and certain reference areas on the monitor image with a light pen. For the analysis of the frontal thermograms three different methods were investigated for the determination of these regions.[5]

1. The region is marked by two horizontal straight lines selected by the observer. The lateral boundaries of the breasts are determined by the computer.
2. The upper limit is again marked by a straight line while the lower limit is outlined with the light pen following the contours of the breasts.
3. The two breasts are outlined individually with the light pen (Fig 2).

Although the first method is the least subjective of the three, the third method yielded better results because it allows for the elimination of the hot areas just below the breasts and of hot or cold spots in the axillae. Consequently, the frontal and the side thermograms are both analyzed by the third method.

The following quantitative parameters based on symmetry considerations appear to be useful in identifying patients with a high risk of having or developing breast cancer[5,6]:

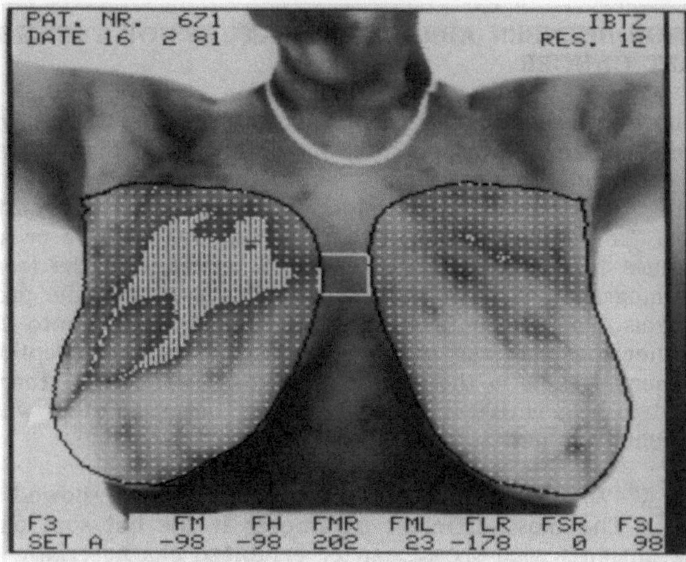

Fig. 2. Microprocessor-evaluated frontal breast thermogram. The analyzed regions are marked with a white point raster; the hot spots are indicated by the white overlay. Patient data and results of analysis appear at the top and bottom of the thermogram.

- Difference between mean temperatures of the two breasts.
- Hot spot parameters: a hot spot reference temperature is defined as the higher value of the two eightieth percentiles of the two breasts. The hot spot parameter is the relative difference between the areas of the two breasts in which the temperature is higher than this reference temperature.
- Difference between the highest temperatures on the two breasts.
- Difference of mean temperatures of the upper 20% of the temperature histograms of the two breasts.
- Difference between the average temperatures of the two areolar regions.
- Difference between the maximum temperature gradients of the two breasts.

Further indicative criteria are:

- Difference between the maximum temperature of each breast and the hot spot reference temperature.
- Difference between the mean temperatures of the upper twentieth percentiles of each breast and the mean temperature of both breasts combined.
- Difference of the mean temperature of each breast and a reference temperature on the sternum.
- Maximum temperature gradients[7] and the variance of the temperature gradient distributions.

CORRELATION BETWEEN ABNORMAL FEATURES OF THERMOGRAMS AND BREAST CANCER

A group of 94 thermograms from patients with a variety of breast diseases was analyzed retrospectively by determining these parameters. The group consisted of 67 patients without proven breast cancer, 26 patients with breast cancer on one side, and 1 patient with breast cancer on both sides. To decide whether a thermogram was positive or negative, threshold values for each parameter were determined in the frontal and the side thermograms. This was done in such a way that the probability of thermograms from patients with breast cancer falling into the class of positive thermograms was as high as possible at an acceptable rate of positive thermograms in the absence of proven cancer. A thermogram was classified to be positive if at least one of the parameters was higher than its threshold values.

Twenty (74%) of the patients with breast cancer showed positive thermograms. The most efficient parameter is the hot spot parameter (Table 1); 12 patients with breast cancer exhibited characteristic hot spot areas with values of the hot spot parameters exceeding the threshold value of 0.75. Only one patient without proven breast cancer was positive by reason of this parameter. The second best parameter is the difference of mean temperatures of the upper 20% of the temperature histograms, ie the temperature difference between the average temperature of the hot spot area on the warmer side and the average temperature of the area of equal size with the highest temperature on the contralateral breast. The threshold value of this parameter is in this study 0.8°C. Again the number of positive thermograms was 12 (not all having a positive hot spot parameter), and the number of positive thermograms from patients without proven breast cancer was 2. Additional parameters of high relevance are the difference between the mean temperatures of two breasts (threshold value 1.0°C) and the difference between the highest temperatures on the two breasts (threshold value 1.2°C). The remaining parameters are less relevant, but all of them, with the exception of the temperature difference between the two areolar regions, were helpful in identifying the thermograms of patients with cancer. All the parameters are calculated for frontal as well as side views.

As can be seen from Table 1, no parameter can identify more than 15 of the 27 patients with breast cancer. But if all of the criteria are used as discussed above, 20 of the 27 are identified as true positive, whereas the 14 additional positive thermograms were recorded on patients with no proven breast cancer. Table 2 gives the correlation between the tumor stage and the result of the corresponding microprocessor analysis. The two breast cancers TO of the patients with negative thermograms were both intraductal carcinomas.

Table 1. Frequencies in which the parameters led to positive and negative classification of the thermograms of patients with and without breast cancer

Patients		With diagnosed breast cancer		Without diagosed breast cancer	
Thermogram		pos	neg	pos	neg
$T_{mean, l}$	- $T_{mean, r}$	10	17	2	65
HSA_l	- HSA_r	12	15	1	66
$T_{max,l,r}$	- T_{HS}	15	12	9	58
$T_{max,l,r}$	- T_{mean}	13	14	7	60
$T_{max,l}$	- $T_{max,r}$	11	16	3	64
$T_{max,l}$	- $T_{max,r}$	12	15	2	65
T_{stern}	- $T_{mean,l,r}$	10	17	4	63
$T_{areol,l}$	- $T_{areol,r}$	4	23	1	66
Total		20	7	14	53

CONCLUSIONS

Although the results discussed above are only preliminary because of the limited number of patients, the microprocessor-assisted analysis of breast thermograms seems to be an aid in the diagnosis of mammary carcinoma. Additional information can be derived from a quantitative analysis of the vascularization patterns of the breasts and by analysing dynamic thermogram sequences recorded during the recovery period after a forced air cooling of the breasts.

This work was supported in part by the Swiss National Science Foundation (grant 3.121-0.77) and was carried out in collaboration with Professor W Schreiner and Dr R Fleiner, Department of Obstetrics and Gynecology of the University Hospital of Zurich.

Table 2. Comparison of tumor stage with result of the corresponding micro-processor analysis

Tumor Stage	Thermograms Pos	Neg	Number of Breast Cancers
T0	-	2	2
T1	5	1	6
T2	12	3	15
T3	1	-	1
T4	1	1	2
Infl	1	-	1
Total	20 74%	7 26%	27 100%

REFERENCES

1. M. Anliker, P. Friedli, Evaluation of high-resolution thermograms by on-line digital mapping and color coding, Appl. Radiol./Nucl. Med. 5;3:114 (1976).
2. M. Negin, M. C. Ziskin, Ch. Piner, M. S. Lapayowker, A computerized breast thermographic interpreter, IEEE Transactions on Biomedical Engineering BME-24:4 (1977).
3. M. Abdel Meged, B. Keith, P. L. Wendel, M. Gautherie, Ch. Gros, Traitement numerique de thermogrammes infrarouges, Proc. Biosigma (Paris) 78:24-28 (1978).
4. G. Leonardi, G. Viganotti, U. Cerchiari, Statistical comparison between human and computerized tele-thermomammography diagnosis, Acta Thermographica 137-141 (1979).
5. P. Bosiger, F. Scaroni, A microprocessor-assisted thermography system for the on-line analysis of thermograms and dynamical thermogram sequences, in: "Biomedical Thermology," Alan R. Liss Inc., New York (1982).
6. M. Gautherie, Ch. Gros, Thermographie und brustkrebs: diagnose, prognose, ueberwachung, Gynakolog. Rundschau. 19:183-227 (1979).
7. R. Haas, H. Wollnik, F. Kassen, R. Kraus, Quantitative tumor identification by differential thermography, Proc. Int. Conf. on Senology, Barcelona (1981).

DIGITAL THERMOGRAPHY SYSTEM

A Tanaka, T Inoue, T Yoshimura, H Horikawa

Fujitsu Ltd.
Kawasaki
JAPAN

Medical-purpose thermography equipment designed for high temperature resolution generally uses a combination of a single infrared detector and mechanical optical scanner. This combination, however, suffers from relatively low scan speeds and tends to cause thermal images on the CRT monitor to have annoying flicker or unsatisfactory brightness or contrast. Consequently, it is very difficult to recognize slight temperature differences or complex thermal distribution patterns, such as breast cancer patterns, from the images obtained on such equipment.

In an effort to improve these drawbacks of conventional thermography systems, attempts have been made to introduce the analog scan converter into a thermography system so as to achieve higher image brightness and contrast. However, this attempt does not seem to be very successful because:

1. The contrast decays with time
2. The picture level drifts considerably.

Against this background, for four years we have been studying a medical thermography system using digital scan conversion techniques. Commercial systems have been available since 1979. Currently, we are studying the multi-infrared detector approach by providing the system with an increased sensitivity and reduced frame time. This report outlines the thermography system using the digital scan conversion technique and discusses the higher sensitivity design for the system utilizing the multi-infrared detectors.

DIGITAL THERMOGRAPHY SYSTEM

Infrared Scanner

The configuration of the optical system is shown in Figure 1. The infrared radiation from the object being viewed (indicated by arrow 1) is scanned by a scanning mirror. A chopper is located at the first focusing point to transfer the infrared radiation from an internal reference source to the detector. The infrared detector is HgCdTe, cooled by liquid nitrogen and has a sensitivity in the wavelength range 8-14 micrometers.

Image Processor and Monitor

The block diagram of the image processor and monitor is shown in Figure 2. A video signal with full dynamic range is digitized into a 10-bit code. Information is extracted from the 10-bit code by specifying the necessary temperature range and level. The extracted 6-bit information is then stored in the internal memory.

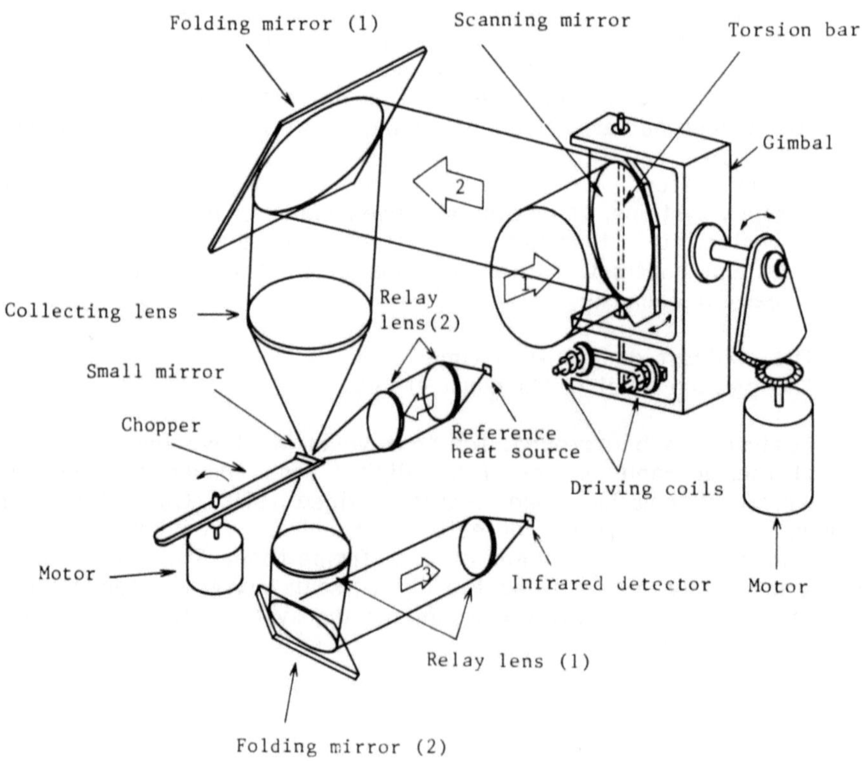

Fig. 1. Optics configuration.

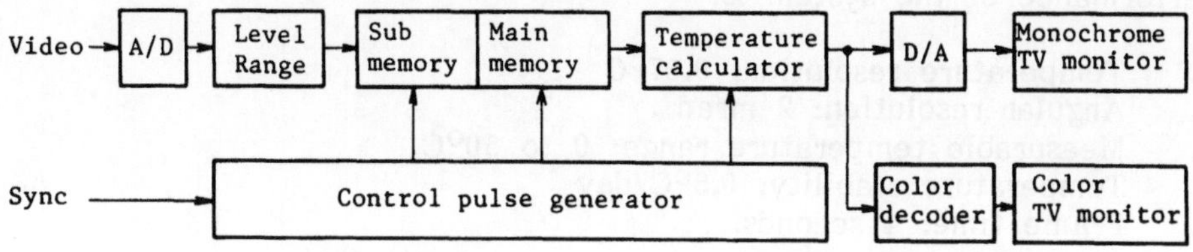

Fig. 2. Block diagram of image processor and monitors.

The video signal coming from the infrared scanner consists of 220 horizontal picture elements and 220 vertical picture elements. In contrast, the image memory has 440 bit capacity for horizontal picture elements to ensure sufficient horizontal resolution. As a result, the total memory capacity is 440 x 220 x 6 = 580,800 bits.

The video signal for individual scanning lines furnished from the infrared scanner is converted in real-time into digital code and temporarily stored in the submemory. It is then transferred to main storage during the blanking periods of infrared scanning and the blanking period of main storage read scanning. This write/read sequence continues until all 220 lines of information are stored in the memory.

Subsequently, the contents of the memory are updated by the information of the next frame line by line until all 220 lines of old information are updated with new information. Memory read and write operations are thus performed continuously in the TV scan mode and infrared scan mode, respectively.

Figure 3 shows a thermographic image obtained from the present system. The number shown just below "ST" indicates the temperature at the intersection of the horizontal and vertical cursor lines. The major

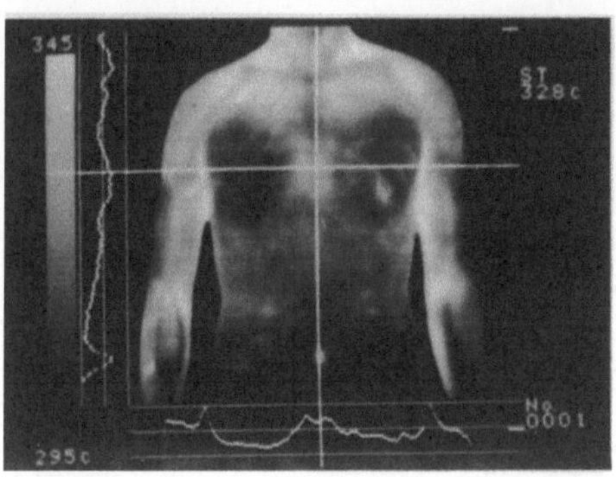

Fig. 3. Thermal image with 0.07°C thermal resolution.

performance of the system is:

1. Temperature resolution: 0.07°C
2. Angular resolution: 2 mrad
3. Measurable temperature range: 0 to 50°C
4. Temperature stability: 0.5°C/day
5. Frame time: 4 seconds.

MULTI-DETECTOR SYSTEM

The temperature resolution of a thermography system may be enhanced by increasing the physical size of its optical system, reducing the angular resolution, and various other approaches. The authors adopted the summation approach using multiple infrared detectors to achieve this target.

In image processing, if signal S for mutually correlated subject patterns is summed n times, the signal level will be increased to nS, whereas random noise level N is increased only to $\sqrt{nN^2}$ (ie $\sqrt{n}N$). Consequently, the signal-to-noise ratio is improved by $\sqrt{n}$ times after the summation.

Based on this fact, the authors contrived a system in which four infrared detectors are arrayed in parallel and are optically scanned simultaneously, the outputs of the four detectors are amplified then subjected to A-D conversion, and the digitized picture elements of each channel are summed and averaged while their relative phases are conditioned. Figure 4 gives the block diagram of the system.

Figure 5 shows a thermal image obtained with the multi-detector system. The four detectors used in parallel configuration give an improvement factor of 2, and hence temperature resolution is improved from 0.07°C with a single sensor to 0.035°C with the multi-detector. The obtained image exhibits a remarkable improvement in visible definition.

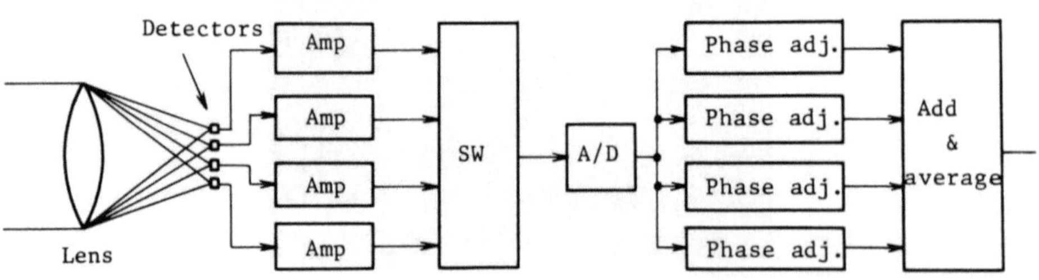

Fig. 4. Block diagram of multi-detector system.

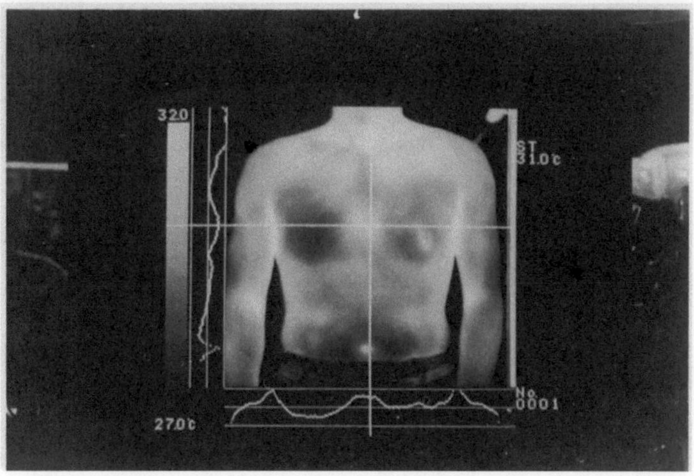

Fig. 5. Thermal image with 0.035°C thermal resolution.

CONCLUSION

The digital scan conversion technique can provide thermal images free of contrast decay or picture level drift on the CRT monitor. In addition, once digitized image information is stored in the memory, the technique enables various post processing of the information, allowing the display of a static image, temperature contours, thermal profiles, and many other display functions. The digitized form of information also facilitates interfacing the system with a computer. This seems to be very helpful in the development of new approaches to thermographic information processing, such as that for skin blood flow studies.

In the meantime, the multi-detector technique has been found to be very effective for improving thermal resolution. It provides an unconventionally high temperature resolution of 0.035°C, and is expected to offer decided advantages in various diagnostic applications such as breast cancer diagnosis, etc.

The other aim of the multi-sensory technique is the improvement of frame time. In this regard, however, examination is now under way to correct the characteristic dispersion among infrared detectors.

DEVELOPMENT OF A THERMOGRAPHIC SKIN BLOOD FLOWMETRY SYSTEM

H Miyake, I Fujimasa, M Iwatani, K Atsumi

Technological University of Nagaoka, Physical Education and Health Care Centre, Nagaoka, and University of Tokyo, Faculty of Medicine, Tokyo, JAPAN

INTRODUCTION

Skin blood flow rates provide important information for the diagnosis and prognosis of peripheral vascular disorders. However, non-invasive blood flowmetry of the skin has not been developed fully. It is evident that medical thermograms include skin blood flow information,[1] and the aim of this report is to develop a method to convert information from the thermogram to local blood flow rates. In order to develop a thermographic skin blood flowmetry system, the following problems must be solved:

1. Construction of a mathematical model to analyze heat conduction of the skin.
2. Evaluation of the model by animal experiments and clinical cases.
3. Installation of the model within a micro-computer system.

PRINCIPLE AND METHODS

According to thermal physiology, the body surface temperature is relatively stable when at rest in an equable thermal environment. In these conditions, the sum of heat flux and heat production of a certain peripheral skin tissue area is equal to zero. In Figure 1, the blood flow heat convection (Qb), the deep tissue conductive heat flux (Qc), and the skin tissue metabolic heat production (Qm) are equivalent to the radiation heat flux (Qr), the convective heat flux (Qf), and the evaporative heat flux (Qe). Thus equation (1) should be established in the stable condition.

$$Qb + Qc + Qm = Qr + Qf + Qe \qquad (1)$$

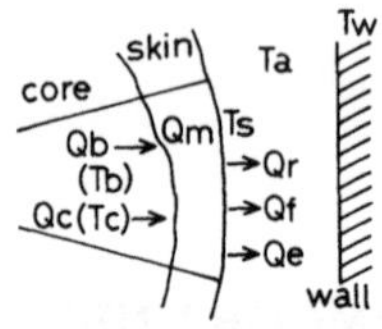

Q: Heat flux and production T: Temperature

(suffix) b:blood flow (suffix) b:blood
c:conductive c:core
m:metabolic s:skin
r:radiative a:ambient
f:convective w:wall
e:evaporative

Fig. 1. Thermal model of the skin and its neighborhood.

and $$Qb = Qr + Qf + Qe - Qc - Qm \quad (2)$$

The skin blood flow (Vs) may be obtained from appendix formula (9), which was constructed for the Figure 1 model, where Ts, Tc, Ta(=Tw), and D are variables and the other parameters are constants.

Ts may be measured by an infrared thermography system. Tc (core temperature or deep body temperature which can be measured locally) is to be measured by a deep body thermometer (Terumo, Core-Temp^{R}). Ta (ambient temperature) and D (radius of the object) are known. A micro-computer system (NEC, PC-8000 series) is employed not only for calculation of Vs but also for image processing.

RESULTS AND DISCUSSION

A photograph of the developed system is shown in Figure 2. The system consists of an infrared thermography system (Fujitsu, INFRA-EYE 150) for Ts measurement, three deep body thermometers for Tc, and the micro-computer system for Vs calculation and image processing. The system is inexpensive and compact in size.

Typical final results for the skin blood flowmetry are shown in Figures 3 and 4 as bar-graphs. Blood flow "profile" patterns are shown in Figure 5. Line printer out-puts by a mini-computer system are also available.

The model evaluation was made by means of animal experiments and clinical cases and was found to correspond well with other methods of measurement.[2]

Fig. 2. Photograph of the developed system - a micro-computer is featured on the top of the color pattern display.

To clarify the contribution of the various physical and physiological parameters to the skin blood flow (Vs), mathematical sensitivity analysis was performed by using the micro-computer system. In Figures 6(a), 6(b) and 6(c), relationships between Vs and Ts with Tc, Ta, and D variations are shown respectively. These results suggest that Ts was one of the most important parameters, whereas D was less important. This analysis was done in a physiological neutral environment with Tb (blood temperature) $37^{o}C > Tc > Ts > Ta$. Seemingly in Figures 6(a), 6(b), 6(c) the skin blood flow becomes zero at a certain point of Ts, where it approaches the basal condition. If Ts falls under this point, the model structure must be altered. Consequently, if the examination for the skin blood flowmetry is done in the well-controlled environmental condition of a physiological neutral environmental temperature, the skin temperature indicated by the thermogram will present information on the skin blood flow itself. The thermogram of peripheral extremities can be converted into pseudo-color patterns of skin blood flow by on-line processing.

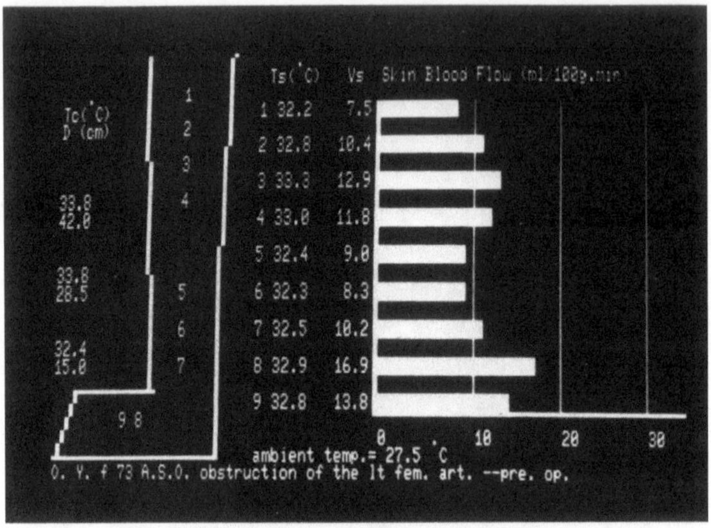

Fig. 3. Bar graph display example (pre-op).

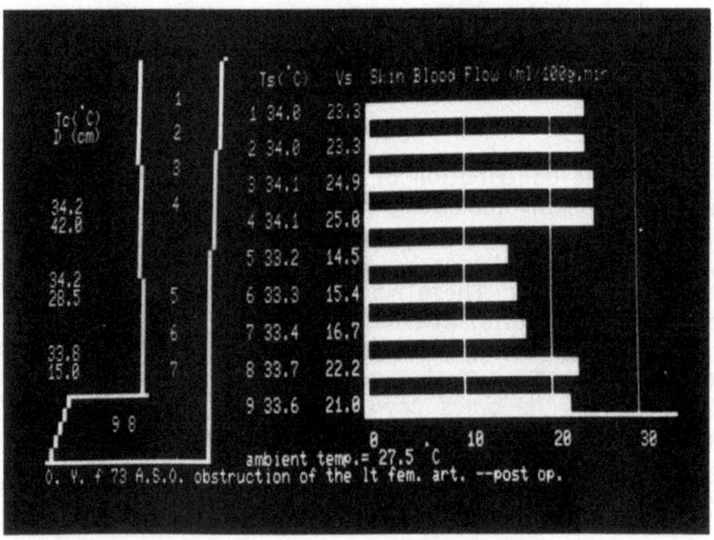

Fig. 4. Bar graph display example (post-op).

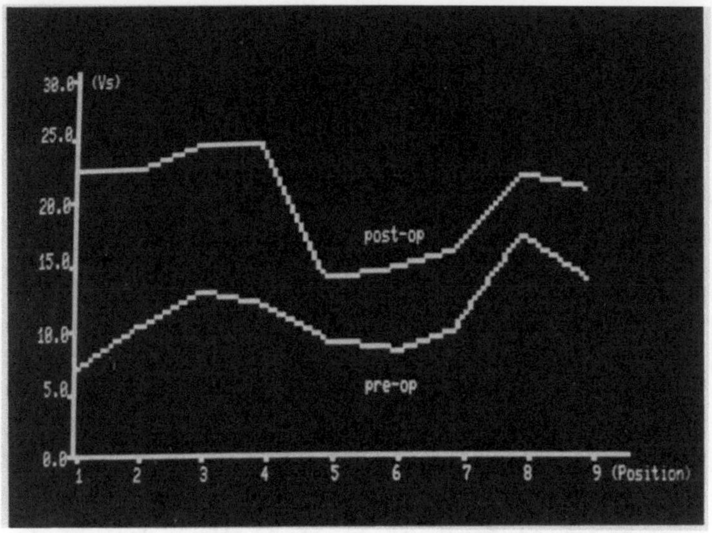

Fig. 5. "Profile" pattern display example.

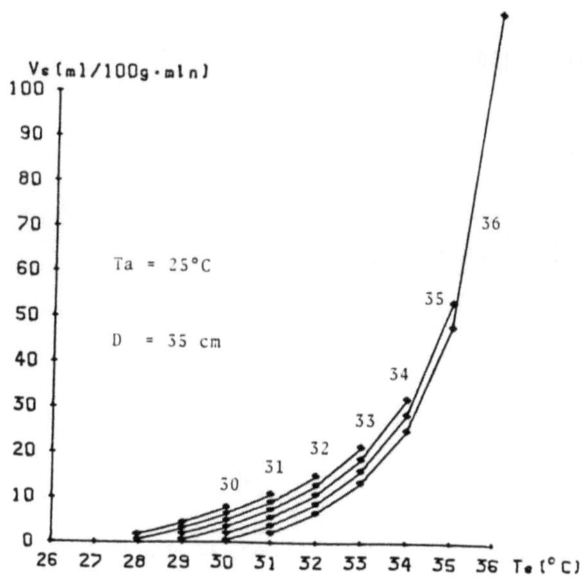

Fig. 6(a). Skin blood flow (Vs) versus skin temperature (Ts) with core temperature (Tc) variation.

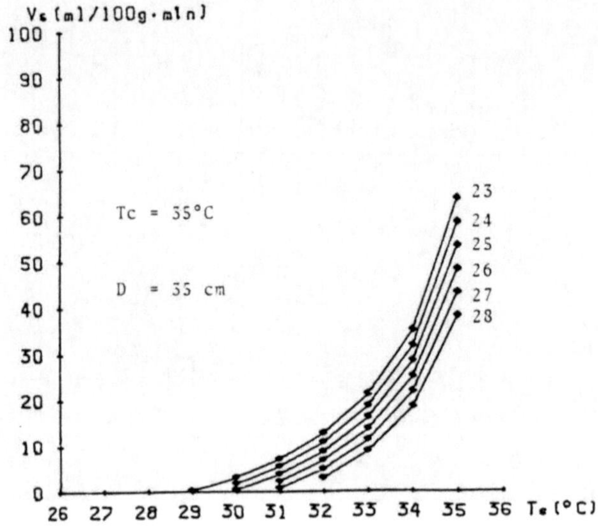

Fig. 6(b). Ambient temperature (Ta) variation.

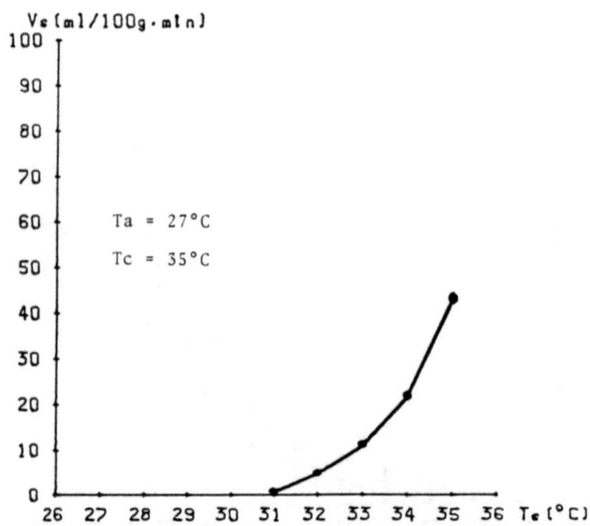

Fig. 6(c). Radius (D) variation of radius D.

APPENDIX

Radiative Heat

1. $Qr = Kr\,\varepsilon\,(Ts^4 - Tw^4)$: $Kr = 4.88 \times 10^{-8} kcal/m^2h^oK^4$, $\varepsilon = 0.99$

$= 0.805 \times 10^{-10}(Ts^4 - Tw^4)$ (cal/cm²min)

Convective Heat

2. $Qf = 1.01 \left(\frac{Ts-Ta}{D}\right)^{0.25} (Ts-Ta)$ (kcal/m²h)

$= 0.53 \times 10^{-2} \left(\frac{Ts-Ta}{D}\right)^{0.25} (Ts-Ta)$ (cal/cm²min)

Evaporative heat

3. $Qe = 11.1$ (kcal/m²h)

$= 1.85 \times 10^{-2}$ (cal/cm²min)

Conductive Heat

4. $Qc = \frac{Kc}{d} (Tc-Ts) \times \frac{1}{3}$ $\quad Kc = 0.168$ Kcal/mh°C

$= 0.93 \times 10^{-2} \left(\frac{Tc-Ts}{d}\right)$ (cal/cm²min)

Metabolic Heat

5. $Qm = 1.44 \times 10^{-2} \times 2^{\frac{Ts-Tm}{10}}$ (cal/cm³ min)

Heat from Blood Flow

6. $Qb = \alpha\rho c Vs\,(Tb-Ts)$ (cal/cm³ min)

7. $Cs.\frac{dTs}{dt} = A.Qc + Qm + Qb - A.(Qr + Qb + Qm)$

8. $\frac{dTs}{dt} = 0$

Blood Flow Rate

9. $Vs = \frac{A.(Qr+Qf+Qe-Qc) - Qm}{\alpha\rho c(Tb - Ts)}$

where,

Kr, Kf, Kc, = constant	Ev = heat of evaporation
Tb = blood temperature	α = counter current rate
Tc = core temperature	pc = density x specific heat of blood
Ts = skin temperature	D = diameter
Tw = wall temperature	d = distance
Ta = ambient temperature	Mo = basal metabolic rate
t = time	Tm = basal metabolic temperature
A = surface area per unit volume	Cs = thermal capacitance of the skin

REFERENCES

1. I. Fujimasa, A. Kouno, K. Atsumi, A method for local blood flow estimation from thermograms under thermal equilibrium, J. J. M. E. 16:(suppl. 351) (1978).
2. M. Iwatani, Y. Mishima, A. Kouno, I. Fujimasa, K. Atsumi, An estimation of skin blood flow and its clinical use, J. J. M. E. 18:(suppl. 670) (1980).

A NEW DEVICE FOR THERMOGRAPHY

A Nagasawa

Metropolitan Hiroo General Hospital
Tokyo
JAPAN

INTRODUCTION

One of the basic requirements in medical thermography is to compare the temperature distribution over body surfaces. In conventional infrared thermography it is sometimes inconvenient to take and observe thermal patterns from different sides of a subject (eg examination times are longer). Developmental thermography, which is the technique for recording thermograms of various parts of a body surface simultaneously has been studied by the author since 1968. The use of this method began with a panoramic thermoprofile[1] with an infrared thermometer, and the basic principles for taking two different types of thermograms were reported at the 2nd European Thermographic Congress.[2] The newly developed device is designed so that a conventional thermocamera can be easily used for developmental thermography. The device has two separate functions: three-sided thermography and panoramic thermography. Developmental thermography has several advantages for medical infrared thermography.

METHOD AND DEVICE

The device is illustrated in Figure 1 and consists of a thermocamera unit (thermocamera with its support pole and H-V conversion mirror) and subject operation unit (a pair of infrared reflector mirrors with its support pole, rotary base, detachable test chair on the rotary base and rotary hand support). The thermocamera can be moved on the base rail connected to the camera unit with the subject unit in order to adjust the correct distance from the thermocamera to target. The subject is viewed sitting on the chair in the case of head or trunk examinations and in a standing position on the rotary base in the case of examination of legs or arms.

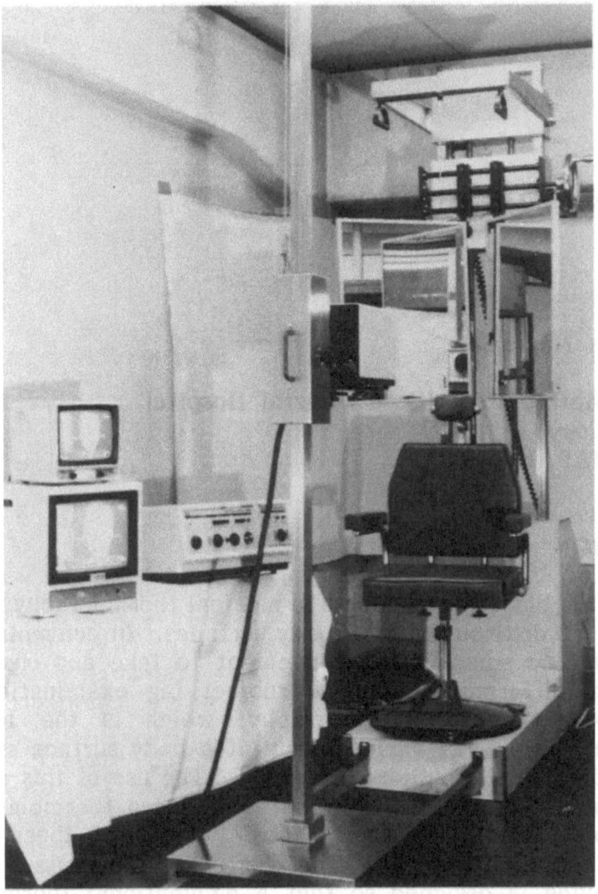

Fig. 1. A new device for thermography.

The subject is supported in the standing test position.

The device is approximately 2 m in height and 3 m in length: it can be operated from the panel on the control box.

The two methods of developmental thermography are used in the following ways:

Three-sided Thermography

This is a technique for taking three-sided thermograms of front, right and left sides of a body surface simultaneously. The two infrared reflective plane mirrors are set symmetrically to a subject's median plane at a selected opening angle to it. The thermocamera scans the front of the subject directly and also both the lateral images on the reflectors, so that three-sided thermograms can be obtained in the same CRT frame simultaneously.

The chair has a head rest and a back rest that can be adjusted to keep the subject's head in the right position for examination, and the seat can be adjusted to the target at the proper level for the reflectors. The two plane mirrors for reflecting infrared radiation from both sides of the head allow the radiation to enter the thermocamera. The mirrors are set symmetrically on a frame on both sides of the patient at an angle to the median plane. The angle of the mirrors is variable and the distance between them can be adjusted to suit the width of a subject with the setting angle kept constant.

Panoramic Thermography

This is a technique for taking a continuous thermogram of a body surface panoramically. The subject is made to rotate by 360° on his center axis with a constant velocity for a full revolution and equal to the frame time of the thermocamera. The thermocamera operating in a line-scanning mode keeps scanning the body surface only on the same vertical line along the rotary axis during rotation, so as to produce a panoramic thermogram of a body surface. In practice, the horizontal scanning of a conventional thermocamera must be connected to scan vertically by a scanning direction convertor, H-V convertor, consisting of a couple of infrared reflective plane mirrors. The motor drives the rotary base to turn the patient whose position can be either standing or sitting. The starting of the thermocamera is synchronized to the starting movement of the rotary base by an interlocking system. The rotation velocity is variable to turn up to 360° in one frame time of the thermocamera. The frame time is approximately 20 seconds for this technique.

RESULTS

Figure 2 shows developmental thermograms of a case of right breast cancer. The temperature distributions on every part of the body surface can be detected in one frame simultaneously and then it is easy to compare them with each other and so reduce the possibility of error.

DISCUSSION

1. In three-sided thermography, coverage of the two oblique thermograms is adjustable by varying the angle of the reflectors to the median plane; also the back of a body can be observed.
2. In panoramic thermography, the surface area covered can be varied up to 360° by changing the rotation velocity of a body.

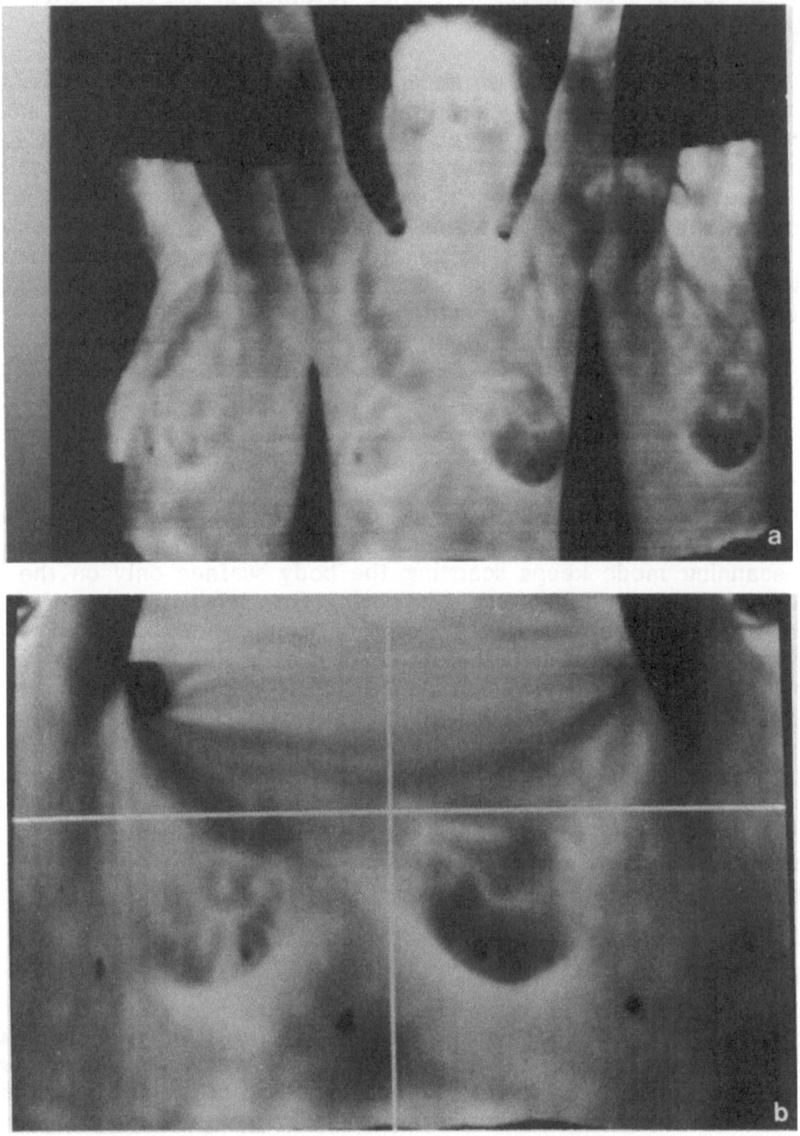

Fig. 2. (a) - Three-sided thermogram of a case of right breast cancer.
(b) - Panoramic thermogram of the same case as (a).

3. The greatest problem in developmental thermography is unavoidable infrared loss caused by the reflector mirrors, but it can be made to be less than 0.2°C by using aluminum mirrors with high infrared reflection coefficients.

CONCLUSIONS

This newly developed device allows two types of developmental thermogram to be taken easily and simultaneously in one frame with a conventional thermocamera. Clinical application has proved its diagnostic usefulness. The advantages of developmental thermography are as follows:

1. Full or wide coverage of target area.
2. Uniform condition for calibration.
3. Reduction of the possibility of error of temperature measurement.
4. It can be used for thermographic screening with a consequential saving in costs for film.

REFERENCES

1. A. Nagasawa, Clinical application of thermography and its values, J. of Jap. Oral Med. 19:182 (1968).
2. A. Nagasawa, System for standardized facial thermography, (abstract) 2nd European Congress of Thermography 128 (1978).

PYROELECTRIC INFRARED SENSORS

D E Burgess

Royal Signals and Radar Establishment
Malvern
UNITED KINGDOM

INTRODUCTION

During the last decade The Royal Signals and Radar Establishment has been involved in the development of two technologies for infrared detection. The first, using quantum detectors cooled to liquid nitrogen temperatures, has resulted in equipments such as those made by AGA which are now in use in the world's hospitals. The second, using thermal detectors based on the pyroelectric effect, forms the basis of this paper.

PYROELECTRICITY

In the electrical equivalent of ferromagnetism, a pyroelectric is an asymmetrical crystalline material having an internal electric dipole moment which is temperature sensitive.[1] The consequent change in this moment when the temperature changes is equivalent to the generation of a surface charge on the material, which if measured describes the material's recent temperature history. This simple operation has two important features:

1. The pyroelectric translation of temperature to surface charge occurs at room temperature so that cryogenics are not required
2. Pyroelectric charge is produced only by a change in temperature, and the material does not respond to a steady flux of radiation.

Pyroelectric detectors may in general be used in one of three ways to present the thermal image of a scene. In the first and simplest case, a single detector is used and a two dimensional mechanical scan of the scene is performed in the optics to build up the picture. A sensible frame rate then requires a rather large detector bandwidth with consequently a high noise level. A reduction in the necessary detector bandwidth is obtained in the second case, where a line of detectors replaces the single element. The detector bandwidth is reduced by a factor equal to the number of detectors and the optics are simplified to accommodate only a one-dimensional scan. Electronic complexity is, however, increased by the need for multiple amplifiers. In the third case, the mechanical scanning is avoided completely by using a two-dimensional detector array with each detector operating at even lower bandwidths, but for any sensible array size the number of amplifiers becomes prohibitive and a different accessing technique must be used. That chosen is electron beam readout in a television camera tube. In the simplest camera tube, the vidicon, the two-dimensional array of detectors (in practice a continuous sheet of material) onto which the scene is imaged is serially scanned in raster format by a focussed electron beam which discharges the signal into a single amplifier. Investigation of all of these techniques has been carried out at RSRE during the past few years with the main emphasis on the last one, the result being the development of the pyroelectric vidicon camera.

THE PYROELECTRIC VIDICON CAMERA

The infrared-sensitive pyroelectric vidicon tube differs from its visible sensitive counterpart in two respects: firstly the normal photoconductive target is replaced by a disc of pyroelectric material, 20 mm in diameter but only 30 microns thick; secondly an infrared transmitting faceplate of germanium is used.

In operation in a television camera, a germanium lens images the thermal scene through the germanium faceplate on to the pyroelectric target, causing localized changes of target temperature.[2,3] Through the pyroelectric effect these temperature changes are converted into corresponding surface charges on the rear face of the target which may be read out in the conventional way by the raster-scanned electron beam.

The video signal from the pyroelectric camera has two important differences compared with the signal derived in a photoconductive tube. The first of these is that the pyroelectric signal is proportional to the rate of change of temperature of the target, so that the tube only responds to changes in the incident radiation level. Once the target has come into thermal equilibrium with the scene (3 seconds) its temperature will no longer change and no signal will be produced. The incident radiation must be modulated in some way to view a static or slowly varying scene, for example, by panning the camera or by using a rotating chopper. The second difference is that the signals from a pyroelectric tube are typically one hundredth of those from its daylight equivalent, and thus a pyroelectric camera requires very careful electrical and mechanical design, especially in the area of the preamplifier.[4]

This design and development work has been carried out and pyroelectric cameras are now available from a number of commercial sources. They are lightweight and portable, easy to operate; working at room temperature they require no cooling, and are relatively inexpensive compared with imagers using cooled quantum detectors. Additionally, being television, accessories such as image processers and tape recorders may be added to produce systems of any desired complexity. On the deficit side the pyroelectric camera has a temperature sensitivity which is worse than that of the best cooled detector imagers, but it has a spatial resolution which is better than many of them. The minimum resolvable temperature of a coarse pattern is typically 0.5°C and the limiting resolution 250 lines (but see below for details of a higher resolution tube).

MEDICAL USES OF THE PYROELECTRIC CAMERA

As early as 1975 at RSRE a pyroelectric camera was used to image medical scenes but the results were not published. More recently in this country and abroad cameras have been used in hospitals in a number of studies. Two of these from the UK are given here as examples. Firstly a pyroelectric camera was used by Dr Wheatley[5] at the Royal Infirmary, Edinburgh, to aid in open heart surgery. The surgeon's task was to insert a by-pass graft across a coronary artery obstruction and then to ensure that blood flowed freely through it, re-establishing an adequate perfusion of the heart tissue. The thermal image of the heart was provided for the surgeon by a camera installed in the operating theater light stand. By a sequence of clamping the graft from the aorta, injecting cold saline and releasing, the surgeon obtained a clear picture of the flow through the graft and into the heart muscle. Individual branches of the coronary tree could be seen, as in some cases could the retrograde filling of one artery via the grafted artery.

Secondly, Mr Newmann at the Southern General Hospital, Glasgow used a hand held camera to scan the buttocks of elderly patients admitted to the hospital[6] and demonstrated that bed sores were not always the result of ignorant nursing, but often of bruising sustained before hospital admission.

In both these trials the temperature and spatial sensitivity of the standard commercially available cameras was sufficient to perform the medical task. The simplicity of operation and portability of the pyroelectric cameras were also important factors in enabling the trials to take place. In particular, the ability of the cameras to operate in any attitude (generally impossible when using cooled detectors with the danger of liquid nitrogen spillage from upturned dewars) ensured the minimum disturbance to patients and to normal operating practices.

FURTHER DEVELOPMENT OF A PYROELECTRIC VIDICON CAMERA SYSTEM

At RSRE during the last five years there has been considerable interest in increasing the performance of the standard pyroelectric vidicon camera, and in adding to it signal processing electronics to aid in image analysis. To this end an improved camera tube has been developed[7] which has a higher resolution pyroelectric target. When the infrared radiation is focussed onto the thin disc of the pyroelectric target and absorbed, it spreads laterally in the target causing a loss of resolution. A higher resolution has been obtained by dividing the target into a matrix of thermally isolated islands stuck on to a supporting plastic backing. This process, known as reticulation, has increased the resolution of the tube from 250 to 350 picture points.

Concurrently with this tube development, the application of electronic signal processing to television-type sensors in order to improve their performance has resulted in equipments being made to perform picture improvements by real time running integration. In these equipments, known as integrating field stores, the random electronic noise, which is different from one television scan to the next, is reduced whilst unchanging picture signals are reinforced. Signal to noise improvements of up to 10 may be obtained in this way at the expense of a time constant for responding to changing scenes of 1.5 seconds - not generally a problem in medical situations. Multiple brightness contouring units producing a different color for each brightness range and spot brightness measuring equipments are also now available to ease image interpretations. Recently such functions have been incorporated into improcessing computers capable of storing and analyzing vast amounts of data.

A MEDICAL TRIAL WITH THE RSRE EQUIPMENT

In order to determine the usefulness or otherwise of the RSRE pyroelectric camera with reticulated high resolution tube and with integrating field store and image contouring unit, the equipment was taken to the Royal National Hospital for Rheumatic Diseases at Bath where it was used to observe a number of patients during a trial kindly organised by Mr E F J Ring. Amongst the patients seen during the day were sufferers from rheumatism in knees and hands and one patient with an ulcer on his ankle. Results were recorded on video tape for later analysis.

RESULTS

Video tape recordings were made of thermal images from five patients, all of whom proved to have measurable skin temperature differences that could be associated with their diseases. Two of these images are analyzed here.

Rheumatic Joints; an Example of Internal Disease

One patient examined during the day had severe pain and swelling caused by rheumatism in his left knee. This type of pain and swelling usually leads to an increased localized skin temperature which was evident in this case. The patient stayed for 15 minutes in a temperature stabilized room without shoes and socks and with trousers rolled up, then stood in front of the pyroelectric camera. Figure 1 is a photograph taken at RSRE from a television monitor displaying the recorded video tape. In this and in the subsequent photographs black corresponds to cold whilst white corresponds to hot. The right (diseased) knee of the picture (the patient's left) is significantly hotter than the normal knee, and a particularly hot area appears to the right and below the patella.

This picture is sufficient to determine the hot areas associated with the disease, but the medical diagnosis of the severity of the disease requires more quantitative measurements of the temperature differences within the scene. Accordingly, the data was replayed through the contouring equipment which was adjusted to give a contour width corresponding to 0.5°C. The contour level was then stepped in intervals of 0.5°C and at each step the monitor was photographed. (The contouring

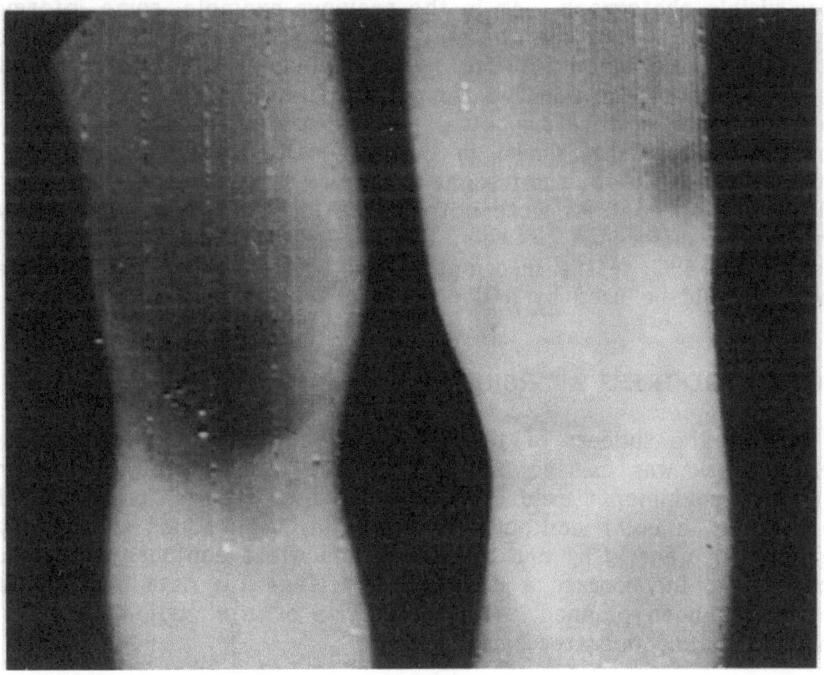

Fig. 1. Thermal picture of a patient's knees.

unit operates by setting to peak white any picture area which lies within the gray shade window set by the operator).

Figure 2 is a composite of a sample of these results. The numbers below each photograph represent the equivalent contour temperature above the temperature of the coldest part of the picture. From an equipment point of view, since the contours are not overlapping the residual electronic noise corresponds to a lower temperature than the 0.5°C contour separation. From a medical viewpoint the maximum temperature difference between the healthy and the diseased knee is 4°C.

Ulcers and Burns; Surface Injuries

In the healing of both ulcers and burns it is of prime importance to determine how well the blood flow to the injured region is returning, since blood flow and healing go hand in hand. Since an increased blood flow near the skin surface will result in an increased skin temperature, then a positive thermal gradient towards such a wound means healing is occurring, whilst a negative gradient shows a failure of the healing mechanism, and requires medical intervention.

One of the patients seen during the day had a small ulcer on his ankle, shown in the upper photograph of Figure 3. The area covered by the lower thermal view is shown by the black corner markings on the upper visible photograph. As in the previous example, some information may be gleaned from the thermal picture. The scab at the center of the ulcer has no blood flow to it and is cooler than its surroundings. The good sign of the positive temperature gradient towards the ulcer is evident. Again, however, a better medical judgement may be made from more quantified data, shown in the composite Figure 4. The top left picture repeats the thermal scene of figure 3 for comparison. Contour width and step controls were both set to 0.5°C. The pictures show the temperature difference between the scene edges and the edge of the ulcer to be 4°C. This information and the shape of the temperature contours could be used by a doctor to chart the healing process.

RECENT PROGRESS AT RSRE

After the success of the limited trial described in the previous section, work was carried out at RSRE to redesign into an integrated unit the equipment used to produce the results presented here. Additionally, a color contouring facility has been added to simplify the procedure of generating multiple black and white contours, whilst future expansion to incorporate a computer interface for data acquisition and control has been planned. The prototype of the equipment has been constructed, and operates satisfactorily.

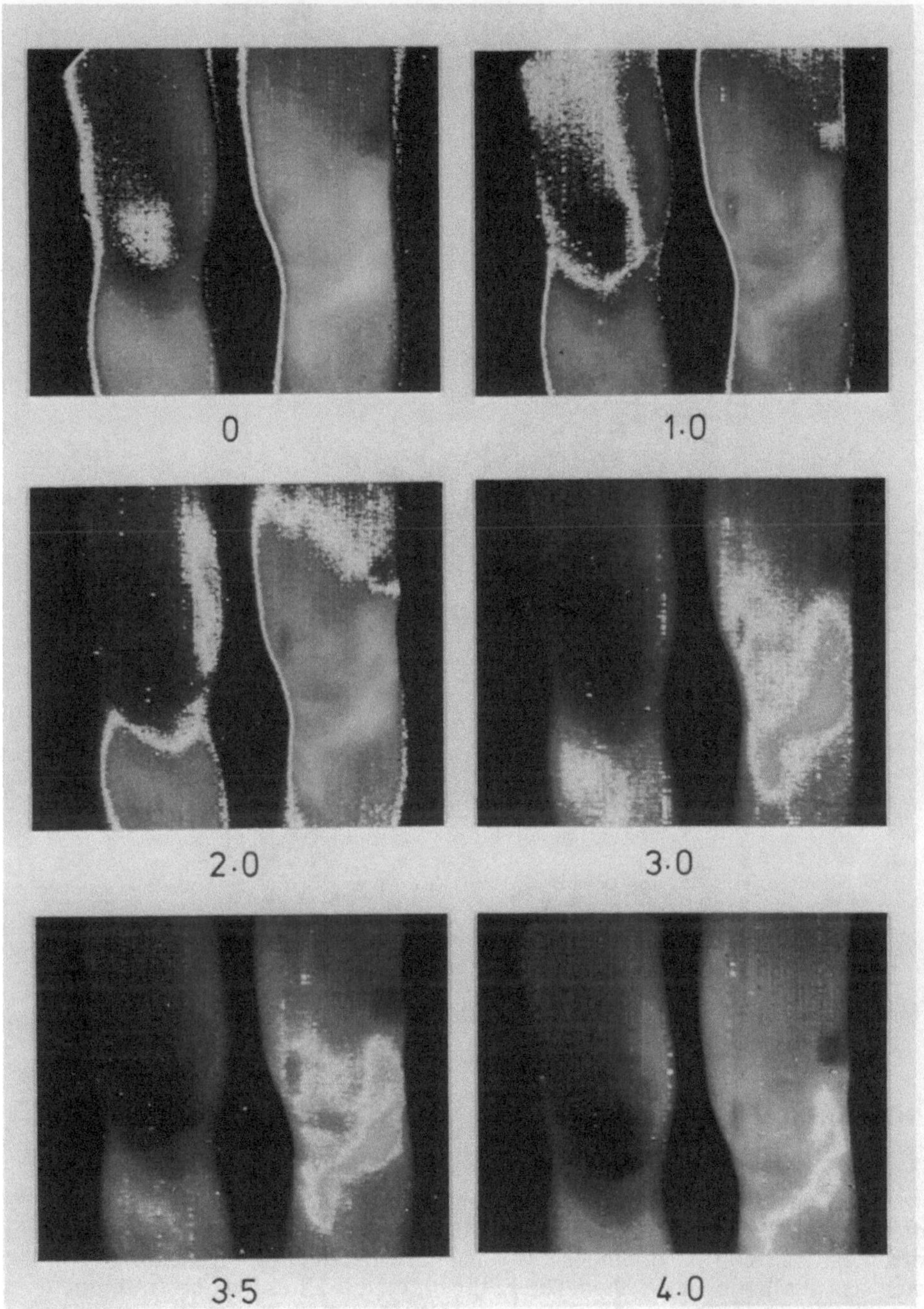

Fig. 2. Temperature contoured thermal pictures of the knees.

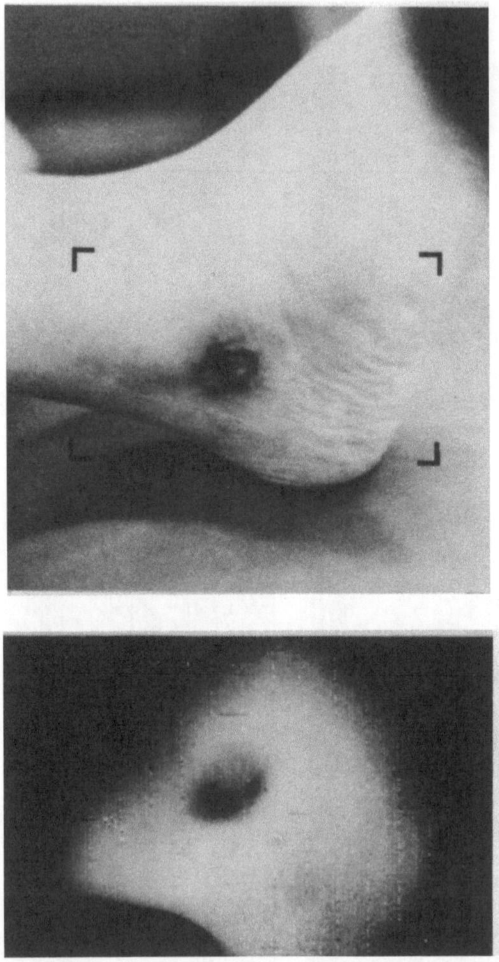

Fig. 3. Visible and thermal pictures of an ulcerated ankle.

CONCLUSIONS

Commercially available pyroelectric vidicon cameras have been shown to be satisfactory for a range of less demanding medical applications, and an optimized camera with its associated signal processing electronics has been used successfully in more demanding situations. The former is available now; hopefully the latter will soon be put into production so that the equipment can fulfil the promise it has already shown in the monitoring of rehabilitation and in the assessment of tissue viability.

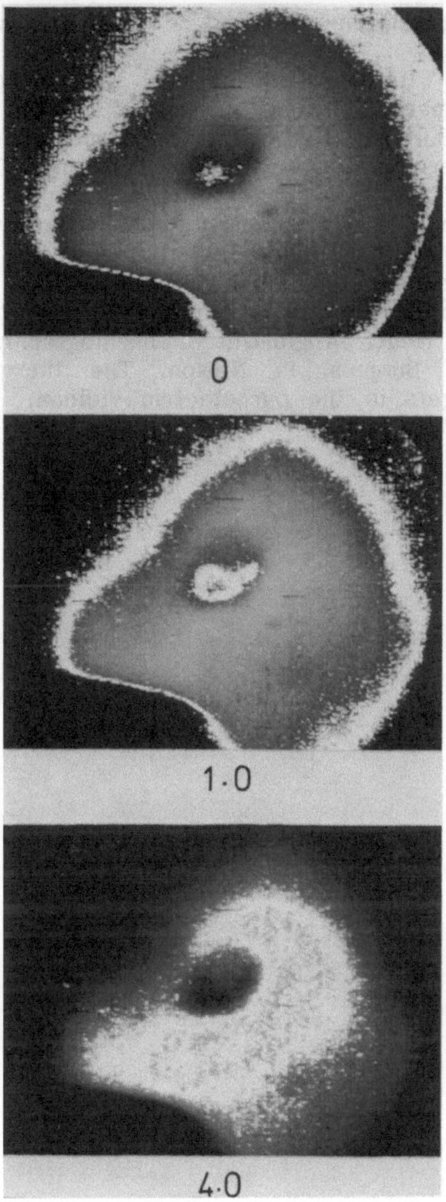

Fig. 4. Temperature contoured thermal pictures of the ulcerated ankle.

REFERENCES

1. E. H. Putley, R. Watton, J. H. Ludlow, Pyroelectric thermal imaging devices, Ferroelectrics 3:263-268 (1972).
2. R. Watton, Pyroelectric materials: operation and performance in

thermal imaging camera tubes and detector arrays, Ferroelectrics 10:91-98 (1976).
3. R. Watton, B. Harper, D. Burgess, An inexpensive portable pyroelectric camera, Infrared Physics 16 (1976).
4. R. Watton, D. Burgess, B. Harper, The pyroelectric vidicon: a new technique in thermography and thermal imaging, J. Appl. Sci. Eng. A 2:47-63 (1977).
5. J. W. E. Brydon, A. K. Lambie, D. J. Wheatley, Thermographic visualization of coronary artery blood flow during by-pass surgery, J. Med. Eng. Technol. 3:77-80 (1979).
6. P. Newman, Applications of a mobile thermograph, in: " Human Metabolism and Heat," Anglo-Dutch Thermographic Society (1980).
7. R. Watton, D. Burgess, P. Nelson, The thermal behavior of reticulated targets in the pyroelectric vidicon, Infrared Physics 19:683-688 (1979).

MICROWAVE THERMAL IMAGING: TECHNICAL ADVANCES AND CLINICAL FINDINGS

M Gautherie, J-L Guerquin-Kern, A Kotewicz

Laboratoire de Thermologie Biomedicale
Universite Louis Pasteur
Strasbourg, FRANCE

INTRODUCTION

In the last few years, prototype equipments of microwave thermography have been developed and used to evaluate subcutaneous temperature distributions in certain medical conditions. The clinical results obtained indicate the potential for microwave thermography for the thermal assessment of certain diseases. Applications include prognosis and follow-up, and the monitoring of tumour temperature during hyperthermia generated for therapeutic purposes.

Clinical experimentation has led to a better understanding of optimal measuring frequencies and the improved forms of analysis and display of thermal distributions. The instrumention developed to date is based on two different technical approaches: (a) remotely-sensing cm- and mm-wave radiometry using focussed reflectors; (b) contact cm-wave radiometry using small matched antennas brought into contact with the skin. Most of the existing devices are combined with scanning and processing systems which allow to display thermal profiles and images.

Within the frame work of co-operative research programmes at national and international levels, we have, over the last six years, carried out several series of clinical investigations using prototypes of microwave thermographs in both (a) remotely-sensing thermographs operating at 68, 33 or 17 GHz (EDRICH's radiometers); (b) contact thermographs operating at 9 or 3 GHz (MAMMOUNI's and LUDEKE' radiometers).

Microwave thermographic examination has been practised in Strasbourg since 1976 on a total of more than 400 patients with benign or malignant disease of various organs: breast, brain, joints, thyroid. Many of these observations have been reported in previous papers.[1-5]

The present work deals with our clinical experience using contact microwave thermographs operating at 9GHz and 3GHz. Two types of instrument have been used successively:

1. Prototypes constructed by the Centre Hyperfrequences et Semi-Conducteurs (CNRS, Lille, France) and the Centre de Technologie Biomedicale (INSERM, Lille, France) were used for preliminary clinical investigations as well as experimental studies on phantoms and animals;
2. Equipment developed by ODAM-BRUKER Corporation (Wissembourg, France) in close co-operation with the previously mentioned research teams, was used in studies on larger groups of patients with different clinical conditions of the breast, brain, bone, joints and muscle.

METHODS AND PATIENTS

Thermal distribution in human tissues is most frequently disturbed in pathological conditions. For example, changes in the local metabolism and circulation associated with the growth of a malignant tumour give rise to thermal changes.[6] These disturbances can be demonstrated by cutaneous thermography using infrared scanners or liquid crystals. However, this only applies if the pathological process is located such that observable changes in the skin thermal pattern occur. Indeed, infrared thermography does not work in some circumstances, for example in intra-abdominal and intra-cranial tumours, because of the barriers to the diffusion of heat presented by fat and bone tissue, both of which have low-thermal conductivity and blood flow. Microwave thermography offers the possibility of overcoming these inconveniences by providing direct access to thermal disorders occurring in the deep tissues.

PHYSICAL PRINCIPLE OF MICROWAVE THERMOGRAPHY

Human tissues spontaneously emit electromagnetic radiation of thermal origin. This phenomena occurs within a very wide spectrum ranging from the near infrared to the radiofrequencies, the radiated energy reaching a maximum around 10 microns (Fig 1). Since the energy radiated at a given wave length is a function of tissue temperature (Planck's law), its measurement indirectly indicates the temperature of the tissue: this is the general principle or radiometric thermometry. However, when the measurements are carried out in the infrared range where human tissues are opaque it is the temperature of the skin that is recorded, since thermal emission can only originate from the very superficial cutaneous layers. By contrast, when the measurements are carried out in the microwave range where the tissues are relatively transparent, the subcutaneous tissues contribute to thermal emission in such a manner that it may be possible to evaluate their temperature.

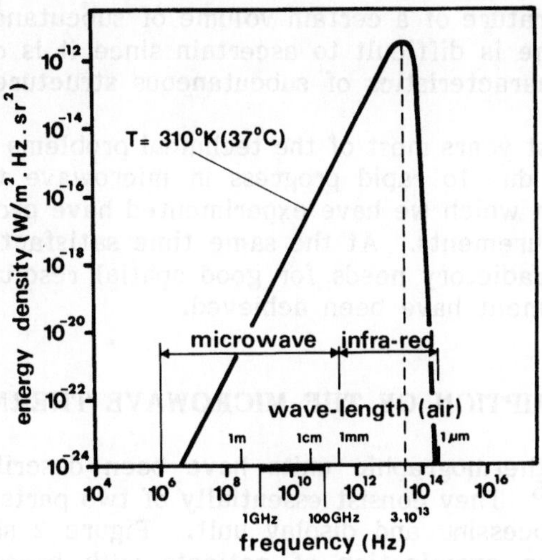

Fig. 1. Spectral distribution of the thermal emission of the human body at a temperature of 37°C (Planck's law).

The physical laws which govern the phenomena of thermal emission and propagation of electromagnetic waves, determine important differences between infrared and microwave radiometry:

1. The energy radiated in the microwave range is approximately 10^6 times weaker than that in the infrared range which requires the use of radiometers of very high sensitivity.
2. Microwave radiometric measurements are much more sensitive to ambient electromagnetic fields; this factor must be taken into account in the design of the receiver and possibly the location of the examination room.
3. Spatial resolution is less favourable with microwave radiometry, since it is limited not only by the characteristics of the receiver but also by the frequency of the microwave transmission in tissue (this varies from a few millimeters to a few centimeters according to the measuring frequency and the dielectric properties of the tissues).
4. Variations in emissivity over the body surface are generally negligible in the infrared range but, by contrast, are large in the microwave range; as a result radiometers are required for the measurement of the emissivity or for compensating for its variation.
5. The depth of measurement with a microwave radiometer depends greatly on the dielectric characteristics of the subcutaneous tissues; experimental studies have shown that it is 5 to 10 times larger for tissues with low water content, such as fat and bone tissue, than for tissues with a higher water content such as muscle.

6. The significance of microwave radiometric measurement is still not obvious because the recorded temperature actually represents the mean temperature of a certain volume of subcutaneous tissue. The actual volume is difficult to ascertain since it is dependent on the dielectric characteristics of subcutaneous structures.

Over the past years most of the technical problems mentioned above have been solved due to rapid progress in microwave technology. The thermographs with which we have experimented have provided significant temperature measurements. At the same time satisfactory compromises between the contradictory needs for good spatial resolution and greater depth of measurement have been achieved.

GENERAL DESCRIPTION OF THE MICROWAVE THERMOGRAPHS

Microwave thermographic units have been described in detail in previous papers.[3,5] They consist essentially of two parts: the radiometer, and the signal processing and display unit. Figure 2 shows one of the units used for the examination of patients with breast disease. The radiometer comprises a hyperfrequency detector of high sensitivity whose noise factor is minimized by the use of a preamplifier FET-AsGa capable of detecting temperature differences of 0.1°C. The probe which receives the electromagnetic signal from the subcutaneous tissue consists of an open waveguide filled with low waste dielectric material, coupled to a coaxial junction. Depending on the type of equipment used, the probe placed directly in contact with the skin may be free at the end of a coaxial cable or mounted on an articulated arm. The latter is provided

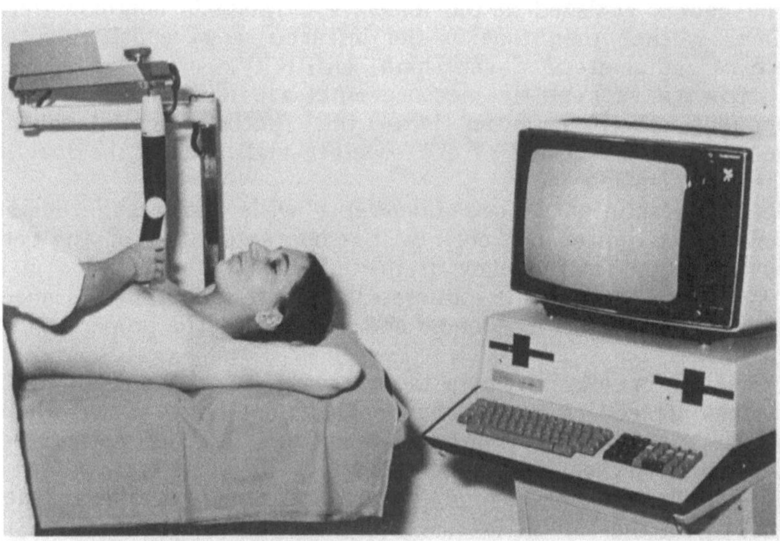

Fig. 2. Microwave thermograph: radiometer and processing and display unit (TMO 3000, ODAM-BRUKER Corp.).

with transducers to define the spatial position of the probe defined by three co-ordinates and three angles.The device for signal processing receives the signal from the radiometer after amplification and digitization. The temperature signals corresponding to the different points of measurement are stored on floppy disk. Following point by point exploration over the body area being investigated, thermal profiles and images can be reconstructed. The practical advantages of this system lie in the possibility of visually following on a screen the formation of the microwave thermogram during the examination of the patient. It also provides the possibility of carrying out additional data processing after the examination, eg for the purpose of enhancing contrast or calculating temperature differences between symmetrical body areas. The use of a colour monitor improves the discrimination between different thermal profiles and the recognition of isothermic areas. A reference scale giving the colour-temperature relationship is displayed together with the thermogram so that the evaluation of isotherms can be easily performed.

CONDITIONS FOR CLINICAL INVESTIGATIONS

Investigations have been carried out using 9GHz and 3GHz microwave thermographs for clinical and technical evaluation of the method. The equipment was installed in clinics which specialize in different pathological areas, in order to have access to all the diagnostic data necessary for detailed analysis of results.

A total of 292 patients underwent microwave thermographic examination. Four groups of patients were represented:

1. Patients with healthy breasts (n=32), benign mastopathy (n=46) or mammary carcinoma (n=96);
2. Patients with cerebral disease, most of whom had a malignant tumour, either primary (n=17) or metastatic (n=21), or a softening process (n=6);
3. Patients with joint disease (n=25) of different locations (shoulder, wrist, spine, hip and knee);
4. Patients with thyroid disease (n=37), of whom 4 had cancer.

In all instances, a careful comparison was made with other imaging techniques, including infrared and possibly liquid crystal thermography and X-ray, isotope and ultrasonic examination. The microwave data were compared with mammography and echotomography for the breast patients, and with arteriography and X-ray CAT-scanning for those patients with cerebral disease.

Prior to each series of examinations, a calibration check and adjustment of the microwave radiometer was carried out. In addition, examinations were carried out in healthy subjects to familiarize clinicians and technicians with the equipment, and to establish a protocol for the measurements. The examinations were performed in a sitting or supine position in normal rooms without electromagnetic protection or air conditioning. Figure 3 shows the examination of a breast patient.

RESULTS AND DISCUSSION

Most of our clinical observations endorse those previously reported either by us or by other authors.[1-5] However, the findings are now more conclusive due to improved presentation of results (thermal profiles and images), and the increased number of cases. Since at this time the study was done on relatively large groups of patients with breast and brain pathology only, the discussion will be limited to these two categories. Special emphasis will be laid upon the critical question of the measuring depth.

Breast Pathology

Approximately 85% of breast cancer give rise to significant changes in the superficial mammary thermal pattern detectable by infrared or liquid crystal thermography.[6] When these techniques show marked hyperthermia, microwave thermography also demonstrates anomalies, except for those carcinomas whose thermal signature consists of vascular hyperthermia with a distorted pattern only. However, the anomalies shown by infrared and microwave temperatures are always found to be higher than infrared temperatures (in most physiological and physiopatholocal conditions, the subcutaneous temperatures are higher

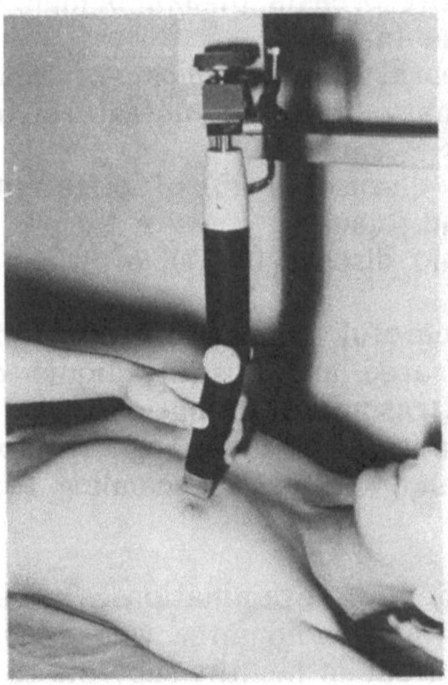

Fig. 3. Microwave thermographic examination of the breast.

than the skin temperature). Figure 4 shows an example of comparative findings with infrared and microwave thermography.

In about 15% of all carcinomas, infrared thermography does not show significant change due to cancer with low tumour thermogenesis (slow-growing cancers), or to restricted heat transfer from the tumour to the skin (low-thermal conductivity, or decreased blood flow, eg in fat tissue). Microwave thermography may reveal significant hyperthermia with variable intensity and extent in the following conditions: (i) in situ carcinomas (low metabolic heat production and moderate thermovascular changes), (ii) diffuse, oedematous cancers with skin infiltration (characteristic skin thickening which acts as a screen to the transmission of heat), and (iii) scirrhous carcinomas with a slow growth rate in a breast with predominantly fat tissue (low tumour thermogenesis and tissue thermal conductivity).

With regard to breast cancers, microwave thermography undoubtedly enables one to cope with most of the insufficiencies of infrared and liquid crystal thermography, especially in the case of non-palpable tumours whose early detection is often uncertain. Furthermore, microwave thermography provides a more accurate thermodynamic characterization of the cancer in those situations where the mean temperature of the tumour volume can be evaluated with sufficient accuracy. The microwave data seem to be more directly correlated with tumour thermogenesis and vascularization whereas, in this respect, infrared data may sometimes be considered erroneous. This may be due to excessive cutaneous thermovascular changes which are "unrelated" to the thermal conditions of the tumour, or by absence of sufficient thermal correlations between the tumour and the skin. This is of great importance from a practical point of view, since the thermodynamic status of the cancer gives essential prognostic data on the relationship between the specific metabolic heat production of the cancer tissue and the doubling time of the tumour volume.[6]

Brain Pathology

Several investigations carried out by microwave thermography in patients with malignant cerebral tumours have given encouraging results. Clear changes in the microwave temperature have been observed in almost all the cases.[3,4] This is remarkable since infrared thermography generally is inefficient, except in the case of superficial tumours such as meningiomas which often provoke a local temperature increase in the scalp which is detectable after shaving.

Our study confirms prior observations since hyperthermia has been shown in most cases, either by measurements at 9GHz (superficial tumors), or 3GHz (deeper tumors). The comparison with other investigative procedures (X-ray, computerized axial tomography, arteriography, scintigraphy) has shown additionally that (i) the hyperthermia corresponds

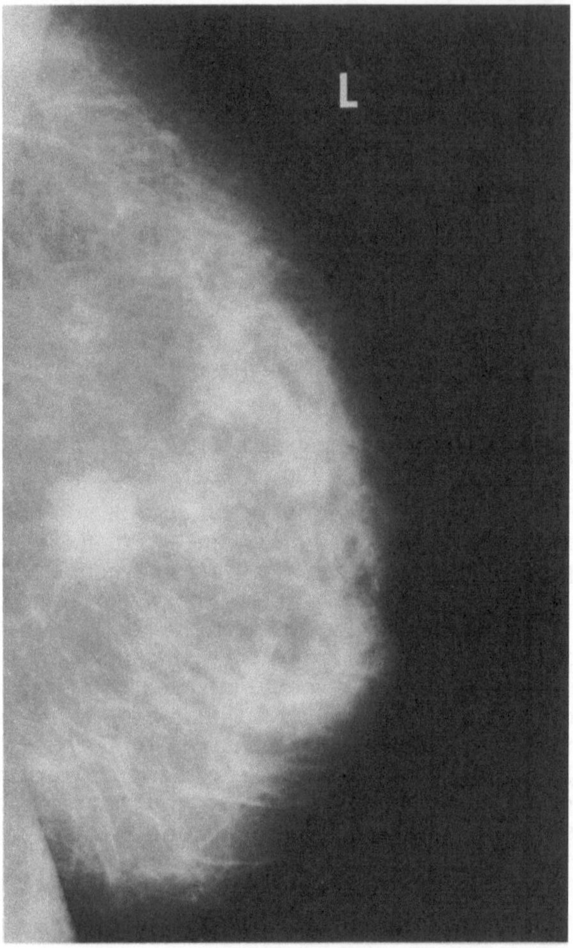

Fig. 4a. Non-palpable mammary carcinoma, left breast (TONOMO), 49-year-old patient. Mammography: deep-seated tumor (4 cm).

topographically with the tumour, and (ii) the intensity of the hyperthermia, often higher than 1°C, is higher when the tumour is superficial (measurements at 9GHz), and when arteriography and scintigraphy show marked hypervascularization and hyperfixation respectively.

Even though the small number of cases in this series does not allow definite conclusions, it seems reasonable to consider the use of microwave thermography for evaluating the thermodynamic status of cerebral tumours for prognostic purposes. There may be a relationship between the growth rate of the tumour and its thermal characteristics, as shown both for mammary carcinomas and malignant melanomas.[6] In the case of

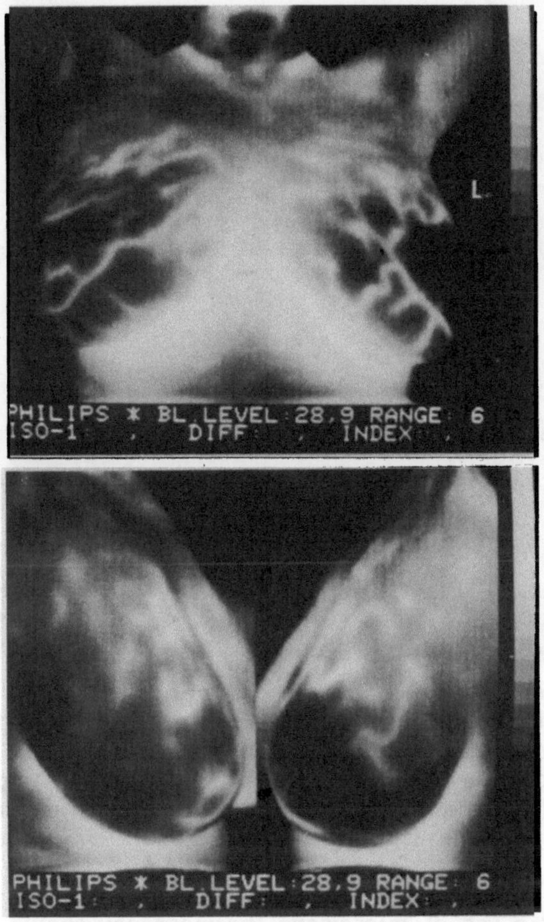

Fig. 4b. Non-palpable mammary carcinoma, left breast (TONOMO), 49-year-old patient. Infrared thermography: symmetrical thermal pattern, no findings.

superficial tumours, peritumoural hypervascularization gives rise to hyperthermia which is more intense than that of the tumour itself. By contrast, oedematous areas often appear as cold areas.

It is also worth mentioning that microwave thermography might have diagnostic potential. In two patients, careful examination has successively enabled us firstly to find a marked hyperthermia in the area of an already known metastasis (localised by CAT-scans) and secondly to discover a second hyperthermic area that on repeat analysis of the CAT-scans revealed a smaller metastatic tumour which at first had passed unnoticed.

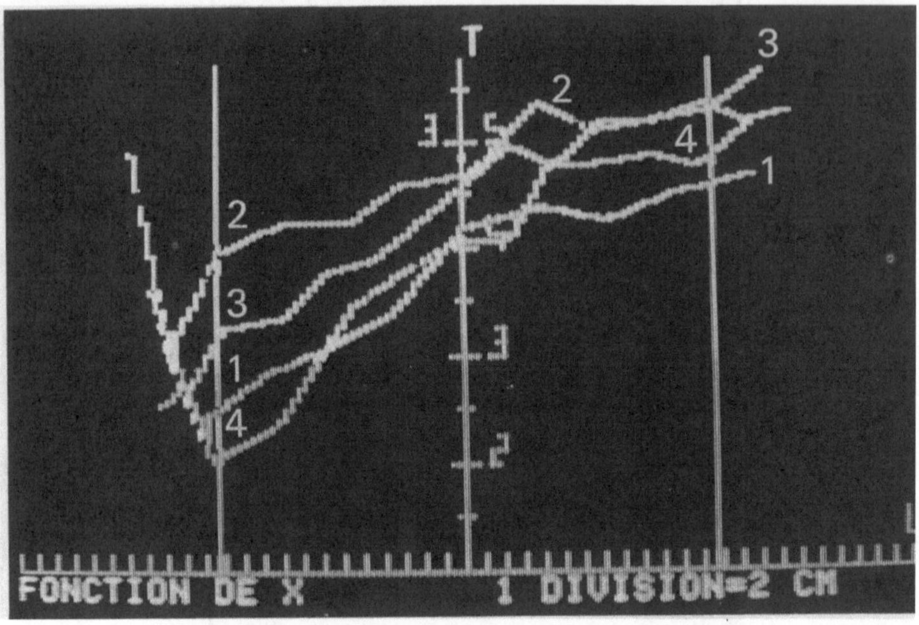

Fig. 4c. Patient as in 4a and b. Microwave thermography (thermal profiles and images reconstructed from the point by point measurement): total hyperthermia of the left breast (mean temperature difference:1.5°C) The profiles 1 to 4 correspond in this order to the following lines of measurement on the two breasts: nipple line, 2 cm below, 2 cm above, 5 cm above.

A systematic use of microwave thermography should be considered by neuroradiologists as an adjunct to the exploration of brain tumours. The technique is non-invasive and the examination quite simple.

Measuring Depth

In general, careful comparison between thermographic and physical data indicates that microwave radiometry effectively analyses subcutaneous temperature distribution. This impression is additionally confirmed by the discrepancies between the results of the infrared and microwave thermographic examinations which become more evident as the pathological process lies at a greater depth.

Our clinical observations show that "measuring depth" depends largely upon the characteristics of the subcutaneous tissue, which was predicted from experimental data from biological tissues. It is obvious that the "measuring depth" is much larger in fat and bone tissue than it

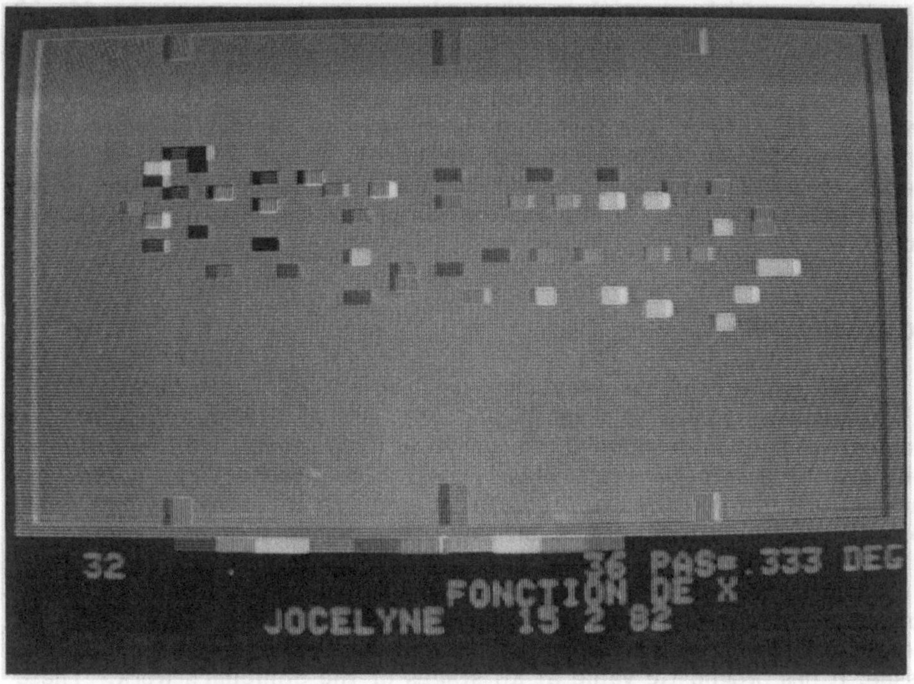

Fig. 4d. Patient as in 4a, b and c. The image has been reproduced from the original color thermogram.

is in muscle tissue. This is why microwave thermography has great potential in intra-cranial and intra-abdominal pathology.

The actual measuring depth is difficult to ascertain, even in anatomically well defined conditions. This is due to the heterogeneity of the biological structures, both from the electromagnetic and thermal point of view. However, our clinical findings clearly demonstrate that measuring depth is often noticeably larger than indicated by experimental studies on mathematical models and excised tissues.

On the other hand, measuring depth also depends upon the measuring frequency. It decreases with increasing frequency, and experimentally has proven to be higher at 3GHz than at 9GHz. This explains, for example, why the 3GHz thermograph give inaccurate findings in most superficial processes and, therefore, why the 9GHz thermograph does not give significant findings in deep-seated processes. This is especially true where subcutaneous tissue has a high water content. In this respect our comparative clinical studies with the 3GHz and 9GHz thermographs have been most fruitful, since they suggest that the operating frequency should be chosen differentially according to the location of the disease. These statements are in keeping with our previous conclusions based on investigations with remotely-sensing microwave thermographs at 68, 33 and 17GHz.[2]

These comments strongly favour the development of multifrequency systems. In spite of the fact that microwave technology becomes more and more difficult with increasing frequency, in our experience, three frequency bands (around 3, 9 and 17 GHz) should be used, the latter being probably optimal for the exploration of the superficial joints (limbs, spine).

CONCLUSIONS

This study confirms previous work carried out in co-operation with other research teams in France and the United States regarding the medical value of microwave thermography. Even though additional studies still are necessary to more clearly define the technology, microwave thermal imaging promises to be useful in many clinical areas, in particular for the thermal assessment of breast, brain, bones and joint diseases.

Microwave thermography is non-invasive and the information obtained is original. Since microwave temperature correlates with the metabolism and vascularization of normal or pathological subcutaneous tissue, it has prognostic value at all stages of the clinician's work. These two characteristics are similar to those of infrared thermography, although the latter gives information only on cutaneous thermal distributions.

Compared with the first generation of contact microwave thermographs, the new equipment obviously represents an improvement which can be used routinely in a variety of medical environments. A particular characteristic of presently available thermographs is the monoreceptor probe which operates at a given frequency; a system proven to work well. It would be advisable, however, to consider the development of new multi-receptor probes or a series of such probes operating at different frequencies, eg 3, 9 and 17 GHz. The clinician would thus have control on the depth of exploration, and perform more significant investigations. Recent technological studies based on the concept of a correlation microwave thermograph using two receptors might be a significant improvement.[7]

The clinical evaluation now in progress at Strasbourg is specialised through clinical departments, ie mammary, cerebral and osteo-articular diseases. Its success will depend on:

1. availability of a variety of physical methods for diagnosis;
2. large number of visiting patients, and
3. systematic long-term follow-up, either pre- or post-therapy.

Other than technical improvement, the objectives should be the following:

1. to define the optimal conditions for examination, in particular those concerning the patient (preparation, posture, ...) and the electromagnetic and thermal environment;
2. to elaborate a specific symptomatology based on the microwave thermographic findings for each clinical application;
3. to determine the value and limits of the method against the cost-effectiveness of the examination.

Apart from the clinical investigations for diagnosis or prognosis, microwave thermography is also used for monitoring tumour temperature during hyperthermic cancer treatment, and for physiological research on thermoregulatory processes (non-invasive recording of brain temperatures) (Fig 5).

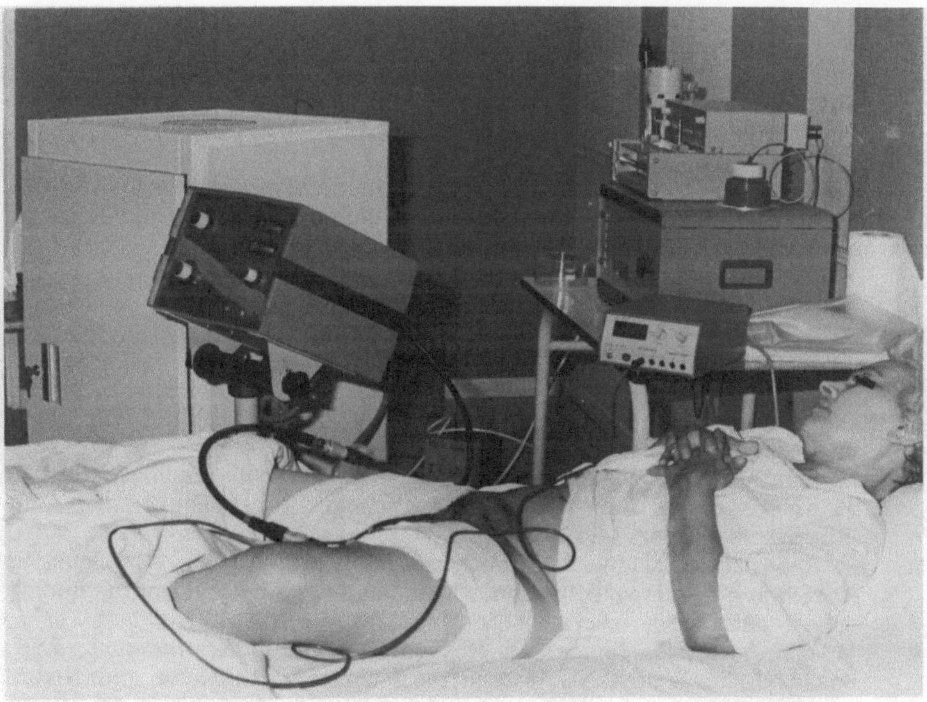

Fig. 5. Hyperthermic treatment of a recurrent cancer of the vulva (equipment: HL1500 SEM, 13.56 MHz).

ACKNOWLEDGEMENTS

The first part of this study was supported by grants from the Delegation Generale a la Recherche Scientifique et Technique (DGRST, nr. 78.7.2271, and 79.7.0812), and from the Caisse Nationale de l'Assurance Maladie des Travailleurs Salaries (CNAMTS) under the auspices of the Institut National de la Sante et de la Recherche Medicale (INSERM). The second part was supported by a grant from the BRUKER Corporation (Wissembourg, France).

The authors wish to express their gratitude to all participants in this research program for their invaluable contribution:

- Y Leroy and A Mamouni (Centre Hyperfrequences et Semi-Conducteurs, CNRS, Lille, France) and Y Moschetto and J P Sozanski (Centre de Technologie Biomedicale, INSERM, Lille, France) who first developed the first prototype equipments of microwave thermographs;

- J C Reithler, V Ringeisen, D Lampert (BRUKER Corporation, Wissembourg, France) who proceeded with the industrial development of microwave thermographs;

- J P Walter, P Haehnel, J P Braun, D Langer, A Wackenheim, D Vallier, Y Dirheimer and B Moyses who allowed us to perform examination on patients, either in the hospital department or in their private practice.

The authors appreciated the invaluable help of Professor L Keith (Northwestern University Medical School, Chicago, USA) in the review of the manuscript.

REFERENCES

1. M. Gautherie, A. Mamouni, M. Samsel, J-L. Guerquin-Kern, Y. Leroy, Ch. Gros, Microwave radiothermometry (9GHz) applied to breast cancer, SPIE 211:154 (1979).
2. M. Gauthèrie, J. Edrich, R. Zimmer, J-L. Guerquin-Kern, J. Robert, Millimeter-wave thermography: application to breast cancer, J. Microwave Power 14:123 (1979).
3. J. Robert, J. Edrich, P. Thouvenot, M. Gautherie, C. Itty, Imaging microwave thermography and some diagnosis applications excluding breast pathology, Acta Thermographica 4:118 (1979).
4. M. Gautherie, M. Samsel, B. Moyses, D. Vallier, A. Mamouni, J. P. Sozanski, Y. Leroy, Y. Moschetto, Imagerie thermique micro-onde appliquee a la detection et au pronostic de tumeurs mammaires, cerebrales et thyroidiennes, in: "Electromagnetic Waves and Biology," A. J. Berteaud ed., CNRS, Paris (1980).

5. J. Edrich, W. E. Jobe, R. K. Cack, W. R. Hendee, C. J. Smyth, M. Gautherie, Ch. Gros, R. Zimmer, J. Robert, P. Thouvenot, J. P. Escanye, C. Itty, Imaging thermograms at cm- and mm-wave lengths, Ann. N. Y. Acad. Sci. 335:456 (1980).
6. M. Gautherie, Thermopathology of breast cancer: measurement and analysis of in-vivo temperature and blood flow, Ann. N. Y. Acad. Sci. 335:383 (1980).
7. A. Mamouni, J. C. Van de Velde, Y. Leroy, New correlation radiometer for microwave thermography, Electronic Letters 17:554 (1981).

PROGRESS IN MICROWAVE THERMOGRAPHY

M Robillard, A Mamouni, J J Fabre, J C Van de Velde, Y Leroy

Centre Hyperfrequences & Semiconducteurs
LA CNRS 287 Universite des Sciences et Technique de Lille I
59655 Villeneuve D'Ascq, Cedex FRANCE

INTRODUCTION

The principle of microwave thermography (MWT) is mainly based on two factors: all bodies emit electromagnetic radiation, the intensity of which depends upon the temperature of the body; living tissues absorb microwaves generated at depths of several centimeters.

A research program on this topic has been carried out in France since 1976; several microwave thermographs have been built in our laboratory[1,2,3] and in collaboration with the CTB Inserm (Lille - France). The value of these systems has been assessed by clinical evaluation of studies on more than 600 patients; these studies were undertaken:

1. In the Laboratoire de Physiologie[4] (Lille)
2. In the Laboratoire de Thermologie and medical centers in Strasbourg[5,6]
3. In the Centre de Medecine Nucleaire (Nancy)[7,8]
4. In the Centre Anti Cancer O. Lambret (Lille)[9,10]

Microwave thermography shows the existence of thermal gradients in the living tissues and since it is atraumatic and non-invasive it is useful for a variety of investigations including prognosis (breast, brain, etc), for brain studies and osteoarticular diseases. Furthermore, in hypertheremia therapy, it allows a non-invasive monitoring of the temperature at a depth.

Collaboration is now taking place with the Laboratoire de Thermoregulation (CNRS Lyon France) to assess a new type of applicator which has been developed.[11] Furthermore, microwave thermography is also being used to study the thermal effects associated with ionizing radiations (Lab. de Radio Biologie appliquee - CEA INRA Jouy en Josas, France). This variety of studies has been made possible by a recent transfer of technology from laboratory to manufacturer (ODAM Bruker).

In addition to these studies, which are continuing in medical centers, we are now tackling, in the laboratory, several physical and technological problems in order to improve and extend the usefulness of MWT. These studies include:

1. Thermal pattern recognition, that is, the quantitative interpretation of the thermal patterns.
2. Correlation microwave thermography, a new method of coherent detection of the thermal noise.
3. The construction of hybrid modules specifically for the medical applications of microwave radiometry.

THERMAL PATTERN RECOGNITION

The different subvolumes of tissue which belong to the volume coupled to the probe contribute to the signal measured by the receiver. A method of computation, based on the reciprocity theorem, has been defined[12] which enables the determination of the thermal patterns. The first rules of thermal pattern recognition can be deduced from this method[13] taking into account the sensitivity of the receiver (for example 0.2°C) and the amplitude of the temperature variations which can be expected in the tissues (for example 5°C). The structure of hyperthermic or hypothermic tissues is assumed to be cylindrical (but, if necessary, more realistic shapes can easily be considered). Using various experimental conditions (frequency of the receiver, characteristics of the probes) computational analysis has been used to determine thermal structure at a depth as well as associated geometric and thermal properties.

This information is being taken into account in the improvement of microwave thermal imaging.

CORRELATION MICROWAVE THERMOGRAPHY[14](CMWT)

CMWT is based on a coherent detection of the thermal signal. The difference between simple MWT and CMTW is similar to the distinctions in radioastronomy, between the Dicke radiometer and the phase switching two aerial interferometer.[15] In CMTW two probes are connected to a microwave correlator. The contribution to the output signal of the thermal signal emitted by tissues depends only on the temperature of the tissues located in the common part of the two volumes coupled to the probes. In practice, the experimenter can modify the volume of tissues under investigation by altering the respective position of the probes. Moreover, the modulation of the phase shift between the two transmission lines of the correlator can provide complementary information by altering the contribution from the different subvolumes.

An experimental set-up, working at 3 GHz, has been built in our laboratory. Preliminary studies have been carried out in which the output signal is recorded as a function of the position of a small noise source embedded in the lossy material. The theoretical aspect of this process is also being studied; computations are in good agreement with the experimental data.

The first conclusions indicate that CMWT is sensitive to thermal gradients (and not to a uniform temperature) and consequently can be used to enhance the spatial resolution in thermal imaging.

THE CONSTRUCTION OF MICROWAVE HYBRID CIRCUITS ESPECIALLY DEVOTED TO THE ACHIEVEMENT OF MWT

This part of our program has not yet been undertaken. At present, the systems incorporate conventional devices such as amplifiers, switches, circulators, hybrid couplers, etc. We think it should be of interest to consider these designs from another point of view: for example, we intend to get together in the same package several switches and several probes in order to achieve an electronic scanning of the probes; the design of other functions such as a correlator or a simplified MWT are planned too. It is intended to undertake this work in our laboratory in the near future.

ACKNOWLEDGEMENT

This work is supported by the Delegation Generale a la Recherche Scientifique et Technique (DGRST - TLBM), by the Agence pour la Valorisation de la Recherche (ANVAR) and by the Etablissement Public Regional Nord - Pas de Calais (France).

REFERENCES

1. A. Mamouni, F. Bliot, Y. Leroy, Y. Moschetto, "A modified radiometer for temperature and microwave properties measurements of biological substances," Published by Microwave Exhibitions and Publishers Ltd., 7e European Microwave Conference, Copenhagen (1977).
2. D. D. Nguyen, M. Chive, Y. Leroy, E. Constant, Combination of local heating and radiometry by microwaves. IEEE Transactions on Instrumentation and Measurements. Vol IM-29, no. 2, 143-144 (1980).
3. Y. Leroy, Microwave radiometry and thermography; present and prospect, Conference Invitee "Symposium Internation de Thermologie Biomedicale" Strasbourg, Publications Alan R. Liss,

New York (1981).
4. A. Mamouni, Y. Leroy, Y. Houdas, Y. Moschetto, Passive subcutaneous temperature measurement for investigation of thermoregulation, 8gh EuMC, Paris (1978).
5. A. Mamouni, M. Samsel, M. Gautherie, Y. Leroy, Radiometrie microonde (9 GHz) : applications aux cancers du sein et a des localisations tumorales diverses, resultats preliminaires, Colloque IMPI, Monoco (1979).
6. M. Gautherie, B. Moyses, Y. Leroy, A. Mamouni, Y. Moschetto, J. P. Sozanski, Imagerie thermique au moyen de systemes radiometriques microondes appliquee a la detection et au pronostic de tumeurs mammaires, cerebrales et thyroidiennes, Symposium International Ondes Electromagnetiques et Biologie Jouy en Josas (1980).
7. J. Robert, J. Edrich, Y. Leroy, A. Mamouni, J. M. Escanye, P. Thouvenot, Clinical applications of microwave thermography. Optics Photonics and Iconics Engineering meeting, Strasbourg (1979).
8. J. Robert, P. Thouvenot, A. Mamouni, Y. Leroy, Detection de lesions intracraniennes a l'aide de la radiometrie microondes, Symposium Internation Ondes Electromagnetiques et Biologie, Jouy en Josas (1980).
9. M. Chive, M. Robillard, J. J. Fabre, G. Giaux, B. Prevost, Hyperthermies Radiofrequence et microonde controlee par thermographie microonde, Colloque Thermologie Biomedicale - Symposium International, Strasbourg, France (1981).
10. P. Besson, M. Chive, M. Robillard, J. J. Fabre, Y. Leroy, Thermographie microonde a 3 GHz: applications cliniques en cancerologie. Colloque Thermologie Biomedicale - Symposium International, Strasbourg, France (1981).
11. A. Dittmar, G. Delhomme, L. Berudon, M. Schmitt, Deep and localized hyperthermia with a new microwave surface applicator, in: "Biomedical Thermology," M. Gautherie, E. Albert eds., Alan R. Liss Inc, New York, (1982).
12. M. Robillard, Contribution a l'etude des sondes et a la reconnaissance d'objet thermique par thermographie microonde (TMO). These de 3e cycle, Lille (1981).
13. M. Robillard, M. Chive, Y. Leroy, Toward an interpretation of the thermal signatures achieved by microwave thermography, Microwave Power Symposium, Toronto, Canada (1981).
14. A. Mamouini, J. C. Van de Velde, Y. Leroy, New correlation radiometer for microwave thermography, Electronics Letters 17;16:554-555 (1981).
15. M. Ryle, A new radio interferometer and its application to the observation of weak radio stars, Proc. Roy. Soc. 211:351-374 (1951).

TEMPERATURE DISTRIBUTIONS BY MICROWAVE RADIOMETRIC MEASUREMENTS

S Caorsi

Instituto di Elettrotecnica
University of Genoa
ITALY

INTRODUCTION

Techniques of microwave radiometry have already yielded satisfactory results in the case of thermographic applications of biological and medical research. In such applications, the main problem is one of determining, by means of microwave radiometric measurements, the temperature distribution inside a body where thermal processes occur.

The aim of this paper is to provide a contribution to the resolution of the above problem.

For this purpose, in the case of a three-dimensional temperature field, the radiative transfer between a volume emitter (ie the body under test) and a surface is considered; this surface can represent the receiving element of the radiometric system. The temperature distribution is then obtained by suitably expressing the problem in matrix form and by using a set of radiometric observations that can be identified either by space-scanning at a given frequency or vice versa, by frequency-domain-scanning in a given position.

PROBLEM FORMULATION

The problem geometry is illustrated in Figure 1. The body under examination occupies the volume V and its characteristics are both space-dependent and frequency-dependent. Moreover, we consider that the body is in local thermodynamic equilibrium. Thus, the power density U that is radiated by the volume V at the wavelength and is incident on the surface element dAn at point P(r) can be derived[1] as follows:

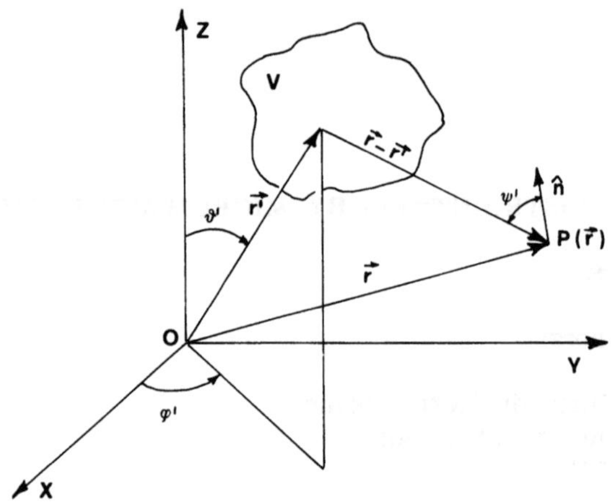

Fig. 1. Problem geometry; non-homogeneous body of volume V in local thermodynamic equilibrium. In P there is a receiving element with orientation n.

$$U_\lambda(\vec{r},\hat{n}) = dA\left[\int_V \alpha_\lambda(\vec{r}')\,\frac{E_\lambda(\vec{r}')}{|\vec{r}-\vec{r}'|^2}\exp\left[-\int_0^{|\vec{r}-\vec{r}'|}\alpha_\lambda\{\vec{r}''\}\,d|\vec{r}''|\right]\hat{n}\cdot\frac{\vec{r}-\vec{r}'}{|\vec{r}-\vec{r}'|}\,dV\right] \tag{1}$$

where:

$\alpha_\lambda(\vec{r}')$ absorption and scattering coefficient

$E_\lambda(\vec{r}')$ black-body emission expressed by the Rayleigh-Jeans function at microwave frequencies

$$E_\lambda = 2KT(\vec{r})/\lambda^2$$

K Boltzmann constant

$T(\vec{r}')$ temperature distribution inside the body.

If we replace dA at point P(r) with a receiving element characterized by the effective area A_e and orientation n, then the power delivered by this area can be written as follows:

$$P_\lambda(\vec{r},\hat{n}) = \frac{\lambda^2}{4\pi}\left\{\int_V \alpha_\lambda(\vec{r}')\,\frac{E_\lambda(\vec{r}')}{|\vec{r}-\vec{r}'|^2}\exp\left(-\int_0^{|\vec{r}-\vec{r}'|}\alpha_\lambda(\vec{r}'')\,d|\vec{r}''|\right)\cos(\psi')\,G_\lambda(\psi')\,dV\right\} \tag{2}$$

where:

$$\cos[\psi'] = \hat{n}\cdot\frac{\vec{r}-\vec{r}'}{|\vec{r}-\vec{r}'|}$$

ψ' angle between $\hat{n}$ and $(\vec{r}-\vec{r}')$

$G_\lambda(\psi')$ gain function of the receiving element in the direction of $(\vec{r}-\vec{r}')$.

The variation of the antenna noise temperature, $T_{A\lambda}$, in the case where the body under test is not present, can be abbreviated to:

$$T_{A\lambda}(\vec{r},\hat{n}) = A_{R\lambda}(\vec{r},\hat{n})P_\lambda(\vec{r},\hat{n}) \tag{3}$$

where $A_{R\lambda}(\vec{r},\hat{n})$ represents the characteristic response of the radiometric system.

By displaying the Rayleigh-Jeans function and by using (2), the above relationship can be written as follows:

$$T_{A\lambda}(\vec{r},\hat{n}) = A_{R\lambda}(\vec{r}',\hat{n})\int_V T(\vec{r}')F_\lambda(\vec{r}'/\vec{r},\hat{n})dV \tag{4}$$

In relationship (4), if $\alpha_\lambda(\vec{r}')$ is a known quantity, the radiometric measurement, $T_{A\lambda}$, turns out to be only dependent on the temperature distribution of the body. Therefore (4) represents the starting points for microwave radiometric detection of thermal distributions.

MICROWAVE THERMAL DISTRIBUTIONS

According to relationship (4), it is possible to derive two different apporoaches based, respectively, on the frequency dependence of the body characteristics and of the radiative transfer, and on the position dependence.

Now we can transform (4) into a matrix equation by using the method of moments.[2] We partition the body into N subvolumes and assume that the temperature $T(\vec{r})$ and the emission characteristics are constant in each subvolume. If we denote by ΔV_n the n-th subvolume located at point $\vec{r}_n'$, then

$$T_{A\lambda}(\vec{r},\hat{n}) = A_{R\lambda}(\vec{r},\hat{n})\sum_{n=1}^{N} T(\vec{r}_n')F_\lambda(\vec{r}_n'/\vec{r},\hat{n})\Delta V_n \tag{5}$$

In the transformation of (4) into (5), we have also performed another matrix reduction:

$$\int_0^{|\vec{r}-\vec{r}_n'|} \alpha_\lambda(\vec{r}\,'',\psi')\,d|\vec{r}\,''| \cong \sum_{l=1}^{L} \alpha_\lambda(\vec{r}_l\,',\psi_n')\ \frac{\Delta r_n}{L}$$

where $\Delta r = |\vec{r} - \vec{r}_n\,'|$

$\vec{r}_l\,'$ position of the 1-th segmentation of Δr_n in the direction of $(\vec{r} - \vec{r}_n\,')$,

ψ_n' angle between $(\vec{r} - \vec{r}_n\,')$ and n.

Therefore, if we perform radiometric measurements at M points $\vec{r}_m$, m = 1,2,...,M, chosen outside the body, we can obtain a linear system of non-homogeneous algebraic equations:

$$A(m,n)\ \tilde{t}(n) = \tilde{r}(m) \tag{6}$$

The discrete temperature distribution $\tilde{t}(n)$, n = 1,2,...N, can then be determined by solving (6).

From another point of view, it is possible to obtain a matrix equation in the same form as (6) by performing radiometric detections for a selected number of working-frequencies of the radiometer at a given point $\vec{r}$ and orientation $\hat{n}$. Thus:

$$T_A(\vec{r},\hat{n},\lambda_m) = A_R(\vec{r},\hat{n},\lambda_m) \sum_{n=1}^{N} T(\vec{r}\,'_n)\, F(\vec{r}_n\,',\lambda_m / \vec{r},\hat{n})\, \Delta V_n \tag{7}$$

If the second approach is taken, a problem that cannot be disregarded arises from the choice of the measurement frequency. In fact, many parameters, both of the body under test and of the radiometric system to be used, are frequency-dependent, and it is possible that they have to meet conflicting requirements.[3]

A first analysis can also be made for simplified geometries, like that in Figure 2. In this figure, the homogeneous object is a body of length D and temperature distribution T(x); in x = 0, there is a receiving antenna. Under these conditions, (2) becomes:

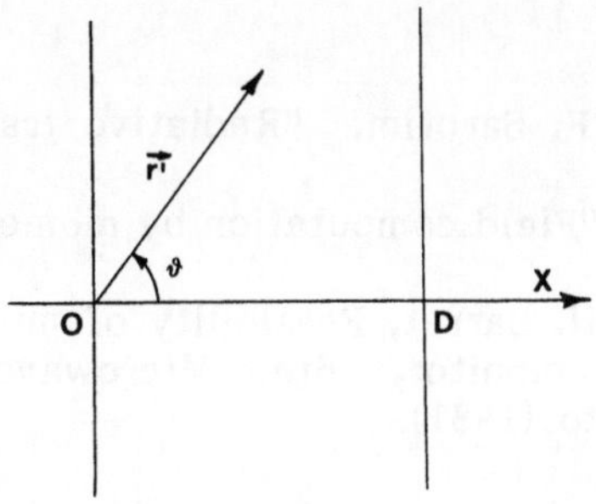

Fig. 2. Thermal radiation by homogeneous layer. In 0 there is a receiving element.

$$P_\lambda(0,\hat{n}) = K\alpha_\lambda A_R(0,\lambda)\left\{\int_{\pi/2}^{\pi/2}\int_0^D T(x)G_\lambda(\vartheta)\exp(-\alpha_\lambda x/\cos(\vartheta)).\cos(\vartheta)\sin(\vartheta)\,d\vartheta\,dx\right\} \tag{8}$$

and then

$$T_A(0,\lambda) = A_R(0,\lambda)\sum_{i=1}^{N}\left\{T_i(x_i)\exp(-\frac{\alpha_\lambda D}{N}(i-1)).(1-\exp(-\alpha_\lambda D/N))\right\} \tag{9}$$

where:

$A_R(0,\lambda)$ now contains also the beam-filling factor,
$T_i(x_i)$ temperature at each point x_i
$x_i = (2i-1)\ D/(2N),\ i = 1,2,\ldots N,\ 0 < x_i < D$

The contribution to T_A of the i-th layer at the temperature T_i and the distance (i-1)D/N is:

$$\Delta T_i = T_i \exp\left(-\alpha_\lambda (i-1)\,D/N\right)\left(1-\exp\left(-\alpha_\lambda D/N\right)\right) \tag{10}$$

Under given geometrical conditions, such a contribution first increases as frequency increases, and then it decreases rapidly, and the frequency corresponding to the maximum value of ΔT_i decreases as the distance increases.

REFERENCES

1. H. C. Hottel, A. F. Sarofim, "Radiative transfer," McGraw Hill, New York (1968).
2. R. F. Harrington, "Field computation by moment method," McMillan, New York (1968).
3. P. C. Meyers, A. H. Barret, Feasibility of microwave radiometer as a hyperthermia monitor, 6th Microwave Power Symposium Digest:171, Toronto (1981).

THE MEASUREMENT OF PERMITTIVITY AND TEMPERATURE OF TISSUES

R J Sheppard

Department of Physics
Queen Elizabeth College, Campden Hill Road
London, UNITED KINGDOM

INTRODUCTION

There is at present much interest in the use of microwave induced hyperthermia.[1] For most applications it is desirable to measure the temperature of the phantom tissue or animal being irradiated without the thermometer interacting with the microwave field. It is also believed that successful microwave hyperthermia requires that the electrical properties of the tissue, ie the permittivity ϵ' and conductivity σ be known. The purpose of this publication is to report ways in which such measurements may be made.

PERMITTIVITY MEASUREMENTS

Until recently it was not possible to measure quickly and accurately, particularly at high frequencies, the electrical properties of tissues, although some pioneering work does exist.[2] Within recent years data have started to become available.[3,4] However, most techniques make measurements in the frequency domain which involves measuring the permittivity at each frequency point of interest, although such techniques are now often automated. The method described here to measure complex permittivity ($\epsilon = \epsilon' - i\epsilon''$) is Time Domain Spectroscopy (TDS) where measurements are made in the time domain. In particular a pulse, considered as a voltage step as a function of time, is applied to the sample and the reflected pulse, after Fourier Transformation, enables the frequency spectrum to be calculated.

Full details of the experimental method have been published[5] and only brief details are given here. The experimental arrangement is shown in Figure 1. A fast rise time pulse (in the order of 35×10^{-12}s) passes

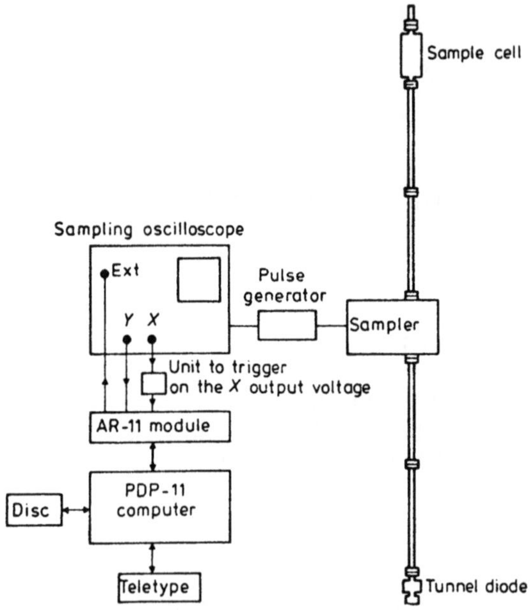

Fig. 1. Diagram of the Time Domain Spectrometer.

along precision 7 mm air lines and is reflected from the sample contained in the sample cell. The sample itself consists of a disc of tissue 7 mm in diameter and of thickness varying between 1 and 4 mm; it is contained between 2 PTFE beads in a precision 7 mm coaxial line. The reflected pulse is detected by the sampler, fed via a sampling oscilloscope to a PDP-11 computer. The computer averages about 100 waveforms, computes the Fourier Transform and then calculates the permittivity as a function of frequency. The actual measurement time is only 6 s and allowing for filling the cell, calibration and temperature stabilization, the tissue is at risk for less than 15 minutes. Data can now be obtained on tissues from about 10 MHz to 10 GHz.

At present the electrical properties of normal and tumor tissues of mice are being measured in a collaborative project with Professor N M Bleehen, Addenbrooke's Hospital, Cambridge. In Figure 2 the permittivity is shown for two mouse tumors, RIF/1 and KHT, and compared with normal mouse muscle. It can be seen that there are no significant differences between the two tumor types but that they are different from the muscle at the lower frequencies. This suggests that there are no gross differences between the free and bound water within the tissues but that structural differences are being seen between the normal and tumor tissues.

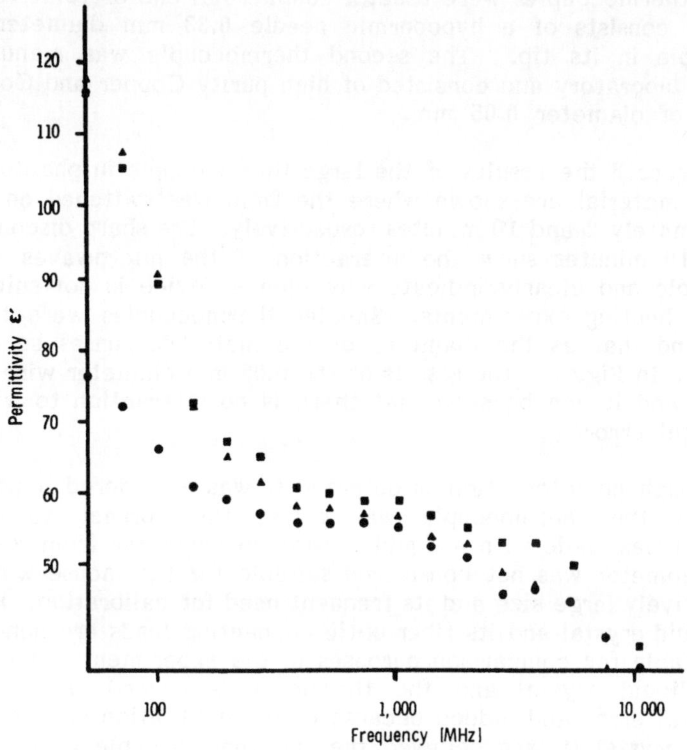

Fig. 2. A comparison of the permittivity of RIF/1 (▲), KHT (■) tumors with normal mouse muscle (●).

TEMPERATURE MEASUREMENTS

For measuring the temperature in a mouse tumor a small thermometric probe is required. Possibly the most convenient thermometer is a thermocouple but being metallic it is not usually considered suitable for use in a microwave field. However, Allis et al[6] suggest that a thermocouple could be employed if shielded leads perpendicular to the field are used but this is not considered suitable in a tumor where in general the polarization of the field is not known. We have performed experiments on the effect of microwaves on a thermocouple and some results are presented below.

The experiments were performed in 2.454 GHz waveguide. The power was applied at one end and an absorbing load placed at the other. Phantom tissue material was contained in a perspex box which had holes drilled in it so that the thermocouples could be placed in various directions of polarization. All results quoted below are for the worst case, ie where the interaction is greatest.

Two thermocouples were used; a commercial one of relatively large size which consists of a hypodermic needle 0.33 mm diameter with a thermocouple in its tip. The second thermocouple was manufactured within the laboratory and consisted of high purity Copper and Constantan wire each of diameter 0.05 mm.

In Figure 3 the results of the large thermocouple in phantom tissue equivalent material are shown where the field was switched on and off at approximately 2 and 10 minutes respectively. The sharp discontinuities at 2 and 10 minutes show the interaction of the microwaves with the thermocouple and clearly indicate why such a device is not suitable for microwave heating experiments. Smaller thermocouples were tried and it was found that as the diameter of the metal decreases so does the interaction. In Figure 4 the results of the 0.05 mm diameter wire junction are shown and it can be seen that there is no interaction to within the experimental error.

Although no interaction is detected it was considered important to check that the thermocouple was giving the correct temperature. Comparison was made with a liquid crystal thermometer from Ramel Inc. This thermometer was not considered suitable for the mouse work owing to its relatively large size and its frequent need for calibration. However, since a liquid crystal and its fiber optic connecting leads are non-metallic it was suitable for comparison purposes in the experiment. It was found that the liquid crystal and the thermocouple agreed well to within experimental error and indeed because of the calibration uncertainties of the liquid crystal it was believed the the thermocouple gave the most reliable temperature measurement.

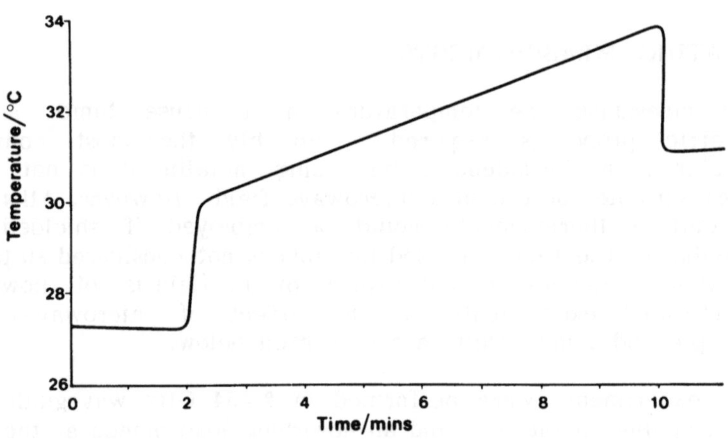

Fig. 3. The large thermocouple in phantom tissue material.

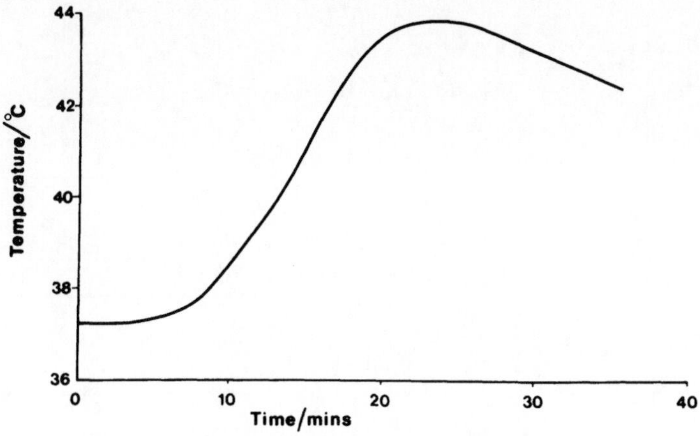

Fig. 4. The small thermocouple in phantom tissue material.

ACKNOWLEDGEMENTS

I wish to thank the Cancer Research Campaign for financial support and Miss J Rogers for making the TDS measurements. Thanks are due to Professor N M Bleehen for supplying the tissues and to Professor E H Grant for helpful discussions.

REFERENCES

1. G. M. Samaras, A. Y. Cheung, Microwave hyperthermia for cancer therapy, CRC Critical Reviews in Bioengineering 123 (1981).
2. H. P. Schwan, Adv. Biol. Med. Phys. 5:147 (1957).
3. K. R. Foster, J. L. Schepps, H. P. Schwan, Microwave dielectric relaxation in muscle, a second look, Biophys. J. 29:271 (1980).
4. M. A. Stuckley, T. Whitt-Athey, S. S. Stuckley, M. Samaras, G. Taylor, Dielectric properties of animal tissues in vivo at frequencies of 10 MHz to 10 GHz, Bioelectromagnetics 2:93 (1981).
5. A. W. J. Dawkins, R. J. Sheppard, E. H. Grant, An on-line computer based system for performing time domain spectroscopy I. Main features of the system, J. Phys. E. 12:1091 (1979).
6. J. W. Allis,. C. F. Blackman, M. L. Fromme, S. G. Benane, Measurement of microwave radiation absorbed by biological systems. J Analysis of heating and cooling data, Radio Sci. 12:1 (1977).

THERMOGRAPHIC TECHNIQUES WITH LIQUID CRYSTALS IN MEDICINE

U Flesch

Rontgendiagnostisches Zentralinstitut of the Rudolf-Virchow-Krankenhaus, Berlin WEST GERMANY

INTRODUCTION

The term liquid crystals consists of two parts which at first sight might appear to be contradictory; crystals consist of fixed elements, atoms and molecules which are fixed in a space lattice whereas elements of fluids are not bound in the same way. However, the spatial distribution of elements is of similar dimensions for both liquids and crystals. On heating, crystals change from the crystalline solid phase into the liquid phase and on cooling from the liquid phase into the crystalline solid phase.

At a definite temperature some organic compounds show an additional phase called the mesophase which occurs between the solid and the liquid phases. In this mesophase some groups of elements within the compounds are ordered in relation to their location and direction, but the groups are displaceable amongst one another. This mesophase is called liquid-crystalline state and the organic compounds which show this property are called liquid crystals.

The honor of discovery and for the early pioneering work goes principally to Lehmann[1,2,3] and Reinitzer[4] in 1877 and 1888: Lehmann introduced the term liquid crystals in 1889.[1]

Liquid crystals have many properties which lie between the liquid and solid states. They show the mobility of liquids and the optical behavior of crystals. The mobility of liquids is limited to one or more degrees of freedom of translation motion and rotary motion. In the following statement only those properties are presented which are important to understand the use of liquid crystal thermography.

Cholesteric liquid crystals are important in medicine. For the most part they are formed from derivatives of cholesterol and have a similar molecular structure. Cholesterol itself does not have a liquid crystalline phase[5] but cholesteryl benzoate which was first discovered by Reinitzer[4] displays a liquid crystalline phase.

THE PHYSICS OF LIQUID CRYSTALS

Figures 1 and 2 illustrate the molecular packing of the cholesteric structure.[13,15] The cholesteric structure of derivatives of cholesteryl shows packing of layers. The direction of the long axis of the molecule is fixed in each layer and there is no order in the molecular position. The molecular layers are very thin. Because of the peculiar shape of the cholesterol molecules, the direction of the long axes of the molecules in each layer is displaced slightly from the corresponding direction in the adjacent layers; the overall displacement traces out a helical path. The molecules themselves are not precisely arranged in layers because the molecules have a space lattice. The molecular thickness in the layer is 0.3 nm and the thickness of one helical path is 550 nm.[6,7]

The most important properties of these liquid crystals when used in thermography are their peculiar optical properties. For instance cholesteryl nonanoate is white in the solid crystalline phase. On heating the substance appears in the mesomorphic phase clear colored and on further heating the substance becomes in the isotrope liquid phase achromatic and limpid. The optical activity is much larger than in any other known substances. The very unusual optical properties are best observed when the material is in a thin sheet, for instance a thin sheet on the skin. Figure 3 shows the principle of measurement of the temperature of the skin with liquid crystals. The sheet will have the properties of a uniaxial negative crystal with the optical axis and the axis of the helix normal to its surface. At a definite temperature - in the mesomorphic phase - a part of the incident beam of white light with a characteristic wavelength is reflected. The other part of the incident beam of white light is transmitted. If a black absorption layer is located between the sheet of liquid crystals and the skin, it will not be possible to reflect the transmitted light upon the white skin and only reflected light from the sheet of liquid crystals will be observed.

It is well known that thin films of oil will scatter light when its thickness is an exact multiple of half wavelengths. The thickness of the cholesteric sheet does not affect the predominant wavelength of the reflected light. The predominant wavelength λ_o of the reflected light of liquid crystals is a function of the index of refraction n and the pitch p of the helix as follows:

$$\lambda_o = 2\ n\ p$$

The band with $\Delta\lambda$, at half-amplitude point is of the order of dimension:

$$\Delta\lambda_o \simeq \frac{\lambda_o}{14}$$

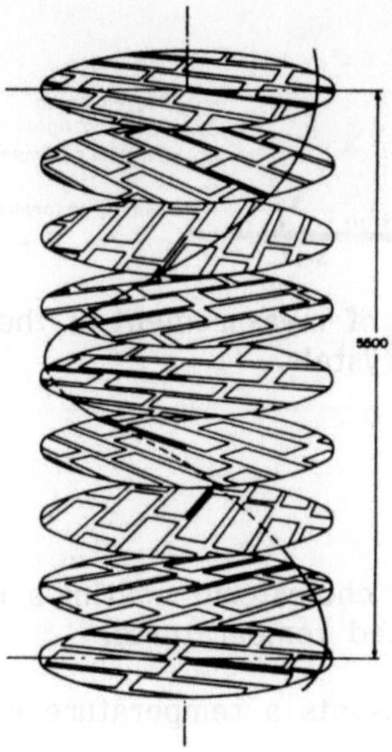

Fig. 1.

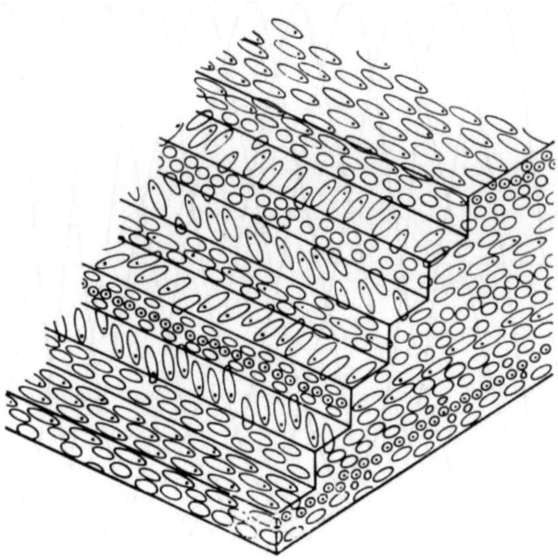

Fig. 2. Both of these figures illustrate the molecular packing of the cholesteric structure.[13,15]

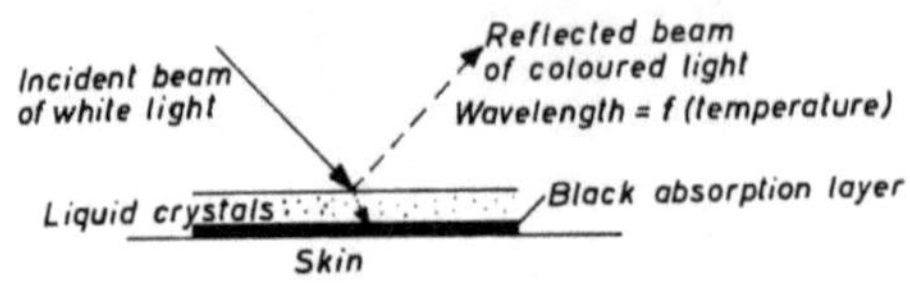

Fig. 3. The principle of measurement of the temperature of the skin with liquid crystals.

Figure 4 shows the cholesteric coating's response to narrow band illumination for 11 selected temperatures.

Each color uniquely represents a temperature and this relationship

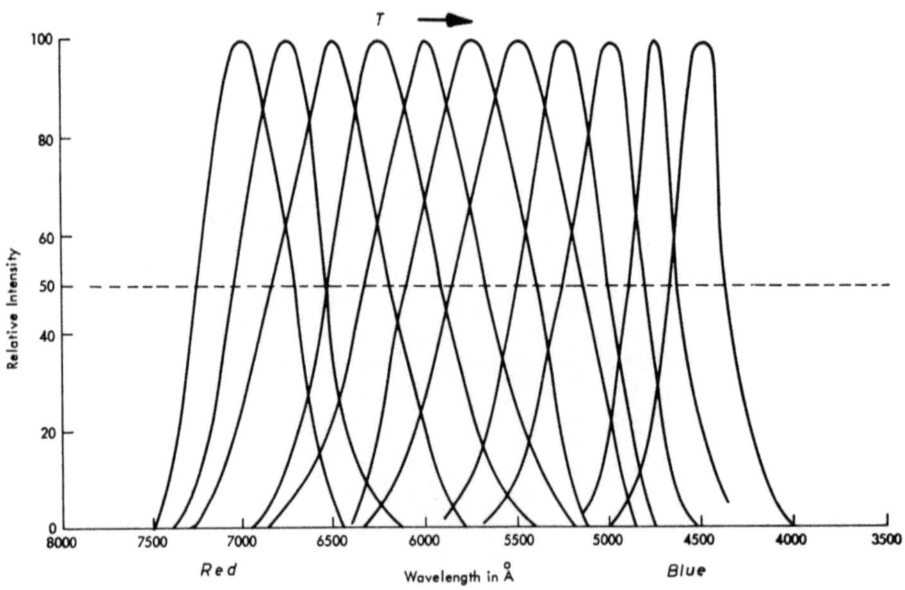

Fig. 4. Reflected intensity of light of a typical medical liquid crystal as a function of the wavelength for 11 selected temperatures.

can be shown by λ-vs-T curve with λ the maximum of the reflected light (Fig 5).

On heating, liquid crystals used in medicine show first the red, then yellow, green, blue and violet colors; but all cholesteric liquid crystals do not respond in the same way to fluctuations in temperature. For example, some substances are initially blue and change to green and then to red when they are heated, others change only from green to red or from green to red and back to red.[5]

The intensity of light reflected is nearly constant for white light. People who are color-blind see almost no differences of intensity in the liquid crystal. Thus black and white photography is in some ways more difficult than color photography. To photograph using black and white film, either a monochromatic wavelength source or a narrow band filter must be used with the camera.

When ordinary white unpolarized light is directed at a cholesteric material, the light is separated into two components, one with the electric vector rotating clockwise and the other with the electric vector rotating counter-clockwise. The reflected light at the maximum of the wavelength rotates in the same direction as the helix of the crystal and the transmitted light in the contrary direction (Fig 6).

Ordinary white light consists of equal parts of left and right polarized light. Therefore 50% of the light is reflected and 50% is transmitted at the maximum of the wavelength. This phenomenon is called circular dichroism. The effect near the maximum of the reflected wavelength

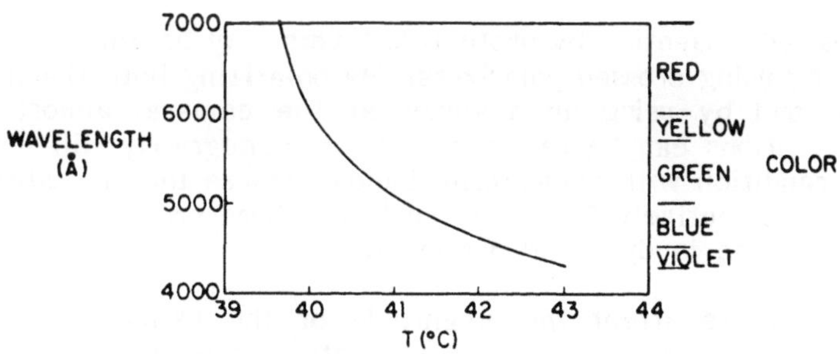

COLOR CALIBRATION FOR VISUAL WORK

COLORLESS-RED	39.5	GREEN-CENTER	40.6
RED-CENTER	39.7	GREEN-BLUE	41.2
RED-YELLOW	40.0	BLUE-CENTER	42.0
YELLOW-CENTER	40.1	BLUE-COLORLESS	43.0
YELLOW-GREEN	40.2		

Fig. 5. Maximum of the wavelength of the reflected light dependence of temperature - the calibration curve.

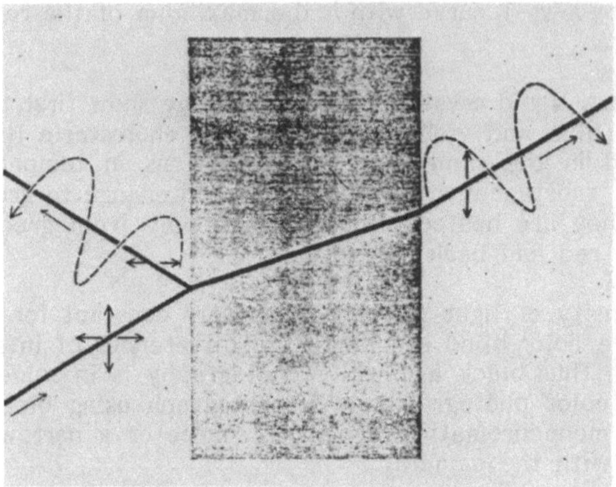

Fig. 6. Circular dichroism. Unpolarized light is directed at a cholesteric substance. The light is broken into two components, one with the electric vector rotating clockwise and the other with the electric vector rotating counter-clockwise; one of these components is transmitted and the other is reflected.

is similar except that the light is elliptical polarized. A right-handed action of the helix of the cholesteric liquid crystal reflects right-handed polarized light at the maximum of the wavelength and transmits left-handed polarized light totally and does not influence the direction of rotation. The reflection of circularly polarized light on a mirror is known to change the direction of rotation.

Undesired reflection by protective layers may be almost completely eliminated by using crossed polarizers. By polarizing both the light from the source and by using an analyzer at the camera, almost all first surface reflections can be removed. Color photography will not give a true color rendition with cholesteric liquids because the cholesteric liquid scatters light selectively from one band and there are no additive colors such as are found in dyes and pigments.

Many factors affect the magnitude of the reflected wavelength. Figure 7 shows alterations of the wavelength of the reflection maximum, λ max, as a function of the angle of incidence and the angle of observation to the normal of the surface. This effect is considerable; the spectrum in visible light may be changed. Perpendicular incidence and perpendicular observation generally yield the greatest wavelength of the reflection maximum. Also the intensity of the reflection spectra changes for different angles of incidence (Fig 8).

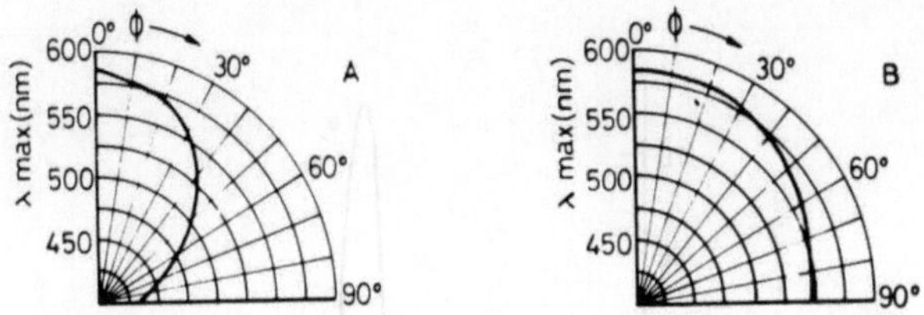

Fig. 7. The wavelength of the reflection maximum, λ max; LEFT: if the angle of incidence = angle of observation, RIGHT: if the angle of observation varies and the film is illuminated normal to the surface.

Cholesterics undergo a progressive untwisting of the helix under pressure because the intramolecular bonding is low (Figs 9 and 10).

The optical effect of the pressure dependence is lower than the effect of temperature.

The spectral half-width of the reflected light is greater in the red than in the blue region (Fig 4). The large overlap of the curves, especially in the red region, shows that if they were simultaneously used, they would not be resolved.

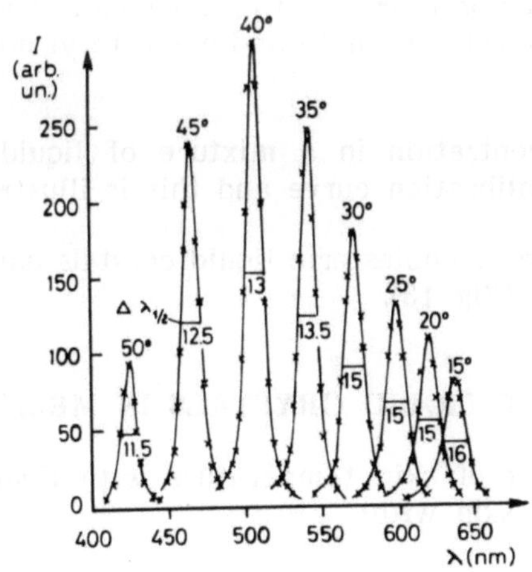

Fig. 8. The intensity I of the reflection spectra versus wavelength λ of cholesterol ethoxyethoxypropionate for different angles of incidence at 27°C (sample thickness 100 μm).

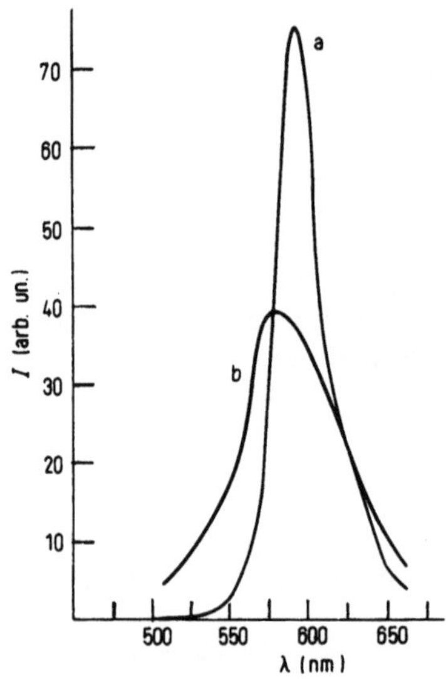

Fig. 9. Intensity of selectively reflected light, I, versus wavelength λ (a) at rest, and (b) under shear (v = 1 cm/min) for a 25 μm cholesteric film.

Mixtures of cholesteric liquid crystals show mesomorphic phases for nearly all ranges of temperature used in medicine. The calibration curve of mixtures of cholesteryl carbonate and cholesteryl nonanoate are shown in Figure 11.

The solute concentration in a mixture of liquid crystals shows a great effect in the calibration curve and this is illustrated in Figure 12.

The color change of cholesteric liquid crystals can also be disturbed by thermal hysteresis (Fig 13).

THE APPLICATION OF LIQUID CRYSTALS IN MEDICINE

The measurement of skin temperature with liquid crystals can be made by painting the skin with:

1. An absorption layer of black paint.
2. A thin layer of liquid crystals.

There are, however, some difficulties:

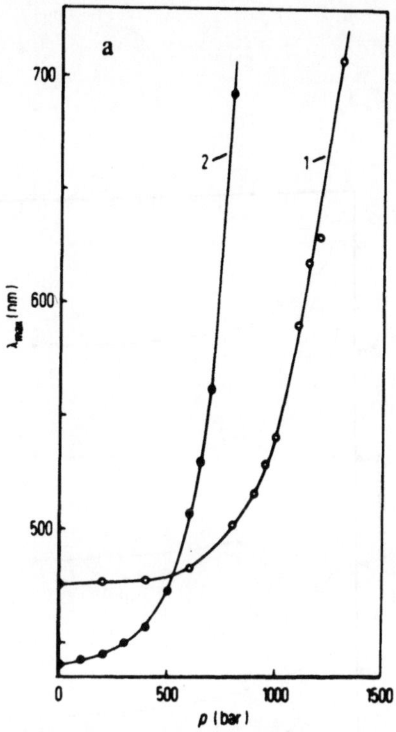

Fig. 10. Pressure dependence of the wavelength of maximum reflection λ max of two cholesteric mixtures of cholesteryl oleyl carbonate.

1. The mixture of liquid crystals on the skin crystallizes only on some areas and the crystals tend to concentrate over these areas. In these circumstances the calibration curve changes.
2. Liquid crystals are oily substances and they combine with the fat of the surface of the skin and this changes the calibration curve.
3. The wet mesomorphic layer can be soiled. For instance, dust is an undesirable nucleus for crystals.

Some of these difficulties were circumvented by Crissey et al.[14] These authors stretched Mylar-film of 0.1 mm thickness evenly in a hoop, coated the film on the skin side with black enamel, and on the viewing side with a suitable liquid crystal mixture. The resolution, however, was much inferior to that obtained with material applied directly to the skin.

The greater part of difficulties were solved by Churchill[15,16] who dispersed liquid crystals as droplets of 5-20 µm diameter in translucent matrix-like polymeric binder material. In principle there are three

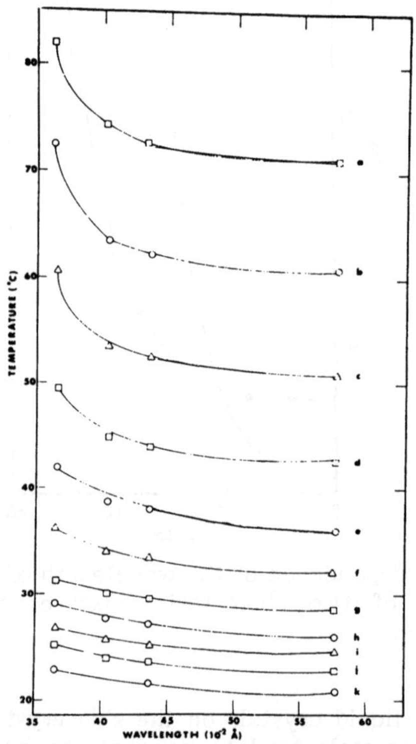

Fig. 11. The wavelength of maximum scattering plotted as a function temperature for various mixtures of cholesteryl nonanoate and cholesterol oleyl carbonate. Curve (a) represents 100% cholesteryl nonanoate. Curve (k) represents 100% cholesteryl oleyl carbonate. Curves (b) through (j) represent changing the content of cholesteryl oleyl carbonate in cholesteryl nonanoate by 10% steps.

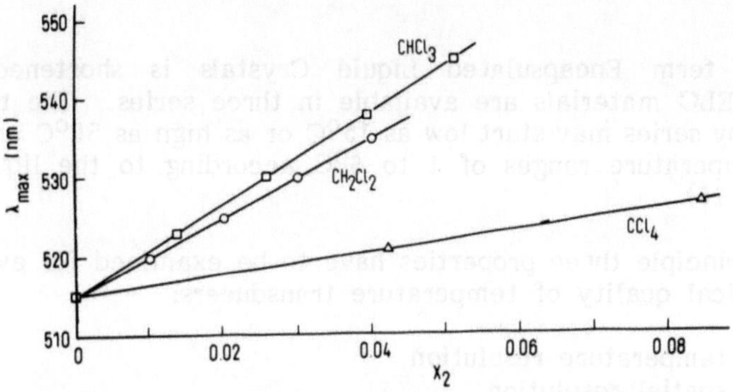

Fig. 12. Wavelength of maximum reflection λ max, vs. solute concentration X_2 in cholesteric mixture at 18°C.

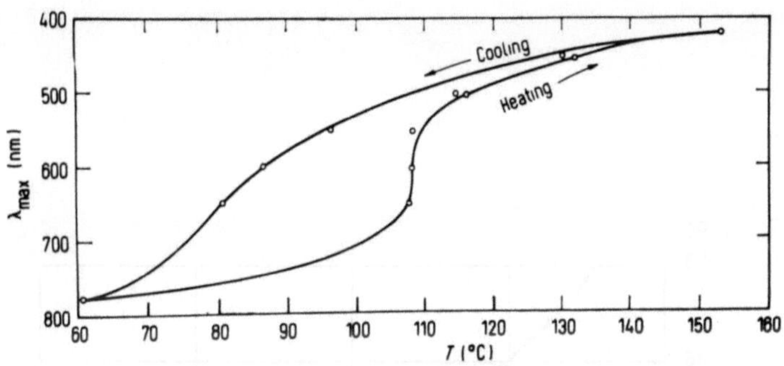

Fig. 13. Wavelength of maximum reflection λ max vs temperature T showing the hyteresis on heating and cooling of a 45:55 mixture of cholesteryl cinnamate and benzoate.

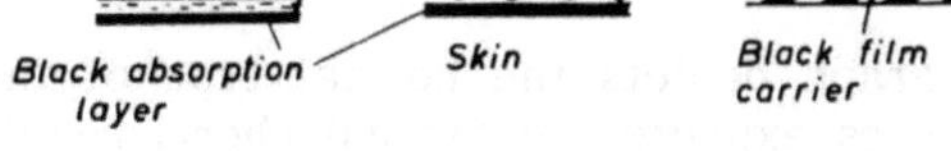

Fig. 14. Construction of films with encapsulated liquid crystals on principle.

possibilities of constructing films with such encapsulated liquid crystals (Fig 14).

The term Encapsulated Liquid Crystals is shortened to ELC. Standard ELC materials are available in three series. The temperature level in any series may start low as 15°C or as high as 50°C and typically cover temperature ranges of 1 to 6°C according to the liquid crystals used (Fig 15).

In principle three properties have to be examined for evaluation of the technical quality of temperature transducers:

1. The temperature resolution
2. The spatial resolution
3. The resolution time.

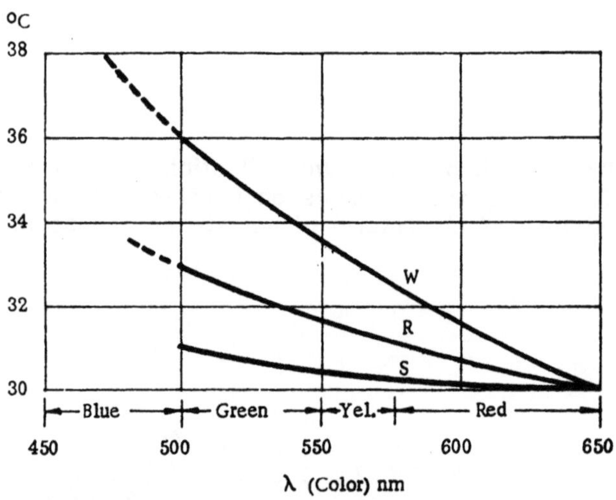

Fig. 15. This graph depicts the normal type of thermal-visible color responses expected of typical short, regular and wide range formulations.

Attention has also to be paid to:

1. The thermal mass of the film which might change the temperature of the surface of the skin,
2. Alterations to the heat exchange between the skin and the surroundings which might affect evaporation
3. The pressure of the film on the skin changes the circulation and the heat exchange in the skin
4. The pressure and shear changes which affect the calibration curve of the liquid crystals
5. The film carrier and the absorption layer which degrade the spatial and time resolution
6. The heat resistance of fat, oil and air between skin and the film with ELC and
7. The heat radiation of the surroundings which changes the temperature and the color of the ELC.

The temperature resolution is dependent on:

1. The range of temperature corresponding to the color spectrum (mixture and purity of the crystals)
2. The spectral half-width of the reflected light and
3. The separation of the overlap of the curves of the reflected light.

In literature for high-sensitivity compounds of 0.1°C total range, resolution up to 0.007°C is implied[17]; but the droplets in ELC do not have exactly the same calibration curve.

The spatial resolutions quoted in the literature of 40 lines/mm[7] are based on a very optimistic view. A line spacing of 0.0025 mm would be of the same order of magnitude as the dimension of the diameter of the encapsulated droplets of liquid crystals.

The rate of thermal change must be sufficiently slow to allow the cholesteric liquid crystal to respond adequately. Thus the rate of change of color with temperature has a time constant which is in the neighborhood of 0.1 sec with a response time of 1/e of the maximum.[7] In most practical cases in medicine this does not represent a limitation.

The theoretical estimation of the different resolutions for measuring the temperature of the skin with liquid crystals is very difficult. It is easier to test this on a phantom.[18,19] Figure 16 shows such a phantom. The subcutis is simulated with a substance of the same heat conductivity (wet sea-sand or wet rock-salt) and the cutis with a fresh excized human skin. A tube circulated with water of suitable temperature is located under the skin to induce a temperature field on the surface of the skin. This apparatus is immersed in a tempered water bath to simulate the body temperature.

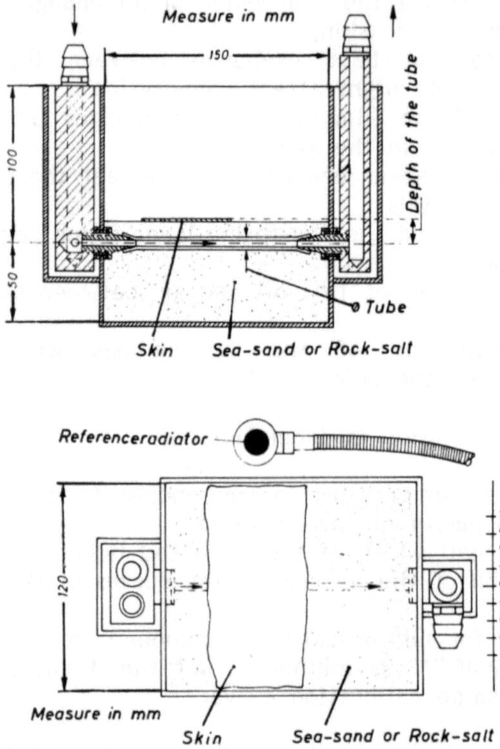

Fig. 16. Phantom of the skin for testing the films with liquid crystals.

The special feature of this phantom is that the evaporation of the skin is simulated too. It is possible to show that the interruption of the evaporation with the film of liquid crystals changes the spectrum of colors of the skin within one minute.

The spatial and temperature resolutions may be examined on a phantom with a grating of heating wires under the surface.[19] The shortest distance of lines which can be separated with the film of liquid crystals lies at 3 mm and the smallest difference of temperature at 0.2°C.

In most practical cases in medicine the temperature of the skin does not always lie exactly within the best range of temperature for the liquid crystal films in stock. It is possible to cool the skin with a blower before pressing the film on the skin or after the film has been positioned on the skin.

Film with many-layered encapsulated crystals has many advantages. It is possible to overlap the temperature-ranges so that mixed colors may be reflected. Such films with many-layered liquid crystals are available.

SUMMARY

Liquid crystals produce good pictures of the temperature field over the skin and are useful for many applications in medicine. However, the errors in measurement are considerable. Many difficulties are solved with the technique of encapsulated liquid crystals. However, some of the difficulties remain unsolved: it is difficult, for example, to deduce from a color the absolute temperature but differences in temperature between neighboring areas of the skin are more readily measured.

REFERENCES

1. O. Lehmann, Physikalische Isomerie, Diss Strassburg (1877).
2. O. Lehmann, Molekularphysik mit besonderer berucksichtigung mikroskopischer untersuchungen und anleitung zu solchen, in: 2 Band W. Engelmann, Leipzig (1889).
3. O. Lehmann, Fliessende kristalle, Z. Physik. Chemie 4:462 (1889).
4. F. Reinitzer, Zur kenntnis des cholesterins, Monatshefte Chemie 9:421 (1888).
5. J. L. Fergason, Liquid crystals, Scientific American 211:76 (1964).
6. J. L. Fergason, Cholesteric structure. I. Optical properties, Mol. Cryst. Liq. Cryst. 1:293 (1966).
7. J. L. Fergason, Liquid crystals in non-destructive testing, Appl. Optics 7:1729 (1968).
8. G. V. Lukianoff, Information content and reoslution aspects in thermal mappings using cholesteric compounds, Mol. Cryst. Liq. Cryst. 8:389 (1969).
9. B. Bottcher, Optische eigenschaften cholesterinischer flussigkeiten, Diss Techn Univ. Berlin (1972).
10. D. F. Ciliberti, G. D. Dixon, L. C. Scala, Shear effects on cholesterics, Mol. Cryst. Liq. Cryst. 20:27 (1973).
11. P. Pollman, H. Stegemeyer, Pressure dependence of the helical structure of cholesterics, Chem. Phys. Lett. 20:87 (1973).
12. D. G. Willey, D. E. Martire, Effect of organic solutes on the selective reflection of visible light by cholesteric mixtures, Mol. Cryst. Liq. Cryst. 18:55 (1972).
13. H. Kelker, R. Hatz, Handbook of liquid crystals, Verlag Chemie Weinheim-Basel (1980).
14. J. T. Crissey, J. L. Fergason, J. M. Bettenhausen, Cutaneous thermography with liquid crystals, J. Invest. Dermatol. 45:3290 (1965).
15. D. Churchill, J. V. Cartmell, R. E. Miller, P. D. Bouffard, National Cash Register Comp. Dayton, Ohio, V St A: A visual display device, in: "Liquid crystals and their applications," Th. Kallard, Optosonic press, New York (1970).

16. National Cash Register Co., Dayton, Ohio (V St A): Temperaturemppfindliche anzeigevorrichtung. DT-PS 1 698 092, 24.2 (1968).
17. Liquid Crystals Ind Inc, Turtle Creek, Pa (V St A): Thermometrische gegenstande oder thermometer und verfahren zu deren herstellung. DT-OS 2 012 493, 17.3.1970, Unionsprioritat 18.3. (1969).
18. U. Flesch, Oberflachenthermometrie der haut mit thermoelement, thermistor, thermovision und temperaturmessfarben. Diss Freie Universitat, Berlin (1974).
19. H. Witt, H. Burger, Mammadiagnostik im rontgenbild. Walter de Gruyter, 2 Aufl, Berlin (1981).

FURTHER READING

1. J. E. Adams, E. L. Haas, J. Wysocki, Optical properties of cholesteric films, J. Chem. Phys. 50:2458 (1969).
2. G. H. Brown, J. W. Doane, V. D. Neff, "Structure and physical properties of liquid crystals," Butterworth, London (1971).
3. J. T. Crissey, E. Gordy, J. L. Fergason, R. B. Lyman, A new technique for the demonstration of skin temperature patterns, J. Invest. Dermatol. 43:89 (1964).
4. R. Dreher, Reflection properties of distored cholesteric liquid crystals, Sol. State Comm. 12:519 (1973).
5. R. D. Ennulat, Selective light reflection by plane textures, Mol. Cryst. Liq. Crist. 13:337 (1971).
6. J. L. Fergason, Electric field device, in: "Liquid crystals and their applications," Th. Kallard, Optosonic press, New York (1970).
7. J. L. Fergason, N. N. Goldberg, R. J. Nadalin, Cholesteric structure. II. Chemical significance, Mol. Cryst. Liq. Cryst. 1:309 (1966).
8. J. H. Flack, P. P. Crooken, Angular dependence of selective reflexion from the blue phase, Mol. Cryst. Liq. Cryst. 69:281 (1981).
9. P. G. de Gennes, "The physics of liquid crystals," Clarendon, Oxford (1974).
10. G. W. Gray, "Molecular structure and the properties of liquid crystals," Academic Press, New York (1962).
11. Hoffmann-La Roche & Co. AG, Basel-Schweiz: Thermochrome Bander. DT-OS 2 014 909, 1.4. (1969).
12. F. Jennen, A. J. Tricoire, Thermografische platte zur messung von temperaturverteilungen, DT-PS 2536:773 (1975).
13. Th. Kallard, "Liquid crystals and their applications," Optosonic Press, New York (1970).
14. Th. Kallard, "Liquid crystal devices," Optosonic Press, New York (1973).
15. P. Kassubek, G. Meier, Optical studies on grandjean planes in cholesteric liquid crystals, Mol. Cryst. Liq. Cryst. 8:305 (1969).
16. G. Meier, E. Sackmann, J. G. Grabmaier, "Applications of liquid crystals," Springer, Berlin (1975).
17. W. Mahler, M. Panar, Cholesteric solids, J. Am. Chem. Soc. 94:7195 (1972).

18. I. Olbricht, Der aussagewert der plattenthermographie im vergleich mit der klinischen und mammografischen untersuchung. Diss Universitats-Frauenklinik Marburg/Lahn (1975).
19. N. Oron, K. Ko, L. J. Yu, M. M. Labes, Chirality in mixed nematics and cholesterics, Liq. Cryst. Ord. Fluids 2:403 (1974).
20. E. B. Priestley, P. J. Wojtowicz, P. Sheng, eds., "Introduction to liquid crystals," Plenum Press, New York (1975).
21. J. Tricoire, La thermographie en plaque. Technique nouvelle d'utilisation des cristaux liquides, Presse Medicale 78:2481 (1970).
22. J. Tricoire, Plaque pour thermographie, Brevet D'Invention Fr. 2.110.505, 20.10.1970 (1970).
23. J. Tricoire, Thermographische platte und verfahren zu deren herstellung, DT-PS 2 152 277, 20.10. (1971).
24. Xerox Corp., Rochester, N Y (V St A) (1971) Flussigkristalline zusammensetzungen und ihre verwendung zur erzeugung von bildern. DT-OS 2 235 385, 13.9.71. (1971).

NON STEROL LIQUID CRYSTAL THERMOGRAPHIC SCREEN

K G Archer

Liquid Crystal Technology
London
UNITED KINGDOM

CHIRAL NEMATIC LIQUID CRYSTALS

The new chiral nematic thermochromic liquid crystals are better in many ways than the hitherto available cholesteryl based liquid crystals. The fundamental advantage the chiral nematics possess is that they are chemically pure and stable. Different batches should therefore be the same, and once a formulation has been established to reflect light at a given temperature, that formulation can be repeated with an accuracy of $\pm$ 0.1°C. This is a considerable improvement over the rather 'hit and miss' formulations of cholesteric liquid crystals.

As reported in another paper, the photochemical stability is also improved substantially; nevertheless, it should be appreciated that devices made with these liquid crystals should not be kept in direct sunlight as they will absorb heat, and the higher temperature will accelerate the photochemical reaction.

A third advantage that the chiral nematics have, is that they do not crystallize readily. Cholesteric liquid crystal formulations usually consist of a mix of three or more cholesteryl esters, at least one of which has a high melting point. If a liquid crystal thermometer or thermographic screen is stored at temperatures below 10°C selective crystallization can occur, leading to erroneous readings. This problem has not been observed with chiral nematics. Given these three advantages it was obviously necessary to investigate the use of chiral nematic liquid crystals to make thermographic screens and thermometers.

PROTECTION SYSTEMS AND COLOR PURITY

Gelatine micro-encapsulation has been the usual method of protecting and applying cholesteric liquid crystals; as far as I know chiral nematics have not been successfully micro-encapsulated, although the method is likely to be perfected soon.

Instead of encapsulation, the liquid crystal can be protected simply by dispersing it in a suitable resin. I have developed an ink consisting of chiral nematic liquid crystal droplets dispersed in a binder polymer. The droplets of liquid crystal are smaller than is usual with micro-encapsulated systems.

It has been found that a film of about 10 μm thickness gives very good color. It is not known exactly why such a thin film shows good color; it may be that the small droplets are packing very efficiently to form an apparently contiguous film. Also the droplets are squashed as the film dries so that they become ellipsoidal; in this shape of droplet it is thought that the liquid crystal is ordered more uniformly and does not have the same disclination lines as a spherical capsule.

The color purity and brightness is to some extent dependent on the shape of the liquid crystal droplet within the dried film.

A film of unprotected pure liquid crystal has a reflectance as shown in Figure 1(a). It reflects back light of a very narrow waveband; this band shifts with variations in temperature. If the angle of view is not normal to the surface of the liquid crystal film it will reflect a shorter wavelength so the color shifts towards the blue end of the spectrum.

A gelatine micro-capsule of liquid crystal will be almost spherical; much of the sphere will be viewed at an angle to the surface and thus the light will be shifted towards the blue. The reflectance band will be wider than the pure liquid crystals in Figure 1(b). The peak intensity of the reflected light will not be so strong and the color not so pure. However, there is an advantage: the film can be viewed from any angle and the color is the same.

In a polymer dispersion there is a compromise between these two states. The droplets are ellipsoidal; as the polymer dries the droplets of liquid crystal are squashed to form a 'flying-saucer' shape. The reflectance is shown in Figure 1(c). Color purity is better and peak intensity is higher than the spherical micro-encapsulated droplet. However, there is a price to pay for this - the color does change to some extent according to the angle of view and this effect has to be controlled when thermographic screens are made.

THERMOGRAPHIC SCREEN CONSTRUCTION

Many thermographic screens in use today for breast thermography

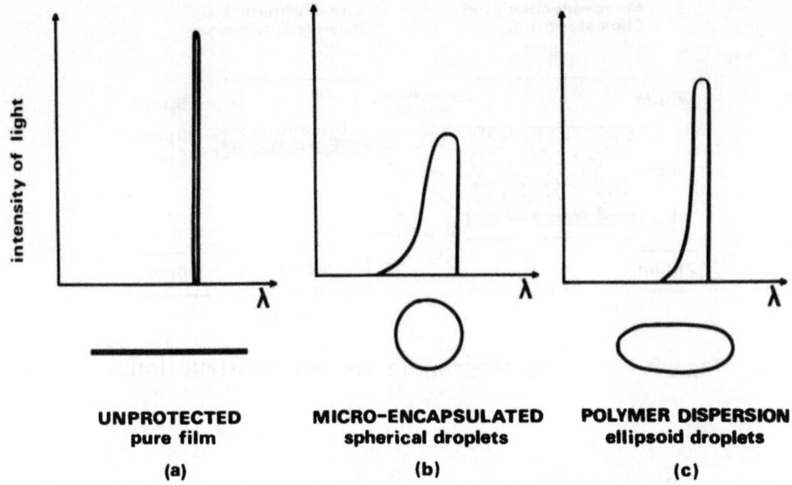

Fig. 1. Reflectance of liquid crystal films.

are 'double spectrum' screens, that is they contain two layers of liquid crystal to give a total of eight different colors. It is possible to use two layers of this polymer dispersion in a similar way.

The construction of a typical screen using micro-encapsulated cholesteryl esters is shown in cross section in Figure 2(a). Behind the plastic carrier film is the liquid crystal, and then a layer of black ink. A protection sheet is usually behind the black ink but this is not physically connected to the liquid crystal sheet. It will be seen that the liquid crystal/gelatine film is about 120 μm thick; the total thickness (excluding the protection sheet) is about 210 μm.

The construction of a screen using the new chiral nematics in a polymer dispersion is shown in cross section in Figure 2(b). The liquid crystal/polymer film is in this case only about 20 μm thick, giving a total thickness of about 110 μm.

This reduction in thickness gives two further improvements. Probably the more important advantage of these new screens is the speed of response. The time taken for the screen to stabilize at skin temperature is considerably reduced. Early tests indicate that these new screens take about half the time of the older screens.

The second advantage due to the reduced thickness is an improvement in the resolution; this has not yet been quantified but it is

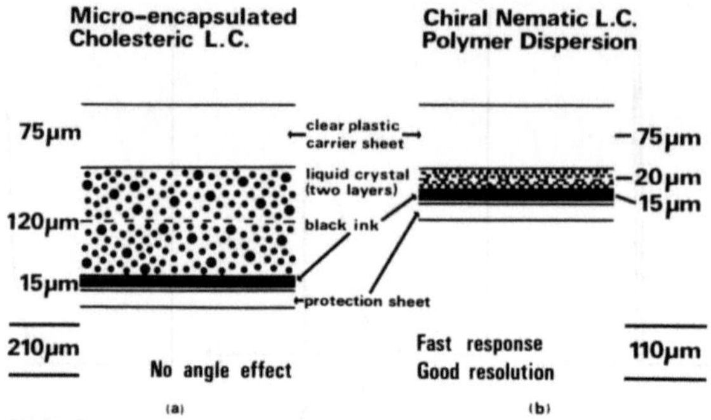

Fig. 2. Thermographic screen construction.

possible to see finer detail than was possible with the other screens.

SUMMARY

Thermographic screens made with chiral nematic liquid crystals in a polymer dispersion have the following advantages over the cholesteric liquid crystal (micro-encapsulated in gelatine) type of screen:

1. An improved accuracy to $\pm$ 0.1°C.
2. Slower photochemical degradation.
3. No selective crystallization.
4. Very fast response time.
5. Improved resolution.

NEW THERMOCHROMIC LIQUID CRYSTALS

D G McDonnell, I Sage

Royal Signals and Radar Establishment, Malvern, Worcs and BDH Chemicals Ltd., Poole, Dorset
UNITED KINGDOM

INTRODUCTION

Liquid crystals are organic compounds which possess ordered fluid phases combining solid-like optical properties with liquid-like flow. The materials exploited for thermometric and thermographic use display a cholesteric (or chiral nematic) phase which has a helical disposition of molecules imposed by an asymmetry in their structure. The less mobile smectic phase, by contrast, is ordered into layers which cannot accommodate such a twisted structure. This smectic phase may appear in the same material as a cholesteric phase, on cooling to a lower temperature. A third liquid crystal phase, the nematic, contains symmetric molecules, has no intrinsic twist in its structure and is used in electro-optic devices.

Although discovered 100 years ago, it has been only in the last twenty years that the unique properties of liquid crystals have found widespread application in temperature sensing[1] and electro-optical display technologies.[2] It is interesting to note that the use of nematic liquid crystals in the latter technology, despite starting ten years after the former has seen a much more rapid development. This may be attributed largely to the discovery of new nematogens and the availability of highly pure, highly stable materials to the electronics industry.[3]

By contrast, until the present work, the only available liquid crystal materials for use in temperature indicating devices were ester derivatives of cholesterol or other sterols. The applicability of these materials has been restricted by poor stability and a limited range of physical properties.

The thermochromic properties of the cholesteric phase arise from the ability of the uniformly aligned helices to selectively reflect incident radiation, the wavelenth (λ) of which is related to the helical pitch (p) by the expression:

$$\lambda = \bar{n}p \qquad \bar{n} = \text{the average refractive index}$$

Furthermore any disturbance to the forces maintaining the helical structure will result in a change in the wavelength of selective reflection. The pitch length is particularly sensitive to temperature in the region of the first order phase transition from a smectic to a cholesteric phase (Fig 1) caused by pretransitional short range ordering effects.

When the pitch length is of the order of visible light, thermochromism is observed. On cooling, the color changes from blue to red. The sensitivity of color play ($d\lambda/dT$) is dependent on:

1. How far the pretransitional effects extend into the cholesteric phase.
2. The purity of the optical activity.

In order to clearly see the selectively reflected light it is important that radiation transmitted through the liquid crystal is absorbed. It is for this reason that all devices using this effect are constructed on a black absorbing layer.

CHIRAL NEMATIC LIQUID CRYSTALS

The helical structure which differentiates the cholesteric from the

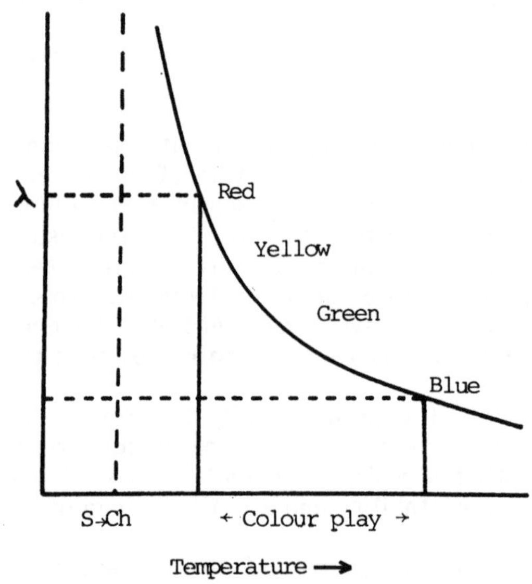

Fig. 1. Selective reflection from thermochromic material.

nematic phase occurs because all or part of the molecules in the liquid crystal are optically active. In recent years[4,5] numerous new optically active (chiral) nematic liquid crystals have been synthesized and some possessing the most interesting properties have been prepared at BDH Chemicals Ltd (Fig 2).

Use of such chiral nematic materials in place of cholesterol derivatives brings several advantages when formulating thermochromic mixtures:

1. The core of the molecule is synthetic and therefore open to variation of structure. Hence a much wider range of physical properties can be obtained than is possible by changing only the terminal group of the molecule as for cholesteryl esters. An extended temperature range can be explored as a result.
2. The optical activity of the material is derived from a branched terminal chain, which can also be prepared in an inactive racemic form. This means that the optical purity of the liquid crystal can be varied without strongly influencing its other properties. The degree of sensitivity of the composition to temperature can be changed conveniently.
3. A greater intrinisic stability of the material to the degrading effects of light and moisture can be achieved by using only chemically stable functional groups in the molecule.

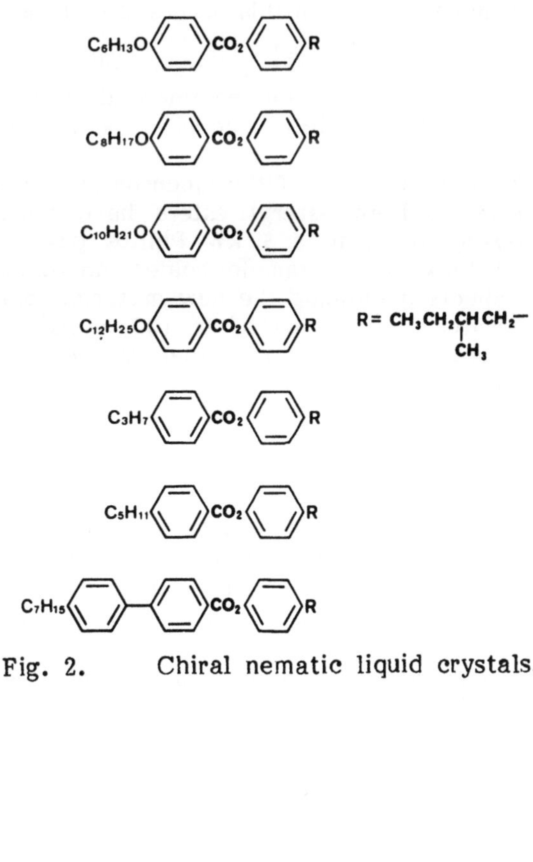

Fig. 2. Chiral nematic liquid crystals.

Using certain of these chiral nematics, a mixture 'system' has been developed[6] which enables the user to obtain selective reflection of colored light varying between 0.5°C and approximately 5°C over the temperature range -30°C to +40°C. The system consists of four mixtures:

TM74A

S_A - Ch	-32.6°C
Ch - I	15.9°C

TM74B

S_A - Ch	-32.3°C
Ch-I	16.2°C

TM75A

S_A - Ch	41.3°C
Ch - I	53.7°C

TM75B

S_A - Ch	42.4°C
Ch - I	54.4°C

B contains 50 wt % of racemic material. TM74A and TM75A are low and high temperature color-play formulations respectively, and TM74B and TM75B are their 50% racemized analogs. The small difference in the thermodynamic properties of the mixtures and their partially racemized analogs is attributed to the difference in their optical activities.

Mixtures TM74A and TM75A are mutually miscible in all proportions and so the point at which color play begins can be tuned by combining the appropriate amounts of each between the limits of -32°C and 41.3°C as shown in Figure 3. The extent of the color play can also be increased by adding the B mixtures to reduce the optical activity of a given composition. The manner in which this occurs is demonstrated in Figures 4, 5 and 6.

These mixtures are now available commercially[7] and make it possible to make devices of higher reliability, accuracy and reproducibility.

The photochemical stability of thermochromic sheets containing the new chiral nematics and cholesteryl esters have been comparatively assessed[8] by exposing them to a 1 kW Philips powerwhite lamp (ref MBF/V) giving 135 klux at the sample holder (approximately 14 times bright sunlight). Sheets containing the new materials only began to lose their color temperature calibration after 100 hours. The cholesteryl esters however started to degrade immediately and lost all color after only 30 hours exposure.

CONCLUSIONS

New chiral nematic liquid crystals, which are suitable for temperature sensing applications, are now available commercially. A mixture system has been formulated to enable a wide range of color plays (0.5°C - 5°C) to be obtained over the temperature range -30°C - +40°C. Devices containing the new liquid crystals show a ten-fold improvement in lifetime compared to devices containing cholesteryl esters.

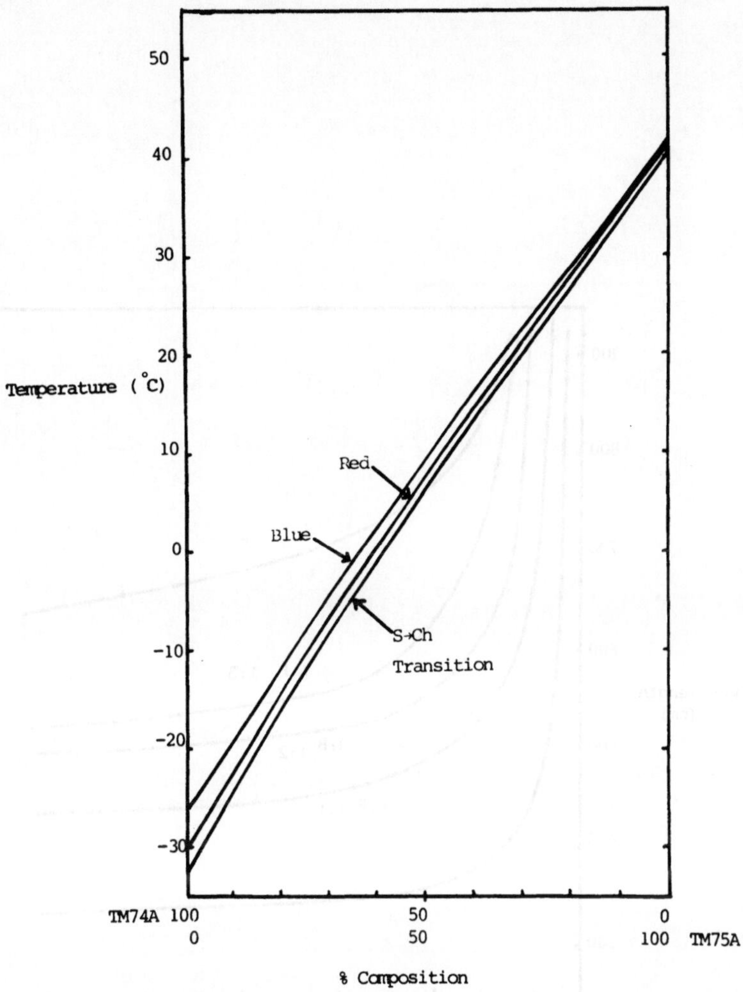

Fig. 3. Color play temperature composition for TM74A/TM75A.

ACKNOWLEDGEMENTS

The authors would like to thank Mr R Smith, RSRE, for spectrophotometric measurements and Mr K Archer of Licril Limited for supplying encapsulated liquid crystals on which stability tests were performed.

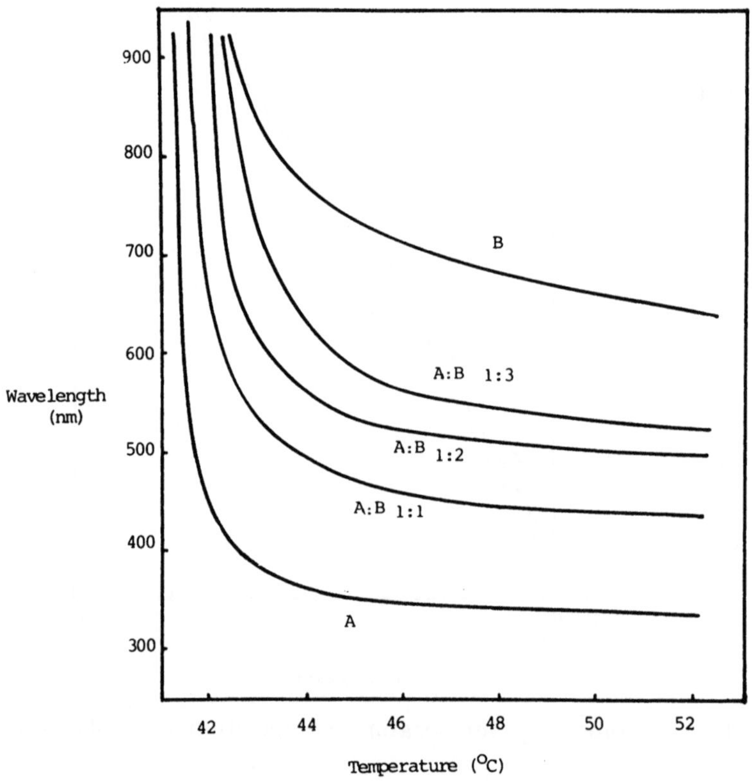

Fig. 4. Wavelength of selective reflection vs temperature for TM75.

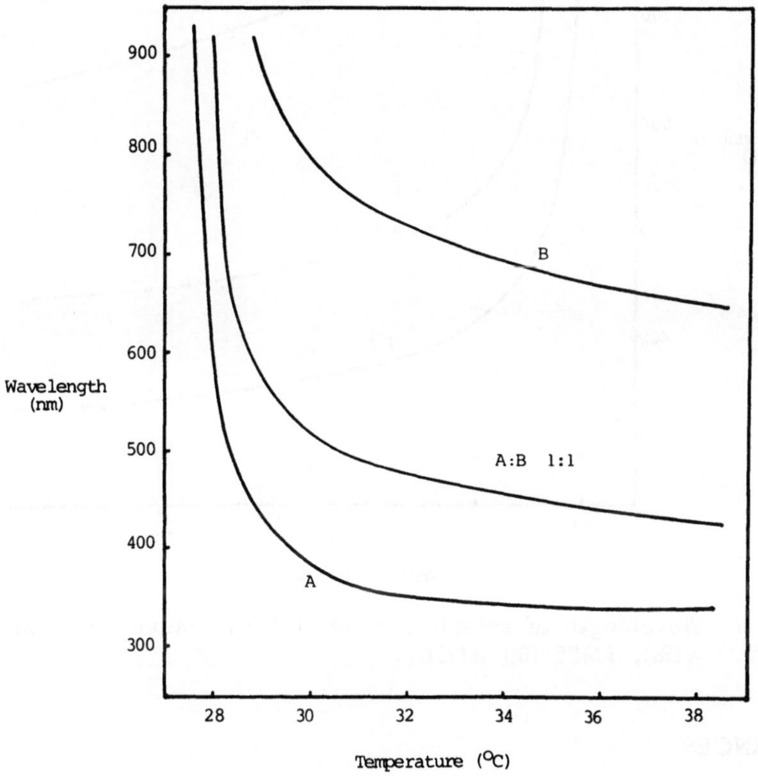

Fig. 5. Wavelength of selective reflection vs temperature TM74 (20 wt%), TM75 (80 wt%).

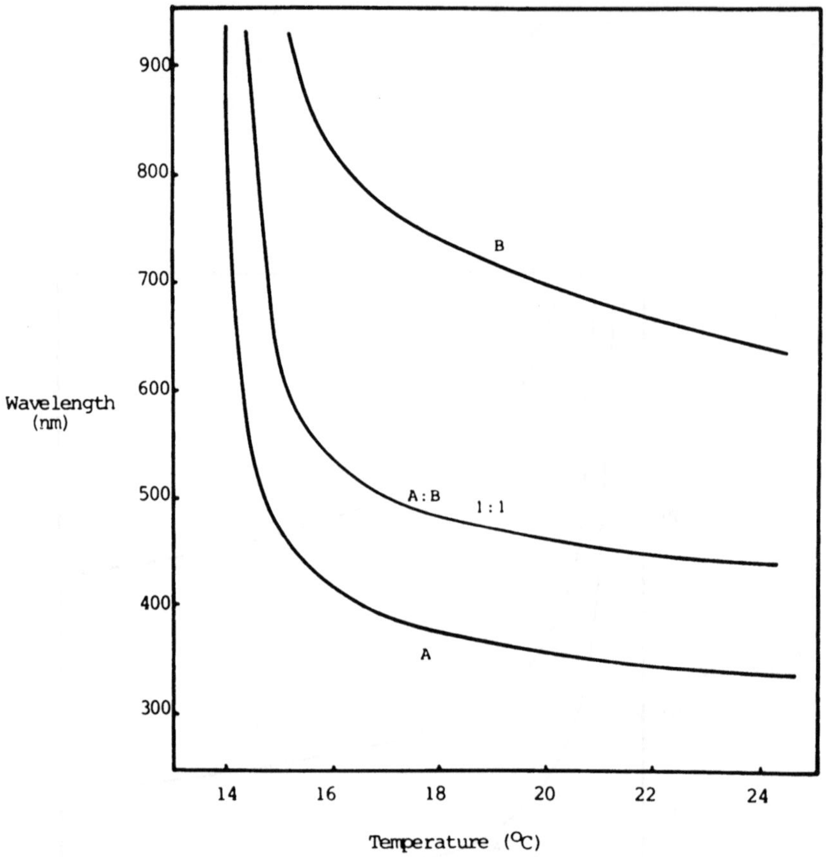

Fig. 6. Wavelength of selective reflection vs temperature TM74 (40 wt%), TM75 (60 wt%).

REFERENCES

1. "Advances in Liquid Crystals," G. H. Brown ed., Academic Press Vol. 2:73-172 (1976).
2. "Applications of Liquid Crystals," G. Meier, E. Sackmann, J. G. Grabaier eds., Springer Verlay (1975).
3. "Liquid Crystals," BDH Publication (BDH Chemicals Ltd, Poole, BH12 4NN, England)
4. G. W. Gray, D. G. McDonnell, Mol. Cryst. Liq. Cryst. 37:189 (1976).
5. G. W. Gray, D. G. McDonnell, Mol. Cryst. Liq. Cryst. 48:37 (1978).
6. D. G. McDonnell, R. A. Smith (In press).
7. For further details contact M. G. Pellatt, BDH Chemicals Ltd, Poole, England.
8. D. G. McDonnell, R. A. Smith (Unpublished results).

BTD: PROGRESS IN PLATE THERMOGRAPHY

A Johnsen

Rohm Pharma GmbH
D-6108 Weiterstadt
WEST GERMANY

Mammography is the most reliable method for the detection of breast cancer, but for several reasons it should not be used for annual screening. Thermography is a useful complementary method for diagnosing breast cancer, particularly since it sometimes indicates the presence of unsuspected cancer. Thermography alone does not lead to a final diagnosis, but indicates the presence of a physiological abnormality. Even this, however, is the first step in selecting asymptomatic women for whom further follow-up examinations are likely to be useful.

THE BTD THERMOGRAPHY SYSTEM

The plate thermography devices on the market work with 3 to 6 single plates, each of which covers a part of the entire range of temperature important for the diagnosis of breast disease. With the foil developed for the BTD Thermography System we have succeeded in covering the clinically relevant range of temperature with only one foil by combining several fluid crystals. This considerably reduces costs; the single examination is quicker and it can be carried out more easily. With this foil it is no longer necessary to find out for each patient which is the most appropriate plate to use for the examination.

Since it is necessary to compare the two breasts a flexible foil with two windows was developed, with which the right and the left breast can be compared immediately. The flexibility of the foil also allows it to be applied to the lateral regions.

For examination in cases of pathological findings and for photo documentation there is also available a BTD thermography plate with a rigid frame. The Polaroid camera system allows the direct control of the picture's quality and avoids problems concerning the co-ordination of the photos and the patient card index.

The BTD Thermography System consists of the BTD Set for the thermographic examination and the Photo Set for the photographic documentation. Additionally offered is an electronic thermometer with a special sensor, to measure precisely the skin temperature.

The American Federal Food and Drug Administration has just approved BTD, which will be marketed to US doctors as Thermascan™ shortly by BCD Products Inc, New York.

A CHEAP AND SIMPLE TEMPERATURE TRANSDUCER: ITS APPLICATION IN MONITORING BREAST DISEASE

U Flesch, H Witt

Zentralinstitut Radiology
Rudolf Virchow Krankenhaus, Berlin
WEST GERMANY

INTRODUCTION

The use of thermological methods in the diagnosis and long term therapy control of breast diseases is a now well established procedure. Thermography is a complementary diagnostic tool that demonstrates the biological activity of the (malignant) pathophysiological process.

Liquid crystal plates only show temperature distributions but cannot easily measure absolute temperatures.

Infrared thermography is a valuable tool for quantitative measurements of absolute temperature and for the display of thermal patterns, but the equipment is expensive.

Both aspects are essential for initial explorations. In follow-up and long term control they can be replaced by a simple and cheap electronic temperature transducer, equipped with a precision thermistor. These fine temperature probes measure absolute temperatures very quickly.

Thermistor contact temperature measurements are cheap and highly reliable procedures in follow up of normal and abnormal patterns, visualized first by liquid crystals or infrared thermography.

After radiation therapy in malignant breast disease we have found that simple temperature control by thermistor proves very useful for early detection of local recurrences of malignancy. This paper describes the design of an appropriate thermistor device.

The tubular container (Fig 1 (3)) which is open at both ends, ie top and bottom, is made of a material which is transparent to thermal radiation and defines a cylindrical air-column having a height h. The bottom of the open cylindrical tube (3) has a rim (8) which carries an annulus (7) made of felt, which is placed in contact with the surface of the skin. The open top of the cylindrical container (3) is held by a handle (10) and communicates with the environment E.

The air-column or cylinder inside this container (3) is insulated against heat flow from the sides. Therefore, the temperature in the cylindrical container falls linearly from the bottom to the top. The bottom end of tubular container (3) is also open, and a miniature thermocouple or miniature thermistor (4) having a height h is located in the center of the bottom of the tubular container (3). Specifically, the thermistor (4) is mounted on a wire mesh (6) which is held in place by plastic thread (5), fastened to the bottom rim (8) of the container (3). Thus the temperature sensing element is suspended across the open bottom of the container (3).

A plug (2) is provided adjacent to the handle (10) as well as to the top of container (3), and connecting wires (9) run to the rim (8) and to the element (4) on mesh (6). A connection can be made from plug 2 to external indicating devices.

The height (h) of the thermistor or the thermocouple (4) to be used can be calculated from the decrease of the temperature over this distance, which is determined by the desired accuracy of measurement of 0.02K. If the radius of the cylindrical tube (3) is small there is no convection in the air-column, particularly when the bottom of the tube (3) is placed directly on the surface whose temperature is to be measured. The felt annulus (7) is in direct contact with the surface, and the wall of the tubular container (3) impedes convective flow in the immediate environment of element (4). Thermal radiation constitutes the greatest part of the heat exchange. The wall of the tube (3) which is the shell of the air-column therein has to be transparent to thermal radiation, so that there is no influence on the thermal radiation inside the cylinder.

The transducer as constructed does not influence the temperature of the skin, and the heat flow through heat conduction, thermal radiation and evaporation is sufficiently low so that the measurement is carried out with an accuracy of 0.02K or better. The probe has a response time of 2 ms. Typically, the cylindrical container (3) is 50 mm high and has a diameter of 30 mm. The wire mesh (6) is provided to integrate the temperature measurement over an area larger than the dimension of the temperature sensor proper.

This apparatus has the following advantages:

1. It is of very simple construction.
2. It is simple to calibrate.
3. It does not influence the temperature of the skin and the heat flow from the surface to the environment (heat conduction, thermal

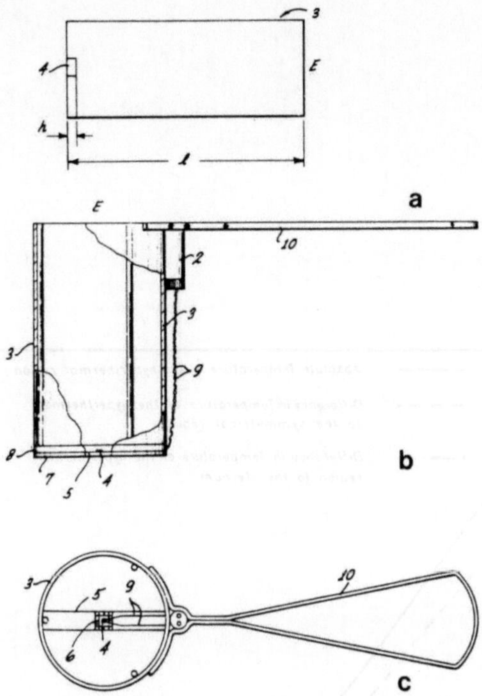

Fig. 1. Schematic diagram of a cylinder with temperature sensing elements, illustrating the geometric relation.
(a) the parameters used in the description.
(b) side view of an apparatus in accordance with the preferred embodiment.
(c) rear view along lines 3-3.

radiation and evaporation) within the accuracy of measurement.

4. It has a small thermal mass.
5. It is not influenced by the skin.
6. It permits measurement of the absolute temperature and of the difference in temperature in relation to the environment (Fig 2).
7. It permits simple and automatic recording of the output and also permits measurement during radiological treatments.

The probe in clinical use is shown in Figure 3.

PATENTS

DT-PS 24 14 048
US 3,938,387

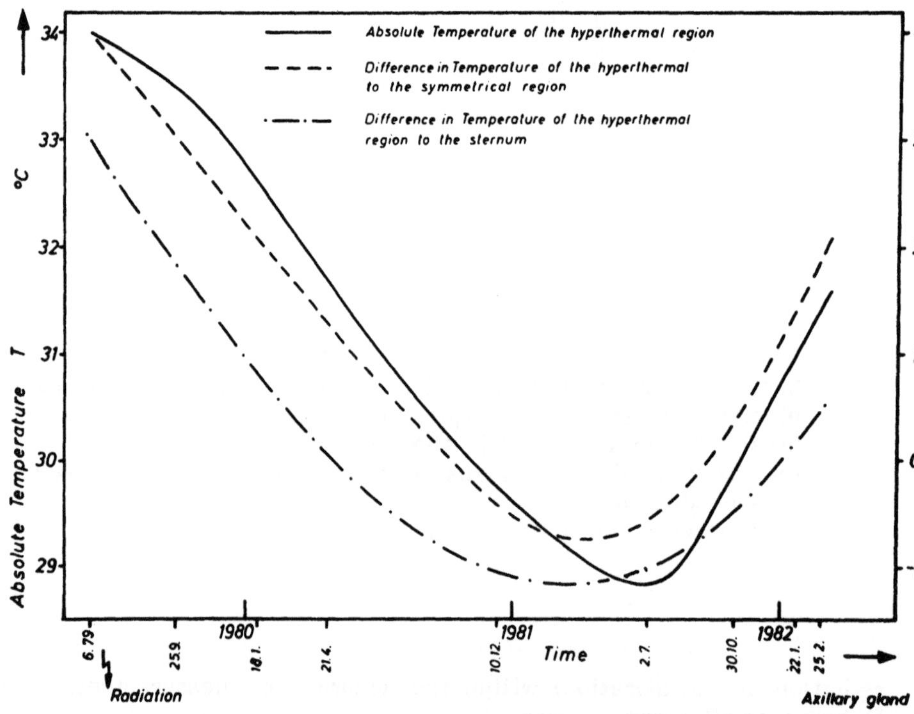

Fig. 2. The digital display device of the equipment shows the absolute temperature of the skin of the hot area.

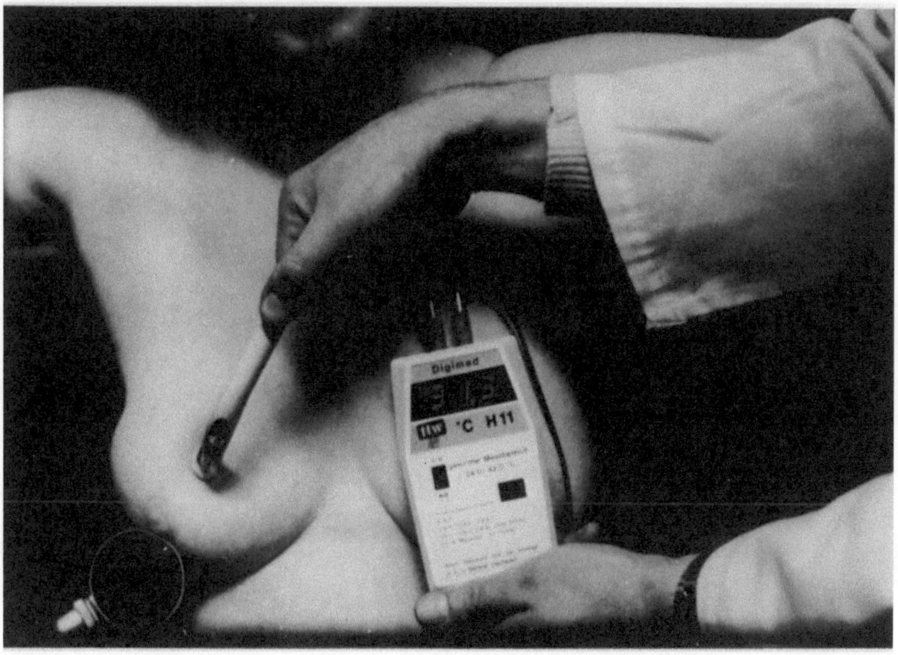

Fig. 3. Patient Fr. E. dob 15.2.26 Thermography of a cancer in the right breast after exclusive radiation as a function of time. Histology: Cancer in the ductus lactifer of the right breast with local infiltration (T1).

MANUFACTURER

TTW Industrie + Medizin Messtechnik GmbH
Am Kohbach 9
D-7808 Waldkirch (FRG)
Phone: 07681-7155 and 9313
Telex: 772 308 ttw

A DEVICE FOR THERMAL CLEARANCE OF THE SKIN

E F J Ring

Royal National Hospital for Rheumatic Diseases
Upper Borough Walls, Bath
UNITED KINGDOM

Thermal conductivity measurements can be made on the skin by a variety of probe designs. The basic essentials are to provide a heated surface and a thermocouple system to monitor skin temperature changes created by the heater. In certain situations the temperature gradient around the probe is found to be related to superficial blood flow. The presence of larger blood vessels and direction of blood flow can be significant factors in thermal conductivity measurements of tissue by this technique.[1] Nevertheless, useful values for 'thermal clearance' of skin can be made, particularly when serial measurements are made from the same area.[2] A special probe has been developed for this procedure which proves to give stable results. It is inexpensive and is constructed from the ceramic base of a potentiometer. The geometric structure of the probe is different from that of other workers.[3,4] The heater is a 270° metal film track, and the thermocouples are placed in the center of the probe (Fig 1). Matched pairs of thermojunctions ranged around the periphery of the heater serve to monitor temperature difference between the area under the probe, and the area 2 cm beyond its circumference.

Infrared thermography of the probe shows that in practice the whole ceramic base achieves an even temperature within a few minutes of applying the heater current. It is therefore unnecessary to construct the probe with a 360° heater in this case.

The technique was simplified by constructing a circuit designed to sense the stability across the thermocouples. The manual procedure required is:

1. Secure the probe to the area of skin, with the subject fully at rest.
2. Wait for a steady state to be achieved, where all the hot junctions are at an even temperature.

Fig. 1. Schematic diagram of the underside of the probe showing the 'C' shaped heater and four inner and outer pairs of thermocouples.

3. Apply the heater current, and monitor the temperature rise.
4. Record the maximum temperature difference required to achieve a new steady state.

These features were incorporated in the control unit. In practice, an LED signal lamp indicates the first steady state. After a predetermined time the heater is automatically switched on. The first lamp is extinguished, and the second (heater) lamp comes on. A second steady state is indicated after some minutes by a visible and audible indicator. A digital display registers the actual rise in temperature difference from the probe (Fig. 2).

EXPERIMENTAL RESULTS

The thermal clearance device has been applied to three different anatomical sites. The mid-forearm has been used avoiding, where possible, any obvious proximity to major blood vessels. A second site over the mid-anterior surface of the tibia has been used. Thirdly, a site under the second metatarsal head, from the plantar foot, was recorded with the subject in a prone position (Table 1).

Changes in the mid-forearm were induced by two methods.

1. Ultra violet radiation
2. Nicotinic acid based aerosol

Ultra violet Radiation

Six hours after a contact exposure of one second from a mercury vapor lamp (Kromeyer treatment lamp), the skin erythema was graded

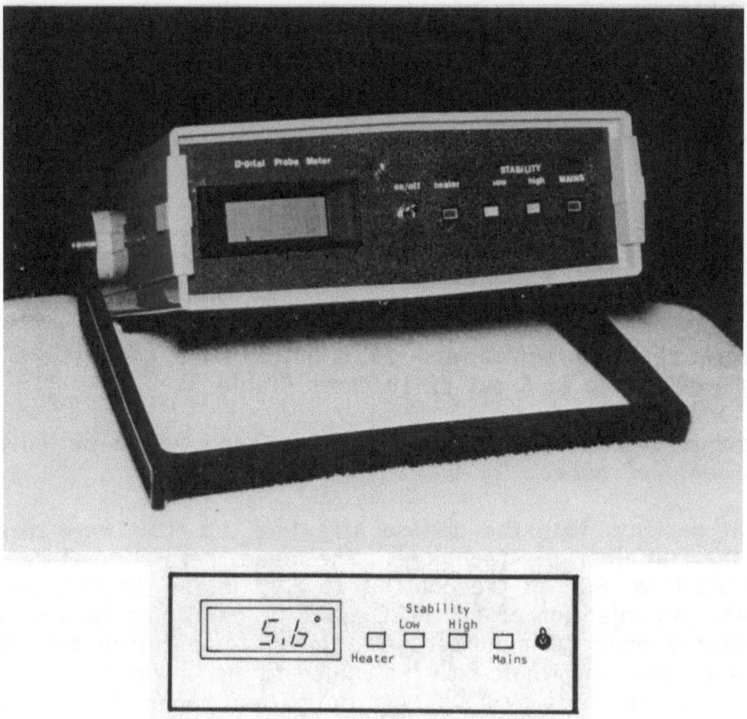

Fig. 2. Thermal clearance display unit.

for severity into one of three categories; +, ++, +++. Twenty-seven subjects were tested, 13 had a minor erythema, 8 a medium and 6 severe +++ erythema (Table 2).

Topical Nicotinic Acid

300 µl from a controlled valve aerosol of a nicotinic based topical agent ('Deep Heat' - Mentholatum), was applied to the forearm. Clearance

Table 1. Results of readings taken at three sites

Mid-forearm ^{o}C	Mid-tibia ^{o}C	Plantar foot 2nd MTP ^{o}C
3.6	8.6	7.3
SD 0.2	SD 0.5	SD 0.9

Table 2. Severity of skin erythema after UV exposure

	Pre UV	+	++	+++
mean Δt	3.4	2.9	2.6	2.3
n = 27°C	SD 0.5	SD 0.5	SD 0.35	SD 0.2

values were obtained before and 20 minutes after application. Skin reddening was visible in 8 out of 10 cases (Table 3).

Changes over the mid-tibial side were recorded in patients with Paget's disease of bone.

Eight patients with the disease affecting the tibia were measured with the thermal clearance technique. The mean thermal clearance value was lowered from 8.3° in the controls to 5.9° in the diseased patients, $p < 0.0005$. An injection of 100 MRC units of calcitonin to one patient was monitored over 70 minutes. The mean temperature recorded by thermography over the whole site from knee to ankle was 32.3°C. After 50 minutes this had risen by 0.8°C, but the patient reported a hot flushing sensation. Thermal clearance changed from 6.2 to 4.8°C, a significant increase. Twenty minutes later the mean temperature by thermography had fallen to 31.8°C and the thermal clearance changed to 6.25°C (Fig 3). Clinical evidence of flushing had ceased. The magnitude of change recorded by thermal clearance was greater than that observed by infrared thermography.

Table 3. Skin reddening after application of nicotinic acid

	Pre Treatment	No Reddening	Skin Reddening
mean Δt °C	3.6	3.0	2.1
	SD 0.2	SD 0.3	SD 0.4
	n = 10	n = 2	n = 8

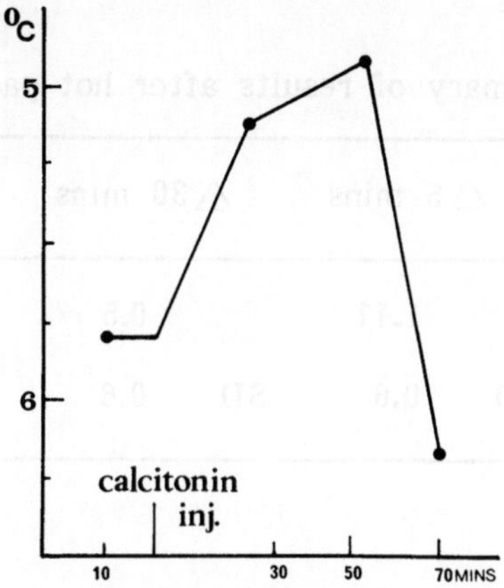

Fig. 3. Change in thermal clearance measured over the tibia in a patient with Paget's disease, following a simple injection of calcitonin.

Reflex Thermostimulation

Nine prone lying patients with rheumatoid arthritis were given hot packs to the knee for 10 minutes. Thermal clearance was recorded over the plantar foot (2nd MTP) before and commenced within two minutes after removing the hot pack from the knee. Further recordings were made 30 and 60 minutes later. Thermal clearance was increased by the hot pack in all subjects, with a return to approximate baseline values by 60 minutes. The experiment illustrates the reflex vasodilation occurring in the foot resulting from the localized hyperthermia to the knee. There was no correlation with severity of joint inflammation. The results are summarized in Table 4.

Despite the multiplicity of factors believed to modify thermal clearance, this technique, when used with care, provides data which can show a greater capacity for change than infrared thermography, where capillary superficial blood flow is likely to be modified by drugs or disease. The device constructed appears simple to use, and to give fairly reproducible results.

Table 4. Summary of results after hot pack application

Baseline		△5 mins		△30 mins		△50 mins
7.3°C		1.11		0.5		0.06
SD 0.9	SD	0.6	SD	0.6	SD	0.22

REFERENCES

1. B. H. Brown, C. Bygrave, P. Robinson, H. P. Henderson, A critique of the use of a thermal clearance probe for the measurement of skin blood flow, Clin. Phys. Physiol. Meas. 1;3:237-241 (1980).
2. N. Britton, J. R. Barker, E. F. J. Ring, An assessment of the thermal clearance method for measuring perfusion, in: "Recent Advances in Medical Thermology," Plenum, New York (1983).
3. G. Holti, K. W. Mitchell, Estimation of the nutrient blood flow using a sequential thermal clearance, Clin. Exp. Derm. 3:189 (1978).
4. W. J. B. M. Van de Staak, A. J. M. Brakkec, H. E. De Riijke-Herweijer, J. Invest. Dermatol. 51:149-154 (1968).

AN ASSESSMENT OF THE THERMAL CLEARANCE METHOD FOR MEASURING PERFUSION

N F Britton, J R Barker, E F J Ring

University of Bath and Royal National Hospital for Rheumatic Diseases, Bath
UNITED KINGDOM

INTRODUCTION

The thermal clearance technique is widely used to give information on skin blood flow. Heat is applied at a known rate at the surface of the skin and the consequent temperature rise is measured. This temperature rise depends on the skin blood flow since heat is convected from the heater by the blood, and a high flow rate therefore leads to a smaller temperature rise. Heat is also conducted from the heater, and so the temperature rise also depends on the thermal conductivity of the skin. It is therefore important that the effects of both parameters are considered, and to do this we introduce a mathematical model of the system.

THE MODEL

The thermal clearance probe which we are modelling is that discussed by Ring and Barker in these proceedings. Its geometry is shown in Figure 1. It is a thin circular ceramic disc on which an annular heating element is fixed. We assume for simplicity that the element is circular, since results show that the heater produces a circular hot spot on the skin. The other assumptions are:

1. The blood perfuses the skin according to Pennes' model.[1]
2. Heat loss from the skin, and the top surface of the heater, to the air obeys Newton's law of cooling.
3. Heat loss from the edge of the heater is negligible.
4. Metabolic heat production is negligible. (In normal tissue it is much smaller than the perfusion effect.)

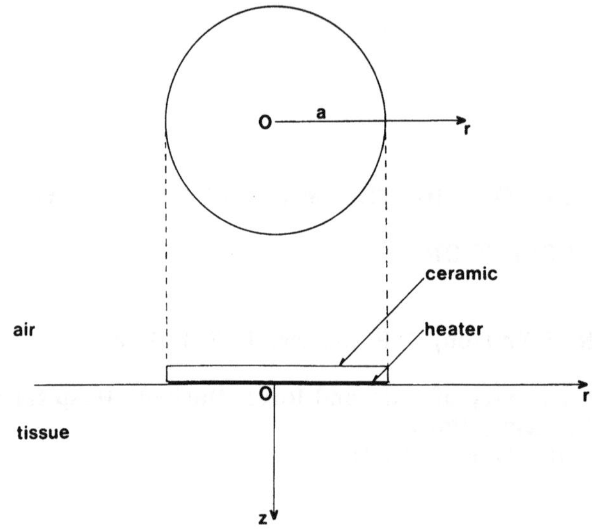

Fig. 1. The geometry of the thermal clearance probe.

5. As depth increases the tissue temperature approaches a constant 'core temperature'.
6. All temperatures have attained a steady state.

Under these assumptions the equation satisfied by the temperature T in the tissue is:

$$k\nabla^2 T - mc_B(T - T_c) = 0$$

$$\underset{\text{conduction}}{\uparrow} \qquad \underset{\text{perfusion}}{\uparrow}$$

where k is the thermal conducitivity of the skin, m is the blood perfusion constant (in kg of blood perfusing each m^3 of tissue per second), c_B is the specific heat of blood and T_c is the core temperature of the body. This equation may be solved with appropriate conditions on the boundaries, as stated in the assumptions above. It is found that the temperature rise at a depth z directly below the heater is given by:

$$\Delta T = \frac{Q}{\pi a k} \int_0^{\infty} \frac{J_1\left(xa\sqrt{mc_B/k}\right)}{\sqrt{1+x^2} + H/\sqrt{kmc_B}} \exp\left\{-z\sqrt{1+x^2}\sqrt{\left[\frac{mc_B}{k}\right]}\right\} dx$$

where Q is the power input of the heater, a its radius and H the heat transfer co-efficient from skin or ceramic to air. This temperature rise is plotted in Figure 2 for various values of thermal conductivity k and blood perfusion m.

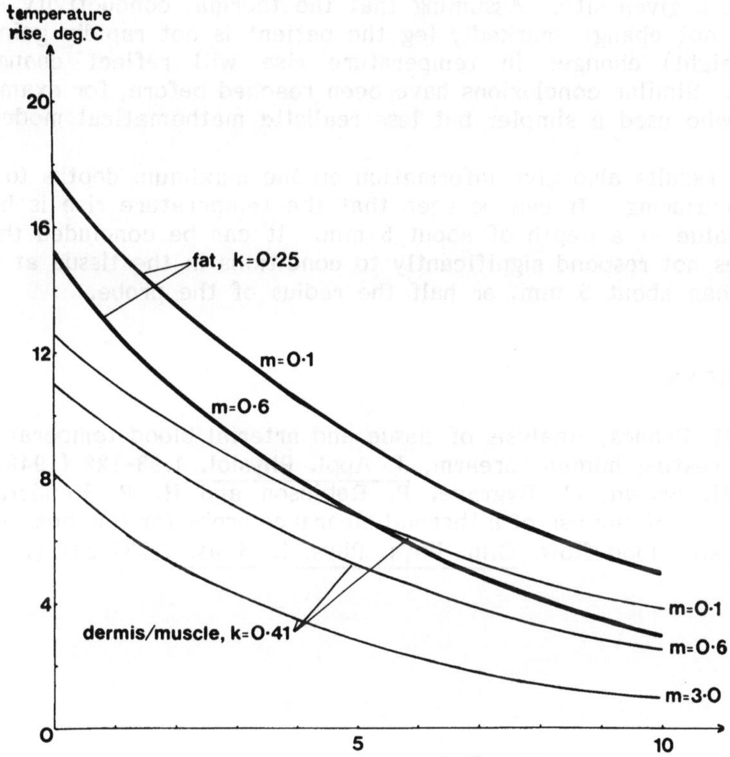

Fig. 2. The temperature rise directly under the center of the heater for various values of thermal conductivity k (in $Wm^{-1}K^{-1}$) and blood perfusion constant m (in kg $m^{-3}s^{-1}$). Parameter values were taken as follows:

H = heat transfer coefficient (skin or ceramic to air) = 13.6 $Wm^{-2}s^{-1}$.
a = radius of heater disc = 1 cm.
Q = power input to heater = 250 mW.
c_B = specific heat of blood = 3800 J $kg^{-1}K^{-1}$.

CONCLUSIONS

It can be seen from Figure 2 that a small difference in the thermalconductivity of the skin makes a large difference to the tempereature rise; for example at a constant value of perfusion, m = 0.6, changing k from 0.41 to 0.25 leads to an increase of 40%. However, changing m from 0.6 to 0.1, a much larger change, leads to an increase of less than 20%. It follows that a very accurate value of thermal conductivity is required in order to estimate the perfusion accurately. This reduces the usefulness of the method of thermal clearance, since accurate values of thermal conductivity are difficult to obtain. However, the technique can be used to monitor changes in perfusion on one given patient at a given site. Assuming that the thermal conductivity at this site does not change markedly (eg the patient is not rapidly gaining or losing weight) changes in temperature rise will reflect changes in perfusion. Similar conclusions have been reached before, for example by Brown,[2] who used a simpler but less realistic mathematical model.

The results also give information on the maximum depths to which we are measuring. It can be seen that the temperature rise is half its surface value at a depth of about 5 mm. It can be concluded that the probe does not respond significantly to conditions in the tissue at depths greater than about 5 mm, or half the radius of the probe.

REFERENCES

1. H. H. Pennes, Analysis of tissue and arterial blood temperatures in the resting human forearm, J. Appl. Physiol. 1:93-122 (1948).
2. B. H. Brown, C. Bygrave, P. Robinson and H. P. Henderson, A critique of the use of a thermal clearance probe for the measurement of skin blood flow, Clin. Phys. Physiol. Meas. 1:237-241 (1980).

VASCULAR DISEASES - INTRODUCTION

E D Cooke

Department of Medical Electronics
St Bartholomew's Hospital
London, UNITED KINGDOM

Despite the number of papers published during the past decade on the use of thermography (THG) in vascular and endocrine disease, the place of this method in these disease entities, at least to this writer, has, with the exception of the diagnosis of venous disease, been somewhat nebulous and ill-defined. Are the results produced in one's own laboratory repeatable and reproducible by other workers? And are there other problems in which THG might be useful but is, as yet, untried? Or have other methodologies such as Doppler ultrasound, isotopic methods and CAT scanning superceded THG. If THG is good, how good is it, ie what is its specificity and sensitivity?

The vascular and endocrine section of the Third International Congress, if not answering all these questions, did much to indicate the present role of THG and pointed the way for future development. It is perhaps useful to state straight away the lasting impressions of the care and scientific accuracy which characterized most but alas not all the papers presented and the fact that thermological methods in vascular and endocrine disease are far from obsolete. On the contrary, the situations in which thermology is useful were more clearly defined. For example, Dr Partsch, who provides a diagnostic service for vascular disease in Vienna, avoids the use of invasive techniques, particularly in venous disease, by incorporating THG into his vascular laboratory - a method which many centers could practice with advantage. It is noteworthy that his other non-invasive methods include foot volumetry, various forms of plethysmography and ultrasound.

Perhaps two of the more controversial areas for the use of THG have been cerebrovascular and peripheral vascular disease. Using THG alone, Yanagi found it of no clinical value in patients with gross cerebrovascular pathology while Karpman and his colleagues, on the other hand, adding Doppler ultrasound, ocular plethysmography and carotid phononangiography to their methodology, found a high degree of specificity and sensitivity combining the methods. Surely this is vindication for the use of several non-invasive techniques rather than depending on one only. The importance of this concept was again evident in a critical review of THG in peripheral vascular disease by Spence. His conclusion that the only use of THG was in end-stage disease to define the level of amputation (skin flap viability) was scarcely surprising since proximal and distal stenosis may produce very similar changes in the distal vasculature. The method of Dewar and his colleagues, who use a thermal mapping technique to define amputation levels, has given promising early results.

This use of thermography in amputation is clearly important considering the difficult decision which often faces the surgeon and the morbidity and mortality which may follow a wrong choice based on clinical findings alone. Unfortunately THG is valueless when the limb is infected. In this situation isotopic methods are at present the only resource. Spence also warns of the difficulties of interpretation of the thermogram produced by the opening of A-V shunts about the area of the ankle in the ischemic limb.

Determining the response to cold stress in vasospastic disease of the hands is an obvious use for thermography. The rewarming patterns of the normal and vasospastic hand described by Kyle and her colleagues are common experience and typical of this condition. Quantification of the response is simple but a standard method of doing so awaits agreement. The work with reflex sympathetic dystrophy continues in Boston. It is disappointing that no advance seems to have been made since the findings were described in 1976. The use of mild cold stress, insufficient to produce pain, may add greater sensitivity to that achieved by the resting thermogram. Stress of the autonomic nervous system is standard in such diseases as diabetes mellitus. Why it should not be so in reflex sympathetic dystrophy is difficult to understand. Cardiac muscle THG, particularly with the popularity of coronary by-pass surgery and now transluminal coronary angioplasty, is an exciting entity of which we seem certain to hear more.

To judge by the papers presented at the Congress and other recent publications, the use of scrotal thermography in the infertile male is increasing. It seems that this must be in response to clinical demand. Could it be that more surgeons are realizing the difficulties of making a diagnosis of varicocele or, just as apposite, diagnosing a varicocele when it is absent? There now seems no doubt that THG is more accurate than the palpating fingers but surely the presence or absence of varicocele is not the sole information available in the scrotal thermogram. In other endocrine abnormalities it seems doubtful if THG has a place, considering the variety and sophistication of the biochemical tests presently available.

What of liquid crystal thermography in vascular and endocrine disease? In this investigator's hands the method is cumbersome and difficult to standardize. Interpretation is subjective and consequently haphazard - enough said.

THE CONCEPT OF A VASCULAR SERVICE

H Partsch

Vascular Service of the Hanusch Hospital
Vienna
AUSTRIA

The concept of a vascular service is based on the fact that the majority of peripheral vascular problems can be diagnosed and managed in an out-patient department.

Diagnosis comprises three tasks mainly:

1. Exclusion of non-vascular disorders.
2. Restriction of the need for angiography.
3. Assessment of vascular function.

EXCLUSION OF NON-VASCULAR DISORDERS

Our own experience shows that of all patients referred to our department, about one half have venous problems, one quarter arterial diseases and one quarter non-vascular disorders (Fig 1). These non-vascular cases have to be ruled out in the first step of investigations.

Arterial Occlusive Diseases

These can be discerned from other conditions presenting similar symptoms by Doppler ultrasound and pulse monitoring methods such as oscillography, rheography and photoplethysmography. It is also possible to localise the flow hindrance approximately. Segmental pressure measurement appears to be a superior method of assessing occlusion in a limb than the determination of flow by venous occlusion plethysmography or radioactive tracer clearance methods.

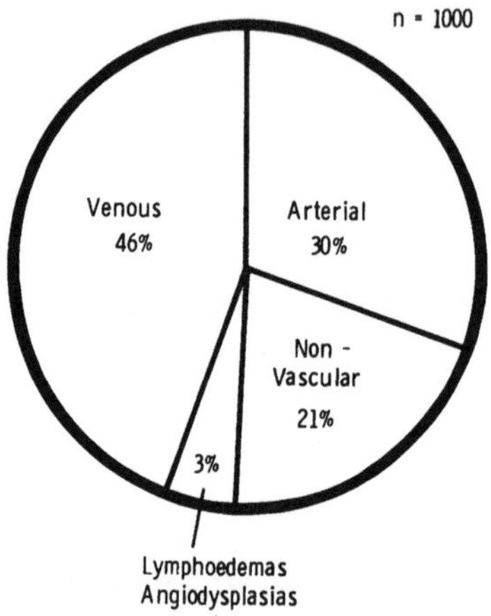

Fig. 1. Patients in a vascular service.

A standardized working stress (ergometer, tread-mill) is helpful for diagnosing border line cases and for assessing the degree of functional compensation of the individual case. It is also a good control measure during follow-up studies, bearing some psychologically positive motivation towards exercises for the patient.

Thermography allows some supplementary information to be gathered regarding the localisation of an arterial embolus, the optimal amputation site or the assessment of the vascular innervation in Raynaud's phenomenon or autonomic neuropathy (see Assessment of Vascular Function). In our opinion this method is not very useful for diagnosing carotid obstructions, in which field we prefer Doppler ultrasound and photoplethysmography as screening methods.

Venous Diseases

Venous diseases of the extremities are easy to diagnose without any instrumental help in cases where the pathological condition is restricted to the superficial tissue layers (varicose veins, venous ulcers, superficial phlebitis). On the other hand, diagnosis may be difficult in patients with incompetent perforating veins and with deep vein thrombosis.

Doppler ultrasound, plethysmographic examinations and thermography are the most important achievements of modern non-invasive diagnosis. Concerning aspects of frequency, late sequelae and socio-economic circumstances these venous disorders are even more relevant than the arterial occlusive diseases.

RESTRICTION OF THE NEED FOR ANGIOGRAPHY BY NON-INVASIVE METHODS

By furnishing every hospital with the screening facilities of a vascular service "normal" arteriograms and phlebograms should be a reminiscence of the past.

By modern non-invasive methods routine angiography can be restricted to cases where the X-ray picture has an immediate therapeutical consequence for the patient:

(a) arterial resting pain and gangrene, severe claudication in young patients.
(b) typical TIA.
(c) deep vein thrombosis if fibrinolysis or surgical treatment is intended.

The vast majority of patients do not belong to this group and can be treated conservatively based on the non-invasive diagnosis.

Thermography, Doppler ultrasound and plethysmography are collectively used in our routine work as the deciding screening methods for deep vein thrombosis. If all three methods give a negative result a DVT is excluded. It is important to keep in mind that this decision can be wrong in up to 8%. On the other hand, phlebography in every consecutive case suspected of having a DVT would reveal no definite diagnosis in about the same percentage range, owing to difficulties in technique or interpretation.

Non-invasive testing, therefore, is able to decrease expenses for the community and discomfort for the individual considerably. At the same time the lack of relevant information is acceptably low.

ASSESSMENT OF VASCULAR FUNCTION

Concerning the functional evaluation of peripheral circulatory disturbances, non-invasive methods are superior even to vascular imaging alone. In the extremities the degree of arterial or venous insufficiency can be assessed. This is of immediate importance for the patient's working and living activities. Such methods include measurement of pressure and flow after a standardized working load by Doppler ultrasound and venous occlusion plethysmography on the arterial level. On the venous side phlebodynamometry in a dorsal foot vein and in the femoral vein, foot volumetry and photoplethysmography are of considerable practical importance. This is also true for selecting patients for varicose vein surgery. Thermography after physical stress seems to offer some valuable information especially regarding an increased risk of getting a DVT. This method is also a very elegant and effective procedure for detecting blood flow changes in the hands and feet induced by local or indirect cooling and heating. Pathological vasoconstriction as in Raynaud's phenomenon and a deficiency of reflex vasodilatation as in autonomic neuropathy (eg diabetic feet) can be proven.

There is no progress in diagnosis and therapy without some clinical research. Thermography, measurement of transcutaneous oxygen tension and photoplethysmography offer promising insights into the pathophysiology of blood flow and nutrition of the skin.

Therapy

Because it also includes therapy the concept of a vascular service is different from the idea of a vascular laboratory. Non-specialized doctors referring patients to a vascular laboratory are quite content to get not only diagnostic but also some therapeutic advice. In patients with arterial occlusive diseases, therefore, conservative management either after surgery or in the large number of patients in whom surgery is neither indicated nor possible can be supervized by a vascular service. Special forms of treatment such as, for example, intra-arterial infusions in patients with gangrene or special compression massage and bandages in venous diseases can be performed according to the local facilities.

Already in our department therapy comprises more than 50% of all our work. Since we regard keeping the patients walking as the most important aim of therapy, especially in our old and partly disabled vascular patients, this ambulatory management has considerable advantages over a hospitalisation. In our experience the medical care of chronic vascular patients before, after and between admissions to hospital is generally on a rather low level. In advanced stages of arterial occlusive diseases or of venous ulcers the benefit for the patient from hospitalization with palliative surgical measures such as profundaplasty, sympathectomy or skin grafting is often very much restricted. Therefore we consider a vascular service as the main institution to guide such patients.

Only desperate cases need admission to the ward. (This view is contrary to the much more expensive, more ineffective and more inhumane concept that the ward is the essential centre of therapy and the vascular laboratory is only a helpful appendix).

In conclusion we think that an acceptable everyday life can be given to the majority of vascular patients by treatment on an ambulant basis consisting of training, medication and a variety of physical aids. To achieve this objective, interdisciplinary co-operation and co-ordination is essential.

AN ASSESSMENT OF THERMOGRAPHY IN ARTERIAL DISEASE

V A Spence, W F Walker

Vascular Laboratory
Ninewells Hospital Medical School
Dundee, SCOTLAND

INTRODUCTION

In the past decade there have been many who have advocated a role for thermography in the routine investigation of patients who have peripheral arterial disease. Much of the evidence in the literature suggested both general and specific applications for the technique in the fields of peripheral and cerebral arteriosclerosis, Raynaud's phenomenon, diabetes, Buerger's disease, amputation surgery and the evaluation of vasoactive drugs. Despite this, thermography is not widely accepted and is often the subject of derision amongst the scientific and clinical establishment working in this branch of medicine. Why should this be so? The answer is complex:

1. There is no evidence in the scientific literature of a confirmed role for thermography in the investigation of arterial disease.
2. Thermographic equipment is not widely available to those involved with vascular disease and will not be so until the scientific evidence establishes a need.
3. The potential role of thermography in other areas of medicine has remained unfulfilled and in many cases has been clearly overstated.
4. Thermography requires a knowledge and understanding of both the physics and physiological concepts of heat exchange between man and his environment, both of which are still not completely understood and afford areas for future research.

Against this background a future for thermography as a valuable asset in the non-invasive vascular laboratory is difficult to recognise. Furthermore, the number of communications of thermography relating to arterial disease has fallen drastically in the past decade (Fig 1) which is further indication of the fall in popularity of the method.

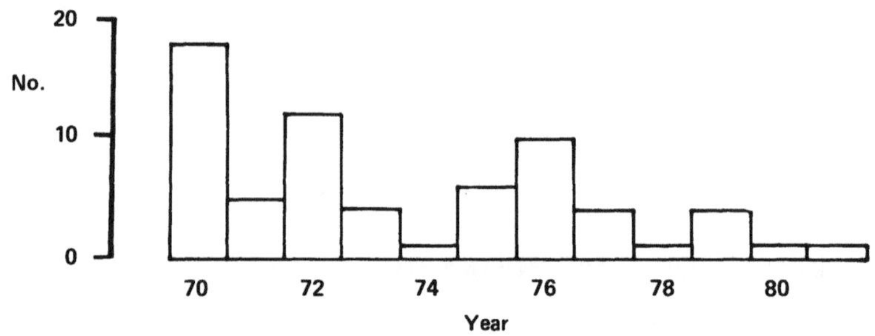

Fig. 1. Publications relating to thermography in arterial disease. Index medicus (English) 1970-1981.

In the early 1970's the majority of publications were concerned with the diagnosis of lower limb and cerebrovascular diseases. Confirmation of a diagnosis relied on areas of hyperthermia at a particular site, which usually depended on the existence of collateral pathways when main arterial branches were occluded. A hypothermic pattern depended on the existence of severe arterial stenosis with a decrease in blood flow.[1] However, despite the fact that asymmetrical temperature patterns are characteristic of specific arterial pathologies, there was generally poor correlation between thermographic and arteriographic findings. At this point it is important to understand the limitations of thermography in relation to arterial disease. The thermographic image is compiled from radiation emanating from the skin surface. Alterations in the thermal pattern of the skin may stem from local superficial changes in skin blood flow or from changes in tissue thermal conductance resulting from structural abnormalities in major arteries, so producing collateral vessels. The variation in skin surface thermal patterns stemming from obstruction of major arteries is enormous. This is so because of the varying severity of vascular disease in individual patients; many have multi-segmental disease rather than a single stenosis with regularly developed heat-exchanging collaterals. Clearly the role of thermography in the vascular laboratory is not one of diagnosis or confirmation of diagnosis. This diagnostic role can be filled effectively and cheaply using the methods of Doppler ultrasound, plethysmography and an exercise treadmill.

SEVERE PERIPHERAL ISCHEMIA

A further objective of a non-invasive vascular laboratory is the investigation of patients with severe lower limb ischemia (Fig 2). These are patients for whom reconstructive vascular surgery is considered unsuitable. Treatment with vasoactive drugs, sympathectomy and amputation are the alternatives.

There are two distinctly different thermographic appearances of the severely ischemic limb. The first is the more common one of hypothermia

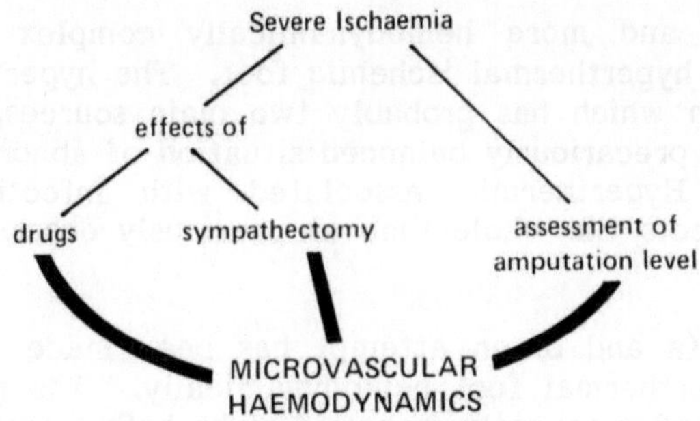

Fig. 2. The role of the vascular laboratory in the investigation of severe peripheral ischemia. The problem is one of assessing the local microcirculatory conditions which exist in the skin and to determine the effects of vasoactive drugs and sympathectomy.

(Fig 3). In this instance blood flow is either absent or greatly reduced in the distal segment of the limb and there is usually a clearly defined temperature gradient which we have taken as the demarcation between levels of viable and non-viable tissue.

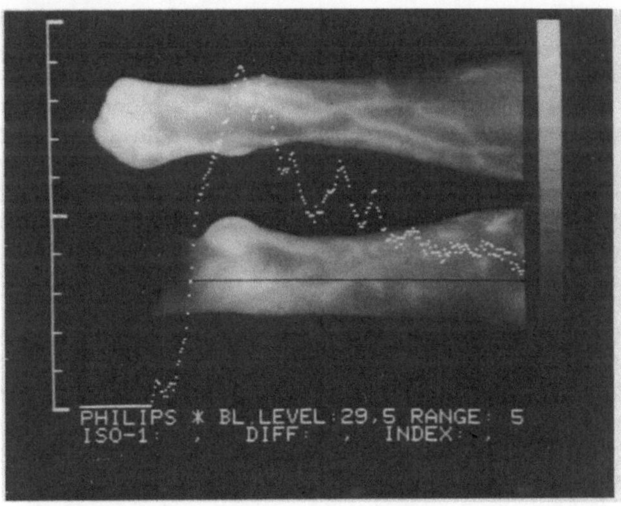

Fig. 3. A thermogram of the lower limbs of a patient with a severely ischemic foot. There is an inflammatory hyperthermal area just proximal to the area of distal hypothermia. The temperature profile along the indicated black line is scaled with a vertical axis in steps of 0.5°C starting at 29.5°C.

The second and more hemodynamically complex thermographic appearance is the hyperthermal ischemic foot. The hyperthermia results from inflammation which has probably two main sources, infection and more commonly a precariously balanced situation of abnormal local flow hemodynamics. Hyperthermia associated with infection is usually extensive and affects the whole limb as previously described by Spence et al.[2]

In figure 4 (a and b) an attempt has been made to explain the inflammatory hyperthermal foot hemodynamically. The patient, a non-diabetic female, had a severely ischemic right hallux without infection. On the thermogram the big toe is hypothermal but a substantial degree of foot hyperthermia exists when compared to the contralateral side. Foot hyperthermia appears to be associated with a significant increase in blood volume flow but at a reduced blood pressure. The lower perfusion pressure accounts for the slightly decreased nutritional skin blood flow despite the larger volume flow. This explanation accords well with the hypothesis put forward by McEwan et al,[3] who explained the inflammatory ischemic foot in terms of the pulse, arterial pressure and volume of blood flow. They questioned the assumption that a regional blood flow of normal or higher than normal volume must adequately perfuse the tissues. A relatively high blood flow in the foot may be due to a reduction in the resistance to flow in the non-nutritional, heat-exchanging A-V shunts following local anoxia.

The importance of an understanding of the microcirculation cannot be overstated if we are to appreciate the subtle changes which take place in the skin circulation following drug treatment or sympathectomy. Thermographic monitoring of such alterations in blood flow can, under strict measuring conditions, prove to be a valuable adjunct to patient management. There is no doubt that the interpretation of thermographic images of ischemic feet is greatly enhanced when taken in conjunction with other hemodynamic measurements, especially if the skin blood flow can be included.

It is also important here to re-stress the need for taking accurate, reproducible thermographic images. This is of particular concern when studying the effects of drugs. If possible, some way of quantifying the temperature distribution either in terms of a region of interest or as a gradient, before and after therapy, makes interpretation so much easier. Precise control of room temperature and the time of day of each investigation is paramount.

The important thermal feature which could be interpreted as a local hemodynamic improvement to the skin circulation after therapy is a positive change in the local thermal gradient, not the actual skin temperature. The skin temperature varies considerably according to many internal and external factors and although the effects of these variables

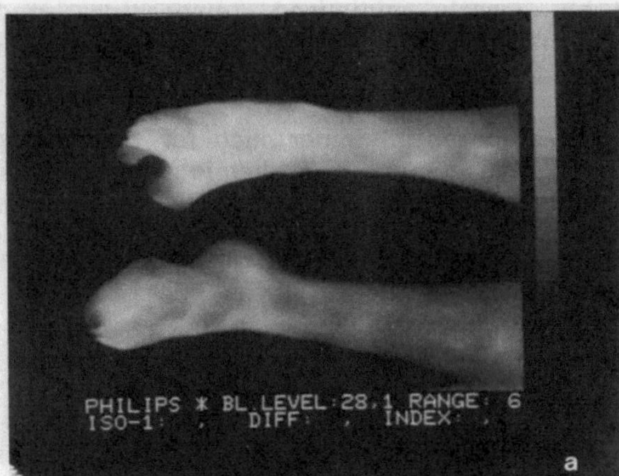

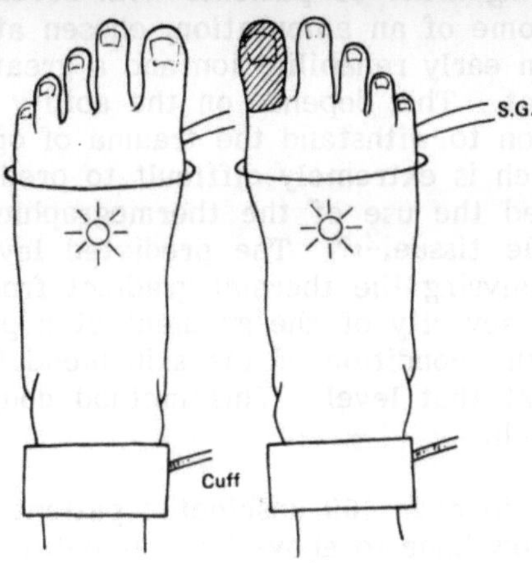

Pressure Index	0.91	0.61	
Arterial Inflow	0.5	2.9	mls/100 ml/mi
Vein Capacity	1.3	0.9	mls/100 ml
Max. Venous Outflow	23	30	mls/100 ml/mi
Skin Blood Flow	9.4	7.1	mls/100 g/min

b

Fig. 4. a) A thermogram of the lower limbs of a patient with pre-gangrene of the right big toe. Apart from the big toe which appears cold on the thermogram the remainder of the foot is hot.

b) Hemodynamic measurements made on the same patient as in figure 4a. The pressure index (ankle/brachial ratio) was recorded by the Doppler method; arterial inflow, vein capacity and maximum venous outflow were recorded by strain gauge plethysmography using cuffs around the ankle and strain gauges around the foot; the skin blood flow was measured by recording the clearance of 50 µl of ^{125}I-4-Iodoantipyrine[4], injected into the skin of the dorsum of the foot, as indicated.

can, and should be, greatly minimised, temperature gradients of an ischemic limb are remarkably stable providing that no further deterioration of the limb circulation takes place.

A further area of potential importance for thermography concerns the investigation and management of ischemic ulcers. Around an ulcer there is, in the majority of cases, a hyperemic response which can be accurately marked out by thermography. The extent and degree of the hyperemic response and its reaction to a thermal challenge may provide an index of the healing potential of the ulcer.

SELECTION OF THE LEVEL OF AMPUTATION

Amputation of the lower extremity is sometimes required as an end result in the management of patients with severe peripheral ischemia. A successful outcome of an amputation, chosen at the most distal level possible, can mean early rehabilitation and a greatly improved quality of life for the patient. This depends on the ability of the skin flaps with marginal circulation to withstand the trauma of operation and eventually heal, a factor which is extremely difficult to predict. For several years we have advocated the use of the thermographic method for assessing the level of viable tissue.[2,5] The predicted level of amputation was ascertained by analysing the thermal gradient from the knee to dorsum of the foot. The severity of the gradient at a particular site provided an indication of the condition of the skin blood flow and therefore the healing potential at that level. This method could be criticised on its qualitative approach.

Nevertheless in over 400 unselected patients with severe ischemia and with a 4:1 below-knee to above-knee amputation ratio we were highly encouraged by our final healing rate of 91%. We have since confirmed our impression that the thermographic pattern at the amputation site correlates well with measurements of skin blood flow.

The thermogram (Fig 5a) is a black and white reproduction of a color thermogram. In the example, the different shades represent isothermal contours 0.6°C apart. The thermal gradient extends from mid-thigh to ankle but a sharp temperature drop of 3.6°C occurs at a level of 12 cm distal to the tibial tuberosity and extends for a further 4 cm. From our previous criteria this gradient would represent a change from viable to non-viable tissue. Measurements of skin blood flow, made on the anterior skin at different thermal levels confirmed the thermographic impression of the level of viable tissue (Fig 5b). The skin of the thigh and leg down to a level 12 cm distal to the tibial tuberosity has a relatively normal skin blood flow of 8.1 ml/100 g/min. The first shaded area has a blood flow of 2.8 ml/100 g/min which is only just consistent with healing. The final shaded area has a blood flow inconsistent with healing. This patient in whom an ankle pressure could not be recorded had a successful outcome of a below-knee amputation,

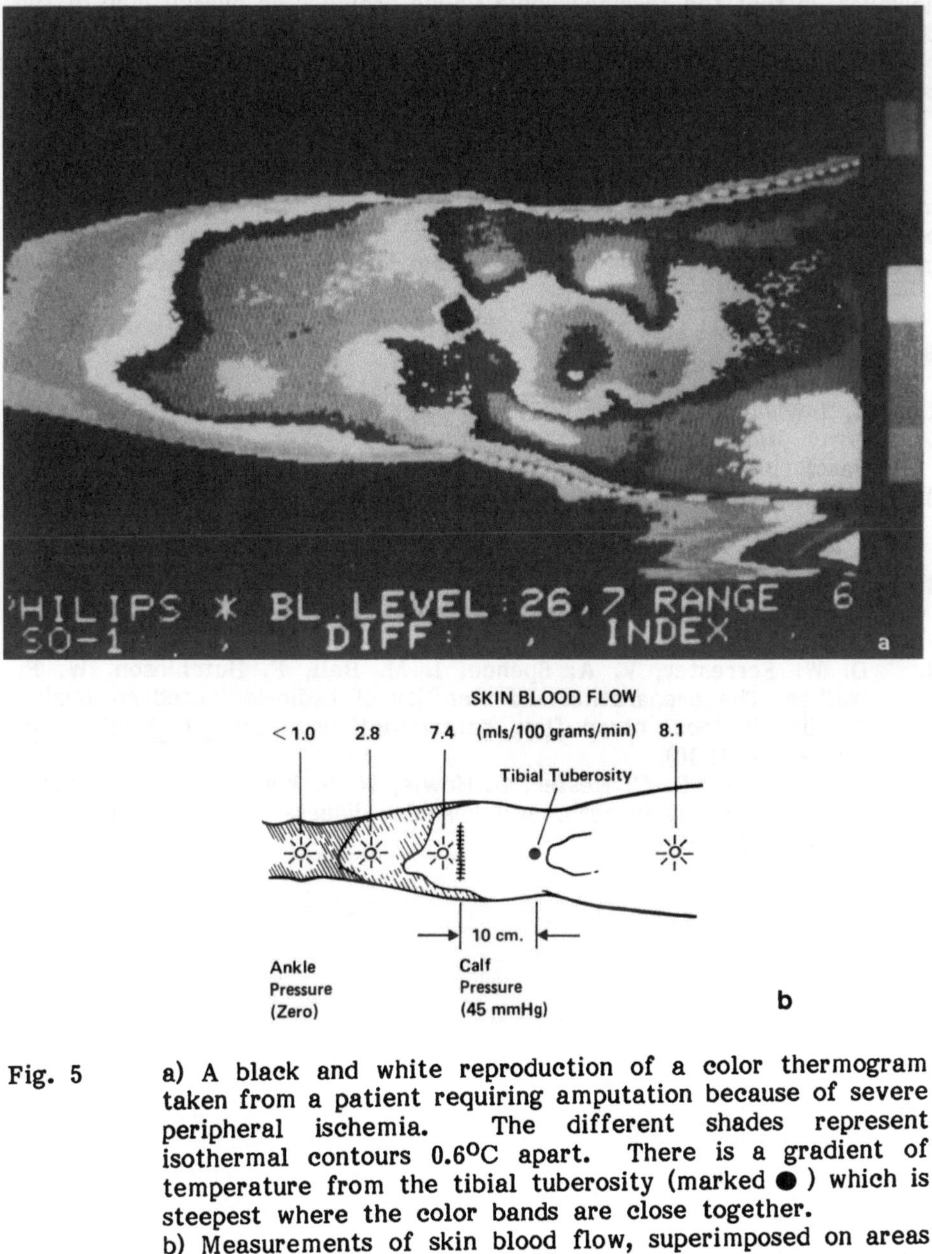

Fig. 5 a) A black and white reproduction of a color thermogram taken from a patient requiring amputation because of severe peripheral ischemia. The different shades represent isothermal contours 0.6°C apart. There is a gradient of temperature from the tibial tuberosity (marked ●) which is steepest where the color bands are close together.
b) Measurements of skin blood flow, superimposed on areas redrawn from the isothermal bands outlined in figure 5a. The pressure at the ankle and calf were assessed by the Doppler ultrasound method.

albeit with a short, modified stump. The important point to make here, however, is that the thermographic image provides an overall map of the different levels of viability of the skin, unlike the measurements of skin blood flow which are representative of a point source only. Thermography is thus an ideal technique for demarcating viable and non-viable skin flaps at the chosen site of amputation.

In conclusion, thermography does have a role to play in the investigation of patients with ischemic disease. The role is not as a primary diagnostic technique but as a method of assessing tissue viability in those patients who reach the end-stage of peripheral arterial disease.

REFERENCES

1. L. Lovisatti, L. Mora, G. F. Pistolesi, Thermographic patterns of lower limb arterial disease, in: Bibl. Radiol. 6:107-114, Karger, Basel (1975).
2. V. A. Spence, W. F. Walker, I. M. Troup, G. Murdoch, Amputation of the ischemic limb: selection of the optimum site by thermography, Angiology 32:155-169 (1980).
3. A. J. McEwan, I. McA. Ledingham, Blood flow characteristics and tissue nutrition in apparently ischemic feet, Br. Med. J. 3:220-224 (1971).
4. D. W. Forrester, V. A. Spence, I. M. Bell, F. Hutchinson, W. F. Walker, The preparation and stability of radio-iodinated antipyrine for use in local blood flow determinations, Eur. J. Nucl. Med. 5:145-146 (1980).
5. V. A. Spence, R. C. Kester, G. Howie, W. F. Walker, Current status of thermography in peripheral vascular disease, J. Cardiovas. Surg. 16:572-579 (1975).

THE ROLE OF THERMOGRAPHY IN THE ASSESSMENT OF THE ISCHEMIC LIMB

M Dewar*, T Coddington, P Jarman*, K Robinson****

*** Bioengineering Centre, University College London Roehampton, London, UNITED KINGDOM**
**** Limb Surgery Unit, Queen Mary's Hospital, Roehampton London, UNITED KINGDOM**

INTRODUCTION

For patients presenting with lower-limb ischemia owing to peripheral vascular disease, the treatment required will depend on the extent and severity of the ischemia and can range from the application of local or systemic vasodilators to lumbar sympathectomy, vascular reconstruction and ultimately amputation. In deciding the most appropriate treatment an assessment of the degree of ischemia must be made.

One area, for example, in which this assessment is of particular value is that of amputation surgery. A proportion of amputations for ischemia fail to heal, largely because of insufficient blood supply to the skin forming the skin flaps. A method which, before amputation, assesses the skin blood flow as an indicator of its viability would be of value.

Skin temperature is commonly used clinically as an indicator of the state of the blood supply to the limb and infrared thermography provides a convenient way of accurately measuring this. It has advantages as a method of assessment in that it is non-invasive, but the question which has to be answered is whether the information that can be provided is (a) reliable and (b) of any practical value.

Although its use in determining the most appropriate level for amputation in cases of peripheral vascular disease has been demonstrated,[1] the detailed information of the skin temperature which thermography can provide has not been utilized.

The aims of this work are:

1 To describe the distribution pattern of the skin temperature in healthy limbs.
2 To investigate the way in which this pattern is changed in ischemia.
3 To compare two amputation techniques, relating the temperature of the skin retained with the proportion of amputations having successful primary healing.

METHOD

Thermograms of the below-knee section of the leg were made for two groups of subjects:

(a) Patients with ischemia caused by peripheral vascular disease for which a successful below-knee amputation was carried out shortly after the thermogram was made.
(b) Volunteers, of the same age group as the patients, who had no symptoms or history of peripheral vascular disease.

Each subject sat with the legs uncovered in a stabilized temperature of $20 \pm 1^{o}C$. After a ten-minute cooling period thermograms were taken, of the affected limb in the case of the patients, and of the right limb in the case of the normals, from four directions: anterior, posterior, medial and lateral. Each of the four aspects of the limb was marked by foil markers placed proximally at the level of the tibial plateau and distally at the level of the malleoli. These defined the centre lines of the four views.

AGA 750 Thermovision equipment was used to give a color isotherm display with ten isotherms covering $5^{o}C$.

Because of the nature of the color display, slight variations in the temperature pattern between individuals appear disproportionately large to the eye and the appreciation of any general underlying pattern is made more difficult. For this reason the general pattern was investigated by converting the isotherms to a standard size and shape and taking an average of all the subjects.

Using the four centre-lines defined by the markers as reference lines the pattern for each subject was transferred to a standard rectangular grid. In effect the temperature pattern was 'unrolled' from the leg (Fig 1). The temperatures at a large number of points on the grid were then taken and combined with those for the other subjects to give mean values for each point. From these data the isotherm pattern was redrawn.

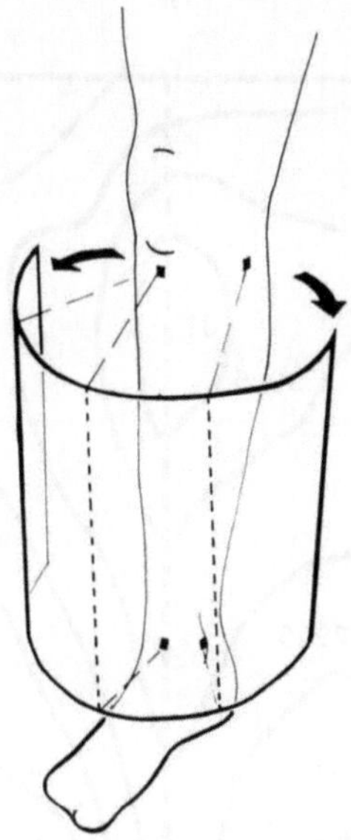

Fig. 1. Derivation of the standard thermograms of below-knee region. Anterior, posterior, medial and lateral midlines are defined by skin markers at the level of the tibial crest and the malleoli.

RESULTS

The mean pattern for the normal limbs is shown in Figure 2. Although there is in general a reduction in temperature of 2-3° from the proximal to the distal end of the limb this reduction is not uniform around the circumference and the pattern appears to reflect the position of the main arteries. Where the femoral artery is close to the surface in the popliteal fossa the skin temperature is high (31°). The path of the posterior tibial artery distally and medially from this point can be detected in the distal curve of the contour lines in this region.

Where the anterior tibial artery appears on the anterior of the limb a few centimeters below the knee the temperature is also high (31°) and the path of the artery down the front of the leg can be discerned in a similar way. The coldest area of skin is over the Achilles tendon.

The mean pattern for the ischemic limbs is shown in Figure 3. There is an overall reduction in temperature compared with the normal limb and the pattern is also different. The temperature gradient along the limb is more even than in the normal pattern and the warm areas extending from the popliteal fossa and on the front of the tibia are reduced or absent.

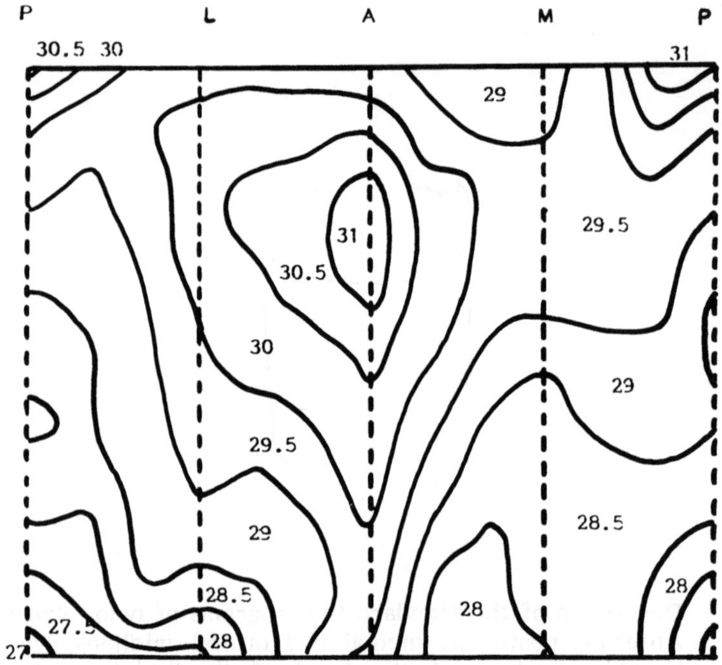

Fig. 2. Average temperature distribution of the below-knee region of the normal leg (number of subjects: 5. Mean age: 63.0 years). The top edge is the level of the tibial crest, the bottom one the level of the malleoli, the five broken lines mark the midlines of the posterior (P), lateral (L), medial (M) and anterior (A) aspects. Numbers are temperature (°C).

In Figure 4 the approximate incision lines for two different types of amputation, the long posterior flap[2] and the skew flap,[3] are superimposed on the mean ischemic pattern.

The long posterior flap incision extends distally and utilizes skin in the colder parts of the pattern. The symmetrical flaps of the skew-flap amputation on the other hand correspond more closely with the temperature contours and therefore enclose warmer areas of the skin.

In a recent series of amputations for peripheral vascular disease performed by one of the authors (KR) using the skew-flap procedure, the rate of primary healing was 71% (n = 24) compared with 49% for previous amputations carried out using the long posterior flap procedure.

DISCUSSION

It is apparent that skin temperature, in isolation, is not an indicator

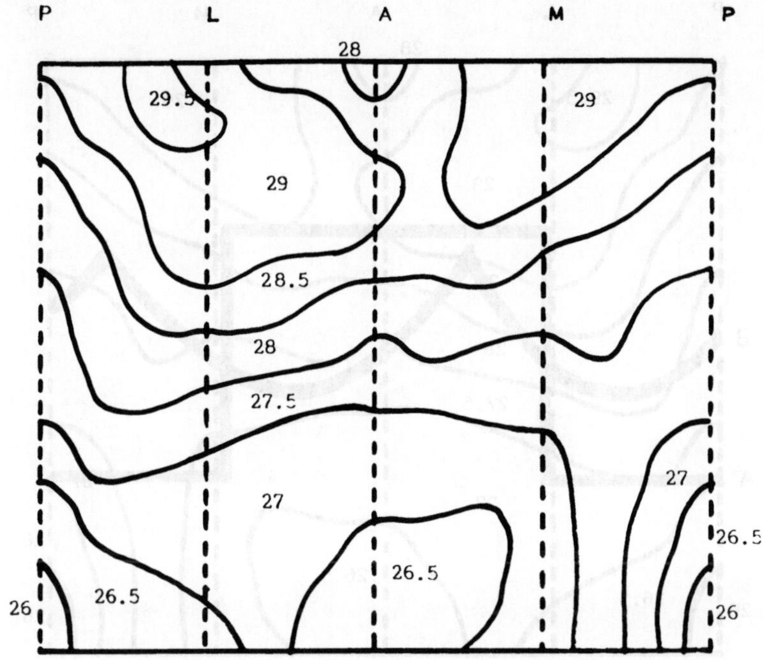

Fig. 3. Average temperature distribution of the below-knee region of the ischemic leg (number of subjects: 5. Mean age: 64.7 years). Details as for Figure 2.

of tissue viability. Factors other than skin blood flow affect it and there are, for example, areas on the normal limb which are obviously viable and yet have a lower temperature than areas on the ischemic limb which are removed during amputation. It is, therefore, the surface temperature in relation to its position on the limb, in other words the temperature pattern, which must be considered.

In relation to amputation surgery the questions which an examination of this pattern might be expected to help answer are:

1 Is an amputation necessary?
2 If so, what level should it be at to maximize the chance of primary healing?
3 Is one particular arrangement of skin flaps indicated which will maximize the chance of healing?

This work so far has indicated that there is a characteristic temperature pattern in the normal limb, that this is signficantly changed in the ischemic limb, and that, in the average ischemic pattern, the warmer areas of skin are associated with greater viability.

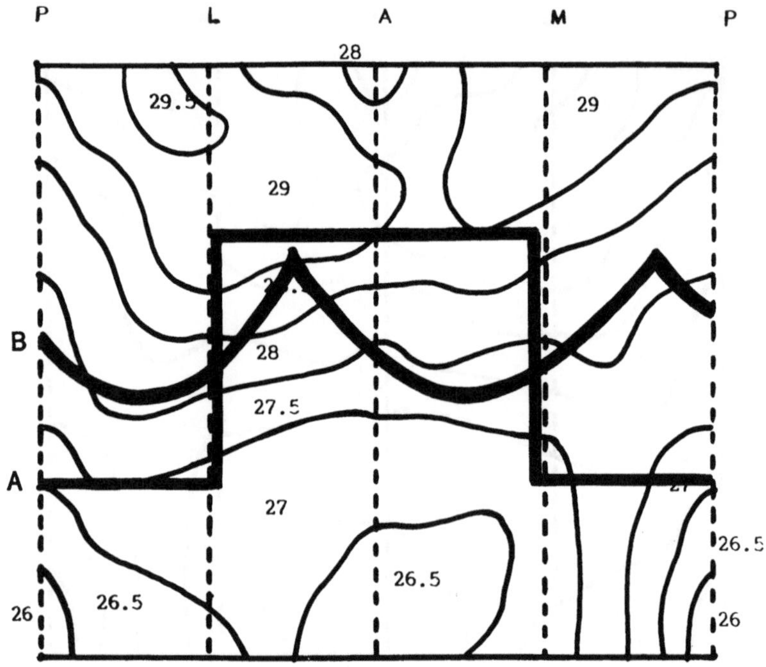

Fig. 4. Diagram of approximate incisions for long posterior flap amputation (A) and skew flap amputation (B) superimposed on the average temperature of the ischemic leg.

Two other groups of patients are yet to be investigated: those presenting with ischemia for which an amputation was not carried out and those who had a below-knee amputation which failed to heal. In addition it will be necessary to consider the variability within each group before it can be decided whether thermography has a role to play in the routine clinical assessment of the ischemic limb.

REFERENCES

1. V. A. Spence, Amputation of the ischemic limb: selection of the optimum site by thermography, Angiology 32:155-169 (1981).
2. E. M. Burgess, R. L. Romano, J. H. Zettl, The management of lower extremity amputations, Publication TR 10-6, Veterans Administration (1969).
3. K. P. Robinson, R. Hoile, T. Coddington, Skew flap myoplastic below-knee amputation - a preliminary report, Br. J. Surg. 69:554-557 (1982).

THE CORRELATION BETWEEN AFTER-EXERCISE THERMOGRAPHY (AET) AND DEEP VENOUS INSUFFICIENCY

T Hallbook, D Bergqvist, A Lindhagen, B Lindroth

Department of Surgery and Radiology
Karnsjukhuset, Skovde
SWEDEN

The technique of after-exercise thermography was described by Cooke and Pilcher[1] and by Henderson et al.[2] They found that abnormal AET indicated a defective calf muscle pump and thus deep venous insufficiency. The available literature, however, contains no reports in which AET was evaluated against some objective method of diagnosing venous insufficiency. Our study was therefore designed for further investigation of AET in deep venous insufficiency assessed clinically and with venous pressure and plethysmographic measurements.

MATERIALS AND METHODS

167 patients were studied for four years or more after major surgery. The mean age was 70.5 years. The surgery performed was hip replacement in 85 cases and abdominal operations in 82 cases. All patients had been postoperatively screened with ^{125}I-fibrinogen uptake test (FUT) to detect deep vein thrombosis. FUT was bilaterally negative in 95 patients, bilaterally positive in 28 patients and unilaterally positive in 44 patients.

The patients replied to a questionnaire concerning clinical symptoms indicating deep venous insufficiency. The clinical status of the legs was recorded and AET, strain gauge plethysmography and measurements of venous pressure were performed. AET was performed with the AGA 680 Medical System. The patient had rested for 10 minutes with legs undressed and raised more than 15 degrees. A grey tone thermogram was taken and the patient was thereafter exercised by walking on the spot for three minutes and another thermogram was recorded. When a network of linear hot spots crossing the anterior tibia was found the result was regarded as positive.

Venous occlusion strain gauge plethysmography and ambulatory strain gauge plethysmography were performed. Venous volume less than 1.5 ml/100 ml tissue and maximum venous outflow less than 30 ml/100 ml/tissue/minute and return time less than 8 seconds were regarded as abnormal. Venous pressure was measured in a dorsal vein of the foot and return time less than 25 seconds was classified as pathological.

RESULTS

After-exercise thermography in 167 patients was positive in 128 legs and negative in 206 legs. Bilateral positive AET was found in 47 patients (94 legs). The patients' replies to the questionnaire are reported in Table 1. The only significant differences between the AET-positive and the AET-negative legs were that varicose veins had developed more frequently and pressure bandaging and use of elastic stockings were more common in the positive group. The findings at the clinical examination are summarized in Table 2. There was greater frequency of varicose veins in the AET-positive legs.

The results of strain gauge plethysmography and venous pressure measurements are shown in Table 3. There were no statistically significant differences between AET-positive and AET-negative legs when the plethysmograms or pressure recordings were considered separately or in combination.

Table 1. Subjective manifestations of venous insufficiency (replies to questionnaire)

Symptoms during follow-up period	AET-positive legs no	AET-positive legs %	AET-negative legs no	AET-negative legs %	Statistical significance
Leg ulcer	4	3.1	4	1.9	NS
Swollen leg	13	10.2	24	11.7	NS
Development of varicose veins	33	25.8	25	12.1	(p 0.01)
Discolouration of leg	16	12.5	17	8.3	NS
Eczema of leg	3	2.3	4	1.9	NS
Nocturnal cramp in leg	26	20.3	35	17.0	NS
Pressure bandage or elastic stocking constantly required	7	5.5	4	1.9	(p 0.05)

Table 2. Findings as clinical manifestations

Clinical findings	AET-positive legs no	AET-positive legs %	AET-negative legs no	AET-negative legs %	Statistical significance
Leg ulcer	1	0.8	1	0.5	NS
Venous eczema	5	3.9	2	1.0	NS
Discolouration	19	14.8	19	9.2	NS
Varicose veins of thigh	21	16.4	6	2.9	(P 0.001)
Varicose veins of leg	55	43.0	53	25.7	(P 0.01)

DISCUSSION

Thermography of the legs performed after exercise has been claimed to show a characteristic pattern in deep venous insufficiency. This study was not able to show any correlation between positive AET and objective or subjective signs of deep venous insufficiency.

Varicose veins were more common in the patients with positive AET than in the AET-negative group. At the clinical examination, however, 57% of the AET-positive legs had no varicose veins. There was also an appreciable incidence of varicosity (26%) in the AET-negative legs. Varicose veins, therefore, do not seem to be the explanation of the characteristic pattern of positive after-exercise thermograms. The causal factors of the characteristic pattern in abnormal AET remain still unclear. The question is raised: if deep venous insufficiency localized to the calf can give this characteristic pattern it has predictive value for postoperative deep venous thrombosis.

Table 3. Pathological plethysmographic and venous pressure measurements

	AET-positive legs no	AET-positive legs %	AET-negative legs no	AET-negative legs %	Statistical significance
Plethysmography					
Reserve venous volume	6	4.7	12	5.9	NS
Maximum venous outflow	9	7.1	17	8.3	NS
Return time	26	35.6	42	38.8	NS
Venous pressure					
Pressure decrease	22	19.1	31	17.1	NS
Return time	24	20.9	29	16.0	NS
Patients with pathological plethysmography and/or venous pressure	40	34.7	61	33.7	NS

REFERENCES

1. E. D. Cooke and M. F. Pilcher, D. V. T. preclinical diagnosis by thermography, Brit. J. Surg. 61:971-978 (1974).
2. H. P. Henderson and M. E. J. Hackett, The value of thermography in peripheral vascular disease, Angiology 29:65-75 (1978).

THERMOGRAPHY AND RAYNAUD'S SYNDROME

M V Kyle, G Parr, R Salisbury, P Page Thomas, B L Hazleman

Rheumatology Research Unit
Addenbrooke's Hospital, Cambridge
UNITED KINGDOM

Thermographs display the pattern of heat distribution of the surface under examination. This pattern depends to a large extent on local blood flow. Thermography therefore seems a particularly suitable medium with which to assess vasospastic disorders such as Raynaud's phenomenon, where symptoms are due to local ischemia. As the condition is episodic, objective assessment is difficult. Cold stress tests (CST) have been used to mimic events when a cold stimulus leads to an attack of Raynaud's. In conjunction with thermography, such tests provide a useful, non-invasive method of quantifying temperature and heat patterns.

Despite widespread use of thermography in breast malignancy and to a lesser extent in inflammatory rheumatic diseases, there is little published information on its use in vasospastic disease. Haberman et al[1] noted almost 20 years ago that thermographs of patients with progressive systemic sclerosis showed a decrease in temperature in distal extremities, especially if Raynaud's phenomenon was marked. Other workers have reported on the presence or absence of hyperemia after mild CST[2] and rewarming times after a more severe stress test in primary and secondary Raynaud's and normal controls.[3] No consistent results have emerged, however, and there has been considerable overlap between normal and abnormal. Nor have other methods of assessing Raynaud's phenomenon, such as thermometry, plethysmography, isotopic scanning or arteriography, proved more successful. Technical difficulties and/or the invasive nature of such investigations have limited their value.

The aim of this study was to use CST and thermographic assessment to examine in detail the response patterns of patients with secondary Raynaud's, primary Raynaud's and normal controls.

PATIENTS AND METHODS

Thirty volunteers with primary Raynaud's, 30 patients with secondary Raynaud's and 30 normal controls had CST and thermographic assessment. Patients' details are shown in Table 1. The criteria required for diagnosis of Raynaud's were a history of cold stimulus resulting in biphasic attacks of numbness and/or pain in association with digital pallor or cyanosis, followed by tingling and/or pain and skin reddening during rewarming. Secondary Raynaud's was considered to be present if there was a known underlying connective tissue disease, or severe digital ulceration, or if gangrenous changes occurred in the fingers, even if a recognisable underlying condition had not been diagnosed. Clinical and laboratory features were also recorded, but will not be discussed in this paper.

Thermograms were taken of the hands at a distance of 1m, using AGA 680M equipment, in a room kept at 20°C. 1° sensitivity (range 10°) was used initially but the majority were recorded at 2° sensitivity with a range of 20°C because of the widespread thermal values. Computer analysis was carried out with an Apple II microcomputer interfaced to the OSCAR unit to measure mean temperature, gradient and grey level distribution. The values were obtained from an area of interest selected from each digit and extending as far as possible throughout the length.

In our cold stress test, both hands of the subject were thermographed on entry and after a 15 minute stabilisation period. The gloved right hand was then immersed in water at 37°C for 3 minutes, after which both hands were thermographed. The right hand was then immersed in water at 20°C for 1 minute; immediately after this cold stress, thermographs were taken of both hands. The right hand was thermographed at 5 minute intervals for 20 minutes and the left after 20 minutes at the end of the test.

Table 1. Age, sex and duration of Raynaud's

	M	F	Mean age (yrs)	Range	Mean duration (yrs)	Range
Normal controls	7	23	33.12	-	-	-
Primary Raynaud's	5	25	40.20	23-68	23.32	6-35
Secondary Raynaud's	6	24	48.36	13-77	12.81	1-50

RESULTS

Characteristic responses were seen in each group.

Normal Controls

Most showed hyperemia during the 15 minute stabilisation period. After the CST, the right hand rewarmed rapidly. Throughout the test, both hands usually maintained a uniform temperature, with the tips tending to be hotter (positive gradient) (Fig 1).

Secondary Raynaud's

These patients usually cooled after entry. After CST the digits failed to show significant rewarming and often cooled further. The mean temperature was usually low on entry (< 27°C) and the range of temperature recorded during the test was high (up to 10°C). A well-defined banded isotherm pattern was seen extending down the digits, with temperature decreasing towards the finger tip (negative gradient). Gradients of up to 14°C were recorded (Fig 2).

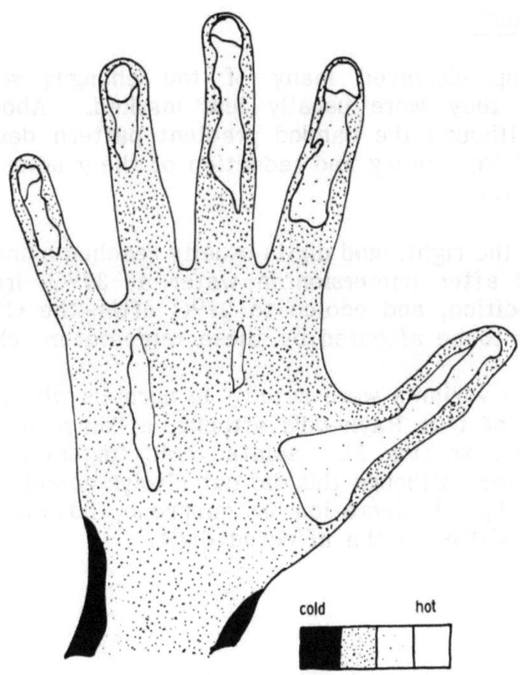

Fig. 1. Normal hand with hot finger tips (positive gradient).

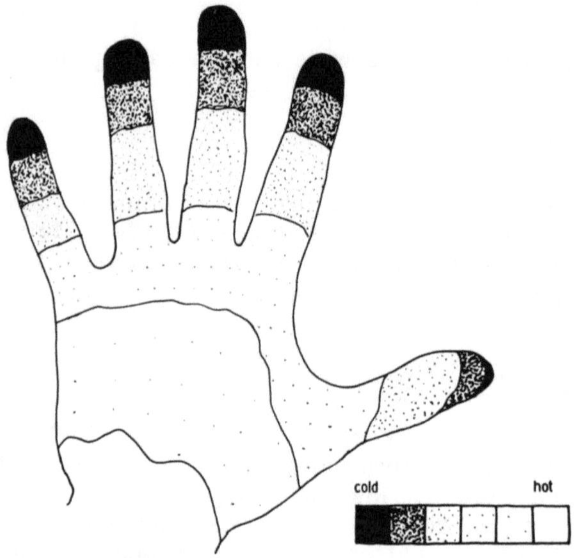

Fig. 2. Raynaud's phenomenon with banded isothermal pattern (negative gradient).

Primary Raynaud's

This group displayed many of the changes seen in secondary Raynaud's but they were usually less marked. About half did show hyperemia. Although the banded gradient pattern described was often seen after CST, rewarming and reduction of the gradient usually occurred by the end of the test.

Although the right hand digits usually reached a uniform temperature of about 33°C after immersion in water at 37°C, irrespective of the underlying condition, and cooled to 27°C after the CST, the left hand did not appear to be affected by these temperature changes.

Interesting variants were noted - in normals who complained of cold hands, but not of true Raynaud's attacks, a "diagonal" gradient pattern was sometimes seen (Fig 3). Isolated cold fingers with gradients were seen in some cases, although this did not always correlate with symptoms. Patients with digital ulceration or gangrene, however, usually showed greater abnormalities in the affected digit.

DISCUSSION

As mentioned earlier, previous attempts to evaluate Raynaud's phenomenon have been difficult to interpret. This may be partly due to

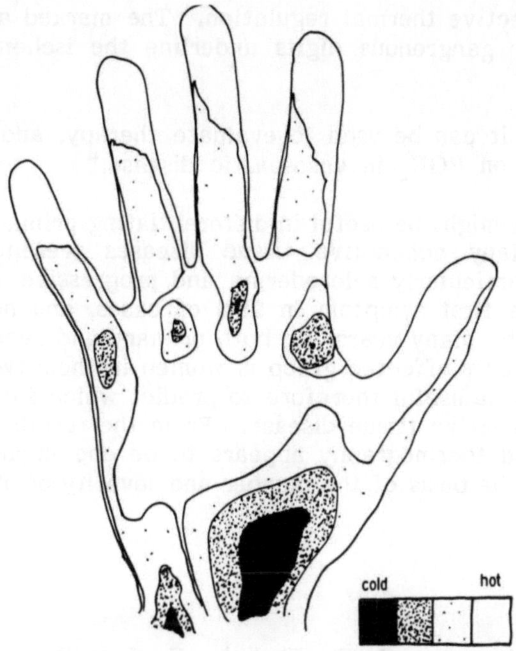

Fig. 3. Normal variant with wedge-shaped thermal patterns.

the different stress tests employed, and the varying times at which thermographs were taken.

In this study, we found characteristic response patterns, most of which have not been previously described. The combination of these appeared to be important, ie the presence of hyperemia, rapid rewarming and positive gradient were typical normal responses; all three never occurred together in patients with secondary Raynaud's and although one or two of these features might be seen in those with primary Raynaud's, all three were rare. Similarly, cooling during the 15 minute stabilisation phase and after the CST, plus the presence of a negative gradient, which in combination were characteristic of secondary Raynaud's, never occurred altogether in normal controls, and were unusual in primary Raynaud's, though one abnormality at least was common.

The value of thermographic assessment and cold stress testing lies in three main areas. Firstly, it may help us to explain the underlying pathogenesis of Raynaud's attacks.

Although the mechanism underlying Raynaud's phenomenon is poorly understood, these results confirm the inability to maintain digital temperature in the face of relatively mild cold stress, and the wide swings of temperature in comparison to normal controls supports the concept of defective thermal regulation. The marked abnormalities seen in ulcerated or gangrenous digits underline the ischemic component to these.

Secondly, it can be used to evaluate therapy, and we have already presented data on PGE_1 in vasospastic disease.[4]

Thirdly, it might be useful in differentiating primary from secondary Raynaud's. Many connective tissue diseases present with Raynaud's phenomenon, particularly scleroderma and progressive systemic sclerosis where it is the first symptom in 90% of cases, and may precede other manifestations by many years. In both primary and secondary Raynaud's, the most commonly affected group is women in their twenties or thirties; it would clearly be useful therefore to predict which sufferers were likely to develop connective tissue disease. From the results we have obtained so far, CST and thermography appears to be one means of making this distinction, on the basis of the number and severity of abnormal responses present.

REFERENCES

1. J. D. Haberman, G. E. Ehrlich, C. Levenson, Thermography in rheumatic diseases, Arch. Phys. Med. and Rehab. 49:187-191 (1968).
2. E. F. J. Ring, A thermographic index for the assessment of ischemia, Acta Thermographica (1981).
3. F. Chucker, R. C. Fowler, T. Motomiya, B. Singh, W. Hurley, Induced temperature transients in Raynaud's disease measured by thermography, Angiology 22:580-593 (1971).
4. M. V. Kyle, G. Parr, R. Salisbury, P. Page-Thomas, B. L. Hazleman, PGE_1, vasospastic disease and thermography, Presented at Heberden Round, February 1982.

COMPARISON BETWEEN ACUTE AND CHRONIC ADMINISTRATION OF A COMBINED ALPHA- AND BETA-ADRENOCEPTOR BLOCKER (LABETOLOL) AND A BETA-ADRENOCEPTOR BLOCKER (PROPRANOLOL) IN HYPERTENSIVE SUBJECTS

E D Cooke, S A Bowcock

Department of Medical Electronics
St Bartholomew's Hospital, London
UNITED KINGDOM

Cold extremities are a well-documented complication of beta-adrenoceptor blockade.[1,2,3,4,5,6] The probable mechanism is digital vasospasm. It is possible that a pharmaceutical with the additional property of alpha-adrenoceptor blockade may attenuate or avoid this phenomenon. This investigation was performed to confirm or refute this hypothesis.

PATIENTS AND METHODS

Acute Study

In a single blind cross-over study 10 hypertensive subjects were given 3 mg of propranolol and 10 mg of labetolol intravenously after measurement of the baseline response of their hands to cold stress (20^oC). Cold challenge was repeated after 12 minutes, ie when the maximum effect of the drug was observed. The investigation was carried out at two different ambient temperatures, 20$^oC \pm 1^oC$ and 24$^oC \pm 1^oC$. Thus each subject was studied on four different occasions.

Chronic Study

The response of the hands to mild cold stress (20^oC) of 30 untreated hypertensive subjects of less than 60 years old was studied at two ambient temperatures, 20$^oC \pm 1^oC$ and 24$^oC \pm 1^oC$.

Cold sensitivity was also studied in 60 treated hypertensives aged less than 60 years after 3 months therapy with either propranolol (30 patients) or labetolol (30 patients) at the same two ambient temperatures.

Excluded were all subjects with prior Raynaud's phenomenon or concomitant disease, ie all subjects suffered from essential hypertension only.

Examples of the normal and abnormal response to cold stress is shown in figure 1.

RESULTS

Acute Study

Half of the subjects showed a variable vasodilatory response 12-20 minutes after the intravenous injection of either substance. No response was recorded in the remainder. Afterwards the effect on cold sensitivity was variable and unpredictable (Table 1).

Chronic Study

Enhanced cold sensitivity was present in 53.3% of untreated hypertensives, in 32.3% of the patients treated with the simple beta-adrenoceptor blocker (propranolol) and in 11.5% of those treated with a combined alpha- and beta-adrenoceptor blocker (labetolol).

DISCUSSION

The finding that half the untreated patients with essential

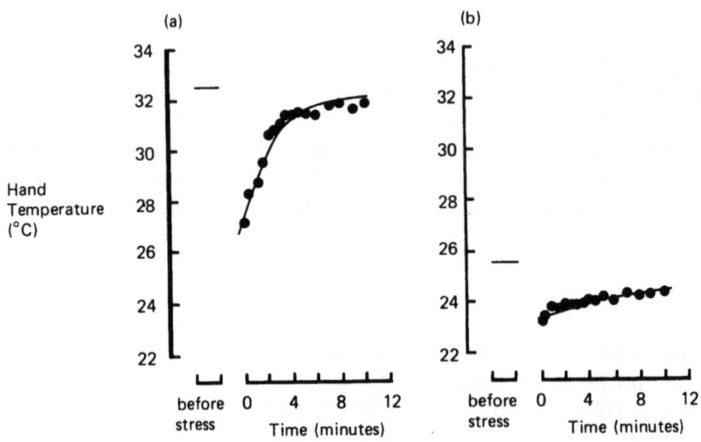

Fig. 1. The normal and the abnormal response following cold stress of the hands. The normal hand is always warmer, the fall in temperature after cold challenge is greater and the recovery time rapid in comparison with the abnormal.

Table 1. The effect on cold sensitivity of an intravenous injection of propanolol or labetolol is unpredictable.

ACUTE STUDY			
	Unaffected	Enhanced	Reduced
Cold sensitivity	16	13	7
β-blocker	7	7	4
α and β-blocker	9	6	3

hypertension showed enhanced cold sensitivity on objective testing is scarcely surprising considering the increase in vascular peripheral resistance associated with essential hypertension. It seems reasonable, therefore, that any treatment which reduces blood pressure may abolish this phenomenon. However, it seems paradoxical that this incidence was reduced during propranolol therapy which is recognized to produce cold extremities in some subjects. Alpha-adrenogenic agonist potentiation has been suggested as a possible mechanism.[3,7] The finding that cold sensitivity is almost abolished in hypertensive subjects treated by the combination of an alpha- and beta-adrenoceptor blocker (labetolol) lends weight to this hypothesis (Fig 2).

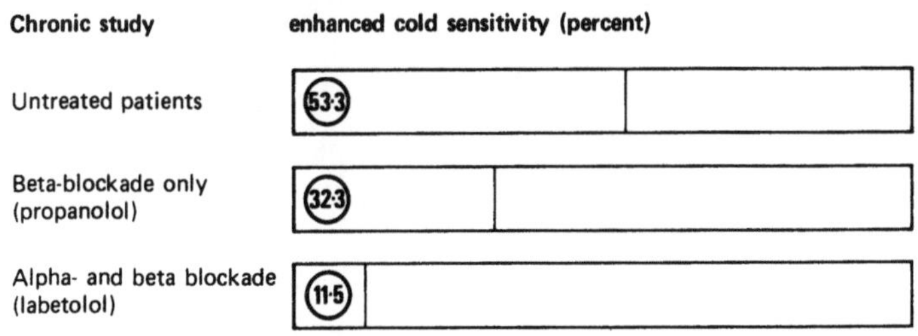

Fig. 2. The reduced incidence of enhanced cold sensitivity following three months of beta-adrenoceptor blockade (propanolol) therapy is further reduced by the addition of alpha blockade (labetolol).

REFERENCES

1. A. G. Marshall, C. J. C. Roberts, D. W. Barrett, Raynaud's phenomenon as side effect of beta blockers in hypertension, Br. Med. J. 1:1498 (1976).
2. D. Trash, M. Grandman, J. Cargill, L. Christopher, Cold extremities and beta blockers, Br. Med. J. 2:527 (1977).
3. C. L. A. van Herwaarden, R. A. Binkhorst, J. F. M. Fennis, A. van T. Laar, Effects of adrenaline during treatment with propranolol and metaprolol, Br. Med. J. 1:1029 (1977).
4. P. D. McSorley, D. S. Warren, Effects of propranolol and metaprolol on the peripheral circulation, Br. Med. J. 2:1598 (1978).
5. R. Gokal, T. L. Dornan, J. G. G. Ladingham, Peripheral skin necrosis complicating beta blockage, Br. Med. J. 1:721 (1979).
6. B. J. Hoffbrand, Peripheral skin necrosis complicating beta blockade, Br. Med. J. 1:1082 (1979).
7. C. de B. White, B. P. Udwadia, Beta-adrenoceptors in the human dorsal hand vein, the effects of propranolol and practolol on venous sensitivity to nor-adrenaline, Br. J. Clin. Pharmacol. 2:99 (1975).

A PILOT STUDY IN THE EVALUATION OF B-BLOCKERS IN HYPERTENSIVE PATIENTS WITH THERMOGRAPHY

N J Aarts, J Rijpert

St Elisabeth Ziekenhuis
Tilburg,
THE NETHERLANDS

INTRODUCTION

The present day treatment of choice in hypertension is beta-adrenoceptor blockade. This therapy is based on Ahlquist's concept of neurotransmission at alpha and beta-adrenoceptors.[1] However, the mechanism by which beta-blockade reduces blood pressure is not fully understood and may be through a variety of mechanisms which include reduction in the rate and force of the heart, direct and indirect inhibition of the effect of noradrenaline on peripheral vessels, and on the renin-angiotensin mechanism.

Cold extremities are a common complication of beta-adrenoceptor therapy.[2] This effect is believed to be less common with the so-called cardioselective ($beta_1$) adrenoceptors than those with unselective activity and may be due to the intrinsic sympathomimetic activity (ISA) of the latter.

This investigation was performed to determine if it is possible to measure these supposed differences objectively using thermography (THG). Further, if this is so, can THG be used to select the appropriate drug for use in individual patients.

MATERIALS AND METHODS

Six untreated hypertensive patients with Raynaud's phenomenon (aged 36-70 years) were selected at random from the out-patient population. All were otherwise asymptomatic and had no serious cardiac condition, asthma or diabetes mellitus. There were 3 males and 3 females. Under standard conditions of temperature and humidity ($19^oC \pm 1^oC$, RH

40-50 per cent) each patient was examined on three separate occasions. Initially, the response of the hands to cold stress (10 minutes immersed in cold tap water) was determined using a Thermographic Index. After 30 minutes a tablet, the nature of which was unknown to the investigator and patient, was given orally and the response of the hands over the next 60 minutes determined, the first reading being taken 30 minutes after ingestion of the tablet. After all six patients had been investigated the code was broken and the tablet identified.

RESULTS

These are shown in figures 1, 2 and 3. Raynaud's response was confirmed in all patients though the response of individual patients varied from one examination to another. Following administration of the drug, some degree of vasoconstriction was apparent in four of the six patients given atenolol and in two patients on acebutolol and propranolol. Well-marked vasodilation was shown in two patients with propranolol and one on acebutolol and atenolol. Symptoms of possible vaso-vagal origin were reported by several of the patients after ingestion of atenolol.

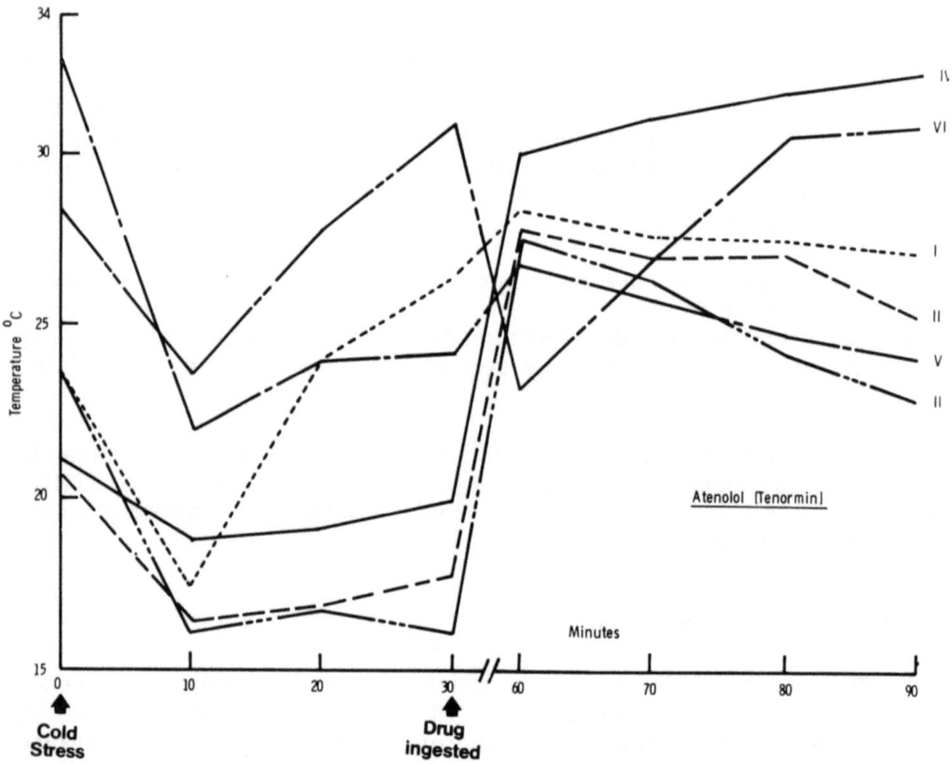

Fig. 1. Influence of atenolol on the recuperative time after cold stress in the region of interest (dorsal hand, MP joints 2-4).

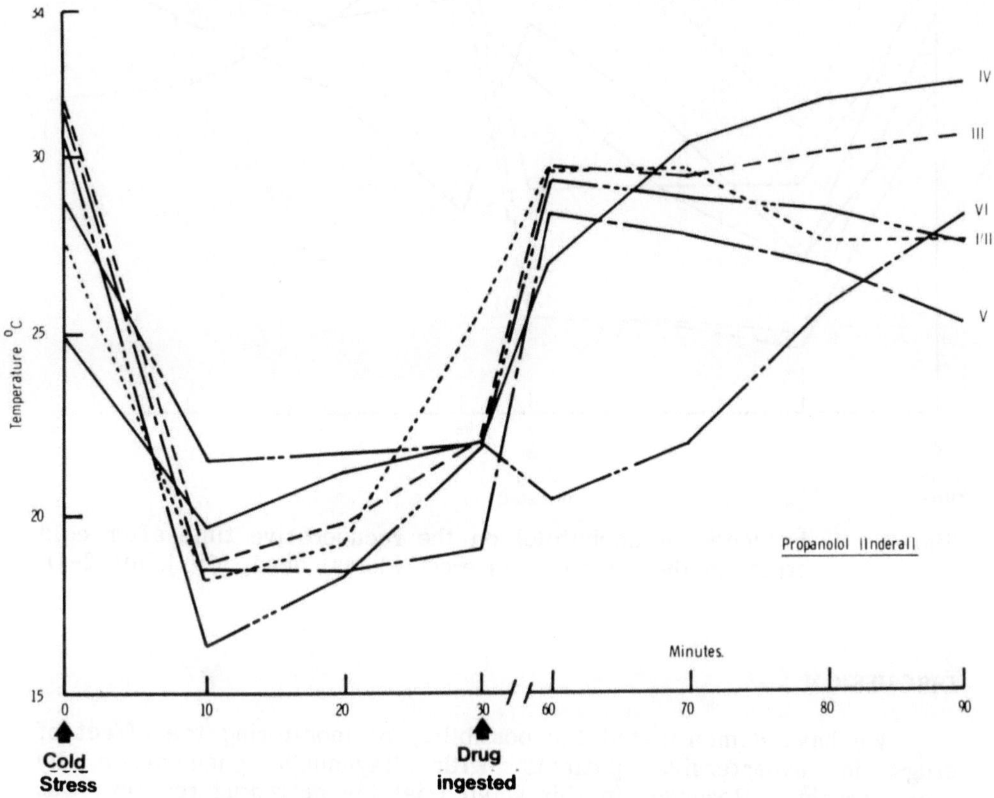

Fig. 2. Influence of propranolol on the recuperative time after cold stress in the region of interest (dorsal hand, MP joints 2-4).

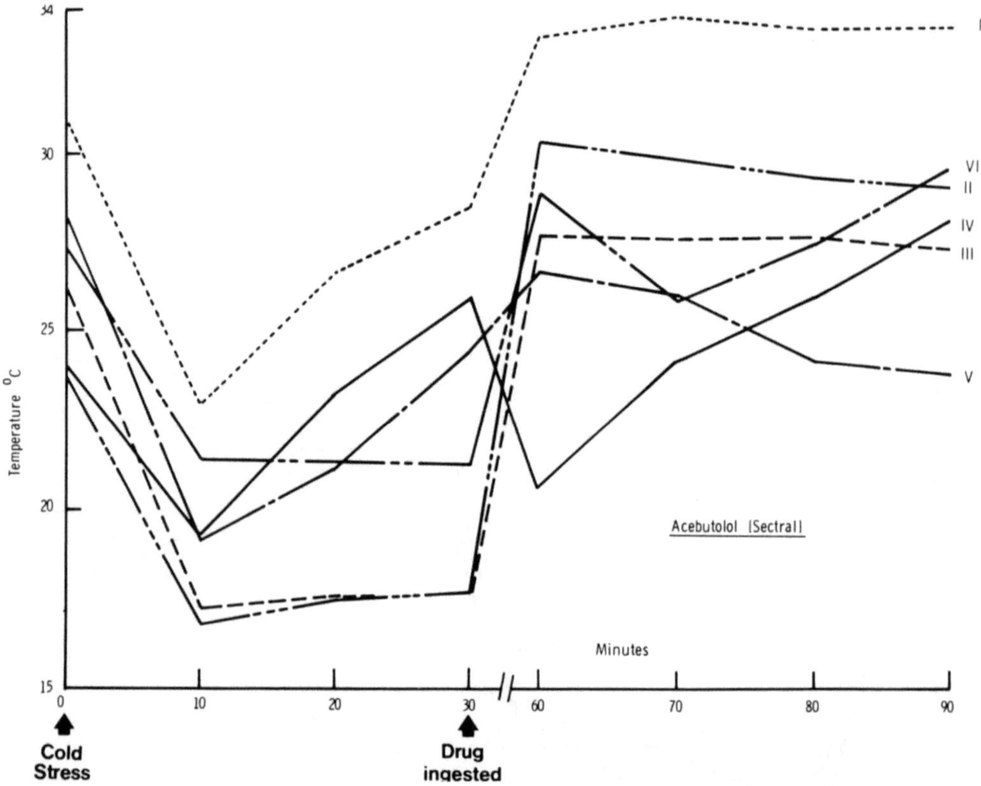

Fig. 3. Influence of acebutolol on the recuperative time after cold stress in the region of interest (dorsal hand, MP joints 2-4).

DISCUSSION

We have demonstrated the possibility of monitoring the effect of drugs in hypertensive patients with Raynaud's phenomenon by thermography. However, in this small trial the post-drug response was variable and inconsistent and could, at least in part, be reasonably attributed to spontaneous change rather than to the effect of the drug, indicating the need for placebo control. Since this is so, no firm conclusion is either possible or justified.

REFERENCES

1. R. P. Ahlquist, A study of the adrenotrophic receptors, Am. J. Physiol. 153:586-600 (1948).
2. J. M. Cruickshank, Beta-blockers, bradycardia and adverse effects, Acta Thermographica 7:309-321 (1981).

THERMOGRAPHY OF FACE AND NECK OF PATIENTS AFTER CEREBRAL APOPLEXY

K Yanagi

Department of Internal Medicine
Towada Municipal Central Hospital
JAPAN

INTRODUCTION

It is reported that the patients after cerebral apoplexy have generally lower skin temperatures on the affected side than on the non-affected side.[1,2] We have found that the limbs of patients are colder on the affected side. However, reports demonstrating such a phenomenon on the face and neck are still lacking.[3]

The author performed thermography on the face and throat of patients after cerebral apoplexy to determine their skin temperatures on the affected and non-affected sides, and analyzed the data according to the type and duration of disease.

SUBJECTS AND METHODS

The subjects admitted in the study were randomly selected from the in- and out-patients. They comprized 31 patients, ie 24 under conservative treatment consisting of 14 males and two females after cerebral infarction as well as four patients from both sexes after cerebral hemorrhage, and two females who had surgical treatment to remove hematoma. Five non-apoplectic patients were studied (two males and three females).

The apoplectic patients were aged 47 to 77 years (mean 62.1 ± 9.1 years) and the non-apoplectic patients 24 to 53 years (mean 36.0 ± 12.0 years).

Thermography was performed to determine the temperatures of the skin of the face and throat. Differences between the affected and non-affected sides were noted and related to the type and duration of diseases.

Urokinase was administered to some patients with cerebral infarction and the changes in the skin temperatures of face and throat before and after treatment observed. The therapeutic effects were studied by CT scanning.

Temperature changes of the face and neck after excision of cerebral hematoma were also studied in patients after cerebral hemorrhage.

The subjects were prohibited from smoking for 30 minutes before thermography and kept at rest. Thermograms were taken while the subjects were seated in an armchair.

The instruments used were Infra Eye 150, an infrared thermographic apparatus (Fujitsu) and a data processor developed by the author. The room temperature was approximately 25°C and humidity approximately 60%.

RESULTS

Skin Temperature of Non-Apoplectic Patients

The mean facial skin temperatures of the five non-apoplectic patients were 33.65 ± 0.26°C (mean + SE) at right side and 33.64 ± 0.19 °C at left side while those of the throat were 33.66 ± 0.25°C and 33.54 ± 0.26°C, respectively.

In both areas, there was no significant difference between the right and left sides.

Skin Temperature of Front Face

The mean facial skin temperatures of affected and non-affected sides in eight infarct patients within one month after stroke were 33.64 ± 0.25°C and 33.63 ± 0.25°C respectively while those of four patients after cerebral hemorrhage were 33.61 ± 0.20°C and 33.62 ± 0.25°C respectively.

At more than one month after stroke, the mean skin temperatures of 12 infarct patients were 33.43 ± 0.19°C and 33.53 ± 0.15°C on the affected and non-affected sides respectively, while those of two patients after hemorrhage were 33.72 ± 0.98°C and 33.43 ± 1.05°C respectively.

The changes of skin temperature after excision of hematoma varied according to the patients, some showing lower levels while the others displayed higher levels on their affected sides.

After more than one month, the temperature on the affected side tended to fall in the infarct patients ($p < 0.1$), which resulted in an increase in the difference of temperature between the affected and non-affected sides ($p < 0.1$). In the patients after cerebral hemorrhage, however, the temperature on the affected side seemed to rise ($p < 0.05$), which eventually brought about a significant increase in the difference between the affected and non-affected sides ($p < 0.05$).

Furthermore, the difference of skin temperatures on the affected and non-affected sides were compared between the infarct and hematoma patients in the periods of within one month and more than one month after stroke. At less than one month after stroke, there was no significant difference between the infarct and the hematoma patients. After more than one month, however, a larger difference between the affected and non-affected sides was noted in the infarct patients ($p < 0.02$).

Skin Temperatures of the Throat

At less than one month after stroke, the mean skin temperatures in eight infarct patients were 33.92 $\pm$ 0.16°C and 33.93 $\pm$ 0.17°C on the affected and non-affected sides respectively. In the four hematoma patients, the mean skin temperatures were 33.46 $\pm$ 0.34°C and 33.36 $\pm$ 0.23°C respectively.

After more than one month, the mean skin temperatures of the throat in the 12 infarct patients were 33.70 $\pm$ 0.14°C and 33.73 $\pm$ 0.14°C on the affected and non-affected sides respectively, while those in the two hematoma patients were 33.83 $\pm$ 0.19°C and 33.82 $\pm$ 0.02°C respectively.

The skin temperatures of the throat were not different between the affected and non-affected sides in the present study.

After urokinase treatment, the cutaneous temperatures seem to have become higher on the affected side than on the non-affected side.

No significant difference in the temperature between the affected and non-affected facial skin was observed, either in comparison of infarct and hematoma patients or in the different periods.

DISCUSSION

This study showed that in a relatively early stage of cerebrovascular accident there is no difference in skin temperature between affected and non-affected sides. Thereafter, the temperatures on the affected side tended to fall slightly in patients with infarction, while in patients with hematoma the temperature on the affected sides is higher than that on the non-affected side.

As described by Senay et al[4] the cutaneous temperatures may be an index for local blood circulation.

In the infarct patients, the temperature on the affected side tended to be lower than on the non-affected side. It is postulated that, in the presence of occlusion or stenosis of the internal carotid artery, intracranial blood flow is supplied by the collateral circulation of the external carotid artery[5] which eventually may cause a rise in the skin temperatures on the non-affected side, ie the side of the lesion, rather than reducing the temperature on the affected side. Further, in cerebral hemorrhage, it is hypothesized that a decrease in the size of hematoma or its disappearance and the subsequent removal of intracranial pressure may induce a development of collateral pathways and the blood flow from the internal carotid artery increases. Consequently, the blood flow in the external carotid artery decreases and eventually may cause a reduction in blood flow in the surface of the non-affected side of the body, ie the side of the lesion, and a consequent decrease of cutaneous temperatures on the non-affected side, rather than increasing the temperatures on the affected side. This, however, is not a definite conclusion.

As for the skin temperature of the throat, there was no significant difference between the affected and non-affected sides. This is ascribed to the fact that patients in the present study had neither occlusion nor stenosis of the common carotid artery. This meant little change in the blood flow of the common carotid artery and therefore no significant difference in the cutaneous temperatures.

CONCLUSION

At more than one month after cerebral apoplexy, the skin temperature of the face in the infarct patients tended to be lower on the affected side than on the non-affected side while in hematoma patients it seemed that the skin temperature on the affected side tended to be higher than on the non-affected side. These changes are too inconsistent and variable to be of clinical significance.

REFERENCES

1. Y. Oike, K. Irinoda, T. Iwaguchi, H. Umehara, K. Onodera, S. Kimura, N. Sasaki, J. Suzuki, S. Tonho, S. Matsuyama, "Clinical study and basic investigation in cerebral apoplexy," Igakushoin p. 39 (1974).
2. K. Atsumi, K. Inagaki, Y. Katsuta, S. Sakamoto, Y. Sakurai, A. Nagasawa, I. Fujimasa, Y. Mishima, "Atlas of Medical Thermography," Igakushoin p. 182-183 (1971).
3. K. Yanagi, The thermograms of patients with cerebrovascular accident, Ikakikai Gaku Zasshi 42:677 (1972).
4. L. C. Senay, Cutaneous vascular responses in finger and forearm during rising ambient temperatures, J. Appl. Physiol. 15:611 (1960).

5. L. Mora, R. Morgante, V. Sbarbaro, Thermographic follow-up of medical treatment in extra-cerebral carotid insufficiency, Acta Thermographica 1:12-17 (1976).

CARDIAC MUSCLE THERMOGRAPHY

I Fujimasa, K Imachi, Y Kajiyama, K Atsumi

Institute of Medical Electronics, Faculty of Medicine
University of Tokyo, 7-3-1 Hongo, Bunkyo-ku
Tokyo, JAPAN

OBJECT

Infrared thermography has been widely used in clinical medicine. Despite the present day availability of small, mobile instrumentation applications are usually limited to the outer surface of the body and internal body cavities are not examined.

The aim of our investigations was to establish measurement technology for surface temperature of internal organs, especially heart surface temperature distribution. Though it was feared that the surface temperature distribution may be homogeneous or the thermogram would be distorted by cardiac movement, our experience has shown that thermography provides a number of useful parameters.

METHOD

At open heart surgery, a small scanning head of a thermograph was fixed above the operation table and heart thermograms were taken at less than 1 second frame time. During experimentation, the chest cavity of a goat was opened by median sternotomy and thermograms were taken by a Thermoviewer MD (JEOL Co). The temperature of the cardiac muscle was measured by two needle thermocouples (Thermo Co). Temperature was measured at a depth of 6 mm and 12 mm. Blood flow through the cardiac muscle was also measured by simultaneous hydrogen gas clearance.

Behind the exposed heart, a cloth was spread for thermal insulation from surrounding tissues and the direction of the axis of the cardiac impulse coincided with the scanning direction of the Thermoviewer.

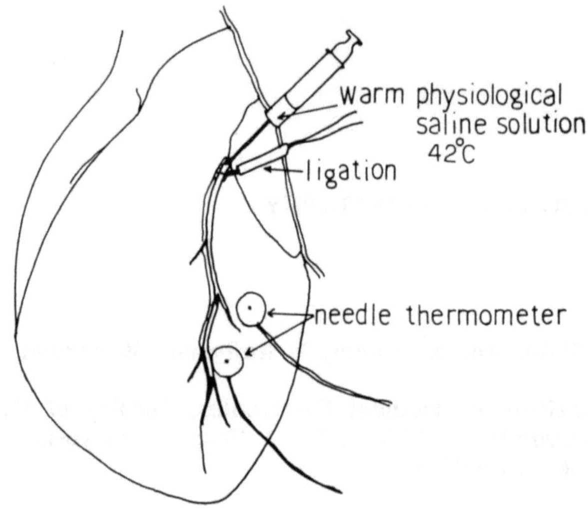

Fig. 1. The experiment method of cardiac muscle thermography.

Three types of experiments were used to analyse the relationships between heart muscle blood flow and its temperature:-

1. Thermograms taken after occlusion of left anterior descending coronary artery.
2. Thermograms taken after injection of 42°C physiological saline solution into the left anterior descending coronary artery (Fig 1).
3. Thermograms taken after 4000 G magnetic field.

The results of the three experiments were analysed and thermal simulation models of heart muscle were constructed.

RESULTS

The quality of cardiac muscle thermograms taken by the Thermoviewer MD was sufficient to enable analysis of the thermologic phenomena of heart muscle. The heart beat of 70 bpm slightly distorted the thermograms but the thermal information provided gave enough data for analysis of blood flow distribution of the heart muscle.

The three types of experiments gave the following results:-

1. Low temperature region of the thermograms indicated the ischemic region of the heart muscle (Fig 2).
2. Warm fluid injected in a coronary artery showed the distribution of the vessel as a hot area.

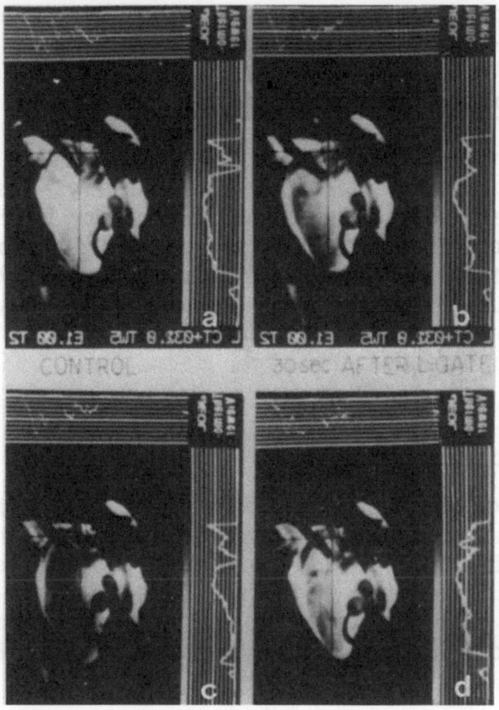

Fig. 2 Effect on cardiac muscle thermogram of left anterior descending coronary artery ligation (a) control, (b) 30 sec after ligation, (c) 3 min after ligation and (d) 3 min after release.

DISCUSSION

The biggest problem with heart muscle thermography was thought to be distortion caused by heart movement. It is impossible to take static heart thermograms because the frame time of modern thermographic instruments is limited to a few seconds. But since the one line scan takes less than 10 msec and frequency response of one picture element is more than 1 MHz, the pattern obtained from one scan line appears as if the heart is still. Actually thermograms taken from the goat heart also showed distortion during the cardiac cycle.

Another problem was the small temperature difference between the heart and other surrounding organs and tissues. If it is possible, a contrast material should be laid behind the heart (for example, cloth was a good contrast medium).

A further problem was that telescopic magnification was essential for heart thermographic technique because the heart is a small organ and the distance between the camera and target was more than one meter.

From the experimental results, the following clinical applications will be considered:-

Differential Diagnosis of Coronary Ischemic Disease

It was clear that decrease of coronary flow caused a temperature drop of the heart muscle after open chest surgery. After 3 minutes ligation of left anterior descending branch of coronary artery, the territory supplied showed more than a 2°C decrease in 27°C room temperature. After the ligation tape was untied, the temperature of the heart surface recovered the pre-ligation temperature immediately. Thus, evaluation of coronary blood flow rate after coronary bypass surgery is appropriate application of thermography.

Thermal Coronary Angiography

After the cardioplegic preservation at open heart surgery, the blood flow distribution is an important parameter to evaluate the prognosis of heart muscle activity. The second experiment showed the possibility of thermal coronary angiography. If the warm solution is perfused into the coronary artery or injected into the aorta, the thermogram of the heart would provide information regarding the flow distribution of cardiac muscle.

Monitor of Heart Surface Temperature and Estimation of Heart Muscle Blood Flow

When monitoring heart muscle temperature at open heart surgery, a needle thermistor is inserted into the ventricular septal muscle directly. But it is not possible to measure the total heart surface temperature and the flow distribution can only be guessed by the color of the heart muscle. In this way thermography can play a useful monitoring role in open heart surgery. However, estimation of cardiac blood flow is completely different from that of blood flow in the skin using thermal methods. We are developing a tool for thermal analysis of heart muscle under open chest conditions.

CONCLUSION

Targetting of the thermographic system on moving internal organs like the heart has scarcely been tried. But, as one line scan takes only 10 msec and the frequency response of one thermographic element is more than 1 MHz, the thermographic instrument has enough capability to diagnose the dynamic blood flow distribution of a moving internal organ.

TELETHERMOGRAPHY IN THE DIAGNOSIS OF THE REFLEX SYMPATHETIC DYSTROPHY

S Uematsu

Department of Neurosurgery
Johns Hopkins University, School of Medicine
Baltimore, Maryland, USA

INTRODUCTION

Reflex sympathetic dystrophy (RSD) is a syndrome. It consists of persistent pain and vasomotor instability. The entity was recognized over a century ago,[1] and subsequent reports have described a number of clinical variants, differing primarily in their associated features and predisposing factors. A clear concept of this disorder has been frustrated by the lack of clear diagnostic criteria, the numerous terms applied to the syndrome, the absence of proven therapeutic modalities and a poor understanding of its pathophysiology.[2]

THE ETIOLOGY OF RSD

Basically, three major theories have been postulated to explain the entity of RSD[3] (Fig 1).

Peripheral Origin

A simple explanation of this theory is that the sympathetic nerve fibers themselves transmit the pain impulses to the central nervous system.[4] This theory is partially supported by electric stimulation studies of the sympathetic nerves during sympathectomy by using "pullout electrodes".[5,6] Stimulation of the central end of the sympathetic nerves after division causes pain. This seems to indicate that stimulation of the sympathetic nerves does cause pain.

A more popular theory was proposed by Douple et al[7] and the theory was supported by experimental evidence conducted by Walker and

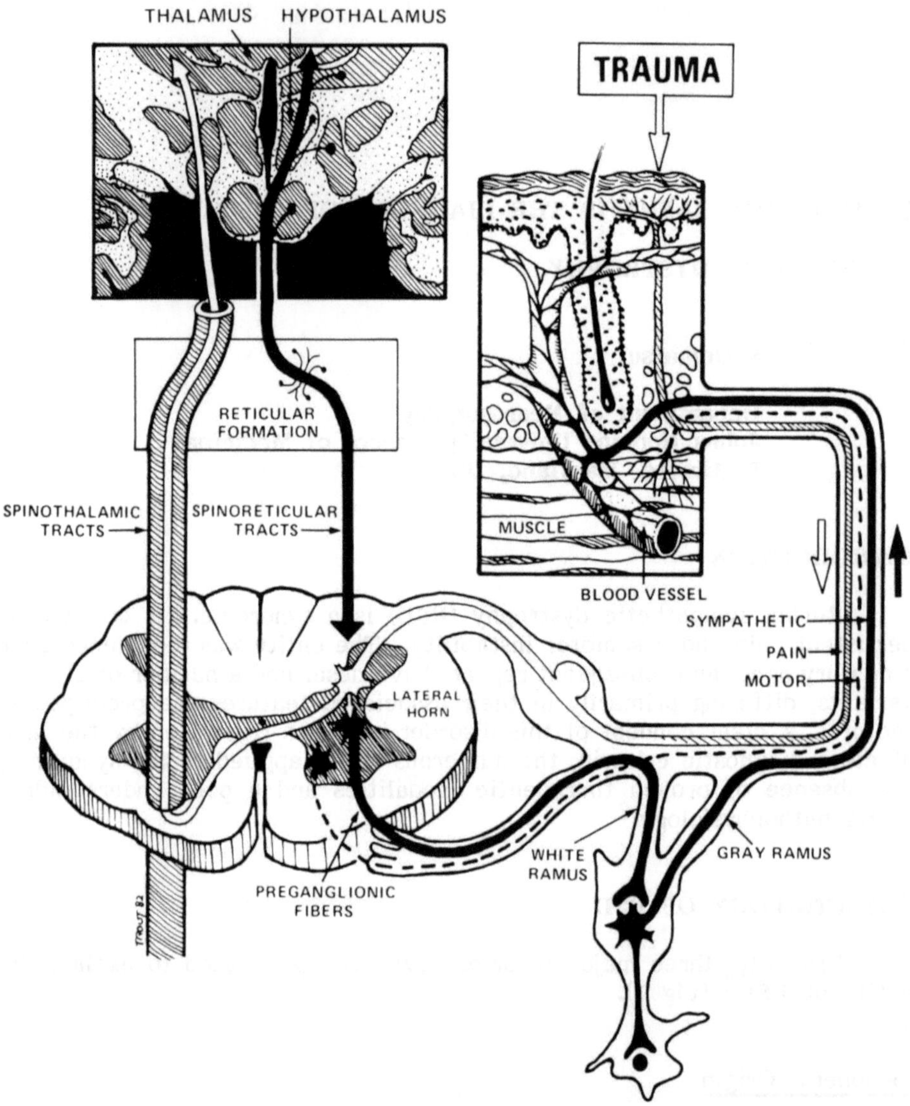

Fig. 1. In the condition of RSD, "cross-fire" between the C-fibers and sympathetic fiber develops at the three levels of the nervous system; (1) peripheral, (2) spinal, and (3) central. All three levels participate in various degrees to constitute a syndrome of RSD. White arrow indicates C-fibers and spinothalamic tracts and black arrow indicates sympathetic fibers and spinoreticular tracts.

Nulsen.[6] In the course of their preganglionic sympathectomies in man, electrodes were applied to the decentralized upper thoracic sympathetic ganglia and on subsequent electrical stimulation, the fully conscious patient reported a variety of sensations. These were referred to the local area in the distribution of the corresponding intercostal nerves. It has been postulated that artificial synapses can develop between the sympathetic efferent fibers and pain carrying afferent fibers. The synapse is the direct anatomical fusion of two different functioning fibers which can be developed during the process of regeneration of the injured nerve. Another possibility is that the overactive fibers can augment neighboring fibers. This results in "cross-stimulation" or "cross-firing" between the afferent pain fibers and efferent sympathetic fibers. Granit and his co-workers[8] demonstrated the cross-stimulation between motor and sensory fibers at a point of nerve injury by recording with cathode-ray oscillography; an afferent "impulse" from the sensory root takes place when the motor root is stimulated. In the absence of any injury to the peripheral nerve, no such returning discharge was observed. They therefore concluded that the axons of the C-group artificially synapse with sympathetic efferent fibers.

The immediate result of the activation of the sympathetic fibers is constriction of the peripheral arteries, causing ischemia of the painful site. A persistent overactivity of the sympathetic fibers leads to a chronic ischemic state resulting in dystrophic changes of the connective tissues (joints, skin, muscle, tendons and others). The dystrophic change, such as a "frozen joint", in turn stimulated C-fibers again in addition to the original trauma. This establishes a chain of events (vicious circle) (Fig 2).

Spinal Cord Origin

Such a chain of events between the efferent sympathetic system and the afferent pain fibers can also occur in the spinal cord. Livingston et al[9] advocated that three main processes take place to establish the vicious circle in the spinal cord level.

(a) Increased production and release of afferent impulses in a peripheral sensory nerve (C-fibers) following injury.
(b) Over activation of posterior horn neurons facilitates activity of internuncial pool and intermediolateral neurons in the lateral horn.
(c) Subsequent discharge of the sympathetic impulses establishes the vicious circle.

Central Origin

In recent years Melzack et al[10] introduced the "gate theory". A similar gate mechanism may be present between the dorsal horn cells and the lateral horn cells. In this condition (RSD) the ascending spinothalamic tract, reticular formation, hypothalamus, limbic system,

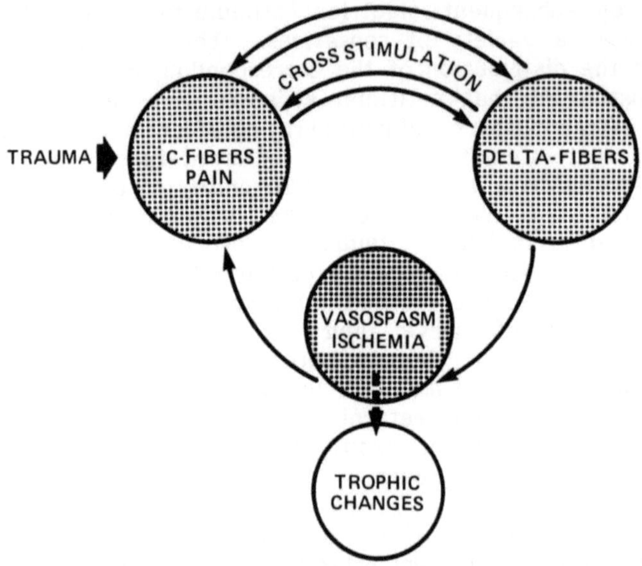

Fig. 2. Persistent irritation of the sympathetic efferent fibers causes chronic ischemic state of painful site, resulting in dystrophy. Dystrophic changes in turn stimulate pain carrying C-fibers to complete the vicious circle.

periaqueductal grey and finally the descending reticulospinal tract can participate to form the complex, at times bizarre, clinical picture of reflex sympathetic dystrophy.

DIAGNOSIS

The clinical diagnosis of reflex sympathetic dystrophy (RSD) is only made when the following two conditions are demonstrated. They are:

1. Burning pain.
2. Vasomotor instability.

Lacking one of these, the condition should not be called RSD. The term causalgia is applied to the same clinical symptoms, typically following major nerve injury. On the other hand, the term RSD is applied to the painful condition after nonspecific tissue injury, such as post-traumatic, post-phlebitic, post-infectious or post-arthritic condition. In these conditions the pain is typically relieved by sympathetic interruption. In the case of RSD, neurological examination can be negative for motor, sensory and/or deep tendon reflexes. The single objective finding in establishing the diagnosis of RSD is the demonstration of vasomotor instability. Unfortunately, the signs of vasomotor instability may not be obvious to the clinician prior to the development of dystrophic changes. It has been stressed that successful treatment of the RSD depends heavily on the early recognition of the condition.

One of the early signs of vasomotor instability is the alteration of skin temperature. White and co-workers[11] observed that the skin in the painful limb was consistently cooler than the opposite normal limb. Rasmussen and Freedman[12] found warmer skin with vasodilation in the first three months, then the skin becomes colder with vasoconstriction in six months. Sylvest and Jensen[13] recently reported significant abnormality in the temperature and blood flow of the brachioradial muscles of the affected forearm with RSD.

Electric skin resistance tests, plethysmography and thermocouple studies have been used to evaluate these clinical observations. However, these techniques have never been widely used in clinical practice owing to their inconsistency. All of these techniques require the application of a measuring instrument on the surface of the skin, a practice which may alter skin temperature or skin resistance. Recently, liquid crystal thermography (LCT) has been introduced. This technique also necessitates direct, firm contact on the tested area.[14] Infrared telethermography overcomes many of the deficiencies of these techniques.

THERMOGRAPHY

Method

Both sides of the body under investigation are exposed to laboratory air, where the temperature is kept relatively stable, in the range of 22°C to 25°C. Black-and-white polaroid film is used exclusively in our laboratory. The lighter tones correspond to warmer areas and the darker grey tones, to the colder areas. Differences in temperature are measured by comparing anatomically identical areas of opposite extremities or areas of the body, using isotherm control. A temperature difference of more than 1.0°C in over one-quarter of the examined major section of the body (example a hand and a forearm or arm and shoulder) is considered abnormal.

Since 1970, we have examined over 2,000 patients with chronic pain. Our accumulated experience indicates that this is not only the most effective technique in establishing early diagnosis of RSD, but also is an effective way of monitoring the treatment. One-third of the patients were referred from our Pain Treatment Clinic. Of these, fifty-four percent had abnormal thermography. Sixty-eight percent revealed colder temperature in the painful area and thirty-two percent revealed warmer temperatures.[15]

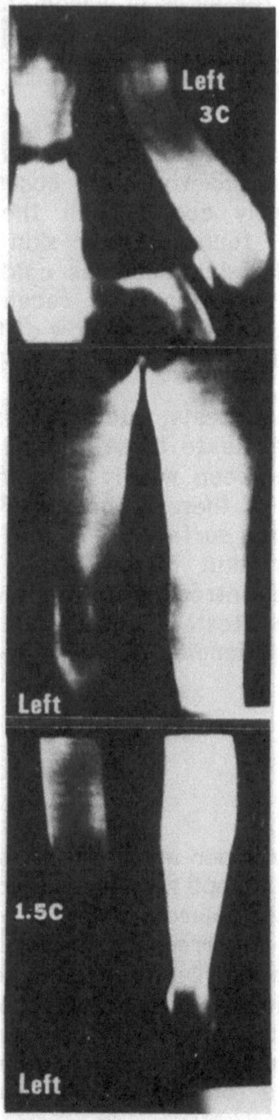

Fig. 3. 73 year old white female with thalamic pain (Dejerine Roussy Syndrome) in left arm for 10 years. Neurologically incomplete loss of sensation on the entire left side of the body with left hemiparesis. The thermography revealed that the painful left hand is 3° colder and 1.5° colder in the left lower extremity.

Thermal (Sympathetic) Dermatome

I propose to use a term, Thermal Dermatome, to describe a topographic change of the temperature of the skin. The skin temperature is directly related to the cutaneous vessel circulation which is regulated by the autonomic nervous system. A defective vasomotor mechanism results in increased or decreased temperature of the skin. The pattern of thermal dermatome may be not specific to underlying causes but varies depending on the involved site in the nervous system. The area of the injury can be: (a) central, (b) spinal cord, (c) spinal roots, (d) spinal nerve, (e) sympathetic chain and ganglion, and (f) vascular.

I should point out that in the majority of cases the most thermal pattern does not match well with classic somatosensory dermatome, except in cases with peripheral nerve injury. The most typical thermal dermatome in RSD is generalized, diffuse coldness in the involved limb, maximum coldness at the site of original injury.

Central. A thermal dermatome seen in the hemiplegic patient is an example. The temperature on the plegic side is usually colder. The coldness is more marked in the distal portion of the extremities. Carmel[16] observed a disturbance in the sympathetic function after stereotaxic lesions placed in the thalamus. The loss of sweating involved half of the body ipsilateral to the thalamotomy (Fig 3).

Spinal cord (pre-ganglionic) origin. Normell[17] observed the segmental thermal dermatome on the paraplegic patients. He describes the helical segmental thermal pattern on the trunk, and the lower extremities (Fig 4). His "sympathetic dermatome" does not match well with the classical somatosensory dermatome. It is probably due to widely overlapping innervation of the preganglionic neurons.

Spinal root (pre-ganglionic). The efferent sympathetic fibers arise in the intermediolateral neurons of the thoracolumbar cord (T1 to L2). They leave the cord at these levels and pass through the ventral root. In other words, the spinal roots at L3, L4, L5 and S1 do not carry efferent sympathetic fibers. The sympathetic fibers join to the spinal nerve distal to the roots outside of the spinal canal at this level. The change in the temperature in the affected limb in the case of disc herniation has been reported.[14] These observations can be explained by the deafferentation or facilitation of the intermediolateral neurons due to the injury of the dorsal root. It should also be noted that there is the possibility that the lower lumbar roots may carry sympathetic efferent fibers, contrary to the generally accepted views.[18] However, I should point out that the change of the skin's temperature is not a consistent finding in the cases of disc herniation, in my experience (Fig 5).

Cranial and Spinal Nerve (post-ganglionic). Thermal dermatome in the section (injury) of the spinal nerve appeared to correspond well with the sensory dermatome. This occurs even in trauma of the pure sensory

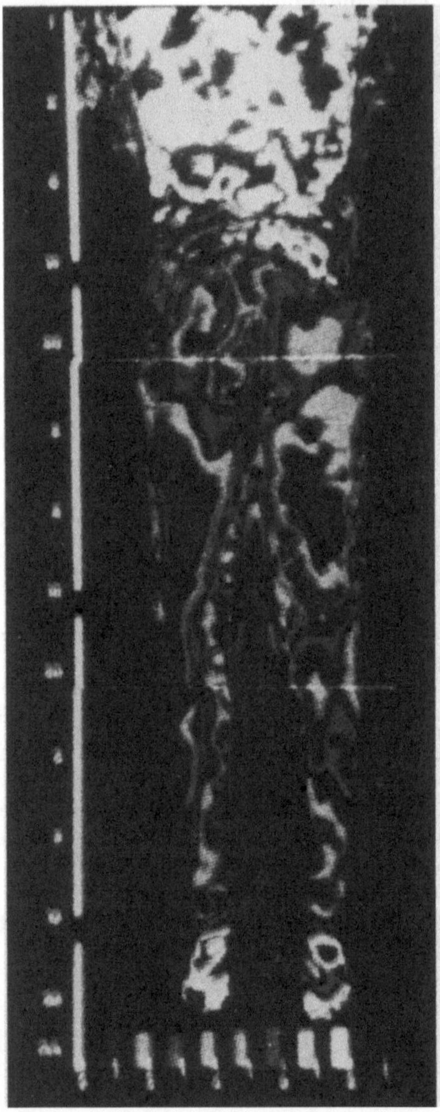

Fig. 4. Segmental thermal dermatome on a paraplegic patient. Decreased temperature below the waist level (corresponding to the sensory level) is obvious. (modified from original publication by Nomell, LA[17]).

cutaneous nerve.[19,20] Demarcated flush is a frequent observation during the radiofrequency trigeminal and spinal rhizotomy (Fig 6). Richter[21] observed an increase in skin resistance in the area of hypoesthesia. These observations can be explained anatomically.

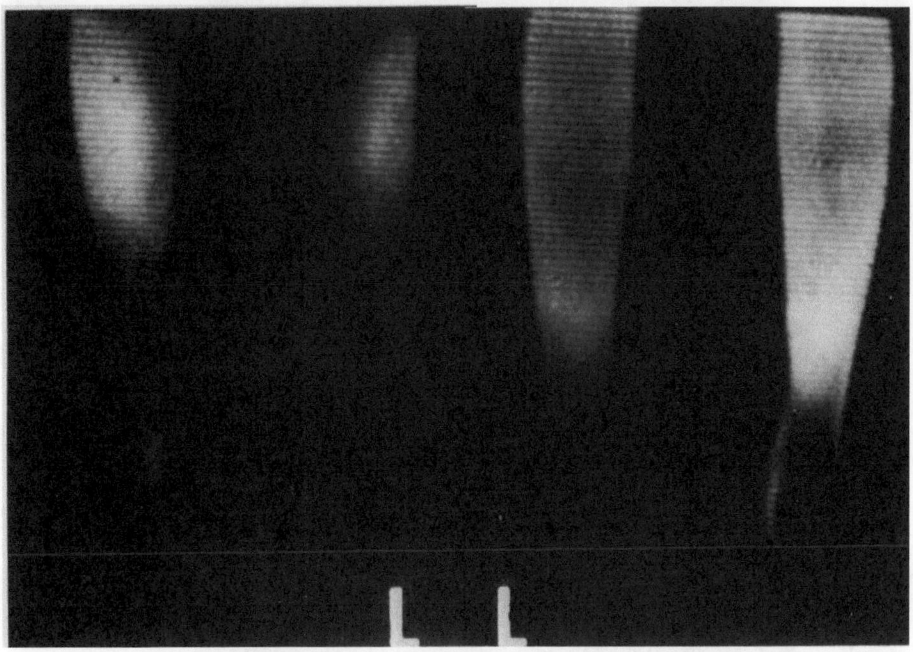

Fig. 5. Thermogram of lower extremities with lumbar disc herniation at L3 and L4 reveals 2.5°C, and 1.0°C coldness in the left foot, medial ankle and posterior aspect of the leg and thigh respectively. The patient had moderate weakness in the anterior tibial and extensor hallucis longus L:left, for 3 weeks at time of the thermogram. A-P view on left, and P-A views on the right.

Postganglionic fibers pass through the peripheral nerves and supply the limb's arteries at various levels. The median nerve supplies sympathetic fibers to the brachial artery and palmar arches. The ulnar nerve supplies the ulnar artery. The radial nerve supplies the radial artery. Similarly in the lower limb, the femoral, obturator and posterior tibial nerves also carry a significant number of postganglionic sympathetic fibers.[22]

Sympathetic Chain and Ganglion and Vascular Origin. The thermal dermatome seen in RSD is the example.

If there is an injury to the tunica media of the vessel and also to the sympathetic system, the sympathetic disturbance should logically be in the area of supply to the artery. The major arterial trunks of the limbs literally supply the quadrants of the body. Thus, in cases of injury of the sympathetic system over a major trunk and/or ganglion the disturbance may be expected in the quadrantal distribution (Figs 9, 11).

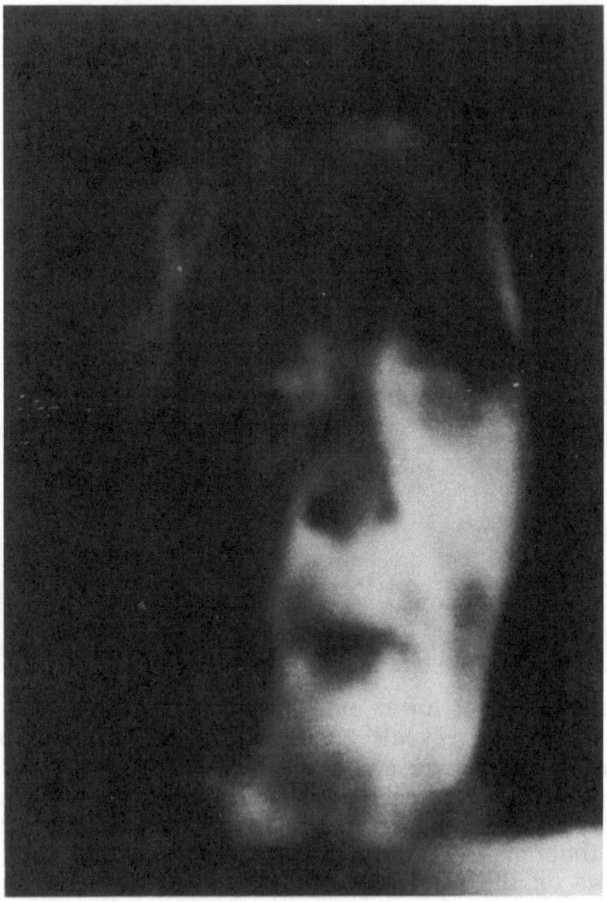

Fig. 6. Thermogram of the face, one hour after trigeminal rhizotomy. White area corresponds to the area of the flush.

Temperature Asymmetry in RSD

The temperature of the painful area was over $2^{o}C$ colder in 67% of those patients who were clinically diagnosed with RSD, and in 42% of chronic pain patients with signs of nerve injury but without clinically detectable signs of vasomotor instability. On the other hand, only 18% of pure chronic pain patients without clinical evidence of nerve injury or vasomotor instability showed thermographic evidence of $2^{o}C$ coldness in the affected area (Fig 7). The degree of the temperature gradient was not correlated with the severity of the nerve injury (Table 1).[15]

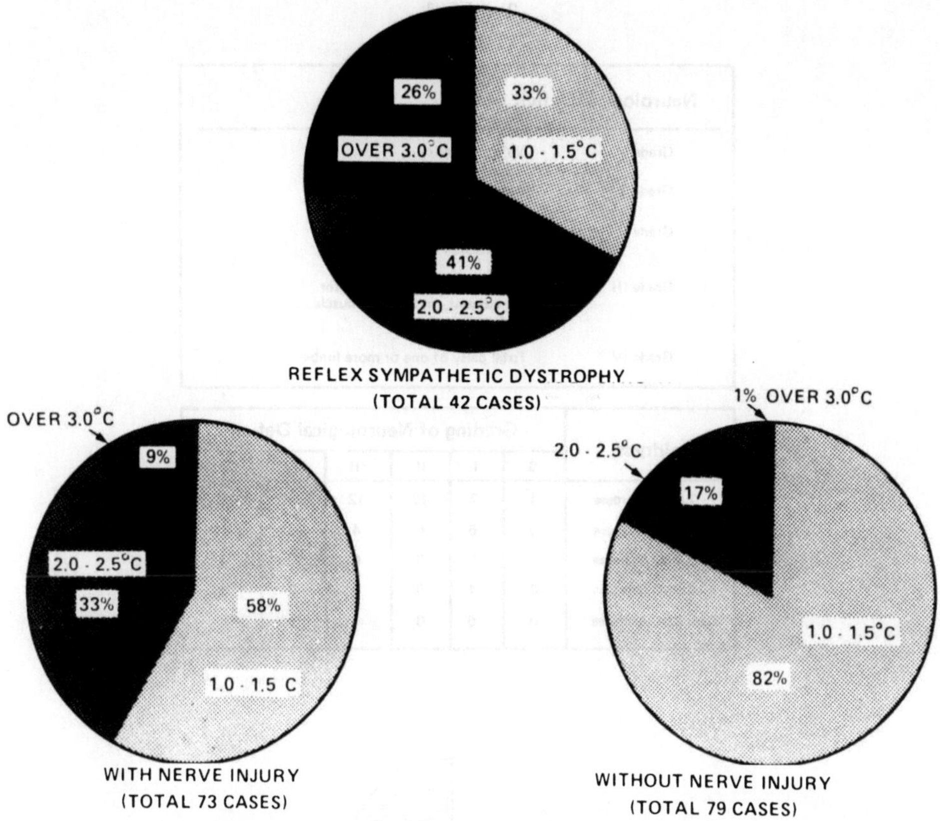

Fig. 7. The temperature of the painful area was over 2° colder in 67% of patients with RSD. Lower percentages were noted in the group without clinically detectable vasomotor instability.

Temperature Asymmetry and Fibrillation Potentials

The percentage of the presence of fibrillation potentials is higher in those cases with a colder painful limb. In cases less than 1.0°C colder, the fibrillations were recorded in only 11% of the total of 94 cases. In cases 1.0°C to 1.9°C colder, the potentials were present in 44% and in cases over 2.0°C colder, the potentials were observed in 66% (Fig 8). In a total of 32 typical cases of RSD, only 16% had fibrillation potentials.[15] These findings indicate that when the painful area is over 2°C colder, motor neurons (and/or muscle fibers) are also likely to be impaired. In the cases of RSD, the motor neuron can also be injured but less frequently. Based on these observations, when thermography is positive, it is highly possible that organic injury has taken place in the delta and/or C-fibers and/or motor neurones. If the thermograph is negative, it may be of functional origin, but one cannot absolutely rule

Table 1.

Neurological Deficit	
Grade 0	No abnormality
Grade 1	Absent deep tendon reflex
Grade II	Suggestive sensory impairment mild weakness
Grade III	Obvious sensory and/or motor impairment, on isolated muscle or dermatome
Grade IV	Total palsy of one or more limbs

Coldness	Grading of Neurological Deficit					
	0	I	II	III	IV	Total
1 - 1.9%C decrease	1	7	19	12	3	42
2 - 2.9%C decrease	5	6	6	4	3	24
3 - 3.9%C decrease	2	2	0	0	0	4
4 - 4.9%C decrease	0	1	0	0	0	1
Over 5%C decrease	0	0	0	1	1	2

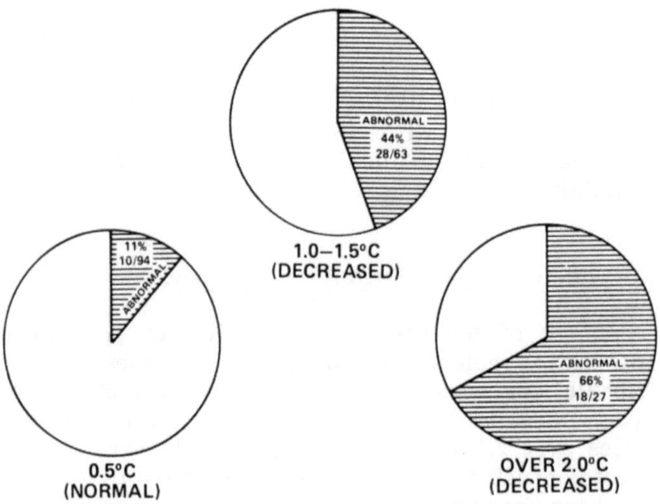

Fig. 8. The fibrillation potentials were more frequently detected in those cases who had colder painful limb than patients with less colder. The chance to detect fibrillation potentials in the painful limb was 11%, 44% and 66% in cases with less than 1.0° cooler (normal), 1.0 to 1.9°C, and over 2.0°C respectively.

out the possibility of nerve injury, particularly in the motor or sensory system.

CASE PRESENTATION

Case JHH 162 16 52 24 YOF

The patient developed a painful foot after wearing new nurse's shoes. Later she could not wear stockings, was unable to put any weight on her toes and had to use crutches. Phlebograms were used to rule out thrombophlebitis. She had a psychiatric evaluation for what the orthopedic surgeons thought were hysterical symptoms. Such things as water splashing on her thigh or wind blowing on the leg caused considerable discomfort.

Physical Findings. Mild oedema of the left foot was found, more marked distally around the toes. The temperature over the entire left leg appeared to be colder than that of the right lower leg. Palpation of the leg was extremely difficult because the patient complained of pain. When asked to stand, the patient could not extend the left knee sufficiently to touch the floor.

Thermography. There was generalized coldness in the left foot extending to the calf. The temperature averaged 2°C colder on the left compared to the right. Follow-up thermography, 2 years after a lumbar sympathectomy, revealed a 2.5°C temperature increase in the left leg compared to the right.

Course. Based on effective lumbar sympathetic blocks, a sympathectomy was performed in March 1976. Since that time she has been free from pain and able to return to work (Fig 9).

Case JHH 166 93 19 42 YOF

Immediately following a hysterectomy, this patient noticed paresthesiae in the fourth and fifth digits of her left hand. These symptoms had increased in intensity, and subsequently, she had developed skin changes with shiny, hairless skin, hypersensitivity and swelling of the hand. Initially, she could close her hand and had a good grip, but this gradually deteriorated. She found her left upper extremity almost useless.

Sensory examination indicated a mixture of hyperesthesia and exaggerated response in the left ulnar nerve distribution. The small muscles of the hand could not be tested adequately because of pain. Deep tendon reflexes were present in the upper extremity and the radial pulses were present. Nerve conduction study and myelogram were negative.

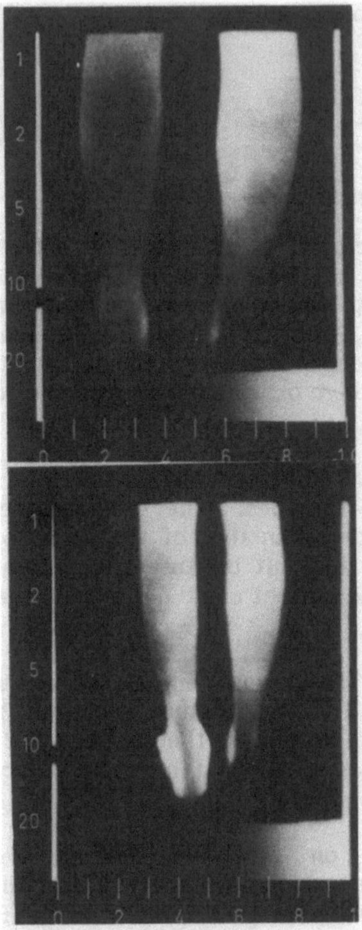

Fig. 9. Generalized coldness in the less painful leg and foot (top). Follow-up examination 2 years after lumbar sympathectomy reveals 2.5°C warmer left foot (bottom).

X-ray. There was marked osteoporosis of the left hand and forearm (Fig 10).

Thermography. The left forearm and hand were 3°C colder than the right. Post sympathectomy examination revealed marked improvement of the temperature pattern (Fig 11).

Diagnosis. Reflex sympathetic dystrophy of the left upper extremity.

Course. Because stellate ganglion blocks of the hand had been repeatedly effective, an upper thoracic sympathectomy was performed on March 9th 1977. Five years after surgery the patient reports that

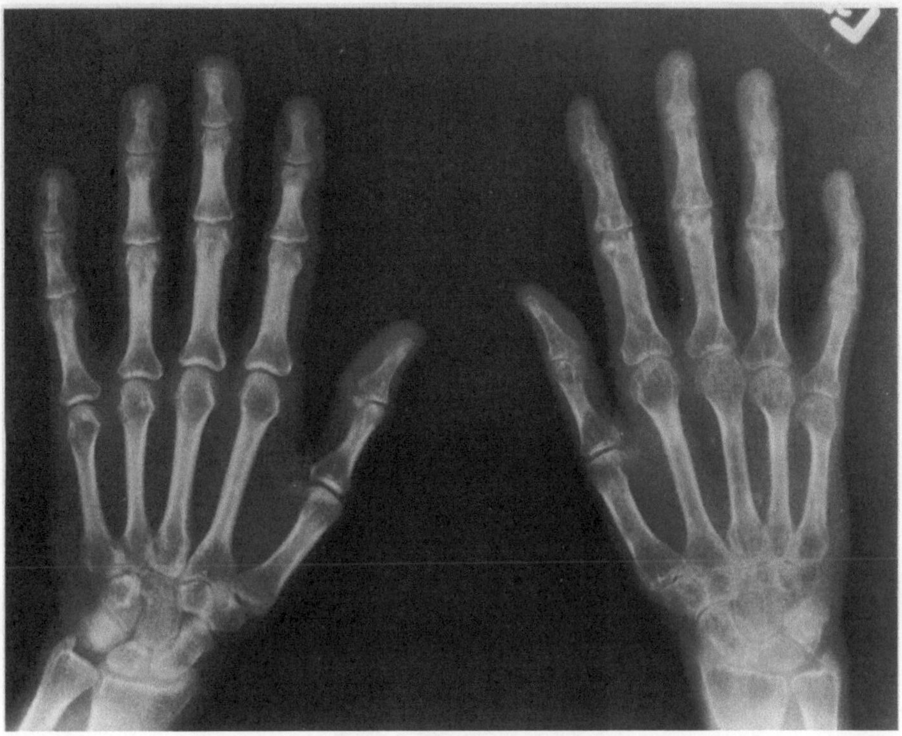

Fig. 10. X-ray of hand. Dystrophic change of painful left hand is apparent.

she no longer has burning pain but still finds it difficult to use that hand owing to contracture of the joints of the hand. She has no doubt that surgery was definitely helpful.

SUMMARY

Telethermography provides a useful tool for early detection of RSD before the development of irreversible trophic changes or myofascial contracture. In the earlier stage, physicians often overlook the condition because of the lack of objective means of assessing the vasomotor instability until the development of obvious clinical signs or deformity of the joints. Temperature asymmetry can be noted in the earlier stage by absolutely non-invasive telethermography. It should be mentioned that thermography has been helpful to physicians in explaining the nature of the pain to the often apprehensive RSD patients by allowing them to review their own thermogram. The technique is also valuable in documenting, as well as confirming, the appropriateness of the sympathetic block or sympathectomy.

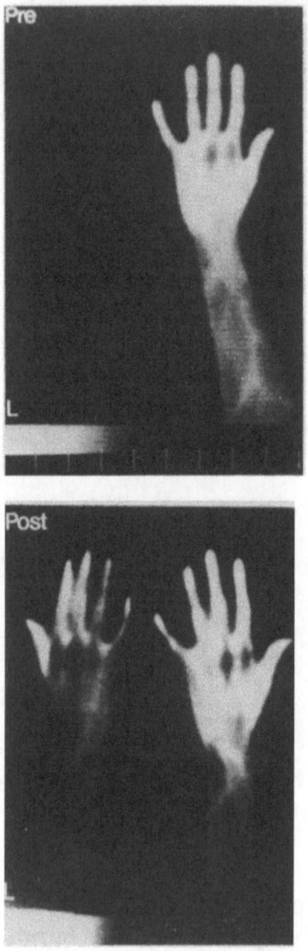

Fig. 11. Markedly colder left arm and hand is barely visible in this thermogram (top). Post-sympathectomy thermogram is shown in bottom of the figure.

REFERENCES

1. S. W. Mitchell, G. R. Morehouse, W. W. Keen, "Gunshot wounds and other injuries of nerves," J. B. Lippincott & Co. 164., Philadelphia (1864).
2. F. Kozin, L. M. Ryan, G. E. Carerra, J. S. Soin, Wortman, The reflex sympathetic dystrophy syndrome (RSD), Am. J. Med. 70:23-30 (1981).
3. J. E. Thompson, The diagnosis and management of post-traumatic pain syndrome (causalgia), Aust. N. Z. J. Surg. 49:3:299-304 (1979).
4. N. E. Freeman, The treatment of causalgia arising from gunshot wounds of the peripheral nerves, Surg. 22:68-82 (1947).

5. J. C. White, B. Selverstone, Pain and related phenomena including causalgia, in: "Peripheral Nerve Regeneration," Chapter 6. B. Woodhall, G. W. Beebe eds., U. S. Government Printing Office, Washington D. C. (1956).
6. A. E. Walker, F. Nulsen, Electrical stimulation of the upper thoracic portion of the sympathetic chain in man, Arch. Neurol. Psychiat. 59:559-560 (1948).
7. J. Doupe, C. H. Cullen, G. Q. Chance, Post-traumatic pain and the causalgic syndrome, J. Neurol. Neurosurg. and Psychiat. 7:33 (1944).
8. R. Granit, L. Leksell, C. R. Skoglund, Fiber interaction in injured or compressed region of nerve, Brain 67:125-140 (1944).
9. W. K. Livingston, "Pain mechanisms. A physiological interpretation of causalgia and its related states," MacMillan Co. N. Y. 83 (1943).
10. R. Melzack, P. D. Wall, Pain mechanisms. A new theory, Science 150:971-979 (1965).
11. J. C. White, W. H. Sweet, "Pain and the neurosurgeon, a 40-year experience," Charles C. Thomas ed., Springfield, Illinois, (1969).
12. T. B. Rasmussen, H. Freedman, Treatment of causalgia - an analysis of 100 cases, J. Neurosurg. 3:167-173 (1946).
13. J. Sulvest, E. M. Jensen, J. Siggard-Anderson, L. Pendersen, Reflex dystrophy resting blood flow and muscle temperature as diagnostic criteria, Scand. J. Rehab. Med. 9:25-29 (1977).
14. R. Pochaczevsk, C. F. Wesler, P. M. Meyers, J. A. Epstein, J. A. March, Liquid crystal thermography of the spine and extremities, J. Neurosurg. 56:386-395 (1982).
15. S. Uematsu, N. Hendler, D. M. Hungerford Long, N. Ono, Thermography and electromyography in the differential diagnosis of chronic pain syndromes and reflex sympathetic dystrophy, Electromyog. Clin. Neurophysiol. 21:165-182 (1981).
16. P. W. Carmel, Sympathetic deficits following thalamotomy, Arch. Neurol. 18:378-387 (1968).
17. L. A. Normell, Recording of normal and impaired cutaneous thermoregulatory vasomotor response by infrared thermography. A methodological study, Scand. J. of Clin. Lab. Invest. 33: suppl 138:3:1-24 (1974).
18. A. Brodal, "Neurological anatomy in relation to clinical medicine," Second edition. Oxford University Press, New York, London, Toronto (1969).
19. C. Wexler, R. Small, Thermographic demonstration of a sensory nerve deficit, J. Neurol. and Orthopedic Surgery 3:73-75 (1981).
20. S. Uematsu, D. M. Long, Thermography in chronic pain, in: "Medical Thermography, Theory and Clinical Applications," S. Uematsu ed., Brentwood Publicating Co., Cal. (1976).
21. C. P. Reichter, Physiological factors involved in the electrical resistance of the skin, Am. J. Physiol. 88:596 (1929).
22. R. Warwick, P. Williams, The autonomic nervous system, in: "Gray's Anatomy" 35th edition. R. Warwick, P. Williams eds., Philadelphia W. B. Saunders, 1065-83 (1973).

INFORMATION FROM SCROTAL THERMOGRAPHY

E D Cooke

Department of Medical Electronics
St Bartholomew's Hospital
London, UNITED KINGDOM

Those of you who have had the temerity, or perhaps inquisitive masochism would state the case more accurately, to embark on the study of fertility will appreciate that my comments in this lecture apply to one aspect only of an exceedingly complex subject as it affects the male.[1] To those who have not, I hope to at least arouse your curiosity.

Before attempting to describe the possible information to be derived from thermography, it is well to clarify our concept of scrotal anatomy with particular reference to the complex venous drainage (Fig 1). Only the left side is shown in the diagram. The small veins are valved and the left internal spermatic vein, in contrast to the vein on the right side, enters the renal vein at a right angle. Compression of the left renal vein between the aorta posteriorly and the superior mesenteric artery anteriorly may produce the so-called 'nutcracker phenomenon' with reflux into the internal spermatic vein producing an increase in pressure in this vein, with the consequent pathological changes of valvular incompetence and thickening of the venous wall. This subject has recently been reviewed in a venographic study by Coolsaet,[2] who has defined other less common causes of venous hypertension in the venous drainage of the testes. For our present purpose it is sufficient to realise that these veins lying within the spermatic cord are normally not palpable but become so if the wall is sufficiently thickened and tortuous with incompetent valves - the so-called varicocele.

The publications of such investigators as Russell[3] and Scott[4] in the early fifties and sixties suggested that varicocele, an almost exclusively left-sided abnormality of the spermatic veins,[5] was a treatable cause of male subfertility. This idea has since been widely accepted, based on the results following surgery, ie improved postoperative sperm quality and pregnancy rate.[6,7] This concept has about it the attraction of

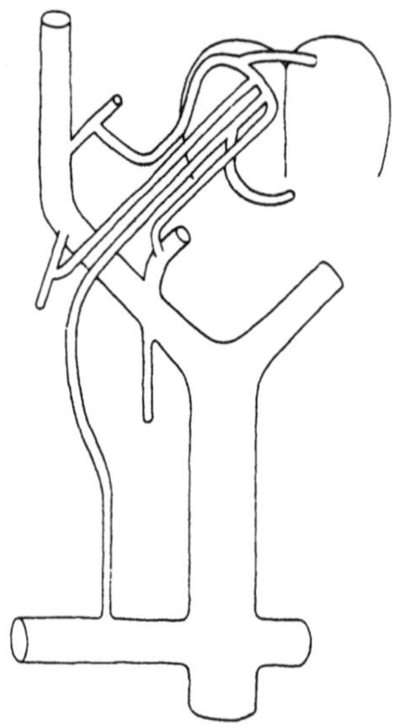

Fig. 1. The venous drainage of the left testicle.

simplicity. Remove the varicocele and your problems are over.

Accepting this then it follows that the clinician should be able to recognise with certainty the presence or absence of a varicocele, but this is not always possible by simple inspection and palpation. For example, analysis of 100 consecutive cases referred by experienced urologists or endocrinologists showed that there was a correlation in 64 cases and no correlation in 36. Of these 36 cases, 25 who were thought to have a varicocele had a normal thermogram, while 11 thought to be normal clinically had a varicocele on thermography (Table 1). There were 7 bilateral cases. Of four cases diagnosed on clinical grounds, three were normal and one had a left-sided varicocele on thermography. Of three diagnosed by thermography only the left-sided varicocele had been detected clinically in each case (Table 2).

This finding alone, that of 100 cases 25 would have had unnecessary surgery and 11 would have had no surgery when this was indicated, would demonstrate the necessity for an objective method of examination such as that provided by thermography of the scrotum. Obviously, all cases following surgery should be examined to ensure that a cure has been affected, ie that the thermogram is normal postoperatively.

But further, it has been suggested that even in the absence of varicocele, if the testicular temperature is raised above normal then spermatogenesis will be affected adversely,[8,9] - a subject which will be touched on again and, if true, a further indication for a thermological method.

Table 1. Correlation between clinical and THG findings in 100 consecutive cases of possible varicocele

Correlation		No correlation
+	-	
36	34	36

Given these considerations it is clear that standardization of the method of investigation is important in order to obtain the maximum information possible and for purposes of comparison of results between investigators. From published papers it is evident that most investigators have accepted the AP view of the scrotum with the patient standing and with the penis removed from the field as the standard view[10,11,12] (Fig 2). This is certainly the simplest view to obtain. In this position the scrotal image is enhanced by use of a belt to hold the penis to the abdomen and a suitable evaginated infrared screen held across the upper thigh. This screen has the advantage of excluding radiated heat from the thigh, but care must be taken that it is placed without pressure on the root of the penis posteriorly and so compress scrotal venous outflow. The resultant thermogram provides a well defined view of the scrotum. Whether this is obtained in grey-tone or color is clearly a matter of

Table 2. Clinical and thermographic findings in 7 bilateral cases of varicocele

BILATERAL VARICOCELE			
Clin +	of which	THG -	
4		3	1 left side only positive
THG +	of which	Clin +	
3		3 left side only	

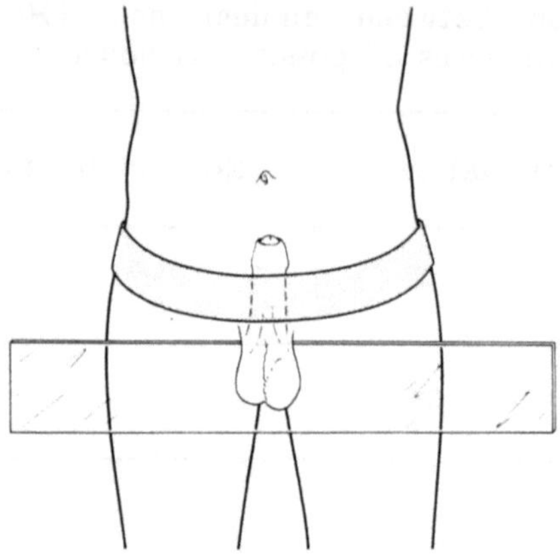

Fig. 2. The position of the patient when performing scrotal thermography in A-P view.

preference and individual usage. In either case the hot streak associated with a varicocele and testicular temperature is usually easily determined.

Early in our investigations of scrotal thermography we recognized that some varicoceles changed in appearance with respiration. From this it was a simple step to investigate the effect of the Valsalva maneuver on scrotal thermographic appearances. The patients were asked to inspire maximally and holding the breath to bear down for 10 seconds after a resting thermogram had been obtained following a period of equilibration (Figs 3a and b). Three effects of the Valsalva maneuver are recognized. In normal individuals the temperature of the testicles falls by 0.5°C or more; in some subjects a varicocele is only apparent following this maneuver; when a varicocele is already apparent the appearance may be enhanced (Table 3). The known tortuosity and thickening of the venous wall with valvular incompetence which is characteristic of varicocele provides an acceptable explanation of the two latter findings. The explanation of the fall in temperature of the testicle in normal subjects is less obvious and presupposes either a paradoxical enhancement of venous return or a fall in arterial input, possibly because of an increase in sympathetic tone. Whatever the explanation it is clear that thermographic examination of the scrotum is incomplete without the subject performing a Valsalva maneuver, as recently shown by Fornage et al[13] working in Rheims.

As a simple inspection of a sufficiently large number of males will readily confirm, the anatomy of the scrotum varies enormously from a firm symmetrical hemisphere with a clearly defined medial raphe to an elongated appendage of redundant flesh hanging centrally and containing at its lower extremity a pair of testicles of variable size. In the latter case partial torsion of the whole scrotum, the right testicle coming to be anterior to the left, is not uncommon. An A-P view in such

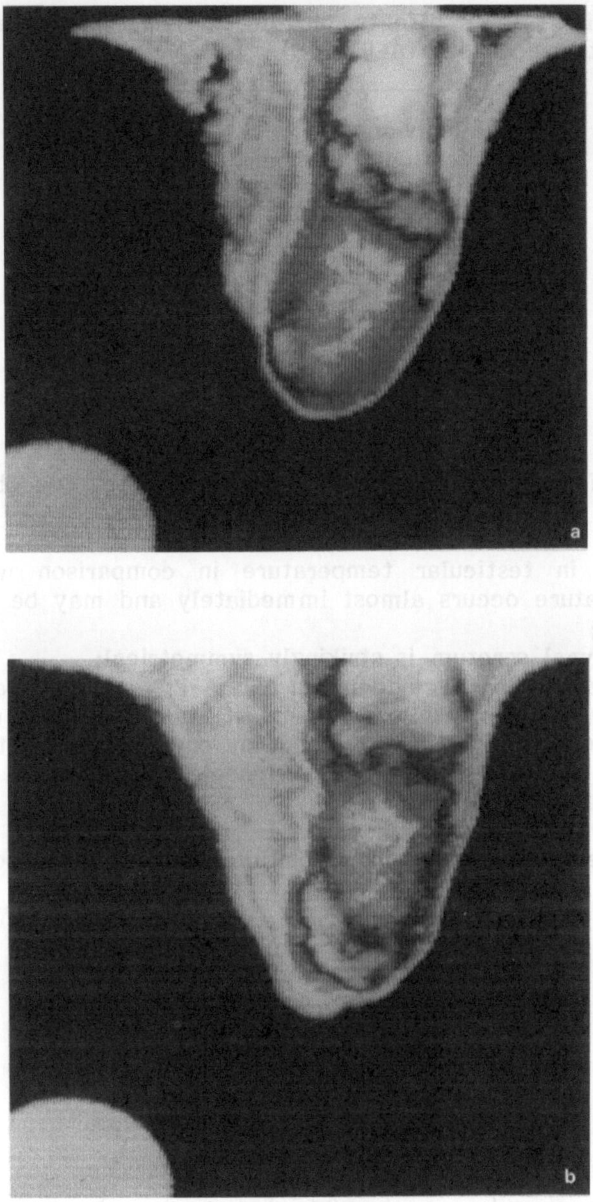

Fig. 3 (a) A-P view of the varicocele. (b) The appearance is enhanced by Valsalva maneuver.

circumstances may not be simple to interpret and relevant information may be lost. For this reason, and simply because study of P-A view seemed an investigation well worthwhile and which to our knowledge had not been previously described, we have examined the posterior scrotum

Table 3. The three effects of the Valsalva maneuver on scrotal thermography (A-P view)

Normal individual: Fall in testicular temperature
Varicocele: (a) Reveals occult varicocele (b) Enhances already apparent varicocele

with the patient in the knee-elbow position (Fig 4). Our observations of the scrotum with the patient in this position are:

1. A fall in testicular temperature in comparison with the A-P temperature occurs almost immediately and may be of the order of 1°C;
2. The normal scrotum is strikingly symmetrical;
3. A varicocele easily evident on the A-P view may be less obvious on the P-A view while, in contrast, the P-A view may reveal a varicocele which was not apparent or doubtful on the A-P view (Table 4). In this view a varicocele may also be enhanced on Valsalva (Figs 5a, 5b, 5c).

This maneuver in the knee-elbow position produces the expected fall of

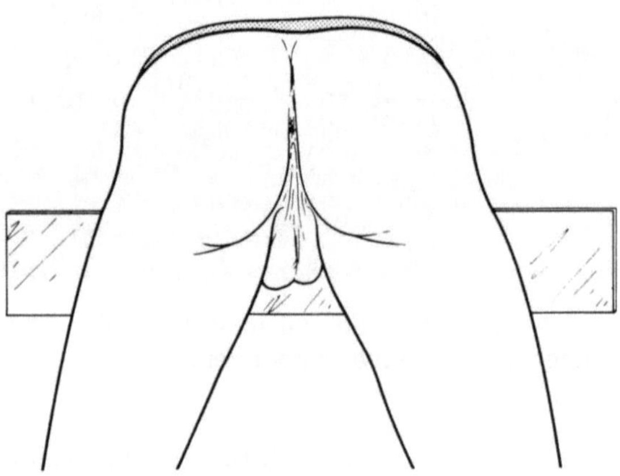

Fig. 4. The position of the patient when performing scrotal thermography (P-A view).

Table 4. Observations on the posterior view of the thermogram of the scrotum

Normal findings:	
A fall in testicular temperature	
Striking symmetry	
Varicocele:	
Possibilities:-	Apparent A-P view, less obvious P-A view
	Diagnostic P-A view, not apparent on A-P view

0.5°C in testicular temperature in normal subjects, just as when the patient is standing erect, ie the A-P view.

The interest in testicular temperature is, as I have said, because of the supposition that a rise in testicular temperature, which is normally held below core temperature by a reflex mechanism, is associated with reduction in both the number and motility of spermatozoa. We have found particular difficulty in defining the upper limit beyond which spermatogenesis is certain to be abnormal. Published results are unhelpful, normality being usually defined in young adulthood, eg male medical students or young cadets (groups which might well be described as being captive for study rather than a yard stick of normality) and not of the age-spread of males who are represented in a clinic for male infertility. It occurred to us that an index between testicular and core temperature might be more productive. In the event, a preliminary analysis of 200 cases indicates that if this figure is greater than 0.83 then impaired spermatogenesis is likely.

In another paper to the Congress we have shown that the appropriate ambient temperature for vascular studies is 24.0°C ± 1°C and this is the temperature used for the present study. Equilibration of scrotal temperature is judged to be complete when there is no further temperature change for a period of two minutes. Both initial, ie pre-equilibration and equilibration testicular temperatures are recorded and, as previously shown, varicocele is recognized as a hot-spot. In other words, the investigation is performed in conformity with previous studies. However, there is serious reason to doubt the validity and hence the value of these recordings. It is apparent that the temperatures recorded are as stated, that is, testicular temperature at the time the individual undresses and then at equilibration in a specific ambient temperature. The latter temperature may be particularly "false" and entirely unrelated to what occurs physiologically when one considers just how little time in a 24-

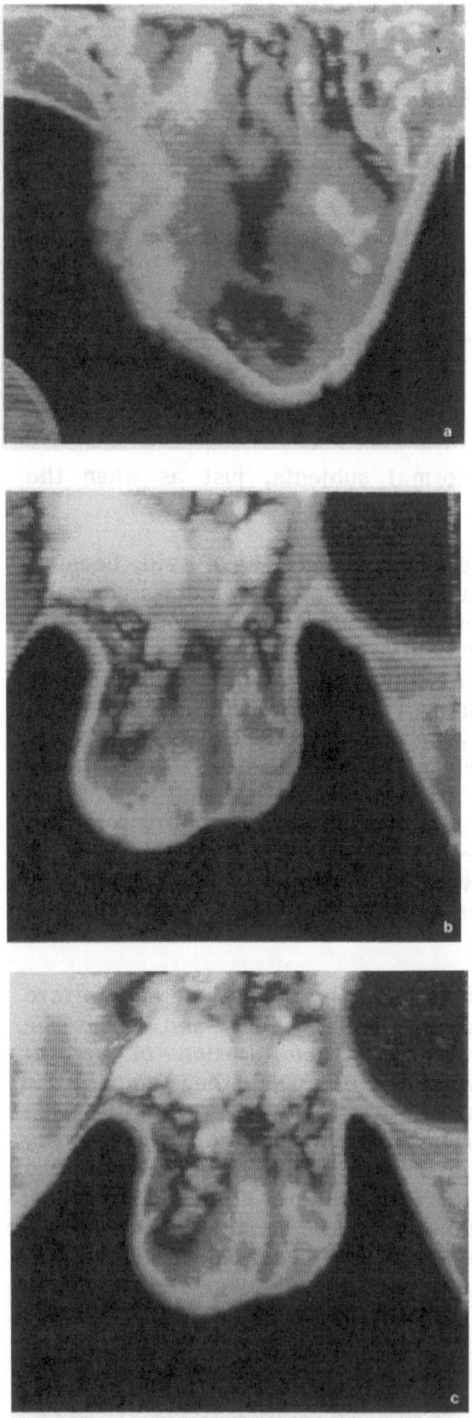

Fig. 5.
A possible left varicocele on the A-P view (a) is easily recognized on the P-A view (b), and is enhanced in this view by the Valsalva maneuver (c).

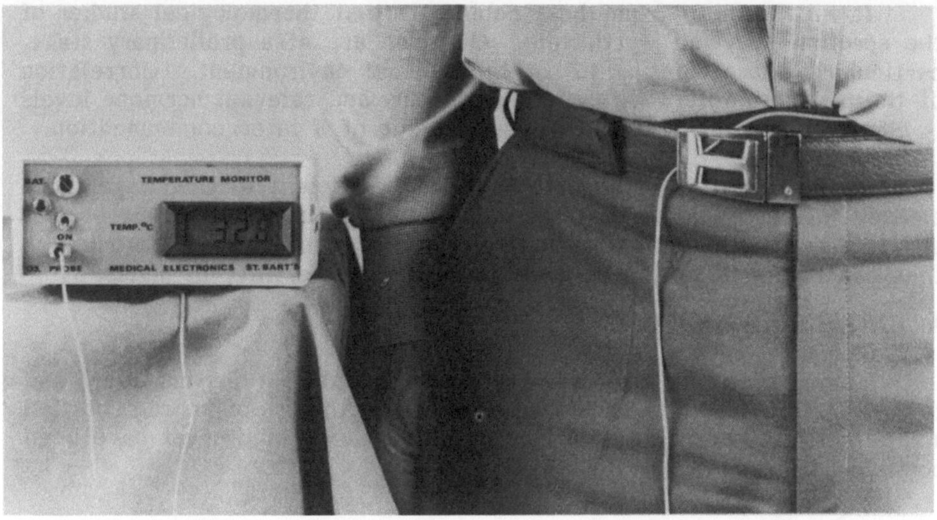

Fig. 6. Determining the thermal environment of the testicles by a thermistor placed between the testicles with the patient clothed.

hour period that an individual is undressed. To extend thought a little further, does the average testicle ever become equilibrated with an environment such as is found under laboratory conditions? In an attempt to provide an answer to this problem we have initiated an investigation into scrotal environment (Fig 6). Our initial studies indicate that the temperature recorded by a thermistor placed between the testicles with the individual fully clothed may be 3-4°C higher than that recorded minutes later when he is undressed. In other words, the testicular temperature falls with great rapidity on exposure and the undressed temperature may bear little relationship to that which prevails ordinarily and thus to spermatogenesis, if this activity is temperature-dependent. Is this finding an objective indication for the infertile male to wear boxer pants? A recommendation that has been hitherto empirical.

These problems have been outlined to indicate that the place of scrotal thermology in the investigation of the infertile male may be more complex than commonly appreciated, and possibly extends far beyond the diagnosis of varicocele. The value of treatment for this condition is not in doubt.[7] The more accurate diagnosis reflected in the figures which I have given earlier, Tables 1 and 2, suggests that a more definitive picture will emerge in the future.

It will be clear from these comments that thermological studies of the scrotum are well worthwhile. Our own are at a preliminary stage, particularly those relating to scrotal thermal environment. Correlation of thermal parameters with spermatogenesis and relevant hormone levels is clearly important and will be the subject of a later communcation.

In the meantime, I hope to have said sufficient to encourage those already interested in scrotal thermology to continue with their studies and to draw the attention of thermologists in general to the contribution that can be made to the complex problem of infertility in males.

ACKNOWLEDGEMENTS

I am grateful to Dr Nick Aarts for his interest in this study and to Sally A Bowcock without whose expertise in the use of instrumentation and pathological and clinical know-how, this study would never have been possible.

REFERENCES

1. W. F. Hendry, Male infertility. Br. J. Hosp. Med. 22:47 (1979).
2. B. L. R. A. Coolsaet, The varicocele syndrome: venography determining the optical level for surgical management, J. of Urology 124:833 (1980).
3. J. K. Russell, Varicocele in groups of fertile and subfertile males, Br. Med. J. 1:1231 (1953).
4. L. S. Scott, Varicocele: a treatable cause of subfertility, Br. Med. J. 5228:788 (1961).
5. N. E. Ahlberg, O. Bartley, N. Chidekel, A. Fritjofsson, Phlebography in varicocele scroti, Acta radiologica diagnostica Stockholm 4:517 (1956).
6. E. F. Kiszka, G. T. Cowart, Treatment of varicocele by high ligation, Journal of Urology 83:713 (1960).
7. L. Dubin, R. D. Amelar, Surgical therapy for male infertility, J. Reprod. Med. 18:211 (1977).
8. D. Robinson, J. Rock, Intrascrotal hyperthermia induced by scrotal insulation: report on spermatogenesis, Obst. and Gynae. 29:217 (1967).
9. H. A. Davidson, Temperature of male subfertility; testicular temperature and varicoceles, Practitioner 173:103 (1954).
10. C. Lafaye, J. Hermabessiere, La thermographie du scrotum, Nuov. Presse Med. 5:1826 (1976).
11. R. H. Gold, R. M. Ehrlich, B. Samuels, A. Dowdy, R. T. Young, Scrotal thermography, Radiology 122:129 91977).
12. R. Monteyne, F. Comhaire, The thermographic characteristics of varicocele: an analsysis of 65 positive registrations, Br. J. Urol. 50:118 (1978).
13. B. D. Fornage, J. L. Valsyre, P. L. Lemaire, B. M. Lardennois, Scrotal thermography of 231 normal young adults, Acta Thermographica 5:38 (1980).

THERMOGRAPHIC EXAMINATION OF THE SCROTUM: PHYSICAL FACTORS

C H Jones, W F Hendry, L Hughes

Royal Marsden Hospital and Chelsea Hospital for Women
London
UNITED KINGDOM

INTRODUCTION

The surface temperature of the scrotum is influenced by that of the testes and the underlying blood vessels. Whereas in the normal healthy male the temperature distribution is symmetrical with a warmer region over the septum, reversal of blood flow in the internal spermatic vein can cause a varicocele which may result in an abnormal temperature increase in the scrotum. IR thermography can be used to detect, measure and record the extent of this thermal abnormality.[1,2] The clinical condition affects about 10% of young men[3] and it occurs in about 20% of subfertile males[4] when it may be associated with a variable degree of impaired spermatogenesis.[5] Clinically, it may be difficult to decide upon the presence or absence of a varicocele and it is even more difficult to be sure whether the abnormality is unilateral or bilateral. Although the mere existence of a varicocele is not by itself an indication for surgery if other findings are normal, thermography can provide accurate objective evidence of disturbed scrotal temperature, which may be important in a subfertile man with evidence of impaired spermatogenesis. The results of surgery are not uniformly good and improvement in semen quality may not occur until the temperature abnormality has been corrected, which may require re-operation.[6]

Thermography can confirm or refute the diagnosis of varicocele associated with a significant alteration in scrotal temperature and can define the extent of the venous abnormality. This information is of help to the surgeon who must decide whether or not to recommend operative correction of the varicocele, and then undertake either unilateral or bilateral surgery. This paper describes a technique which has been found to produce consistent results and permits examination of the patient before and after corrective surgery.

TECHNIQUE

Thermographic examination of the scrotum is best carried out with the patient positioned with his legs slightly apart and penis held out of the field of view. The imposition of making the patient assume an unnatural, uncomfortable position should always be avoided. It is important to view the front, side, underside and posterior of the scrotum. Usually this can be achieved by taking anterior and underview thermograms (Fig 1). The anterior view thermogram is recorded after a period of at least 10 minutes cooling of the trunk, scrotum and legs with the subject standing in the examination position in an ambient of 19° - 20°C. It is helpful to observe the temperature distribution over the scrotum throughout this period so that any gross changes in the temperature pattern can be noted. Additional time should be allowed if the scrotal temperature continues to change significantly. Thermal imaging is accomplished with AGA Thermovision (680 Medical). Once equilibration has been attained, the maximum temperature of both sides of the scrotum and the temperature of the septum are measured by means of isotherms. In addition, the AGA integrator is used to display a square or rectangular area over the scrotum and the maximum and mean temperatures within this area are recorded. It is sometimes advantageous to view the patient at a slight angle so that the lateral contour of the scrotum can be observed

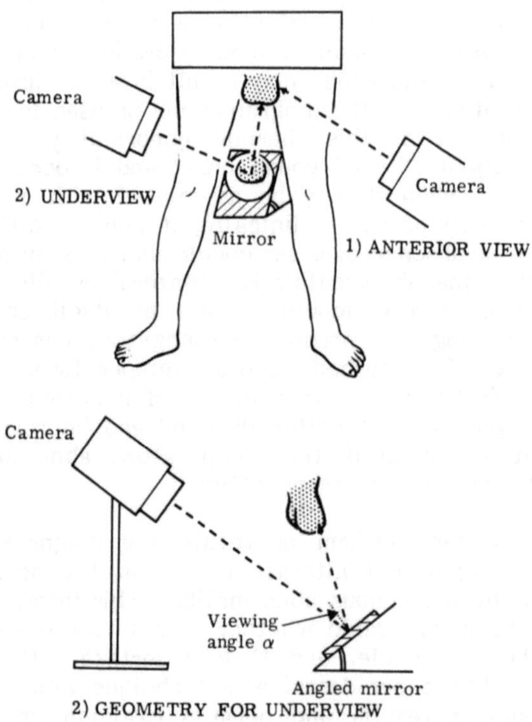

Fig. 1. Geometry for anterior and underview scrotal thermography.

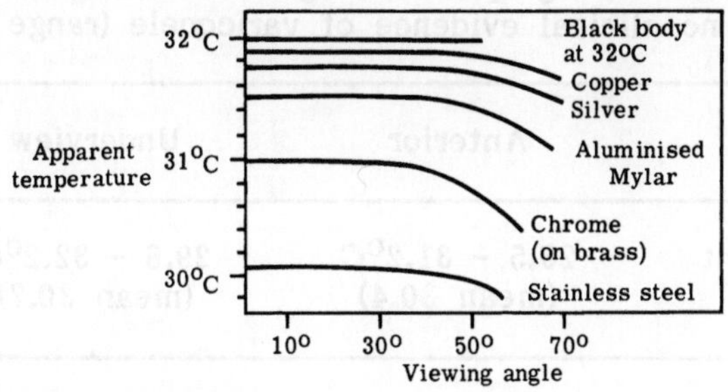

Fig. 2. Apparent temperature changes due to variations in reflectivity and viewing angle.

and when this is the case care must be taken to view the patient symmetrically from either side of the midline. An underview of the scrotum is obtained with the aid of a circular mirror 5" diameter made from aluminized Mylar. The mirror is supported at an angle of about 45° to the horizontal so that the thermal image of the under-surface of the scrotum is reflected into the field of view of the THV camera. The angle of the mirror can be adjusted so as to provide optimum viewing conditions. The patient stands with his legs slightly apart over a small stool on which the angled mirror is positioned at 45° to the viewing axis of the camera and between 20-40 cm beneath the scrotum of the patient. If precise temperature measurements are required, corrections must be made to allow for the effects of surface emissivity and viewing angle.[7] However, these corrections are small compared with the effect of the reflection coefficient of the mirror and its variation with viewing angle. In practice using a highly polished aluminium, or aluminized Mylar mirror a correction of about 0.5°C has to be made (Fig 2). A similar correction has to be applied to integrator values measured by the underview technique.

The maximum and mean temperatures are recorded on the anterior and underview thermograms with the THV integrator (Fig 3). It has been found convenient to make all temperature measurements with a Mid-Temperature-Level of 32°C with a THV sensitivity of 1° per division so that temperatures between 27-37°C can be recorded. The integrator is set so that 50 integrator units correspond to a uniform temperature of 32°C (0 integrator units correspond to 27°C, 100 integrator units correspond to 37°C). Thermograms are recorded on Polaroid film usually in inverted mode (black hot, white cold) or in color.

Table 1. Scrotal thermographic findings in 10 fertile control subjects with no clinical evidence of varicocele (range and mean)

	Anterior	Underview
Right	29.5 - 31.2°C (mean 30.4)	29.6 - 32.2°C (mean 30.7)
Left	29.5 - 31.5°C (mean 30.5)	29.6 - 32.2°C (mean 30.7)
Maximum difference between the sides 0.3°C		

NORMAL THERMAL PATTERNS

The muscular fibers of the dartos muscle divide the scrotal pouch into two partitions by forming a septum, which can be seen on the thermogram as a warm linear zone. After a cooling period of 10 minutes in an ambient of 19-20°C the mean surface temperature of the scrotum is usually within 28-32°C and the temperature difference between each hemiscrotum is less than 0.5°C (see Table 1). The normal thermal pattern is characterized by (a) a uniform temperature distribution, (b) a warm linear zone over the septum, (c) a scrotal temperature which is usually lower than the thigh temperature (Fig 3). It has been found that the average temperature of the scrotum is less than 32°C although small areas such as those over the septum or those caused by the influence of elongated folds of skin can cause temperatures up to 34°C.

EFFECTS OF VARICOCELE ON THERMAL PATTERN

A varicocele is caused by reversed blood flow due to incompetence of valves in the internal spermatic vein which causes pooling of blood around the testicles and an increased scrotal temperature, the extent of which depends on the size of the varicocele (Figs 4 and 5).

An abnormal thermogram is characterized by one or more of the following features:

(a) a very high localized temperature over one or more quadrants of the scrotum; the temperature difference between the affected quadrant and a symmetric point on the contralateral side can vary between 1° and 5°C.

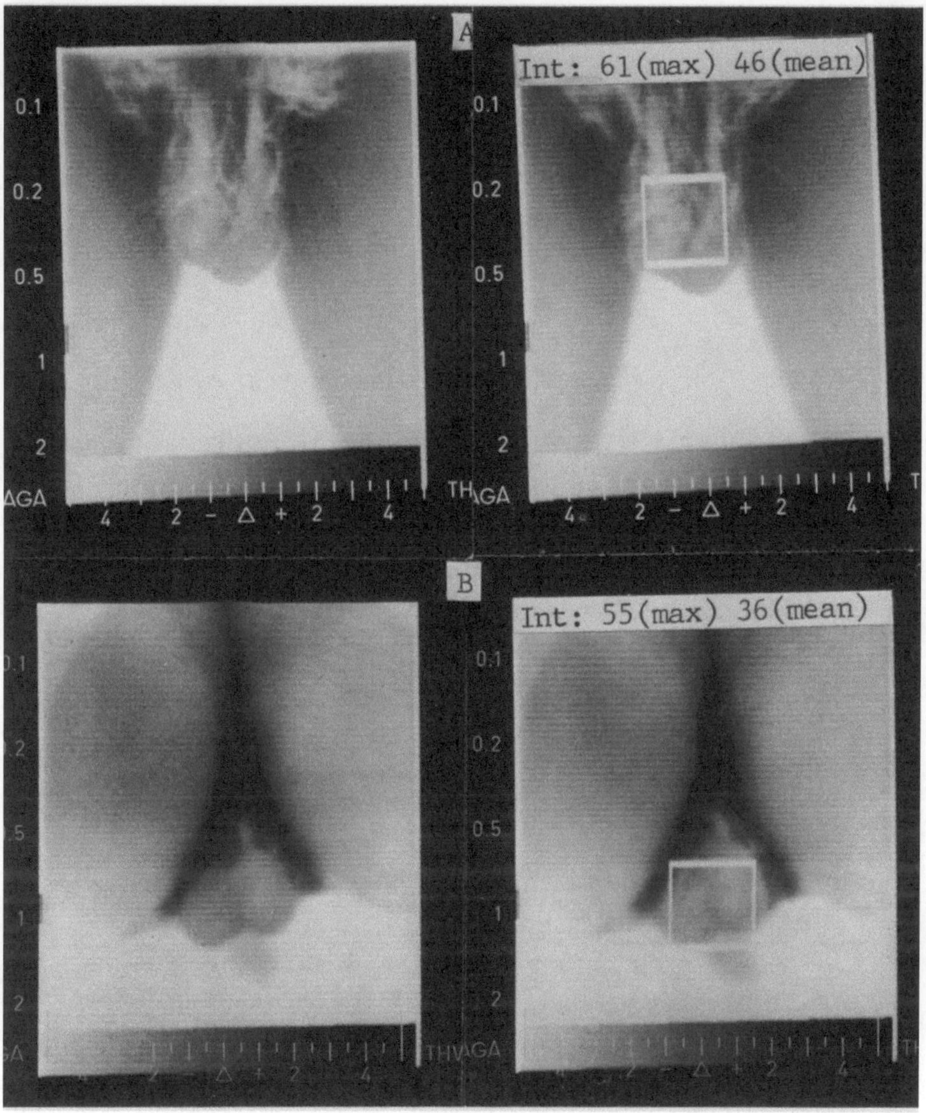

Fig. 3. Anterior and underview thermograms of healthy fertile males (subjects A and B). (MLT 32°C: Int 50).

(b) a mean scrotal temperature greater than 32°C; this temperature is usually greater than the thigh temperature.
(c) unclear delineation of the septum.

Bilateral varicocele can cause the scrotal temperature to increase symmetrically and in such cases the magnitude of the mean temperatures

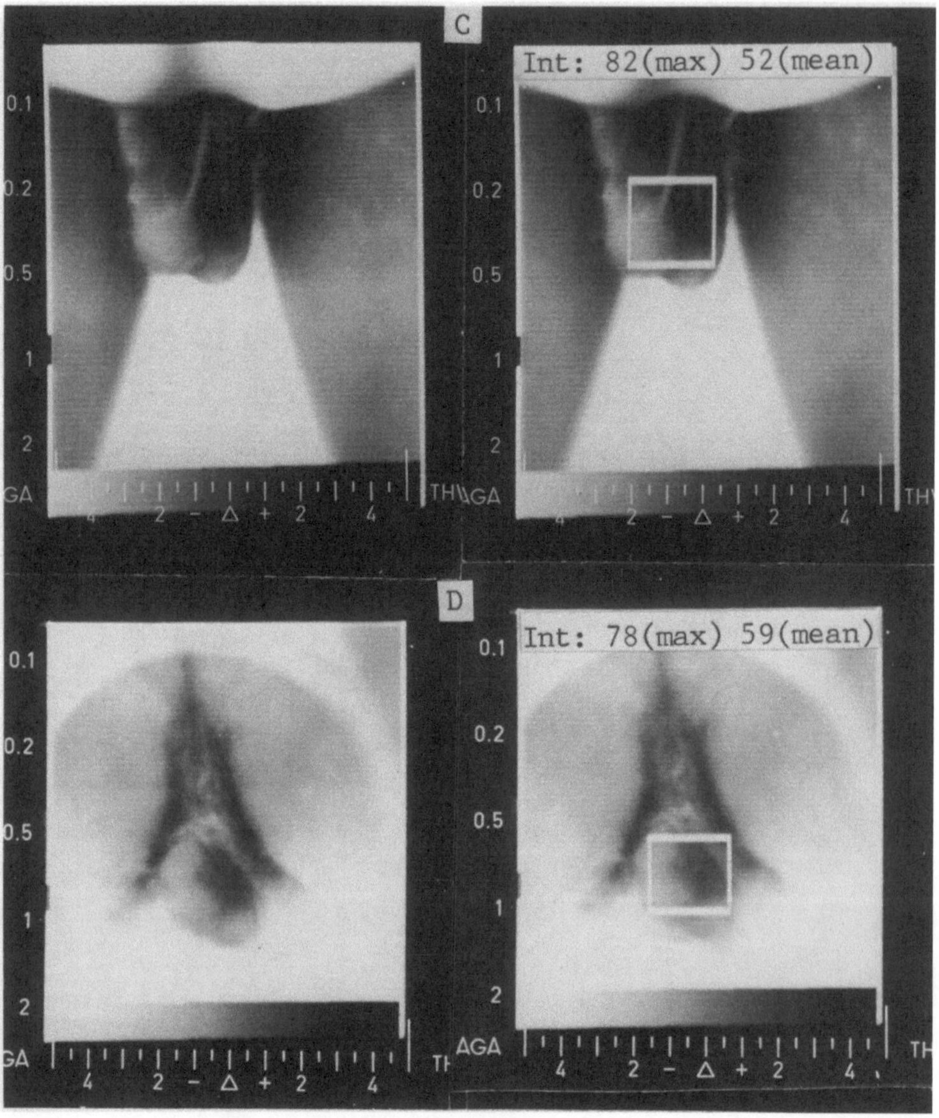

Fig. 4. Abnormal anterior and underview thermograms (subjects C and D). (MLT 32°C: Int 50).

of the scrotum becomes a key factor. The THV integrator is particularly useful for measuring average temperatures. We have found that the normal scrotal temperature distribution is such that under the conditions described above the number of integrator units for the mean temperature is usually less than 50 (corresponding to 32°C). It is helpful to use the integrator for anterior and underview thermograms, particular attention

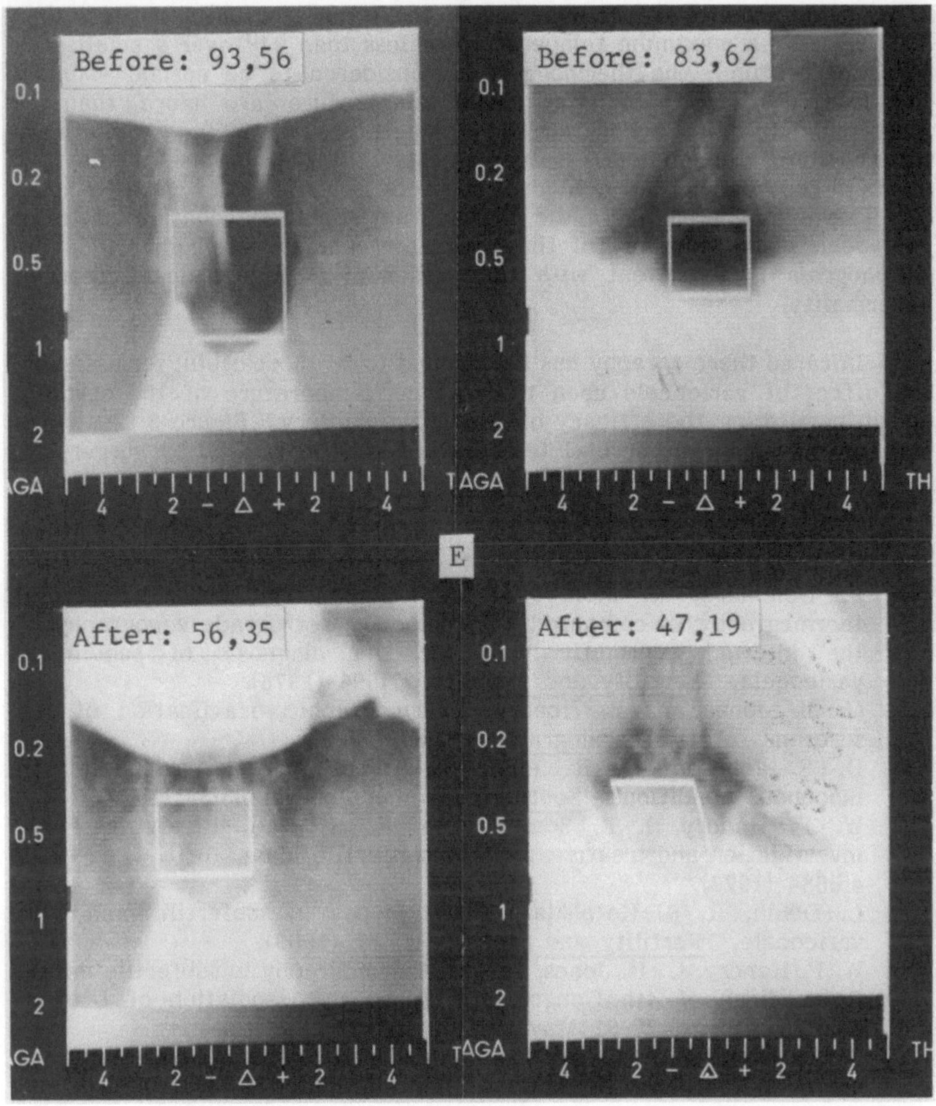

Fig. 5. Thermograms of subject E before and after ligation of varicocele, illustrating changes in temperature and how Integrator can be altered to accommodate scrotum. (MLT 32°C: Int 50).

being paid to the maximum and mean integrator readings. Thus for each examination four relevant integrator numbers are obtained. Since the integrator records the maximum temperature over a very small area it is sometimes useful to identify where this area occurs by means of a

THV isotherm. A small area of high temperature over the septum is often of no clinical significance and if the integrator units are less than 70 (recording a maximum temperature of less than 34° over a small area over the septum), the thermogram is considered to be normal - unless the integrator units for the mean temperature are more than 50 (corresponding to a mean surface temperature of 32°). Preliminary assessment of the clinical results suggests that if only one of the four numbers recorded for an individual subjects is higher than either a mean of 50 or a maximum of 70 the thermogram should be considered to be equivocal: if two or more of the integrator numbers are higher then the thermogram is consistent with the presence of a signfiicant thermal abnormality.

Infrared thermography has been found to be most useful for assessing the effect of varicocele upon the surface temperature of the scrotum and in monitoring the efficacy of corrective surgery. Figure 5 illustrates the way in which the scrotal temperature can be altered by surgery.

REFERENCES

1. F. Comhaire, R. Montyene, M. Kunnen, The value of scrotal thermography as compared with selective retrograde venography of the internal spermatic vein for the diagnosis of subclinical varicocele, Fertility and Sterility 27:694 (1976).
2. C. H. Jones, W. F. Hendry, Thermographic examination of the scrotum, Acta Thermographica 4:38 (1979).
3. D. E. Johnson, D. R. Pohl, H. Rivera-Correa, Varicocele: An innocuous condition? Southern Medical Journal 63:34 (1970).
4. W. F. Hendry, I. F. Sommerville, R. R. Hall, R. C. B. Pugh, Investigation and treatment of the subfertile male, Brit. J. of Urol. 45:684 (1973).
5. L. Dubin, R. S. Hotchkiss, Testis biopsy in subfertile men with varicocele, Fertility and Sterility 20:51 (1969).
6. W. F. Hendry, C. H. Jones, Scrotal thermography in subfertile males. Proceedings of 4th Congress of European Association of Urology held in Athens, 28-31 May (1980).
7. J. A. Clark, Effects of surface emissivity and viewing angle on errors in thermography, Acta Thermographica 3:138 (1976).

THERMOGRAPHIC CHARACTERISTICS OF THE SCROTUM IN THE INFERTILE MALE

V Vlaisavljevic

Department of Gynaecology
General Hospital Maribor
YUGOSLAVIA

Varicocele is a frequent cause of male infertility, and small varicoceles may be just as pathological as large ones. Slight changes in blood flow into the internal spermatic vein (ISV) are frequently difficult to detect by clinical examination only. These inpalpable vascular changes can be the cause of "idiopathic" infertility. In the last eight years we have tested some methods (thermography, plate thermography, phlebography and angioscintigraphy)[1,2] which have been found valuable in the diagnosis of venous insufficiency and varices of the ISV.

MATERIALS AND METHODS

During the period of our study between 1979 and 1981 thermographic recordings of the scrotum were made in 500 patients who were attending regular treatment sessions at the Outpatient Clinic for Infertile Couples. Thermographic examination was performed by means of AGA - Thermovision equipment. The patient was allowed to stand for 15 minutes in a constant temperature environment of 20°C. Thermograms were obtained from the anterior aspect of the scrotum with the patient in the erect position as described by Kormano et al[3] at a distance of 1 meter. The diagnostic criteria applied in our study are based on observations described by many other authors.[4,5,6,7,8]

RESULTS

Temperature over the scrotal surface is usually distributed symmetrically. Varicocele is the manifestation of retrograde blood flow into the ISV. The surface temperature of the skin covering the side with vascular abnormality is higher than the rest of the scrotum. To demonstrate these circulatory changes we observed the cooling effect over the scrotal surface during the examination. Absolute values of temperatures measured over left and right testis were registered in 28 males with normal clinical and thermographic findings during the cooling process (Fig 1). The effect of cooling a scrotum with clinically palpable varicocele under the same conditions was registered in 26 patients (Fig 2). During the period of observation, heat loss over the varicocele surface tends to remain high because warm blood is continuously supplied from the deeper parts of the body. Initial absolute values of temperature over both sides of the scrotum in men with varicocele were higher than in normal males. During the cooling process the close connection between the two regions disappeared because the patient was undressed; the slope of the curve depicting the right hand side of the scrotum with left sided varicocele was steeper than the left side.

The warm areas on abnormal thermograms differed in location and extent. In 158 abnormal thermographic registrations of temperature, distribution varied from local hyperthermia in the upper part of the scrotum to generalized hyperthermia (Table 1).

Temperatures measured over left and right pampiniform plexus and left and right testis in normal thermograms (n = 152) were 30.47 ± 1.26°C, 30.44 ± 1.22°C, 28.78 ± 1.65°C, 28.84 ± 1.61°C respectively. Values in the same areas with abnormal thermograms (n = 95) were 32.53 ± 1.75°C, 30.35 ± 1.43°C, 30.19 ± 2.0°C, 29.44 ± 1.34°C (Fig 3).

Raised temperature of the scrotum in patients with normal clinical

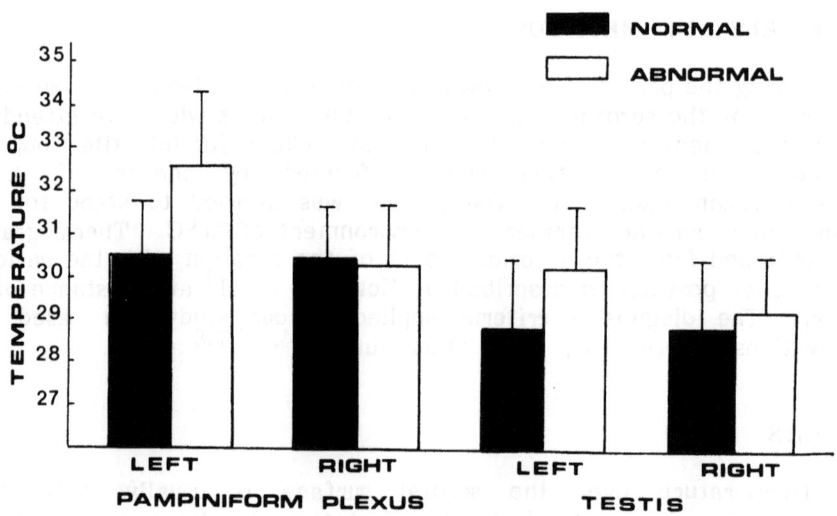

Fig. 1. Measured temperatures ($\bar{X}$ ± SD) over left and right pampiniform plexus and left and right testis in 52 patients with normal and in 95 patients with abnormal thermograms.

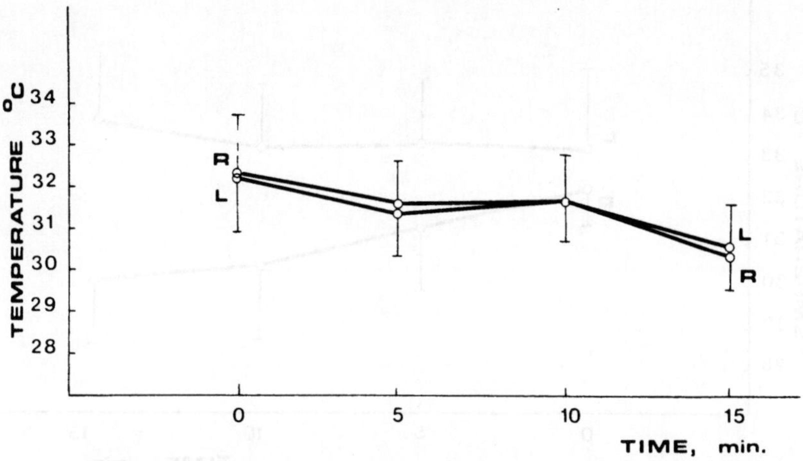

Fig. 2. Plots of temperatures ($\overline{X} \pm SD$) measured over left (L) and right (R) pampiniform plexus as a function of time for 28 patients with normal clinical finding and normal thermogram. Ambient temperature was stable at 20°C.

Table 1. Comparison of thermographic and clinical examination in 500 patients

THERMOGRAPHIC FINDING	CLINICAL FINDING			
	NO VARICOCELE	POSSIBLE VARICOC.	PALPABLE VARICOC.	CONTROL AFTER SURG.
NORMAL (342)[1]	257	14	19	52
ABNORMAL (158)	87	5	52	14
HYPERTHERMIA OVER PAMPINIFORM PLEXUS ($\Delta T \geq 1.5°$ C)	27	2	16	9
HYPERTHERMIA OVER PAMPINIFORM PLEXUS ($\Delta T > 1.5°$ C) AND IPSILATERAL TESTIS ($\Delta T > 1.0°$ C)	38	1	30	4
HYPERTHERMIA OVER TESTIS ($\Delta T \geq 1.5°$ C)	12	1	4	1
ELEVATED TEMPERATURE OVER SEPTUM SCROTI COMPARED WITH PAMPINIFORM PLEXUS ($\Delta T \geq 1.0°$ C)	7	1	2	
HYPERTHERMIA OVER WHOLE SCROTUM	3			

[1] number of patients

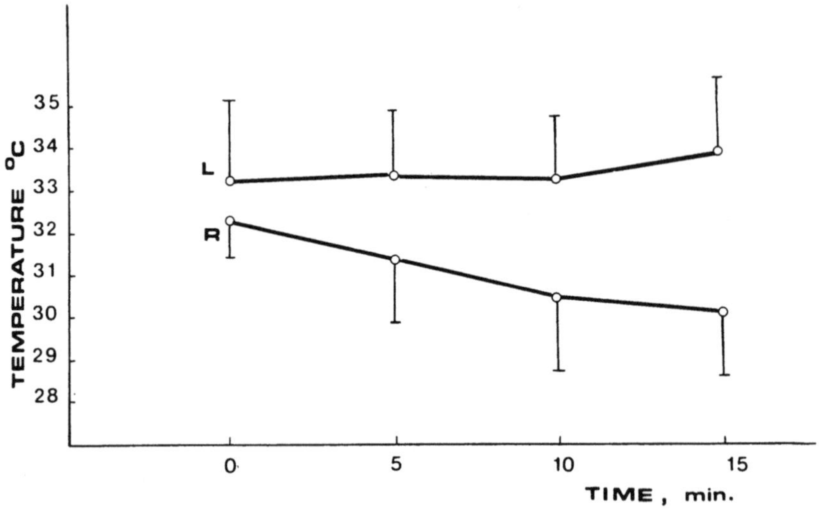

Fig. 3. Plots of temperatures ($\bar{X} \pm SD$) measured over left (L) and right (R) pampiniform plexus as a function of time for 26 patients with varicocele and abnormal thermogram. Ambient temperature was stable at 20°C.

findings occurs significantly more frequently parallel to a spermiogram of lower quality (sperm density $< 20.10^9$/L, $\leqslant$ 60% motile sperms) (chi square test, $p < 0.001$).

In cases where some disagreement between clinical and thermographic findings was observed, we applied additional, more invasive diagnostic methods. The most interesting group was the group of patients with poor semen quality and no clinically evident varicocele, but with a presence of abnormal thermographic registration. Among 44 patients with such a combination of findings vascular abnormality was certified in 32 (72.7%) by means of retrograde phlebography and radionucleotide angioscintigraphy.

CONCLUSION

Disturbances in temperature regulation in patients with varicocele have been thermographically demonstrated by Kormano et al.[9] Our results obtained by the observation of the scrotum during the period of cooling suggest the same conclusions. The close connection of varicocele and central body circulation is the reason why the temperature over the scrotal surface on the varicocele side changes only slightly during the examination.

A comparison of absolute temperature values in patients with normal and abnormal thermograms shows us that absolute temperature values alone do not suffice in diagnosing varicocele with reasonable accuracy. Absolute values of temperatures can be useful only in diagnosing bilateral varicocele.[8]

The results of our study suggest that scrotal hyperthermia can be caused by different types of circulatory disturbances: reflux of warmer blood, distention of pampiniform plexus combined with venous reflux and with other circulatory, clinically unrecognizable disturbances. Increased temperature over the scrotal surface in patients with varicocele is in close connection with varicocele size.[7] Hyperthermia over the septum scroti was observed more frequently in patients whose varicocele was localized behind the testis. In such cases the heat transmission through the septum was the most visible sign of its presence. Hyperthermia over the whole scrotum was observed only in few cases. In one of them, the presence of varicocele was certified by phlebography. In two cases such type was registered just before collapse. The problem of diagnosing minimal circulatory changes of the scrotum and testis has not been fully solved.[10] After a period of preliminary testing we included thermography as a screening method for more invasive diagnostic procedures (angioscintigraphy and retrograde selective venography). Thermography is an objective adjunct to clinical examination for the diagnosis of varicocele.

REFERENCES

1. V. Vlaisavljevic, B. Gorisek, M. Picajt, The thermographic diagnosis of varicocele, Zdrav Vestn 48:415 (1979).
2. V. Vlaisavljevic, B. Zakotnik, J. Subic, R. Breznik, E. Borko, Use of radioisotope angioscintigraphy in diagnostics of male infertility, Zdrav Vestn 51 (in press) (1982).
3. M. Kormano, K. Kahanpaa, U. Svinhufvud, E. Tahti, Thermography of varicocele, Fertil. Steril. 21:558 (1970).
4. J. P. Amiel, L. Vignalou, J. Tricoire, B. Jamain, J. H. Ravina, Thermographie du testicule, J. Gynecol. Obstet. Biol. Reprod. (Paris) 5:917 (1976).
5. F. Comhaire, R. Monteyne, M. Kunnen, The value of scrotal thermography as compared with selective retrograde venography of the internal spermatic vein for the diagnosis of "subclinical" varicocele, Fertil. Steril. 27:694 (1976).
6. H. R. Gold, R. M. Erlich, B. Samuels, A. Dowdy, R. T. Young, Scrotal thermography, Radiology 122:129 (1977).
7. R. Monteyne, F. Comhaire, The thermographic characteristics of varicocele: an analysis of 65 positive registrations, Br. J. Urol. 50:118 (1978).
8. C. H. Jones, W. F. Hendry, Thermographic examination of the scrotum, Acta Thermographica 4:38 (1979).
9. M. Kormano, K. Kahanpaa, E. Tahti, Thermographic recording of the reaction of normal and varicocele scrotum to increased temperature, Andrologia 5:201 (1973).

10. A. V. Hirsh, M. J. Kellett, G. Robertson, P. J. Pryor, Doppler flow studies, venography and thermography in the evaluation of varicoceles of fertile and subfertile men, Br. J. Urol. 52:560 (1980).

THE ROLE OF LIQUID CRYSTAL THERMOGRAPHY IN THE DIAGNOSIS OF TESTICULAR AND THYROID DISORDERS

P Goblyos, I Szabolcs, I Szilvasi, Z Kopcsanyi, G Vydra, G Irsy, M Goth, G Szilagyi

Postgraduate Medical School
Department of Radiology, Urology and Internal Medicine, 1389 Budapest, POB 112, HUNGARY

Liquid crystal thermography (LCT) has been used, predominantly, in the diagnosis of tumors and inflammation. Less information is available about the thermographic features in endocrine disease. The thyroid gland as well as the testicles is superficial and easily accessible to this method.

Besides the use of LCT in the diagnosis and post-operative assessment of varicocele, we investigated a number of endocrine disorders causing male infertility.

The place of LCT in the diagnosis of thyroid disease is not defined; therefore, comparable data provided by physical examination, hormonal, isotope and ultrasonographic investigation as well as by LCT has been accumulated in patients with a number of thyroid disorders.

MATERIALS AND METHODS

LCT of the testicles was performed in 23 males with oligo-azoospermia of genetic, hormonal, genetic-hormonal or unknown origin. Determination of pathogenesis included investigation of the chromosomal picture and blood levels of follicle stimulating hormone (FSH), luteinizing hormone (LH) and prolactin (PRL), measured by commercially available radioimmunoassays. Testosterone levels were measured by the competitive protein binding method. The patients were divided into groups; Klinefelter's syndrome (3 cases), primary hypogonadism of other origin (7 cases), oligo-azoospermia of unknown origin, hyperprolactinemia (2 cases) and pituitary hypogonadism (6 cases). Eleven fertile men of the same age group (20-40 years) served as a control group.

In another study 24 patients with differing thyroid disorders were investigated and classified according to the clinical features, physical examination, routine hormonal analysis and findings on Technetium-99 scintigraphy and ultrasonography. All the patients were women aged 16-71 years (mean 32.7 years). Three female control subjects were also investigated.

With the patient sitting or standing liquid crystal thermography was performed with application of thermosensitive flexible plates over the thyroid gland or over the testicles with the subject standing. The temperature of the thyroid was 33-34°C, that of the testicles 32-34°C. The testicles appear on the plate as areas of different color. Relating to the thyroid, the differentiation of hot and cold parts of the gland is also possible and blood vessels become visible. Sometimes interpretation can be assisted by local cooling of the skin with alcohol or cold air.

RESULTS

Normal fertile men may have one or both testes cooler than usual or a previously unrecognized varicocele could be detected. In primary hypogonadism due to either Klinefelter's syndrome or another cause the testicles were usually atrophic but in some cases were at the normal expected temperature. On the other hand in some patients with oligo-azoospermia but with a normal hormonal and genetic status, the testicles had decreased heat production. In these cases LCT showed better correspondence with the sperm count than did physical examination. The cases of hyperprolactinemia and pituitary hypogonadism showed normal testicular temperature. The patient's state was more accurately defined by LCT than by clinical examination, eg the colder testicle as shown by LCT could be selected for biopsy.

Using LCT we were not able to distinguish either hypo- or hyper-thyroidism. However, small areas of reduced isotope uptake are better detected by LCT; in cases of cystic degeneration with great colloid content the temperature of the gland was usually lower than normal and in cases of tumor or tumor-like solid areas the temperature was higher. Large solid cysts were colder by LCT.

DISCUSSION

These data indicate that LCT of the testicles may be a useful additional method in the diagnosis and subsequent treatment of hypogonadism. On the other hand, there appears no clear indication for the use of LCT in thyroid disease.

LOCOMOTOR DISEASES - INTRODUCTION

D P Page Thomas

Rheumatism Research Unit E6
Addenbrookes Hospital, Hills Road
Cambridge, UNITED KINGDOM

This session by its very nature embraced a wide variety of different applications of thermological investigations which studied the reaction of the locomotor system to trauma, non-articular inflammation, joint disease, and the effects of physical therapy and drugs.

The contributions of the invited speakers emphasized either newer applications of thermography to various locomotor conditions or reviewed and updated its current role in well established applications.

Schmitt and Guillot reviewed the usefulness of serial thermographic examination of musculotendinous lesions in sports medicine, and found that it was of particular value not only in the assessment of the severity of acute lesions, but also for monitoring their progress through to the resumption of full activity. In injuries of long standing, it was also of value in diagnosis and in following recovery after surgery. They stressed the value of differentiated thermal measurement with the comparable area in the normal contralateral limb, and also the regional hypothermia which may overlap the lesional hypothermia after severe muscle lacerations, or tendinous tears at the point of any insertion.

Physical methods of treatment have been used to treat painful conditions of the locomotor system from time immemorial, but their effects upon the body have been relatively little studied by objective thermological methods, as Rusch reported in a wide-ranging and detailed review of the literature. He laid particular stress on the need for further work in this field, and that it was important to draw a clear distinction between thermal measurements made in the hours immediately following treatment, and the longer term therapeutic effects, if any, of repeated treatments upon the thermal state over the lesion.

The value of asymmetric thermal changes was further stressed in Wexler's experiences in an extensive series of spinal thermograms and also the hypothermic changes along the course of dermatomes. He also found that the thermogram correlated well with electromyographic studies and myelography in herniated lumbar intervertebral discs, and had a similar statistical accuracy in these patients.

Engel reviewed in depth the relationship between abnormal thermograms and the underlying complexity of physiopathologic processes in joint disease, which could influence thermal changes.

The Seminar session was concluded by an update of the value of thermal quantification in pharmacological assessment in locomotor disorders by Bacon and Ring. Numerous examples were given of their extensive experience in this field. However, they stressed the need for rigidly controlled techniques, if it is to be profitably used in longitudinal studies, and concluded that thermal change associated with arthritis, whatever the exact underlying mechanism, is a valid reflection of clinically important inflammatory change and is the best objective assessment method presently available.

The contributions to the poster session emphasized further many areas of current interest previously mentioned in the seminars. Sports injuries and the role of pain in altering thermal patterns occupied a number of contributions. The thermographically difficult problem of analysis in ankylosing spondylitis was also covered by other workers. The importance of pattern recognition in both articular and non-articular disease was stressed by a number of contributors, and the increasing need and use of the microcomputer in the numerical assessment of pattern changes will undoubtedly prove to be of increasing importance in future years.

THERMOGRAPHY IN RHEUMATOLOGY

J M Engel

Staatliches Rheumakrankenhaus
Baden-Baden
WEST GERMANY

As a medical discipline, rheumatology has come of age. A remarkable transition has occurred from empiric spa treatment to the multidisciplinary specialty of today. Scientific efforts of the last decades have led to a better understanding of inflammatory, metabolic and immunologic processes, even if the causes and primary pathogenic pathways are still obscure. A diversity of medical disciplines now help the rheumatologist in disease management and differential diagnosis. Technical progress in X-ray techniques and scintigraphic scanning provide detailed views of structural and functional changes in various rheumatic diseases. Enhancement in biochemical methods continues to enlarge the spectrum of laboratory tests available to monitor the systemic course of disease, and secondary changes in different organ systems. Among the more recent technical methods, temperature measurements, particularly infrared thermography (IRT), is contributing to this progress. Even if IRT is not yet a routine procedure in every rheumatology unit, recent technical progress is such that thermology will soon leave the laboratory and experimental stage. The time has come to establish thermological methods as a routine for clinical rheumatology.

In rheumatology the common goal is "to keep the patient going". Most of the rheumatic diseases are systemic, non-linear and life-long processes. Therefore, patient management calls for dynamic, non invasive procedures to monitor the course of disease and the influence of therapeutic measures. Body core temperature measurements are well established as a check for fever, and a sign of central temperature regulation disturbance. The measurement of surface temperatures and temperature distribution has remained a science for enthusiasts. The physiology of the nervous system controls almost every functional system of the human body. There must, therefore, be some interference within the pathophysiology of rheumatic diseases. Skin microcirculation is well controlled by sympathetic nerves. Surface temperature measurements made under standard conditions or in response to well defined stimuli must present evidence of pain and inflammation, two of the classical signs in clinical rheumatology. Both symptoms are difficult to measure objectively. This diagnostic gap can be closed by thermological methods such as IRT or liquid crystal plate thermography. The latest technical equipment includes IRT-cameras measuring absolute temperatures with image processing facilities (Fig 1) and flexible liquid crystal plates of high temperature resolution and short response/reaction times.

SKIN MICROCIRCULATION - THE OBJECT OF THERMOGRAPHY

The skin is an essential regulating mechanism in human physiology and is not just the barrier against the outer world. Within the cybernetic circles of blood pressure and body temperature the skin has important functions. The effector is the capillary network within the deeper skin layers. With regard to temperature regulation, peripheral skin areas act as a mediator for heat loss. Thus skin surface temperatures and infrared emission depend on convective heat transport from the body core according to microcirculatory blood flow. Since the skin capillaries form a narrow

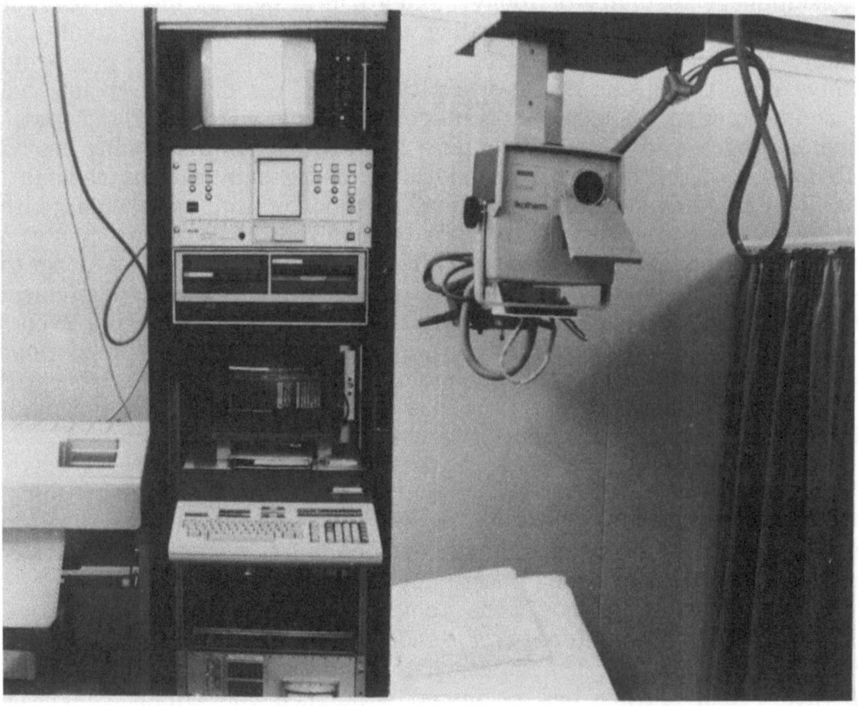

Fig. 1. Ikotherm - IR - Camera (Cell Zeiss, West Germany) connected to a Hewlett-Packard 9845 computer using special software (Thermotom - System) for rheumatology.

network along the whole body surface, it is by no means clear that heat conduction from deeper layers or local metabolic heat production should be of importance to surface temperature, while the microcirculation in the skin is working properly. Skin microcirculation prevents man from being "cooked" in the Finnish sauna and from freezing in the cold, both being external influences. Why then should this system work in only one direction?

Several experiments in the past have shown that the convective heat transport via skin microcirculation is the predominant factor in building up skin temperature at the surface. When applying thermographic methods for measuring skin temperature distribution, it is therefore essential to know of the factors and mechanisms influencing skin microcirculation. The pre-capillary sphincter is responsible for controlling blood flow in the microcapillary network. Changes in blood flow can be as high as 3000% (!) by changing the diameter of the pre-capillary sphincters. The pre-capillary sphincter is not an anatomically well defined arteriole, but is a function of the vessels which can react to different nervous, humoral, chemical and other stimuli - even when smooth muscle cannot be detected in the capillary wall. Nevertheless, the sympathetic nervous system plays the key role in controlling local skin blood flow.

This is the basis on which thermography of the skin surface measures convective heat transport from body core to skin microcirculation. Pathological changes can only be detected thermographically when they affect one or more factors regulating skin microcirculation. Standard physiological conditions during and preceding thermographic measurements must be observed in the detection of pathological changes.

ALTERATIONS IN SKIN TEMPERATURE IN LOCOMOTOR DISEASES

The locomotor diseases consist of about 450 different diagnoses, pathologies and symptoms. Differential diagnosis in such a poly-combination of signs and symptoms with only a handful of well defined clinical entities must be a discriminating process in which many different methods can be useful. Surface temperature measurement is one of these complementary methods. Its advantage is that it is a non-invasive procedure, repeatable at any time interval required for monitoring the course of disease. A disadvantage may be the short time constant in thermo regulation at skin surface level. In contrast, radiography has a very long time constant (bone mineralization/destruction process being relatively slow). Radiation hazards exist if radiographs of all joints and structures involved in a rheumatic disease are repeated too often.

There are three different patho-physiological situations in rheumatology that can be successfully detected and followed by thermographic methods:

1. <u>Inflammation</u> with secondary, mediator induced vasodilation.

2. Pain with nocioceptive vasoconstriction.
3. Vasculitis with vaso-obstruction/vasospasm.

In rheumatic inflammation mediators eg prostaglandins, kinins etc are liberated. Extensive vasodilation proportional to the degree of inflammation will result. Thus induced alterations in local blood flow increase surface temperature above normal levels.

Functional lesions in joint movement lead to a nocireaction because of the many free nerve ends in joint synovium. At a spinal level this nociceptive input is switched to a sympathetic - vasoconstrictive output as well as to a motor output. This pathophysiological pathway is especially applicable to the small spine joints, sacroiliac joint and other small joints of the trunk (sternocostal, costovertebral, costochondral).

Direct mechanical nerve (root) irritation can in an early stage also result in segmental hypothermia because of vasoconstriction in the skin microcirculation. The more the mechanical compression irritates and blocks neural structures the more thin, less myelinated fibers of the sympathetic system are destroyed, thus lowering the vasoconstrictive action with resulting vasodilation. In complete lesions there is complete loss of neural vasomotor control. This pathomechanism is involved in degenerative spine disease (herniated disks) or inflammatory disease (carpal tunnel syndrome).

Vascular involvement in the course of rheumatic disease may also result in changes in surface temperature, either by mechanical obliteration (vasculitis) or by vasospastic reactions (Raynaud's syndrome).

All these different pathomechanisms may lead to changes in "normal" surface temperature distribution. A very carefully standardized recording technique is thus essential to get valuable results. Standardization must also include careful preparation of the patient to avoid errors caused by nutrition, medication, physical work, blood pressure, or smoking etc.

NORMAL SURFACE TEMPERATURE DISTRIBUTION

Normal surface temperature distribution has no fixed value in terms of absolute temperature. However, an individual temperature pattern is observed, the absolute value of which is dependent on the patient's temperature regulation, metabolism, emotional situation and ambient conditions. The short time constants of skin microcirculation regulation may result in changing surface temperature patterns if any of the different parameters is in a transient stage. Careful bedside patient preparation and constant monitoring of the technical equipment and the ambient conditions are the chief influence on skin surface temperature measurement. The term "normal surface temperature (pattern)" thus must always be quoted with the ambient temperature.

Surface temperatures at 18°C ambient are very different from surface temperatures at 24°C ambient. In the first case regulation stabilizes at vasoconstriction level, whereas at 24°C there is a slight vasodilation. The physiologist also shows that there is a difference in the level of regulation, a regulatory gradient from central to more peripheral areas of the body surface. Nevertheless, it is true that "normal", healthy people will present a normal thermogram with symmetrical temperature pattern. To detect pathophysiological changes it is necessary to adapt the ambient conditions to the mechanism to be detected: vasodilation or vasoconstriction.

THERMOGRAPHIC DETECTION OF INFLAMMATION

To detect inflammatory changes in the locomotor system, ambient conditions of 18.0 ± 0.5°C are selected to cause a vasoconstrictive response in all areas not affected by the inflammatory process. A time of at least 15 minutes is necessary to achieve an almost stable temperature situation. The transient process of adaptation to the low ambient temperature will last longer (up to 120 minutes). But the most significant changes occur within the initial period. Adaptation is dependent on the patient's metabolism, temperature regulation, circadian rhythm and other intrinsic factors. Thus temperature measurements for monitoring disease and inflammation should be taken when the temperature change is less than 0.1°C/minute.

After careful adaptation to 18°C ambient, increased convective heat transport from the body core can be quantified by several parameters. Maximum temperature in a region of interest, thermographic index (mean temperature difference to a constant), heat distribution index have all been used (Fig 2). These quantifying parameters provide a method for monitoring inflammation and anti-inflammatory therapy. This will only apply if calculated from thermograms taken under the same standard conditions. Apart from figures in °C the temperature distribution pattern is also important to detect inflammation, and to establish the thermal differential diagnosis.

All joint inflammation - independent of pathogenesis - may be detected and monitored by thermography in a cold ambient temperature. Paget's disease of bone, bone tumors and osteomyelitis can be monitored, as well as inflammatory muscle or tendon disease such as myositis, tenosynovitis and tendinitis. In some cases of clinical swelling of joints or tendons thermography may be the only method to differentiate the inflammation still present from the secondary edema or hygroma.

Apart from static thermographic measurements, as described above, which have been clinically recognized for several years, modern IR-camera systems can be used for dynamic studies to measure the "thermal relaxation time" after cold or heat challenge. With more mediators present, vasodilative activity may lead to shorter relaxation times and

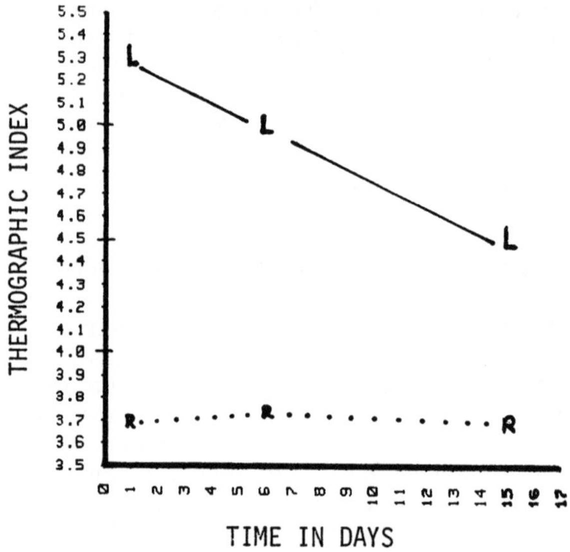

Fig. 2. Thermographic monitoring: 47 year old female with monoarthritis of the left knee joint. Ambient temperature 18^oC; thermographic index according to Ring et al.[1]

less reaction to the challenge. These thermographic procedures still have to be tested in detail, before they can be recommended for clinical routine.

The thermographic procedures available for clinical routine can be used for monitoring joint inflammation in the course of any arthritic involvement of peripheral joints. Inflammation of small joints of the spine cannot be detected since the articular blood drainage is not directed towards skin surface. An exception to this rule may be the sacroiliac joints where a correlation between skin temperature, clinical signs of acute inflammation and, for example, scintigraphy may be shown. A lack of correlation with radiographic signs of inflammation has often led to a negative opinion on the clinical value of thermography. However, thermography and radiography are complementary methods, measuring different pathologies with different time constants. Bone erosions and other aspects of chronic inflammation occur later, and last longer than the inflammatory process with mediator induced vasodilation. For this reason the clinical value of thermography should not be assessed by radiographic comparisons. Thermography provides another dimension in diagnosis.

The detection of inflammation in the joints of the spine by

microwave thermography seems very promising. However, this procedure is not yet ready for routine clinical use. There are several techniques which are readily available for clinical studies.

In the interpretation of thermograms demonstrating inflammatory lesions, one must be aware of possible complicating factors. Skin lesions or skin diseases such as psoriasis, erythema nodosum or tinea pedis may interfere with the result. When skin inflammation is present local blood flow may be increased due to inflammatory mediators or decreased due to edema in the skin, thus masking the rheumatic thermal pathology. Reading thermograms without knowing the patient in detail is unwise and will lead to misinterpretation of results. This is as true for thermal diagnosis as it is for differential diagnosis. Following diagnosis and the establishment of quantitative thermal parameters, the disease progression can be monitored by thermography. Under standardized conditions, thermography is a valuable tool for monitoring inflammatory diseases of the locomotor system, most of which are chronic, and need non invasive procedures. Apart from primary inflammation, the secondary inflammatory reactions as in osteoarthritis, can also be monitored by thermography. The effects of systemic anti-inflammatory medication or physical therapy can be followed at the site of the disease process. This is a particular advantage of thermography, in contrast to laboratory findings or measurements based on subjective clinical symptoms.

THERMOGRAPHIC DETECTION OF FUNCTIONAL LESIONS

To detect the sympathetic nocireaction in functional lesions of joint movement - especially in small spine joints - ambient conditions of 24 - 25°C should be used. This is necessary to achieve basal vasomotor tone in the skin microcirculation in the non-affected areas. Even if the temperature difference between the ambient temperature in the thermographic department and skin temperature beneath the patient's clothing is small, an adaptation time of about 10 minutes is recommended. This is needed for a stable surface temperature distribution and to eliminate discrete "mini-tourniquet" effects at microcirculation level caused by clothing.

After adaptation to the ambient temperature of 24°C there is normally a symmetrical thermal pattern on the patient's surface, especially on the back. If a nocireaction is present due to a functional lesion, segmental vasoconstriction in the affected dermatome can be detected, the degree of which can correlate with the intensity of pain, and the acuteness of the lesion (Fig 3). In chronic malfunction and pain of small spine joints a polysegmental reaction can occur with less definition. This clearly detectable vasoconstrictive hypothermia of 0.2 to more than 1°C is related to the efferent sympathetic nerves and their surface representation. A clear understanding of the sympathetic dermatomes is essential in order to interpret these thermograms. For example, functional lesions of C_7 are represented in the hand (finger II/III)

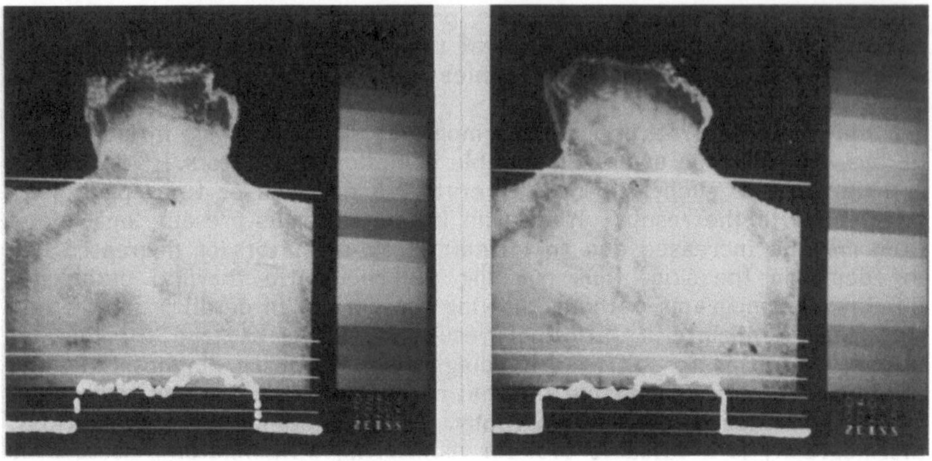

Fig. 3. Thermograms of the back and neck of a 50 year old female with chronic back pain. The left at AT 25°C shows marked hypothermic in the dermatome C4 on the right. The right shows increased hypothermic (ΔT 0.4 to 0.6°C) after rotation and lateral flexion accompanied by pain.

and functional lesions of the sacroiliac joint are represented at the level S_1 - S_3 at the back of the upper leg.

The thermographic examination is not specific: it is known that the internal organs (heart, lungs, liver, gall bladder, etc) are also represented at the skin surface. However, here the anatomy of the sympathetic nervous system always causes a polysegmental representation, which clearly constrasts with the monosegmental lesions of small spinal or costovertebral joints.

However, thermographic measurements are very sensitive: healthy people always present a symmetrical thermal pattern. Asymmetry is a sign of internal or external derangement. For this reason the thermogram can only be read by the doctor who knows the patient's symptoms, history, and possible differential diagnosis. To avoid misinterpretation due to artefacts during thermographic recording, a well trained assistant or even the doctor himself should take the thermograms. He is then able to ask the patient to move the spine in the painful direction, thus provoking more marked hypothermia in the dermatome ("thermatome") affected.

The technique is very sensitive in detecting functional lesions of this kind. Direct mechanical compression of peripheral nerves in an early stage before there is clinically evident loss of sensory or motor function may present as a hypothermal zone on the skin surface related to the peripheral nerve affected. This mechanism applies to degenerative spinal disk disease and peripheral compression syndromes, ie carpal/tarsal tunnel syndrome. In later stages - when motor lesions are present - the compression of sympathetic fibers causes a lack of vasoconstrictive impulse with secondary vasodilation and hyperthermia of skin surface. This mechanism is well known from sympathectomy (Fig 4).

Quantitation of thermal images in these diseases is desirable but not essential, as the thermal diagnosis is dependent on thermal patterns which are detected visually. Therefore these conditions with segmental hypothermia can be studied using flexible plate thermography, despite problems with the human anatomy and the limited flexibility of liquid crystal plates. As the ambient temperature of 24°C is easy to maintain, this procedure can therefore be utilized in almost any rheumatologist's practice. The painful, functional lesions and nerve compression syndromes are particularly common. Until the introduction of this technique, the rheumatologist was without an objective method by which to examine the patient's complaints, or to monitor the effect of manipulative treatment.

THERMOGRAPHIC DETECTION OF BLOOD VESSEL INVOLVEMENT

Many inflammatory changes and diseases in rheumatology also can affect small and large blood vessels, eg in collagen disease vasculitis is a common complication. Raynaud's syndrome - the vasospastic reaction to cold - is also more common in systemic connective tissue disease and rheumatoid arthritis. These changes in the fingers and toes can be objectified by a simple thermographic cold stress test. The ambient temperature should be kept at 24 - 25°C for normal, basal vasomotor tone in skin microcirculation. The hands are then immersed in water at 16°C for one minute. To prevent later evaporative cooling the hands must be protected by thin plastic gloves. The rewarming reaction can be monitored by thermography. Quantification of thermal images provides a more accurate measure of the time constant for peripheral hyperemia and hyperthermia.

There are several types of possible thermal reaction after this simple cold stress test. The normal reaction is a rapid rewarming within 3 to 10 minutes, after a reactive hyperemia which exceeds the baseline temperature. All fingers react symmetrically which is modified by central temperature regulation. In patients (mostly women) with chronically cold fingers, due to the regulatory process for central body core temperature, a slow onset of this normal hyperthermia can occur. After glucose uptake or a hot drink the reaction can revert to the normal pattern. In patients with Raynaud's syndrome (primary, idiopathic or secondary, symptomatic) the normal hyperthermia is never detected due to the vasospastic reaction in both hands and fingers. In these cases very slow rewarming of the fingers occurs. After 30 - 60 minutes the pre-test temperatures are not re-established, due to the severity of the vasospastic reaction. This

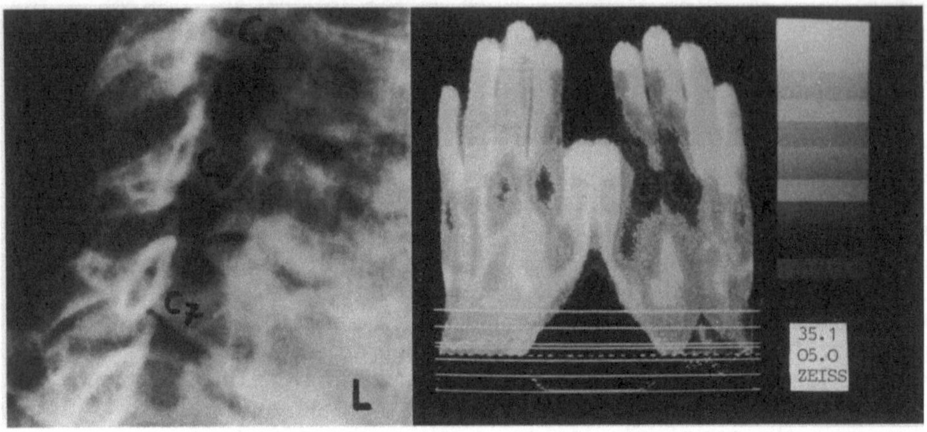

Fig. 4. X-ray and thermogram of hands of a 58 year old female with chronic pain in the left shoulder. The X-ray shows an arthrosis affecting C6/C7 on the left. The thermogram (AT 25°C) shows marked hypothermic over second and third fingers (median nerve: dermatome C7).

pathological reaction is usually symmetrical (Fig 5).

If asymmetrical reactions are detected the distribution pattern of normal and abnormal thermal values is significant for differential diagnosis. If the fingers relating to the segmental distribution of peripheral nerves (eg median, ulnar) show a retarded thermal reaction the cause of this may be a peripheral nerve compression or a sympathetic nocireaction. If the fingers do not relate to the neural pattern, occlusive or inflammatory peripheral artery disease may be the cause of the restricted blood supply to the finger. Immunogenic vasculitis is often localized in only one or two fingers. For full interpretation of this cold stress test, other clinical signs and symptoms must be taken into consideration. This thermal test is very sensitive, and also relatively specific if carried out carefully. It can be used to study objectively the common complaint of cold fingers in patients with rheumatic disease.

CONCLUSIONS AND PERSPECTIVES

The different thermological methods have reached a technical stage where they can be applied to clinical rheumatology. Thermographic procedures are now well established and tested. A more widespread application in rheumatology can be recommended. Even if quantitative infrared thermography will be the method of choice, all other methods such as contact thermography, bolometry, and contact thermometry with

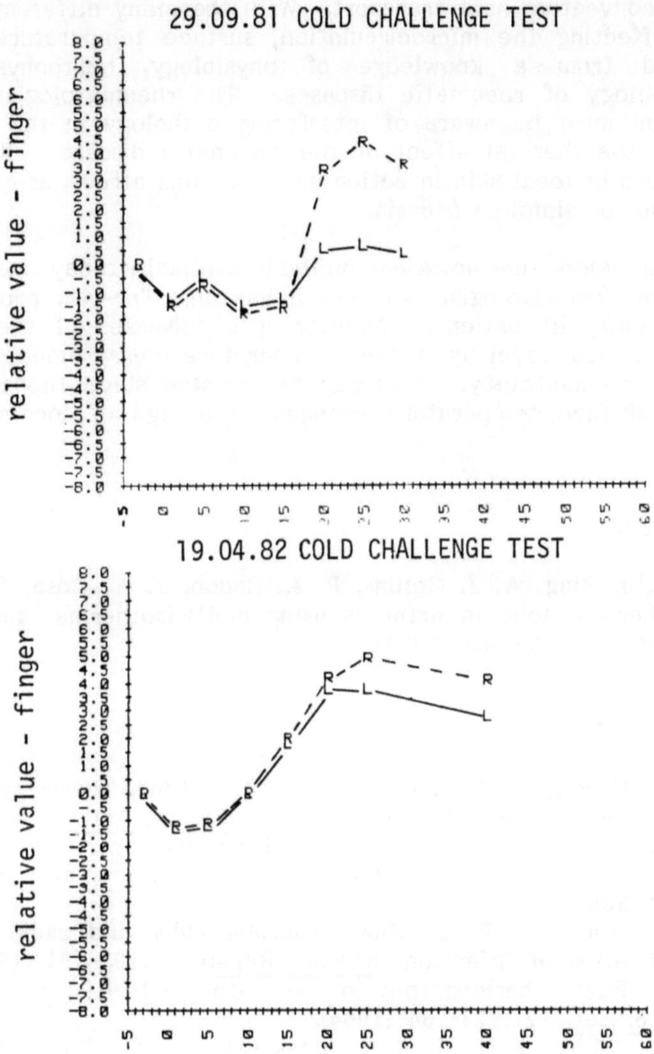

Fig. 5. Cold challenge test of the hands of a 22 year old female with acute iritis of the left eye. Vasoconstriction improved with treatment of the iritis. (Pain fibers from the eye interface with the sympathetic nervous system at the left of the upper cervical ganglia).

thermistors can be used for special clinical situations and diagnostic problems.

Interpretation of thermograms and thermometric results is not easy.

Skin surface temperatures are the result of skin microcirculation blood flow and convective heat transport. With the many different regulatory systems affecting the microcirculation, surface temperatures can only interpreted from a knowledge of physiology, neurophysiology and pathophysiology of rheumatic diseases. The rheumatologist reading a thermogram must be aware of interfering pathology in the skin, which may mask the thermal effect of the rheumatic disease. Skin diseases like psoriasis or local skin infection can have this effect, as can phlebitis, lymphedema or simple varicosis.

Nevertheless thermological methods available today are a valuable tool to the rheumatologist, and are a harmless "no-risk procedure" for the chronically ill patient. Monitoring of diseases of the locomotor system at a local level by surface temperature measurements fills a gap in clinical rheumatology. This gap has existed since Hippocrates first described surface temperature changes as a sign of locomotor (joint) disease.

REFERENCE

1. E. F. J. Ring, A. J. Collins, P. A. Bacon, J. A. Cose, Quantitation of thermography in arthritis using multi-isothermal analysis, Ann. Rheum. Dis. 33:353 (1974).

BIBLIOGRAPHY

1. P. A. Bacon, E. F. J. Ring and A. J. Collins, Thermography in the assessment of anti-rheumatic agents, Rheum. Arthr. 105-110 (1977).
2. P. A. Bacon, J. Davies, E. F. J. Ring, Benoxaprofen - dose range studies using quantitative thermography, J. Rheumatol. Suppl. 6 48-53 (1980).
3. H. A. Bird, E. F. J. Ring, Thermography and radiology in the localization of infection, Rheum. Rehab. 17:103-106 (1978).
4. N. F. Boas, Thermography in rheumatoid arthritis, Ann. of the N. Y. Ac. Sci. 121:223-234 (1964).
5. A. J. Collins, E. F. J. Ring, Measurement of inflammation in man and animals by radiometry, Br. J. Pharm. 44:145-152 (1972).
6. A. Collins, E. F. J. Ring, P. A. Bacon, J. D. Brookshaw, Thermography and radiology complementary methods for the study of inflammatory diseases, Clin. Radiol. 27:237-243 (1976).
7. L. Daenens, Thermology in a rheumatological private practice, Medicamundi 26:107-111 (1981).
8. J-M. Engel, J. A. Cosh, E. F. J. Ring, D. P. Page, T. Van Waes, D. Schoenfeld, Thermography in locomotor diseases. Recommended procedure. Thermographic terminology, Acta Thermographica 2:299-306 (1978).
9. J-M. Engel, Quantitative infrarot-tele-thermographie, Verh. Dtsch. Ges. Rheumatol. 6:280-283 (1980).

10. J-M. Engel, THERMOTOM - Ein Software-Paket fur die thermographische Bildanalyse in der Rheumatologie, Z. Biomed. Technik. 26:53 (1981).
11. J-M. Engel, Thermographie - die andere Dimension der Diagnostik, DMW 106:27:879-81 (1981).
12. K. H. Friedrich, Assessment criteria for infra-red thermography systems, Acta Thermographica 4:68-76 (1980).
13. I. Fujimasa, Y. Sakurai, K. Atsumi, Digitalizing approaches to thermogram analysis, in: "Medical Thermography," K. Atsumi ed., University of Tokyo Press, (1973).
14. J. R. S. Hales, M. Iriki, Differential thermal influences on skin blood flow through capillaries and arteriovenous anastomoses and on sympathetic activity, Bibl. Anat. 189-191 (1977).
15. T. Hashimoto, Y. Katsuta, Clinical application of infrared thermography to the study of Raynaud's phenomenon in collagen diseases, in: "Medical Thermography," K. Atsumi ed., University of Tokyo Press (1973).
16. S. Katz, T. Masterfano, A preliminary report on the clinical testing of a new elastomeric liquid crystal film, in: "Thermal characteristics of tumors. Applications in detection and treatment," Annals of the New York Academy of Sciences, 335 New York (1981).
17. Y. Mishima, K. Hara, S. Oohashi, Thermography in peripheral vascular diseases, in: "Medical Thermography," K. Atsumi ed., University of Tokyo Press, (1973).
18. B. H. Phillips, E. F. J. Ring, Quantitation in medical thermography, I. Mech. Eng. 6:107-111 (1977).
19. E. F. J. Ring, Quantitative medical thermography, SPIE Industrial Civil Appl. Infrared Techn. 110:78-84 (1977).
20. E. F. J. Ring, C. Watson, J. Barker, Infrarot-Thermographie und thermische Clearence der Haut, in: "Thermologie Fachberichte," 2 J. M. Engel ed., Baden-Baden (1981).
21. E. F. J. Ring, Kontaktthermographie der planta pedis, in: "Thermologie Fachberichte," 2 J. M. Engel ed., Baden-Baden (1981).
22. D. Rusch, Dynamische Quantitative Thermographie, in: "Thermologie Fachberichte," 2 J. M. Engel ed., Baden-Baden (1981).
23. D. Rusch, Thermographische Befunde bei ankylosierender Spondylitis, Akt. Rheumatol. 7:102-108 (1982).
24. R. S. Salisbury, M. De Silva, B. L. Hazleman, G. Parr, D. P. Page-Thomas, Thermography and the assessment of inflammation, Ann. Rheum. Dis. 40:524 (1981).
25. G. Strobel, J. M. Engel, Dynamische Thermographie bei peripheren Zirkulationsstorungen, in: "Thermologie Fachberischte," 1, J. M. Engel ed., Baden-Baden (1981).
26. G. Stuttgen, U. Flesch, T. Siebel, Erythementwicklung im thermographischen Bild, Der Hautarzt. 32:622-628 (1981).
27. E. R. Tichauer, The objective corroboration of back pain through thermography, J. Occup. Med. 19:727-731 (1977).
28. S. M. Viitanen, A. L. Laaksonen, Thermography in juvenile rheumatoid arthritis, Acta Rheum. Scand. 16:91-98 (1970).
29. C. E. Wexler, Thermographic evaluation of trauma (spine), Acta Thermographica 5:3-5 (1980).

THERMOGRAPHY AND MUSCULAR INJURIES IN SPORTS MEDICINE

M Schmitt*, Y Guillot**

*** Laboratoire de Physiologie, Faculte de Medecine A Carrel Lyon, FRANCE**
**** Centre de Medecine du Sport (Dir Professuer Guillet) Hopital Edouard-Herriot, Lyon, FRANCE**

Among all the injuries which can be observed in sports medicine, muscular injuries are the most frequent after osteoarticular lesions. Non-muscular lesions have been studied in more depth. As they generally have an important inflammatory component (tendinitis, synovitis) they are more easily detected. Muscular pathology is often responsible for very disabling sequelae, especially for highly trained athletes. If diagnosis of a muscular traumatic injury is generally easy, its functional importance is much more difficult to assess. It will, nevertheless, determine the treatment and therefore the date of sport resumption. Resumption will be allowed only when complete anatomical healing has been achieved. Patients' follow-up must, then, be very strict. This is particularly important in professional and high competition level athletes. The classification of muscle injuries remains obscure and it is therefore rather difficult to make a clear distinction between the different degrees of injury, ranging from the single pulled muscle to heavy muscle strain, laceration or rupture.

Among the paraclinical means available to the sport clinician, thermography has been used with increasing frequency over the last 15 years. However, only a few papers have been published on that subject and results are often contradictory.[1,2,3,4]

Muscle injuries occur mainly in the lower limbs (95%) and especially in the following muscles: ischio-tibial, rectus femoris, triceps surae. The upper limbs are injured in only 3% of cases (biceps). In our study, we used infrared thermography. Its main advantage is its harmlessness. It is obvious that, as for any complementary examination, thermal patterns have to be interpreted according to the clinical context. As cutaneous temperature varies considerably from one region to another, interpretation of the results must be based on the comparison of symmetrical regions. To be significant and to affirm an eventual hypo or hyperthermia, differences in temperature of more than 0.5°C are necessary.

MATERIALS AND METHODS

More than 200 patients were examined using thermography, from very occasional to high level and professional athletes. No distinction was made between men and women, nor between the different kinds of sport. The site of injuries was variable but mostly the lower limbs. For acute injuries a first thermogram was performed during the week following the injury. Patients were followed up every week or every fortnight until complete normalization of the thermal pattern. In chronic and neglected lesions the delay between injury and thermographic examination was generally unknown. Thermograms were performed before and after appropriate (surgical) treatment.

Before thermographic examination the undressed patient was allowed to equilibrate at ambient temperature for at least 15 minutes. Room temperature was maintained at 20°C-21°C in order to provide a better contrast on the thermal image. No clinical examination had been done in the half hour preceding the thermogram. No ointment or cream, which could modify skin temperature or skin emissivity, had been applied since the day before. In order to avoid any error related to an upright position (hyperthermia over varicose veins for example) the patients were examined reclining. Thermograms were performed through a 45° inclined mirror.

Two types of infrared camera were used: AGA 680 and Philips Digital Thermograph. In each case, color and black and white thermograms were recorded. As previously pointed out, only thermal gradients greater than 0.5°C, ie twice the detector sensitivity, were considered.

RESULTS

A distinction between the two main lesion types must be made: acute accidents, whatever their cause (external contusion, internal laceration, elongation) and chronic lesions corresponding to bad healing or to secondary functional complications.

In acute and recent lesions, three kinds of thermal patterns may be observed:

Normothermia

That is thermal symmetry between the two examined limbs. Sports resumption is allowed within one or two weeks.

Hyperthermia

Localized to the injured area: most elongations, hematomas or lacerations are responsible for such localized hyperthermia. The difference in temperature with the symmetric region or the surrounding regions is variable, from 1 to 4°C (average 1.5-2°C) (Figs 1 & 2).

Such hyperthermia confirms the severity of the injury, though the muscular lesion is generally painless and without important functional disorder. Sometimes hyperthermia may overlap the injured area and diffuse around it. Such thermal images have already been described by several authors.[1,2,4] According to Karpman,[1] hyperthermia is due to an increase in heat production secondary to muscular spasm. As hyperthermia appears just a few hours after trauma and may remain over a month, it seems more credible to relate it to an inflammatory process. Moreover, its early development after injury enables one to also evoke a local release of vasodilating substances, secondary to trauma and disrupted muscle fibers. These substances will subsequently increase local blood flow and heat convection from the underlying tissues to the skin. Increased thermal conductivity by injured muscular tissue or due to hematoma is probably not involved. In fact, blood thermal conductivity is very close to the thermal conductivity of intact and normally vascularized muscle (about 12.10^{-4} Cal/cm°Csec).[5,6] The same local hyperthermic thermal patterns are observed in the weight-bearing points of the body in muscular crush-syndromes (rhabdomyolysis) following intoxications and comas.

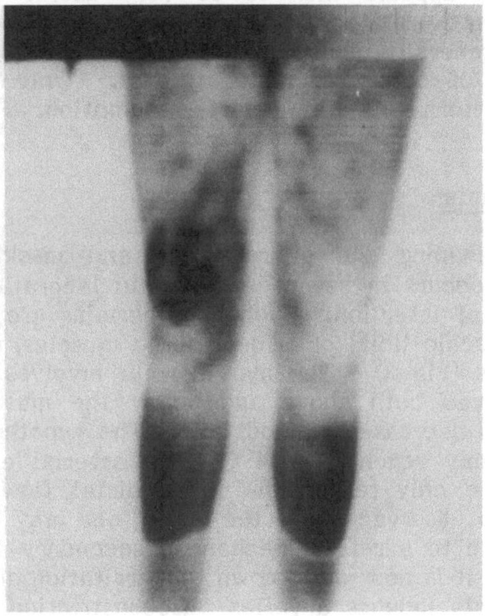

Fig. 1. Localized hyperthermia secondary to a laceration of the left rectus femoris. ΔT:3.5 - 4.0°C. (10°C scale : black is hot).

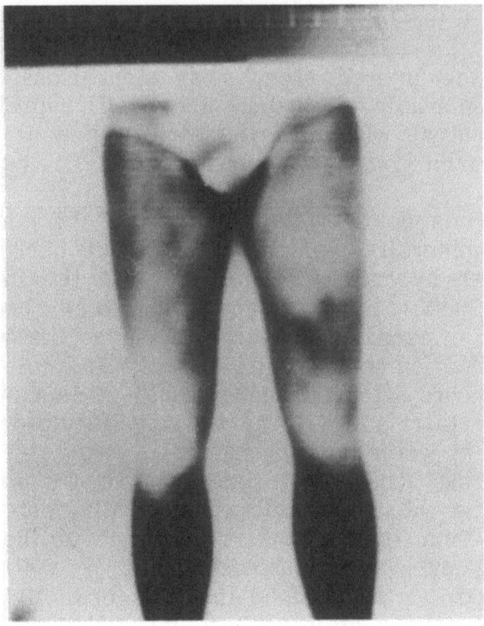

Fig. 2. Hyperthermia over injury to right quadriceps. Δ T: 2.0°C.

Thermography appears then to be very useful in following the evolution of sports related injuries until the disappearance of hyperthermia. It generally occurs within three or four weeks, later than the disappearance of clinical symptomatology. Only the return to a normal thermal pattern will allow sports resumption.

Regional Hypothermia

Largely overlapping the injured area and masking the lesional hyperthermia. It occurs in severe and painful lacerations or in muscle tears at the point of insertion. Such hypothermias are mainly observed in injuries to the ischio-tibial or triceps surae muscles, ie in inextensible aponeurotic regions (Fig 3). The hypothermia involves the entire limb and is thus observed both above and below the muscular injury. It indicates a regional decrease in blood flow. The hypothermia is variable in intensity and may reach 3 or 4°C. An arterial compression by a hematoma would be only responsible for a distal flow reduction. As hypothermia is also observed above the lesion, one may invoke a regional vasoconstriction due to a reflex mechanism secondary to pain or nerve irritation. In fact, it is now well known that irritation of sensitive nerves or dorsal nerve roots induces a reflex vasoconstriction and therefore a cutaneous hypothermia in the corresponding area or "dermatome".[7,8] In some cases a relative hot spot, when compared with the surrounding areas, may be observed over the lesion itself, contrasting with the

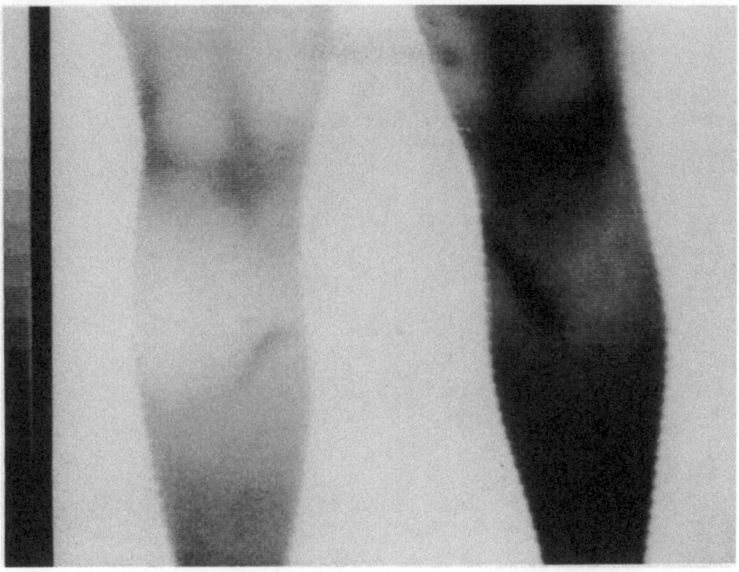

Fig. 3. Hypothermia of the whole left limb due to reflex vasoconstriction. Secondary to an external contusion of the calf. Δ T: 2.5°C.

hypothermic background (Fig 4). The reflex hypothermia disappears after three or four weeks and then gives way to the characteristic lesional hyperthermia. Most often a complete thermographic normalization occurs in the six to seven weeks following the injury and allows sports resumption.

In chronic states of pain, discomfort or helplessness: in most cases the thermogram presents a cold, hypothermic area corresponding to the painful region. Such a cold spot evokes either a decrease in local blood flow or a decrease in muscular or subcutaneous thermal conductivity. It is observed in old untreated injuries and corresponds to a bad muscle healing. Surgical exploration generally shows a fibrotic or sclerotic callus corresponding to the hypothermic area. Thermography appears then to aid in the diagnosis and is a good guide for eventual surgery. This is especially important for neglected lesions of ischio-tibial muscles where clinical signs are often rare. Furthermore, after surgery and excision of the fibrotic area, thermography is a very good follow-up method. Post surgical healing will, of course, determine a local hyperthermia, which will progressively decrease with anatomical repair. In our experience, an approximate delay of 3 to 6 months appears necessary to recover a normal, perfectly symmetrical thermogram. Once again, apparent clinical recovery generally occurs earlier than thermal symmetry. Thermography is then an accurate method in the assessment of complete anatomical repair. Despite its usefulness in the diagnosis and follow-up of muscle injuries, thermography has its limits, responsible for false negatives.

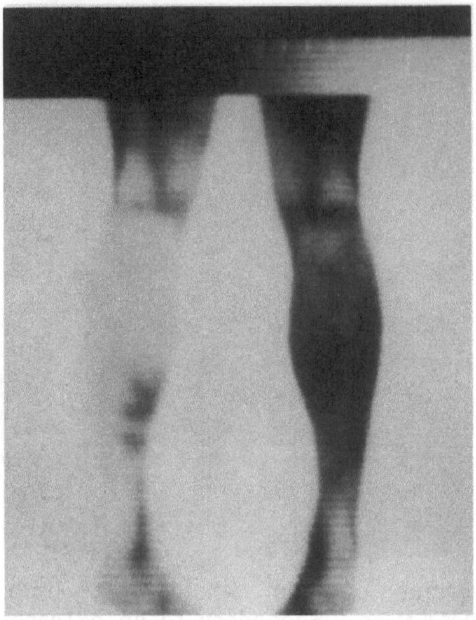

Fig. 4. Laceration of the left calf showing hypothermia due to reflex nociceptive vasoconstriction. Note relative hyperthermia over the lesion.

These limits are:

1. Depth of the lesion: it is obvious that a very localized and deep muscle laceration will have no thermographic translation. In the same manner, ossification of a deep hematoma may be responsible for either a hyperthermia or a hypothermic skin area, depending on the metabolic level of the ossification process.
2. Repeated micro-traumas: this is particularly important in professional and semi-professional athletes (football, rugby players for example). Repeated injuries induce in these subjects a mottled pattern in the thermogram of the inferior limbs. Lack of symmetry renders difficult the interpretation of an eventual dysthermia. In any case thermography remains useful for following up these patients, providing that the clinician or thermographist is in possession of a "normal" thermogram before any new trauma.

In conclusion, thermographic examination of a recent injury allows one to:-

1. Estimate its severity.
2. Monitor its evolution.
3. Estimate the time of healing and eventual resumption of sports activity.

and in chronic cases of old and neglected injury:

1. To help reach a correct diagnosis.
2. To monitor an eventual post surgical recovery.

REFERENCES

1. H. L. Karpman, A. Knebel, C. J. Semel, J. Cooper, Clinical studies in thermography. II. Application of thermography in evaluating musculoligamentous injuries of the spine, Arch. Environ. Health 20:412-417 (1970).
2. C. Zuinen, Interet de la thermographie dans les lesions des tissus mous chez les sportifs, Med. Medicale 142:103-106 (1977).
3. R. Guillet, J. Genety, M. Schmitt, E. Brunet-Guedj, P. Meffre, Apport de la thermographie infrarouge au diagnostic des lesions musculaires de la pratique sportive, Lyon Chir. 15;1:7-10 (1979).
4. P. Rochcongar, A. Belossi, Les images d'hyperthermie dans les accidents musculaires du sportif, Med. Medicale 232:7-8 (1980).
5. K. E. Spell, The thermal conductivity of some biological fluids, Phys. Med. Biol. 5:139-153 (1960).
6. H. Hensel, J. Ruef, Fortlaufende registrierung der muskeldurchblutung am menschen mit einer kalorimetersonde, Pflugers. Arch. 263:259-267 (1954).
7. C. E. Wexler, Lumbar, thoracic and cervical thermography, J. Neurol. Orthop. Surg. Charter issue Nov 37-41 (1979).
8. A. A. Marinacci, Thermography in the detection of non neurological imitators of peripheral nerve complexes, Bull. Los Angeles Neurol. Soc. 30;1:1-11 (1965).

THERMOLOGY IN THE ASSESSMENT OF PHYSICAL THERAPY

D Rusch

Justus-Liebig-Universitat Giessen
D-6350 Bad Nauheim
WEST GERMANY

Medical research is often directed to observing acute therapeutic effects. Physical therapy differs from other therapy in having its effect through biological or physiological reactions in response to physical stimuli. The latter are often thermal in their nature, or connected with thermal phenomena. This makes thermology the means of choice in the study of the said reactions. But there are many difficulties to surmount.

Thermal processes and thermal reactions to physical stimuli change chronobiologically in different periods, circadian, circaseptan and seasonal.[1,2] The immediate effects of physical therapy are often very weak and cannot be intensified as the therapeutic dose may not be increased without changing the effect totally. Heat therapy may be considered as an example. A hot pack may be up to 50°C. This is about 15°C above skin temperature and will be tolerated by the skin. An increase to 65°C is impossible without burning the skin.

The long term effects of repeated treatments are even more difficult to detect. We need to know more about the adaptive character of the reactions on repeated treatment with physical therapy. Serial or continuous physical stimuli over a period of time lead to a marked time structure of the adaptational process.[2] To follow the thermal changes of long term physical treatment we need even better measuring instruments. Instruments with better reproducibility and dedicated quantification of the temperature results. Two possible solutions were given by Bosiger et al[3] and Rusch.[4]

Most thermologists deal with immediate thermal effects of physical therapy. But this is only one side of the question. The other side is represented by resulting therapeutic alterations in the patient's disease which follow repeated application. We must encourage more research in this field. We do not want, for example, to substitute tissue warmth by external therapeutic heat, we want the body to react to thermal stimuli applied with heat therapy. We must learn from the study of immediate effects, but our main interest should be directed towards the evaluation of real therapeutic results.

The results published on thermology in the assessment of physical therapy will be reviewed in three parts.

THERMOMETRY, RADIOMETRY, THERMAL CLEARANCE AND RELATED METHODS

Studies have been carried out in the more common types of therapy: galvanization, iontophoresis, cryotherapy (ice packs), hot packs, UV irradiation, diathermy and magnetic fields.

Maurer[5] reviewed many thermal effects of different therapies on the knee joint. She showed that in galvanization the higher skin temperatures are found under the cathode. Iontophoresis with histamine chloride leads to an additional increase in skin temperature in the anode region. She also studied radiometrically skin temperatures over the knee joints of healthy persons before and a relatively long time after intense cryotherapy. She found a distinct two phase rewarming. The initial rapid increase of temperature was finished in 15 to 20 min. It was followed by a four hour period of slower temperature increase with 1.5°C per hour. Some patients with arthritic knees observed radiometrically showed significantly lowered knee temperatures after seven days of cryotherapy.[6,7]

Franke and Engel[8] studied radiometrically temperatures in a series of patients with degenerative and inflammatory joint diseases several times before, during and after cryotherapy. They only found in some patients the temperature pattern described by Maurer.[5] The results obtained by Franke and Engel[8] showed mostly a reactive hyperemia within 10 to 20 minutes after cryotherapy and temperature increase above initial levels. An earlier paper of Cosh and Ring[9] reported on the diagnostic application of standardized ice packs. The authors found much faster rewarming after the ice pack in arthritic knees than in normal knees. Perhaps joint inflammation together with cautious cooling may produce the hyperemia observed by Franke and Engel[8] but not by the other authors.

Haimovici[10,11] reported on intra-articular temperature measurements in normal, arthrotic and arthritic peripheral joints. Some of the joints were treated with diathermy and some with low frequency magnetic field therapy. It was shown that the main advantage of magnetic field over other physical therapy is the lack of local heating.

Ring et al[12] examined UV erythema of the skin with a special thermal clearance probe and IR thermography. They found a linear correlation between the results of thermal clearance and the degree of erythema. Thermography showed elevated skin temperatures only in high grade UV burns. Adams et al[13] gave a detailed description of a new thermal probe for quantification of the microcirculation which is useful in the measurement of physical therapy effects.

Law and Pettigrew[14] used the temperature rise of a small copper disc on the skin to calculate the density of heat flow rate they needed to know for their method of whole body hyperthermia. Kleinschmidt and Schnizer[15] measured the density of heat flow rate directly with a specially designed probe. The paper cited here deals with experimental application of hot packs. Among others an interesting new design of hot packs was examined.

INFRARED THERMOGRAPHY

Scientific literature surveys reveal a number of publications of experimental results with electrotherapy, ultrasound, magnetic fields, exercise therapy, hot packs and a combination of hot pack and galvanization.

Danz and Callies[16] showed thermographically that in ultrasound therapy the skin is only heated at high doses of ultrasound. This is due to cooling by the metal parts of the transducer. The authors could differentiate the various intensities of ultrasound on the basis of the course of skin temperature. The same authors dealt with different forms of electrotherapy.[17,18] Galvanization with current densities of 1 to 2 mA/10 cm^2 gave uniform elevation of skin temperature under the electrodes. With higher current densities skin temperature under the cathode rose significantly higher than under the anode. The form MF of the diadynamic currents was the only one to show the same difference in skin temperature increase.

Scott et al[19] reported on exercise therapy in ankylosing spondylitis. Besides other objective and clinical parameters they measured the TI over the sacroiliac joints as an objective criterion for disease activity. However, quantitative thermography could not assess the distinct functional improvement of the patients.

Tauchmannova et al[20] investigated thermographically temperature changes after local mud packs and after remedial exercise in rheumatoid arthritis. They found the TI remained higher than initial values for 5 hours after the pack. No change of TI was seen after therapeutic exercise.

Warnke and Altmann[21] measured skin temperature thermographically before and during magnetic field application. They postulate a strong vasoactive effect on skin vessels. Unfortunately, they did not use quantitative thermography and no temperature reference is shown in their thermograms.

Bergthold[22] described a thermographic temperature assessment of the insensible foot. His method seems to be similar to the contact thermography device Ring demonstrated at the September 1980 meeting of the Anglo Dutch Thermographic Society.

Rulffs[23] combined a hot pack with electrotherapy. He showed in a qualitative comparison by thermography that skin temperatures rise higher after a combination of hot pack and galvanic current than after hot pack alone.

MICROWAVE THERMOGRAPHY

As microwave thermography is not yet a routine method there are fewer published papers. The only reports of physical therapy assessment by microwave thermography are those of the Lille/France group.

N'Guyen et al[24] and Chive et al[25] described the possibility of simultaneous microwave local heating and microwave thermography with examples of its clinical application.

Mamouni et al[26,27] reported on passive subcutaneous temperature measurement for investigation of thermoregulation. The authors showed the advantage of direct deep temperature measurement by microwave thermography compared with infrared radiometry in the field of ergonomics or exercise therapy.

Thermal measurement and imaging techniques have a useful role in monitoring objectively the changes in body temperature caused by physical therapy. This is an area of medicine which needs more assessment and evaluation for a better understanding of the clinical and therapeutic effects of these treatment regimens.

REFERENCES

1. G. Hilderbrandt, D. Crynjak, Zur tageszeitlichen ordung der hydrotherapie, Z. Phys. Med. 1:51 (1970).
2. G. Hilderbrandt, Chronobiologische grundlagen der pravention und rehabilitation, Z. angew. Bader. - und Klimaheilk 25:326 (1978).
3. P. Bosiger, F. Scaroni, Ein mikroprozessor-unterstutztes thermographie-system zur quantitativen on-line-analyse von statischen und dynamischen thermogrammen, Thermologie Fachberichte 2:15 (1981).
4. D. Rusch, Dynamische quantitative thermographie, Thermologie Fachberichte 2:13 (1981).
5. R. Maurer, Zum verhalten der hauttemperatur am kniegelenk nach physikalischer therapie und bei rheumatischen entzundungen.

Dissertation, FB Humanmedizin, Giessen (1980).
6. K. L. Schmidt, R. Maurer, D. Rusch, Zur wirkung ortlicher warme - und kalteanwendungen auf die hauttemperatur am kniegelenk, Z. Rheumatol. 38:213 (1979).
7. K. L. Schmidt, R. Maurer, D. Rusch, Zum verhalten der hauttemperatur uber entzundeten kniegelenken unter taglicher kryotherapie, Z. Phys. Med. 9:58 (1980).
8. M. Franke, J-M. Engel, Thermal reactions in cryotherapy of rheumatoid joint diseases, XVth International Congress of Rheumatology, Paris (1981).
9. J. A. Cosh, E. F. J. Ring, Techniques of heat detection used in the assessment of rheumatic diseases, J. Radiol. Electrol. 48:84 (1967).
10. N. Haimovici, Die magnetfeld-therapie bei der behandlung der degenerativen gelenkerkrankungen, Z. Phys. Med. 11 (in press) (1982).
11. N. Haimovici, Wert der direkten, intraartikularen temperaturmessung in der diagnostik der gelenkerkrankungen, Z. Phys. Med. 11 (in press) (1982).
12. E. F. J. Ring, C. Watson, J. Barker, Infrarot-thermographie und thermische clearance der haut, Thermologie Fachberichte 2:18 (1981).
13. T. Adams, S. R. Heisey, M. C. Smith, M. A. Steinmetz, J. C. Hartman, H. K. Fry, Thermodynamic technique for the quantification of regional blood flow, Am. J. Physiol. 238:H682 (1980).
14. H. T. Law, R. T. Pettigrew, Heat transfer in whole-body hyperthermia, Ann. N. Y. Acad. Sci. 335:298 (1980).
15. J. Kleinschmidt, W. Schnizer, Experimentelle untersuchungen zu thermischen vorgangen bei der anwendung von heisspackungen, Physiother. (Lubeck) 71:95 (1980).
16. J. Danz, R. Callies, Thermometrische untersuchungen bei unterschiedlichen ultraschallintensitaten, Z. Physiother. 30:335 (1978).
17. J. Danz, R. Callies, Infrarotthermometrie bei differenzierten methoden der niederfrequenztherapie, Z. Physiother. 31:35 (1979).
18. J. Danz, R. Callies, Thermographische objektivierung einer differenzierten gleichstromtherapie, Z. Physiother. 32:15 (1980).
19. D. G. I. Scott, B. Schlecht, E. F. J. Ring, The effect of inpatient exercise therapy on disease activity in ankylosing spondylitis, IXth European Congress of Rheumatology, Wiesbaden (1979).
20. H. Tauchmannova, M. Tauchmann, T. Mistina, Use of thermography for the evaluation of physical therapy, Acta Thermographica 4:129 (1979).
21. U. Warnke, G. Altmann, Die infrarotstrahlung des menschen als physiologischer wirkungsindikator des niederfrequent gepulsten schwachen magnetfeldes, Z. Phys. Med. 8:166 (1979).
22. H. T. Bergthold, Temperature assessment of the insenstivie foot, Am. J. Phys. Ther. 58:18 (1979).
23. W. Rulffs, Potenzierende wirkung simultaner thermo- und elektrotherapie, Z. Phys. Med. 11 (in press) (1982).
24. D. D. N'Guyen, A. Mamouni, Y. Leroy, E. Constant, Simultaneous

microwave local heating and microwave thermography. Possible clinical applications, J. Microwave Power 14:135 (1979).
25. M. Chive, D. D. N'Guyen, Y. Leroy, Hyperthermie locale. Vers un controle atraumatique de l'echauffement fonde sur la thermographie micro-onde, Bull. Cancer (Paris) 68:293 (1980).
26. A. Mamouni, Y. Leroy, Y. Houdas, Y. Moschetto, Passive subcutaneous temperature measurement for investigation of thermoregulation, 8th European Microwave Conference, Paris (1978).
27. A. Mamouni, D. D. N'Guyen, M. Robillard, M. Chive, Y. Leroy, Thermographie micro-onde - nouvelles methodes diagnostiques fondees sur la mesure du bruit thermique emis par les tissus vivants, L'onde Electrique 60:30 (1980).

HEAT DISTRIBUTION OVER JOINTS:- THE NORMAL AND ABNORMAL PATTERN

R S Salisbury*, G R Parr, B L Hazleman, D P Page Thomas

Addenbrooke's Hospital, Cambridge, UNITED KINGDOM
*** Department of Rheumatology, Hope Hospital, Salford UNITED KINGDOM**

SUMMARY AND CONCLUSIONS

Measurement of inflammation by absolute temperature methods has proved unreliable owing to the many physiological factors which affect skin blood flow and temperature. We have identified patterns of heat which are present over normal knees, elbows, wrists, shoulders, and ankles. In the presence of inflammation these patterns are altered or lost due to the abnormal skin blood flow.

Sixty healthy controls and 85 patients with rheumatoid arthritis were thermographed at different times of day over a 1 year period. No patient with clinical evidence of synovitis had a normal thermogram of that joint. Only 4 out of 618 thermograms from the healthy controls were abnormal.

The extent of the abnormal pattern can be quantified by several mathematical models based either on temperature, gray level frequency distribution, or alteration of the temperature gradient.

HEAT PATTERN

We have identified regular thermal patterns present over normal wrist, elbow, shoulder, knee and ankle joints. A negative thermal gradient extends from the center of the joint to the periphery (Fig 1). This negative gradient is distorted or reversed when active inflammation is present in the joint (Fig 2). These "heat patterns" or "spatial arrangements of temperature" remain relatively constant during the course of the day, although the absolute temperatures which make up the heat patterns may change. Thus, the heat patterns of a single joint may be moved up or

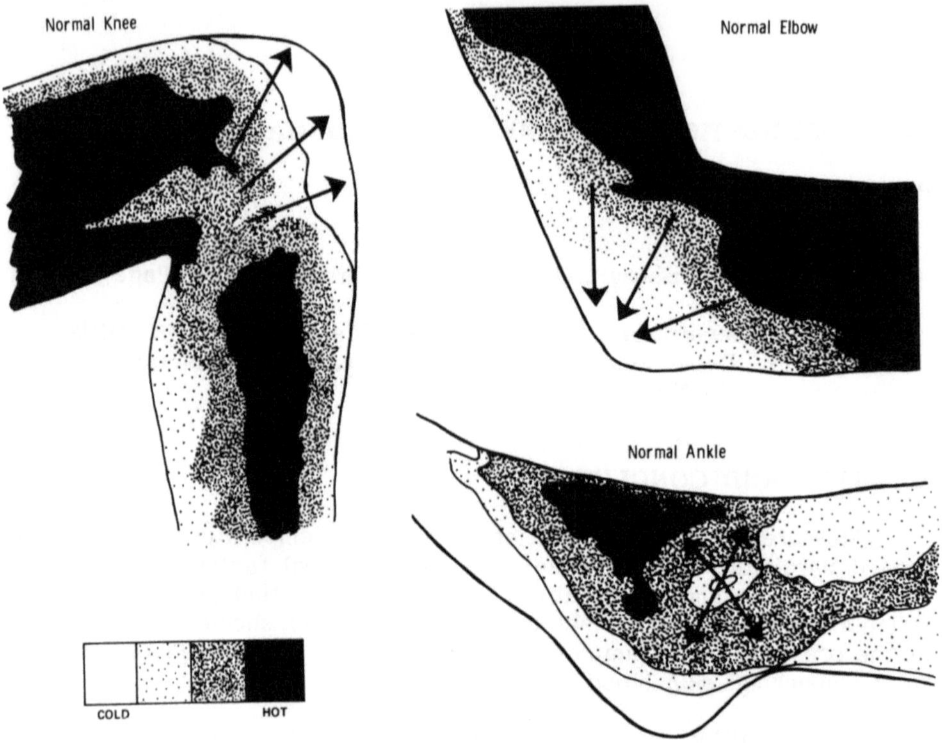

Fig. 1. Thermal gradient from center of the joint to periphery (normal).

down a temperature range depending on the individual's physiological or environmental changes, whilst the actual pattern of heat remains constant.These diurnal/circadian changes in surface temperature illustrate the inaccuracies in methods of assessment of inflammation based on absolute skin temperature, whilst the relative constancy of these heat patterns confirms the usefulness of quantification based on relative temperature distribution.

METHODS AND RESULTS

To test the reproducibility of heat patterns to distinguish normal from inflamed joints, 60 healthy controls and 85 patients with classical rheumatoid arthritis (ARA criteria) were thermographed at different times of the day. Standard methods of subject preparation and thermographic views as described by Engel et al[1] were used.

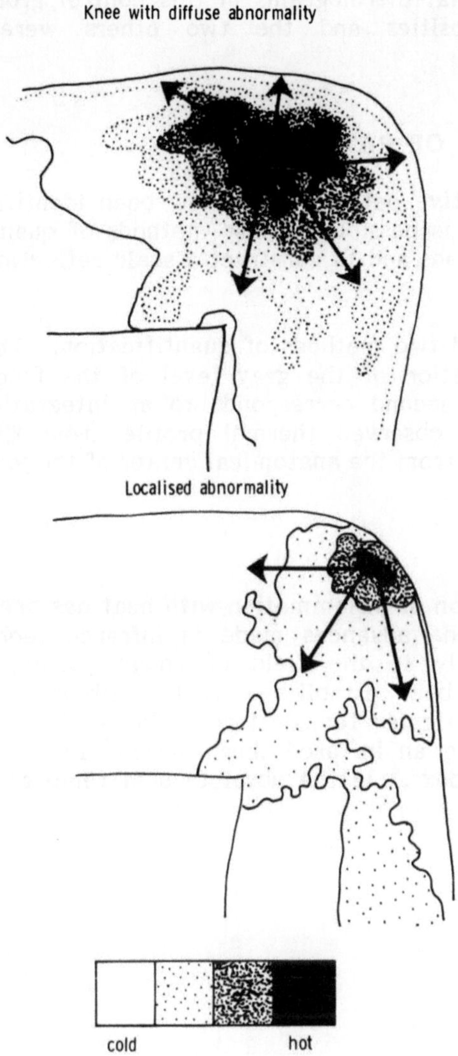

Fig. 2. Distortion of normal thermal pattern resulting from active inflammation.

Thermography was carried out using an AGA 680M Thermovision system and thermograms were digitized for storage and analysis by an on-line micro-computer.

No joint with clinical evidence of active synovitis had a normal heat pattern, irrespective of time of day. 614 out of 618 joint thermograms from healthy controls showed a normal heat pattern. Two of the four abnormal thermograms in this control group were attributed to surface varicosities and the two others were both apparently idiosyncratic.

QUANTIFICATION OF PATTERN

Since distinctive patterns have now been identified for normal and abnormal joints, it is possible to base methods of quantification on these abnormal distributions and to construct a scale reflecting their differences from the normal.

We have used two methods of quantification. The first of these is the standard deviation of the gray level of the frequency distribution (Figs 3, 4). The second corresponds to an integration of the thermal difference of the observed thermal profile from the negative linear gradient (Fig 5), ie from the anatomical center of the joint to its periphery.

DISCUSSION

The association of inflammation with heat has been recognized since Hippocrates and the advances made in infrared technology lend these techniques naturally to the field of rheumatology. The foreground appeared to have been established by Horvath and Hollander[2] in their studies on intra-articular temperature. The early surface temperature measurements using an infrared thermograph[3] led to the proposal of a "Thermographic Index",[4] ie the absolute skin temperature minus an

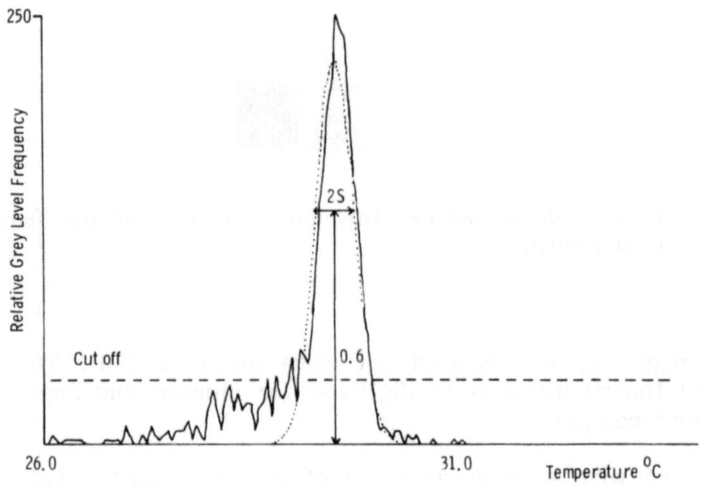

Fig. 3. Temperature histogram - normal.

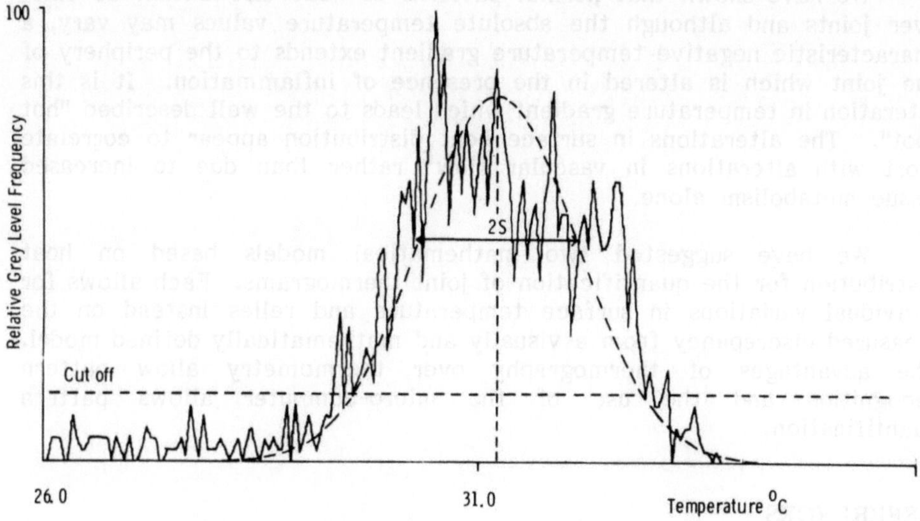

Fig. 4. Temperature histogram - abnormal.

arbitrary constant. Because of the many individual, physiological, diurnal, circadian and environmental factors affecting skin temperature,standardization and conditions of operation reached a high degree of complexity and many workers and observers found both difficulty in separating normal from pathological values and in achieving reproducible results. Many of these variables appear less critical when measuring intra-articular temperature. Since those involved in rheumatological thermography could often visually identify "abnormal" thermograms despite "normal" surface temperature values, and vice versa, a search was made to identify distinguishable patterns, although these proved elusive.[5]

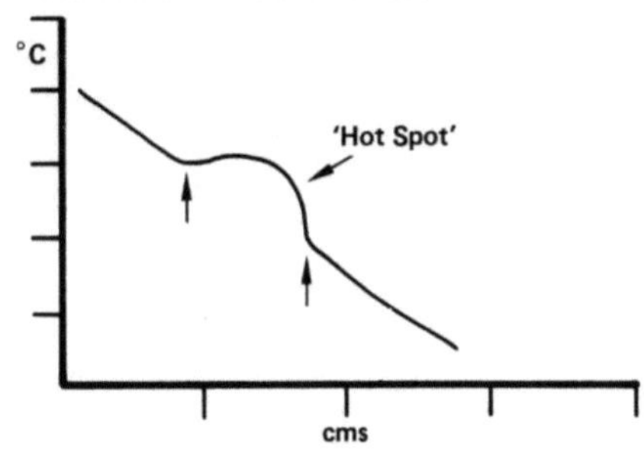

Fig. 5. Thermal profile showing a "hot spot" over the joint.

We have shown that normal patterns of heat distribution do exist over joints and although the absolute temperature values may vary, a characteristic negative temperature gradient extends to the periphery of the joint which is altered in the presence of inflammation. It is this alteration in temperature gradient which leads to the well described "hot spot". The alterations in surface heat distribution appear to correlate most with alterations in vascular flow[6] rather than due to increased tissue metabolism alone.

We have suggested two mathematical models based on heat distribution for the quantification of joint thermograms. Each allows for individual variations in surface temperature and relies instead on the measured discrepancy from a visually and mathematically defined model. The advantages of thermography over thermometry allow pattern recognition and the use of the micro-computer allows pattern quantification.

REFERENCES

1. J. M. Engel, J. A. Cosh, E. F. J. Ring, D. P. Page Thomas, P. Van Waes, D. Schoenfeld, Thermography in locomotor diseases:- recommended procedure, Group report of the Anglo-Dutch Thermographic Society (1980).
2. S. M. Horvath, J. L. Hollander, Intra-articular temperature as a measure of joint reaction, J. Clin. Invest. 28:469-473 (1949).
3. J. A. Cosh, E. F. J. Ring, Thermography and rheumatology, Rheum. and Phys. Med. X;7:342-8 (1980).
4. A. J. Collins, E. F. J. Ring, J. A. Cosh, P. A. Bacon, Quantification of thermography in arthritis using multi isothermal analysis. i. The thermographic index, Ann. Rheum. Dis. 33:113-115 (1974).
5. T. J. Love, Thermography as an indicator of blood perfusion. Thermal characteristics of tumors. Applications in detection and treatment, Ann. N. Y. Acad. Sci. 335:416-428 (1980).
6. S. K. Nielsson, S. E. Gustafsson, L. M. Torrell, Skin temperature over a heat source. Experimental studies and theoretical calculations. Thermal characteristics of tumors. Applications in detection and treatment, Ann. N. Y. Acad. Sci. 335:416-428 (1980).

THERMAL PATTERNS OF OSTEOARTHRITIC KNEES AND HANDS

P A Dieppe, E F J Ring, B Cadge

Royal National Hospital for Rheumatic Diseases
Upper Borough Walls, Bath, and
Bristol Royal Infirmary, Bristol, UNITED KINGDOM

INTRODUCTION

Osteoarthritis (OA) is a disease characterized by destruction of articular cartilage and increased activity of the underlying bone. There is an associated inflammatory component in most cases. Increased vascularity is a feature of both the subchondral bone and the synovium.[1]

Previous work has shown that thermograms are often abnormal, and that anti-inflammatory drugs can reduce infrared heat emission from OA joints.[2] The purpose of the present work was to examine the pattern and frequency of abnormal thermographic images in OA, and to relate these to radiological changes.

METHODS

Patients attending rheumatology out-patient clinics at Bristol Royal Infirmary and the Royal National Hospital for Rheumatic Diseases, Bath were included in the study. 70 consecutive patients with knee or hand pain, thought on clinical and radiological grounds to be due to OA, have been examined. AP radiographs of the hands (including wrists) and AP and lateral views of the knees were taken. AP thermographic images were recorded in standardized conditions, after 20 minutes cooling at 20°C. The radiographs and thermographic images were then examined independently by two observers.

Osteoarthritic changes were graded from 0-3 for the medial, lateral and patello-femoral compartments of the knee, and for individual joints in the hand. Thermographic images were reported as normal or abnormal, and the site of increased temperature recorded; for this the knee was divided into five zones (superior, inferior, medial, lateral and central [patella]), and when possible individual joint areas in the hand were isolated. Correlations between abnormal X-ray and thermographic images were tested by the correlation coefficient (r).

RESULTS

By definition, all patients had some radiological evidence of OA at at least one joint site, and all were symptomatic.

Areas of increased temperature were recorded on the thermographic image in 53% of hands and 37% of knees examined (Table 1).

In the hands the two areas most frequently abnormal on the thermographic image were the distal end of the index finger (presumed first distal interphalangeal joint - 22% abnormal) and the base of the thumb (presumed carpo-metacarpal joint of the thumb - 28% abnormal). This correlated well with radiographic changes, as these two joint sites were those most frequently abnormal on the X-ray. Many patients had abnormal radiographs and normal thermographic images, but no area of increased temperature was radiologically normal.

Table 1. Incidence of abnormal radiographs and thermograms in different joint areas of osteoarthritic hands and knees.

	% with OA Radiographic Changes	% with Hot Area on Thermograph
KNEES		
Overall	100	37.0
Medial	82	19.2
Lateral	61	17.8
Superior	-	3.3
Inferior	-	11.1
Patella	-	0
HANDS		
Overall	100	53.0
Index DIPJ	94	11.4
Middle DIPJ	80	6.9
Ring DIPJ	90	8.5
Little DIPJ	63	4.8
PIPJs	33	1.4
1st CMC	82	14.8

Knee thermograms usually showed a normal pattern, or an isolated area of increased heat emission in the medial or lateral area. The patella (central) area was always cold, and a "synovitis pattern" (hot medially, laterally and inferiorly) was only seen in two cases.

The correlation coefficient between radiological and thermographic abnormalities in the medial and lateral compartments of the knee was 6.1 ($p < 0.02$). As in the hand, many abnormal radiographs were associated with normal thermograms. Only one patient had an area of abnormal heat in a radiologically normal joint area.

DISCUSSION

Thermography is an established technique for recording inflammation of joints, and measuring treatment responses, in rheumatoid arthritis (RA). There have been few thermographic studies of OA.

Our data show that between one-third and one-half of symptomatic OA joints have an abnormal thermographic image. The pattern is different from that seen in RA. Small isolated areas of increased heat emission, rather than a generalized "hot joint", is the usual abnormality in OA. Many patients with severe pain and X-ray changes have normal thermograms, but the converse is not true. There is some correlation between the areas of greatest radiological change and the chances of the thermographic image being abnormal.

Increased heat emission could arise from increased subchondral bone blood flow, or from the soft tissues. Evidence that anti-inflammatory therapy reduces the thermographic index in OA suggests an important synovial element.[2]

OA includes a heterogeneous group of diseases. In this limited study no special features were apparent in patients with hot thermographic images, as opposed to those with normal thermograms. However, further studies might help us to understand why a proportion of radiologically abnormal joints are hot, and to define OA subsets.

SUMMARY AND CONCLUSIONS

Seventy patients with symptomatic and radiographic evidence of OA of the knees and hands have been examined thermographically. An abnormal pattern of heat emission was seen in 53% of hands and 37% of knees. There was some correlation between the radiological sites of damage and the isolated areas of increased heat emission seen. The pattern of abnormality is different from that seen in other joint diseases, and only occurs in a proportion of radiographically abnormal joints. These paradoxes are being explored in further studies.

REFERENCES

1. "The Aetiopathogenesis of Osteoarthritis," G. Nuki ed., Pitman Medical Publishing Co. Ltd., England (1980).
2. P. A. Dieppe, B. Sathapatayavongs, H. E. Jones, P. A. Bacon, E. F. J. Ring, Intra-articular steroids in osteoarthritis, Rheum. and Rehab. 19:212-217 (1980).

THERMAL IMAGING IN ASSESSMENT OF DRUGS IN RHEUMATOLOGY

P A Bacon, E F J Ring

Department of Rheumatology, University of Birmingham and Royal National Hospital for Rheumatic Diseases Bath, UNITED KINGDOM

INTRODUCTION

In a disease with no known cure, drug treatment has to be rigorously assessed, but assessment in rheumatic diseases is not a simple process. This is attested to by the extensive literature on the subject and the large number of tests available (for review, see [1]). It is clear that no one single test adequately reflects the complex changes which occur in the whole patient with an inflammatory arthritis. For this reason, most drug trials include a battery of tests which measure joint pain, joint tenderness and joint function. In addition, the patient's preference and various laboratory tests are subjective and will be strongly influenced by the patient's state of mind and his relationship with the drug trialist. The problems of placebo response in chronic diseases, such as rheumatoid arthritis, are well known. There is, therefore, strong need for an objective test which would indicate the changes in joint inflammation, independently of pain or pain threshold. Ideally such a test needs to be simple, cheap, non-invasive, repeatable and easily reproducible and acceptable to the patient. The only two methods which fulfil even some of these ideals are the use of radioisotope scanning of joints and infrared thermography. Both of these reflect chiefly the increased vascularity of inflammation and are not influenced by pain. Isotopes suffer from the disadvantages of any radioactive material and are invasive. Thus, thermography seems to be a logical choice. Extensive work in Bath over the last decade has suggested that thermography does indeed have many advantages and correlates with a number of aspects of joint pathology. These have included intra-articular temperature, vascularity of the joint as measured by both plethysmography and isotope studies, inflammatory changes in the synovial fluid as reflected by the white count, the total volume, and the protein content, and the synovial pathology as assessed microscopically at operation and histochemically using intracellular cathepsin levels.

These correlations have previously been extensively reviewed.[2]

It is fair to point out that not all authors have found thermography easy to use in the clinic, probably due to minor differences in technique.[3] More seriously, the validity of correlating the surface temperature over a joint with intra-articular temperature has recently been seriously questioned[4] and it is clear that further work is needed on this subject. Nevertheless, the purpose of this paper is to suggest that thermography can be used with great advantage in the assessment of drug therapy in rheumatology and appears to reflect the clinical changes and correlate with some immunological tests as well. This suggests that it may reflect drug induced changes in the disease pathogenesis. It is thus a useful empirical tool for use in the drug clinic.

THERMOGRAPHIC TECHNIQUES IN RHEUMATOLOGY

Compound isotherm pictures of arthritic joints provide a graphic illustration of the site of inflammation, together with a qualitative impression of the degree of inflammation. The use of a color filter to each isotherm lends artistic value to the pictures. However, to be useful in assessing the results of therapy, it is essential that the thermographic image is quantified. This has been achieved in Bath using an on-line computer to provide an integrated figure for the area of each isotherm within a standardized box. This is, in effect, an integrated temperature measurement for the joint surface.[5] It can hardly be stressed sufficiently that rigid standardization of technique is an essential prerequisite if such a thermographic index (TI) is to be applied to either cross-sectional or longitudinal studies[6] (Fig 1). With such standardization we have shown that the thermographic index of normal healthy joints varies somewhat according to the joint area, but is usually 2.5 or less. Assessment of large numbers of joints in patients with rheumatoid arthritis, has shown much greater variation with thermographic index values ranging from 3 to 6. It is thus reasonable to look for changes in the elevated thermographic index of disease, in particular, changes induced by drugs. However, it is clear that because of the intrinsic variation, all studies should be based on the change in thermographic index induced by a drug within each individual patient. For groups of patients, the mean of these individual changes is used rather than the mean of the absolute thermographic index.

Other authors have different approaches. In particular, it has been advocated that changes in pattern rather than in the absolute thermographic index, more closely reflect joint inflammation.[7] However, this involves a reversion to visual assessment patterns and neither quantification of this pattern, nor its use in drug assessment, has yet been described.

We have used the thermographic index extensively to assess drug therapy in arthritis. This paper illustrated its use in the assessment of

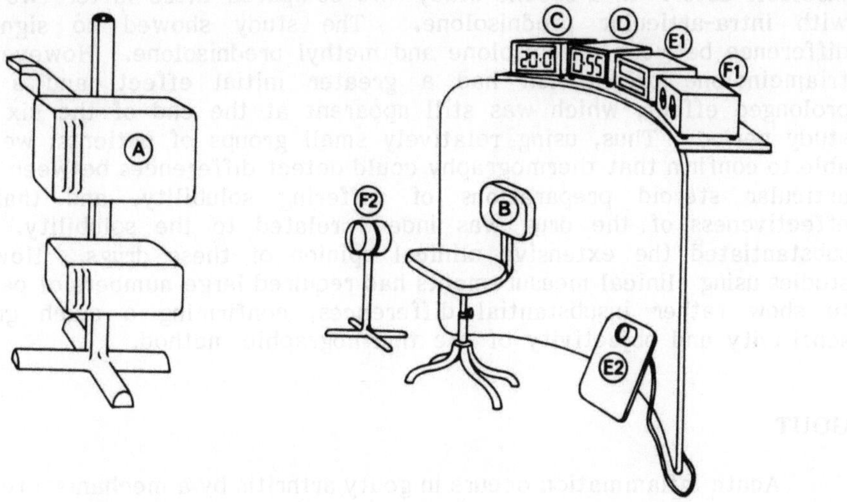

Fig. 1. Quantitative thermal imaging requirements. A - the thermograph with parallax free stand; B - height adjustable seat for the patient; C - large easy to read indicator of room temperature; D - time clock; E - the control unit (1) and thermal reference source (2) (portable) for upper limbs; F - control units (1) and reference source (2) for lower limbs.

single joints, both after an intra-articular injection in rheumatoid arthritis and in the acute inflammation of gout after oral non-steroidal drugs. In addition, the changes induced in a multiple joint compound, TI by both non-steroidals and slow-acting anti-rheumatic agents will be described.

INTRA-ARTICULAR STEROIDS

The use of steroid injections directly into a joint is a long-established practice in rheumatoid arthritis. Multiple preparations are available including a number of less soluble formulations which are said to have a longer duration of action. We directly tested this in comparing three analogues of prednisolone of different solubility.[8] This showed that the speed of action was greater with the most soluble preparation but the duration of action was greatest with the least soluble preparation. The changing profile of the TI from the injected knee correlated closely with the profiles of serum levels of prednisolone taken simultaneously, the latter reflecting solubility of the steroid and its escape from the joint. The use of two different doses shows that there was a dose effect with the most soluble steroid, which had a more prolonged action with the bigger dose. However, with the less soluble steroids, no such effect was seen. Other preparations of steroid are commonly advocated, particularly methyl prednisolone and more recently triamcinolone hexacetonide, a very insoluble ester. In a second study[9] we compared these latter two drugs with intra-articular prednisolone. The study showed no signficant difference between prednisolone and methyl prednisolone. However, the triamcinolone preparation had a greater initial effect, and a more prolonged effect, which was still apparent at the end of the six week study period. Thus, using relatively small groups of patients, we were able to confirm that thermography could detect differences between intra-articular steroid preparations of differing solubility, and that the effectiveness of the drug was indeed related to the solubility. This substantiated the extensive clinical opinion of these drugs. However, studies using clinical measurements had required large numbers of patients to show rather insubstantial differences, confirming a much greater sensitivity and objectivity of the thermographic method.

GOUT

Acute inflammation occurs in gouty arthritis by a mechanism relating to phagocytosis of urate crystals. The clinical response to effective doses of non-steroidal agents is remarkably rapid. However, very few studies of objective changes have been made. Gout is a subject with a strong historical link with Bath and its waters, and in more recent years Dr G D Kersley has made extensive evaluation of the more modern drugs for this disease. It was a logical progression to look at this further using thermography. With the co-operation of local general practitioners, we managed to collect a number of cases of gouty arthritis in the acute stage before therapy had been administered. The patients were divided into two groups, one receiving standard indomethacin therapy and the other receiving a new agent. These studies confirmed that indomethacin induces changes in the inflammation of gout within hours and indeed a maximum fall in thermographic index occurs within 24 hours which is not improved later at 48 hours. In addition, it showed that the changes induced by indomethacin were larger than those with the new non-steroidal agent. Subsequent extensive experience with the new agent has confirmed that its anti-inflammatory effect is somewhat less marked than indomethacin.[10]

These studies were extended into the use of thermography to assess urate-induced inflammation after intradermal injection of crystals. Again the reduction of inflammation by indomethacin was greater than that achieved with the new agent, confirming the results in gouty patients and suggesting that this might be a useful method for screening potential anti-gout agents.

RHEUMATOID ARTHRITIS - NON-STEROIDAL ANTI-INFLAMMATORY DRUGS

The inflammation of rheumatoid arthritis is much more chronic than that of gout, and the disease is more common. For these two reasons, it is much easier to accumulate large numbers of patients with active inflammatory arthritis for therapeutic studies in rheumatoid arthritis and we have extensive experience of this subject. Several studies have shown that the pattern of response to non-steroidal anti-inflammatory drugs (NSAID's) in rheumatoid arthritis (RA) is very different from that of gout. Firstly, the rate of response is very much slower in RA, with the maximum effect not being visible for two or three weeks, as compared with the 24 hours of gout. This must reflect the different pathogenetic mechanisms working in the two diseases. However, the effect of NSAID's, although delayed, is not due to change. We consistently see an increase in thermographic index, reflecting deterioration in the arthritis, when the patient is switched from a non-steroidal to a pure analgesic drug, such as paracetamol. Another important point is that different joints respond at different rates and to differing degrees in RA (Fig 2). The response of the small joints of the hands, (wrists and MCP's) is both more rapid and of a greater magnitude than is the response of larger joints, such as knees and elbows. We have consistently found that a further improvement appears between two and three weeks in the large joints, although the maximum has already been achieved by two weeks in the small joints. For this reason it is important to standardize the joints used in any drug trial. To overcome this and to facilitate the comparisons between trials, we have over the last few years consistently used a compound thermographic index which involves six joint areas. These are the hands, including the small MCP joints; the knees and the ankles. These are all joint areas frequently involved in rheumatoid arthritis and we feel that this compound thermographic index, although not assessing all the joints involved in any one patient, provides a clearer reflection of the total patient state than any measurement of a single or only a few joints. There are some disadvantages in that not all the joints assessed will necessarily be acutely inflamed at the time, so that only smaller changes can be expected and the sensitivity of the method will thereby be reduced.

Student t

n = 60	HANDS	ANKLES	KNEES	
PARACETOMOL 1 WEEK ▲	<0.004	<0.025	<0.25	P
ORAL NS. 2 WEEKS ▼	<0.0005	<0.005	<0.01	

Fig. 2. The change in compound TI of hands, ankles and knees in 60 patients with rheumatoid arthritis after one week on paracetamol followed by two weeks on oral non-steroid. A TI increase during analgesic therapy is followed by a decrease with anti-inflammatory preparations.

Using the compound thermographic index we have compared six standard non-steroidal agents, given to groups of ten patients each, and assessed the change in TI over a two week period. It is clear that not all patients will respond to all of these drugs and indeed, over twenty per cent of patients will not respond to the agent used first. However, the mean change within the group, whether the total change is considered, or only the change within responding patients, shows no signficant differences between any of these drugs. This suggests that the minor differences in patient preference between NSAID's noted in some drug studies reflect subjective assessments such as pain and the incidence of side effects, rather than any anti-inflammatory effect of these drugs.[11]

In studies with non-steroidals, an overall correlation between changes in thermographic index and changes in other clinical assessments, is noted, but the degree of correlation is never very great. This again reflects the dual action of the NSAID's, acting as both analgesic and anti-inflammatory drugs. Using an analysis of covariance in six different drug studies, we found little correlation between thermographic index and either joint pain or joint tenderness. However, when the index for the TI for an individual hand was compared with the grip strength of that hand, a signficant correlation was found in the majority of studies. This suggests that the TI does indeed reflect the important changes in function induced by inflammation within that individual joint.

Another important aspect of non-steroidal drugs is that of the optimum dose. An analgesic effect may be observable at a low dose, as with aspirin. Quite often higher doses are needed to achieve anti-inflammatory effects. The limitations in clinical methodology often mean that extensive clinical experience of the drug is necessary before the ideal dose is established. Using compound TI we have been able to produce a dose response curve on several non-steroidals. Although the dose per body weight is not the same, the shape of these curves is otherwise remarkably similar in human rheumatoid arthritis and in induced animal arthritis, when the same agent is considered.

The correlation between TI and clinical indices and the ability to produce a dose response, has been most useful in assessment of new drugs as they become available. This is illustrated by the results with one new agent, benoxaprofen, a non-steroidal which appears to have a slightly different mode of action to some of the others. With this drug, we were again able to show dose response curve, this time produced by comparing different groups of patients with a matched control group in each case.[12] Differences between 400 and 600 mg per day doses (the two higher doses) are particularly interesting. There was no difference between these doses when considering pain or the articular index of joint tenderness. However, there were marked differences in the degree of inflammation as assessed by the thermographic index between these two doses (Fig 3). In addition, changes in hand function, assessed by grip strength, were also apparent between the 400 and 600 mg dose so that the correlation between grip strength and the individual hand TI, held up in this study (Fig 4).

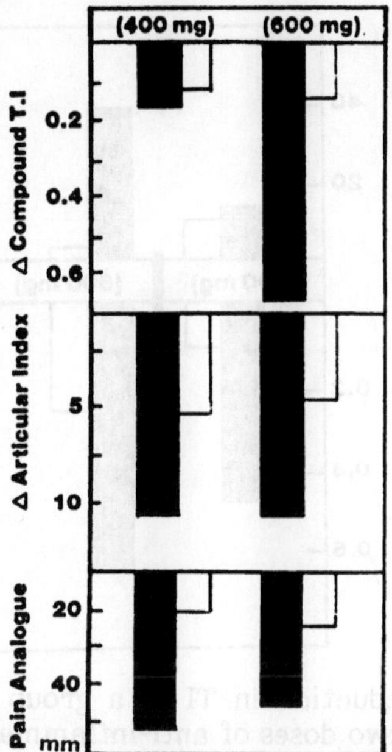

Fig. 3. The mean fall in TI in four groups of patients treated for three weeks with an anti-inflammatory drug. The solid histograms represent the change from 400 and 600 mg. The open histograms represent the change in two matching placebo groups.

In conclusion, extensive experience with TI assessment of NSAID's has confirmed that these drugs have an analgesic effect which appears early and at a lower dose but, their anti-inflammatory action appears much more slowly and at a higher dose. Because of the important differences between individual joints, particularly between large and small joints, compound thermographic index involving multiple joints is necessary for adequate evaluation.

RHEUMATOID ARTHRITIS - LONG-ACTING DISEASE MODIFYING AGENTS

The studies with NSAID's were all conducted over a short period, usually up to four weeks total. However, it is known that clinically, the so-called disease modifying agents used in rheumatoid arthritis, classically gold and penicillamine, take much longer to show an effect but that the magnitude of that effect is usually greater. We have used thermography both to examine new agents for potential long-acting effect and to look at established disease modifying agents.

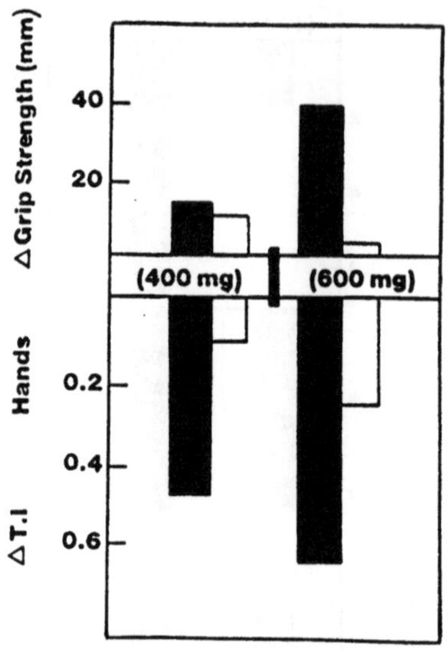

Fig. 4. The mean reduction in TI in a group of patients treated (as above) with two doses of anti-inflammatory therapy, compared with their placebo response. Grip strength improvement matches the increased temperature fall in the higher dose group.

The use of TI to assess a new agent, is illustrated by a study comparing a new drug from ICI (Clozic), in comparison with a standard NSAID (flurbiprofen).[13] When pain assessments were considered, there was little difference between the two drugs initially, although by two months, a difference was apparent. Using thermography a difference between them was apparent immediately. The changes in compound TI induced by Clozic showed a consistent improvement from two weeks onwards, with statistically significant improvement by four weeks. The consistency of Clozic induced changes were in marked contrast to those induced by the non-steroidal drug, where haphazard and non-significant changes occurred reflecting the fluctuating nature of inflammation in rheumatoid arthritis. The changes in thermographic index were reflected by some changes in laboratory parameters, as illustrated by the marked fall in the elevated C-reactive protein, which occurred in the patients on the Clozic, but not on the non-steroidal. Thus, thermography agreed with the clinical and the laboratory changes indicated that the new drug did indeed have a disease modifying effect. Most importantly, this change was detected early, and quite consistently, using compound TI.

Several studies with established drugs, such as D-penicillamine, in our unit have shown that its effects are different from those of non-steroidals. Firstly, only a minimal change in compound TI is apparent at six weeks. However, at six months, the maximum effect is considerably greater than that achieved with the non-steroidal, confirming that the disease modifying agents really do have a different mode of action. As with the non-steroidals, a dose effect can be seen, as illustrated by serial studies with D-penicillamine.[14] Extended studies over a twelve month period have shown that for D-penicillamine as well as for gold, little or no effect on the compound TI is seen at one month, but a pronounced effect is apparent at three months and seems to reach its maximum at six months. Thereafter, little further improvement occurred. This is in accordance with long established clinical observations that these drugs take two to three months to produce an effect but have their maximum effectiveness by six months.[15]

The pathology of the synovial membrane in rheumatoid arthritis shows a striking infiltration with lymphocytes. There has been extensive interest in changes in peripheral blood lymphocyte populations in rheumatoid arthritis over the last few years. We have tried to correlate the changes in thermographic index induced by long-acting anti-rheumatic drugs, with some changes in lymphocyte function assessed in the laboratory. Using different doses of cyclophosphamide to treat severe rheumatoid arthritis, we were able to show a highly signficiant correlation between changes in TI and changes in the leukocyte migration index.[16] This held good both for patients who were improving and for those who were deteriorating while under treatment. More recently, we have been particularly interested in our laboratory in changes in a sub-population of activated lymphocytes, or immunoblasts, which we detect in a ficoll density gradient. In a further group of patients treated with intravenous cyclophosphamide we were able to correlate serial changes in TI with the percentages of activated low density lymphocytes.[17] In addition, serial studies over a four month period have been performed on fresh groups of patients treated with either gold or D-penicillamine.[18] In both cases, the maximum fall in thermographic index, in some patients to within the normal range, occurred between sixteen and twenty weeks. Changes in clinical parameters were very variable and did not always correlate with the changes in compound TI. However, the changes in low density lymphocytes closely parallelled those of the compound TI. Several of our laboratory studies have now shown that low density lymphocytes are particularly associated with active synovitis in rheumatoid arthritis. Therefore, it is of considerable interest that changes in these low density lymphocytes are closely related to the changes in thermographic index, suggesting that the latter accurately reflects important aspects of synovial inflammation.

OTHER DISEASES

Vasospastic conditions are commonly associated with the connective tissue diseases seen by rheumatologists. A number of studies have been used to demonstrate the effect of vasodilating drugs, such as the ability of the drug piribedil to produce increased temperature from increased blood flow, shown by color thermograms.[19] Vasospasm is by its nature intermittent and usually related to cold. The development of a thermal stress test for the hands for use in vasospastic disease, therefore represents a considerable advance.[20] This has been applied to serial studies of a vasodilator drug in Raynaud's disease.[21]

Metabolic bone disease also presents to rheumatologists. Paget's disease of the bone can present with bone pain which may be locally warm where superficial bone is involved, as over the tibia or the skull. Not surprisingly this is reflected by an increased surface temperature which can be detected and quantified using computerized thermography. Again drug induced changes can be assessed and our group has shown a rapid decrease in temperature occurring within a few weeks with treatment with calcitonin. However, on cessation of therapy rapid rebound occurs which predates clinical deterioration often by a long period.[22]

Both Raynaud's and Paget's disease, coupled together with other vascular diseases, appear to be suitable areas for expanding thermographic assessment of the effects of drugs.

CONCLUSION

Extensive studies have suggested that computerized thermography, provided it is rigidly standardized, provides a useful index for joint inflammation. The method is reproducible, quantitative, and objective. This last is particularly important and contrasts with the majority of assessments made in rheumatic diseases.

The most important use of thermography is in the longitudinal assessment of drug induced changes in joint inflammation. It can be used to detect changes within a single joint after intra-articular injection, or in the acute monoarthritis of gout. However, in the polyarthritis of rheumatoid and similar diseases, multiple joints need to be assessed. The use of compound thermographic index diminishes the problems of differential responses between different joint areas, particularly between large and small joints. The compound thermographic index can be shown to measure both the short term effect of a non-steroidal drug and is useful for comparative purposes and the production of a dose response curve. It can also measure the more slowly occurring changes induced by disease modifying agents which are eventually of greater magnitude.

The changes in compound thermographic index reflect functional changes induced by inflammation in the same joint area, but they do not correlate with measurement of pain. More importantly, several correlations have been shown in long term studies between changes in compound thermographic index and changes in a lymphocyte function. The latter may be of fundamental relevance in the pathogensis of rheumatoid inflammation. The correlation with TI suggests that thermography really does correlate with the important disease producing mechanisms in arthritis, lending credibility to its use as a clinical tool for assessing drug action. In a complex multi-faceted chronic disease, such as RA, it is not entirely clear as to precisely what thermography is measuring and doubt has been cast on its relationship to intra-articular temperature. Nevertheless, our studies suggest that it is a valid reflection of clinically important inflammatory changes and is certainly the best objective assessment method available in arthritis at the present time.

REFERENCES

1. P. A. Bacon, Assessment in rheumatic diseases, in: "Scientific Basis of Rheumatology," G. Panayi ed., Churchill Livingstone, London (1982).
2. P. A. Bacon, A. J. Collins, E. F. J. Ring, J. A. Cosh, Thermography in the assessment of inflammatory arthritis, in: "Clinics in Rheumatic Diseases," M. I. V. Jayson ed., W. B. Saunders Co. Ltd. (1976).
3. J. Paterson, W. Watson, E. Teasdale, P. Newman, W. James, D. A. Pitkeathly, Assessment of rheumatoid inflammation in the knee joint: a reappraisal, Ann. Rheum. Dis. 36:288 (1977).
4. A. C. Kennedy, Joint temperature, in: "Clinics in Rheumatic Diseases," P. Hasselbacher ed., W. B. Saunders Co. Ltd. (1981).
5. E. F. J. Ring, A. J. Collins, P. A. Bacon, J. A. Cosh, Quantitation of thermography in arthritis using multi-isothermal analysis. II Effect of nonsteroidal anti-inflammatory therapy on the thermographic index, Ann. Rheum. Dis. 33:353-356 (1974).
6. Anglo Dutch Thermographic Society, Thermography in Locomotor Diseases: Recommended Procedure, Eur. J. of Rheumatol. and Inflam. 2:299-306 (1979).
7. R. S. Salisbury, M. De Silva, B. L. Hazleman, G. Parr, D. P. Page Thomas, Thermography and the assessment of inflammation, Ann. Rheum. Dis. 40:522 (1981).
8. W. Esselinckx, P. A. Bacon, E. F. J. Ring, D. Crooke, A. J. Collins, D. Demottaz, A thermographic assessment of three intra-articular prednisolone analogues given in rheumatoid arthritis, Br. J. Pharmac. 5:447-451 (1978).
9. H. A. Bird, E. F. J. Ring, P. A. Bacon, A thermographic and clinical comparison of three intra-articular steroid preparations in rheumatoid arthritis, Ann. Rheum. Dis. 38:36-39 (1979).
10. P. A. Bacon, E. F. J. Ring - unpublished data.
11. P. A. Bacon, E. F. J. Ring, A. J. Collins, Thermography in the assessment of anti-rheumatic agents, in: "Rheumatoid Arthritis," J. L. Gordon, B. L. Hazleman ed., Elsevler, Amsterdam (1977).
12. P. A. Bacon, J. Davies, E. F. J. Ring, Benoxaprofen - dose range studies using quantitative thermography, J. of Rheumatol. (suppl. 1 no 6) 7:48-53 (1980).
13. J. Davies, P. A. Bacon, N. Hall, Comparative study of Clozic and Flurbiprofen in RA, Submitted for publication, (1982).
14. D. R. Blake, E. F. J. Ring, R. C. Bucknall, P. A. Bacon, D-penicillamine in rheumatoid arthritis: fixed versus variable regimes,

in: "Inflammation - Mechanisms and Treatment," D. A. Willoughby, J. P. Giroud ed., MTP Press Ltd. (1980).

15. Empire Rheumatism Council, Gold therapy in rheumatoid arthritis: final report of a multi-centre controlled trial, Ann. Rheum. Dis. 20:315-334 (1961).
16. N. D. Hall, H. A. Bird, E. F. J. Ring, P. A. Bacon, A combined clinical and immunological assessment of four Cyclophosphamide regimes in rheumatoid arthritis, Agents and Actions 9;1:97-102 (1979).
17. G. M. Papadimitriou, P. A. Bacon, N. D. Hall, Circulating activated lymphocytes in rheumatoid arthritis: a marker of synovial inflammation, (In press)
18. G. J. M. Alexander, F. Cavalcanti, D. R. Blake, P. A. Bacon, N. D. Hall, Low density lymphocytes, anti-rheumatic therapy and rheumatoid arthritis, Ann. Rheum. Dis. 40:523 (1981).
19. J. Huys, H. Brens Vaerenhergh, P. M. Van, Telethermographic demonstration of the increase in arterio-capillary perfusion induced by injectable piribedil, Pharmatherapeutica 1;4:221-228 (1976).
20. E. F. J. Ring, Computerized thermography for osteo articular diseases, Acta Thermographica 1;3:166-175 (1976).
21. E. F. J. Ring, P. A. Bacon, Quantitative thermographic assessment of inositol nicotinate therapy in Raynaud's phenomenon, J. Internat. Med. Research 5:217-222 (1977).
22. E. F. J. Ring, J. Davies, J. R. Barker, Thermographic assessment of Calcitonin in Paget's disease, Bone Disease and Calcitonin 39-48. Armour Pharmactic Co Ltd, Eastbourne (1977).

THERMOGRAPHY IN ANKYLOSING SPONDYLITIS

H Tauchmannova, V Svec, D Zlnay

Research Institute of Rheumatic Diseases
Piestany
CZECHOSLOVAKIA

INTRODUCTION

The early diagnosis of ankylosing spondylitis (AS) in the absence of clearly defined radiological changes may present clinical difficulties. The incipient inflammatory manifestations cannot be revealed radiologically, the inflammatory activity in region of sacroiliac joints (SIJ), however, may be detected by scintigraphy. The disadvantage of this method lies in its invasiveness and there is hesitation whether to use radioisotopes in young individuals (frequently under 20 years). From prognostic and diagnostic aspects it is just this group where it is necessary to find out the cause of low back pain. In the last few years, studies of Ascarelli,[1] Cosh and Ring,[2] Sadowska-Wroblewska,[3] Schlecht et al,[4] Scott[5] and Zorzin[6] have drawn attention to the possibility of differentiation of temperature changes in the region of SIJ by means of thermography (T), thus detecting the inflammatory manifestations by a non-invasive method. The aim of this study was to evaluate the usefulness of T as a suitable complementary diagnostic method in detecting sacroiliitis.

MATERIALS AND METHODS

In forty out-patients with clinical symptoms of incipient AS (fulfilling the first four New York criteria for diagnosis of AS), radiological, immunological (presence of HLA B27 antigen) and thermographic examinations were made. In the present study the temperature distribution over the SIJ was recorded by means of a thermoprofile focused on three selected levels where inflammatory manifestations had been expected. For this purpose, we have chosen three lines interconnecting two opposite formations between:

1. Spinal iliac dorsalis cranialis
2. Spinal iliac dorsalis caudalis and
3. Line passing in the midway between the two connecting lines (Fig 1).

Thermographic examination was made under standard conditions (patient sitting on the chair in riding position, equilibration time - 20 min, the three selected lines were indicated by cooled needle).

As "normal" we considered the thermoprofile with one temperature peak situated directly over the spine, due to the rich vascular network in this area (Fig 2). Figure 3 shows symmetric increase of temperature over the SIJ, especially on the levels II and III. Asymmetric increase of temperature over the right SIJ is shown in Figure 4.

RESULTS

Out of forty patients with clinical symptoms of AS, ten had 4 positive signs (clinical, X-ray sacroiliitis, presence of HLA B27 and "hot" symmetric thermoprofile). In 22 patients without radiological signs of sacroiliitis, thermopathological curve of profile and the presence of HLA B27 were found. Two signs (clinical and radiological) were present in one patient, another combination of two signs (clinical and thermographic) was found also in a single patient, still another combination of two signs (clinical and HLA B27) - in two patients; one sign - the clinical one -

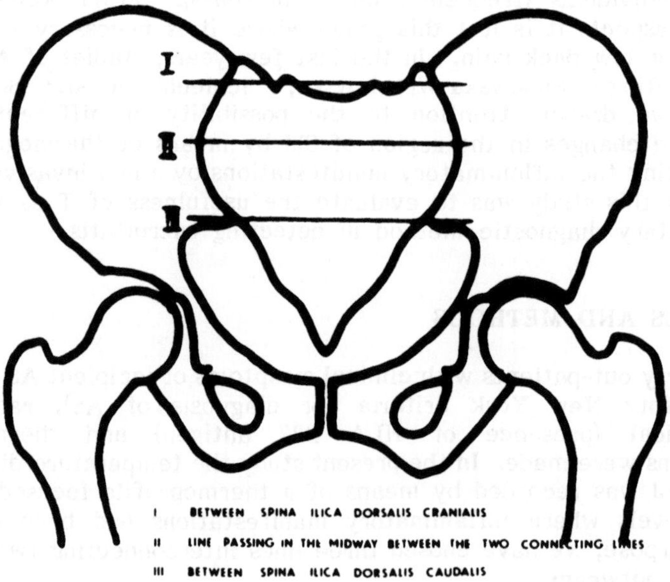

Fig. 1. Localization of linescan for ankylosing spondylitis.

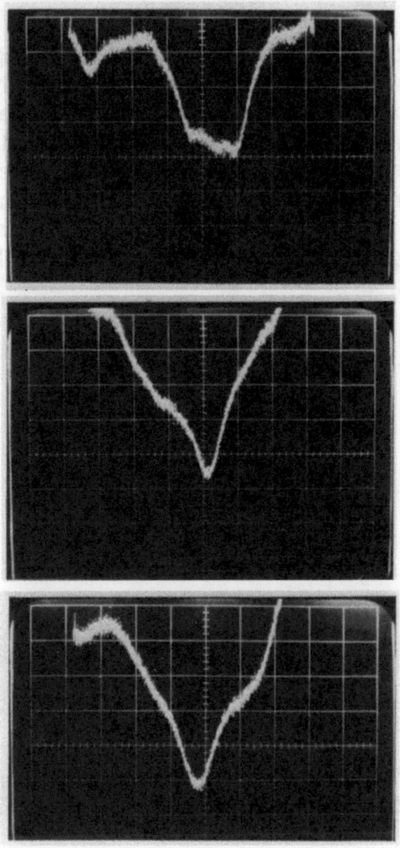

Fig. 2. Normal thermoprofile over sacro-iliac joints (SIJ's).

was found in four patients. In one patient an advanced form of sacroiliitis without finding of HLA B27 antigen nor a "hot" thermoprofile was observed. The results are summed up in Table 1.

DISCUSSION

Various authors approached studying inflammatory changes in the region of SIJ in various ways determined by technical possibilities and parameters of their equipment. There is no doubt that, as yet, the best way of evaluating the temperature changes is a quantitative one. Not having the means to perform the evaluation in this way we have used the thermoprofile to record temperature distribution over the SIJ. We have considered as thermopathological only curves with a symmetrically "hot" thermoprofile; the asymmetrically "hot" profile was regarded only as suspect. In a high percentage of our patients the presence of HLA B27 antigen and a "hot" symmetrical thermoprofile was found. In our opinion, this finding can be explained only by the fact that all the patients referred to the Institute did fulfil the four diagnostic criteria for ankylosing spondylitis and, therefore, there was a high probability they had also inflammatory changes in SIJ, but undetectable by X-ray. Radiologically confirmed sacroiliitis of II degree did not manifest the increased temperature pattern, not even in the patient with definite AS. Thermographic examination of the sacroiliac joints in the three suggested lines seems to be a sensitive indicator for detecting inflammatory changes in this region and might therefore be a useful complementary method in early diagnosis of AS.

Table 1. Summary of results

Combination of signs	No of patients
4 signs (clinical, radiological, HLA B27, thermographic)	10
3 signs (clinical, HLA B27, thermographic)	22
2 signs (clinical, radiological)	1
2 signs (clinical thermographic)	1
2 signs (clinical, HLA B27)	2
1 sign (clinical)	4
	40

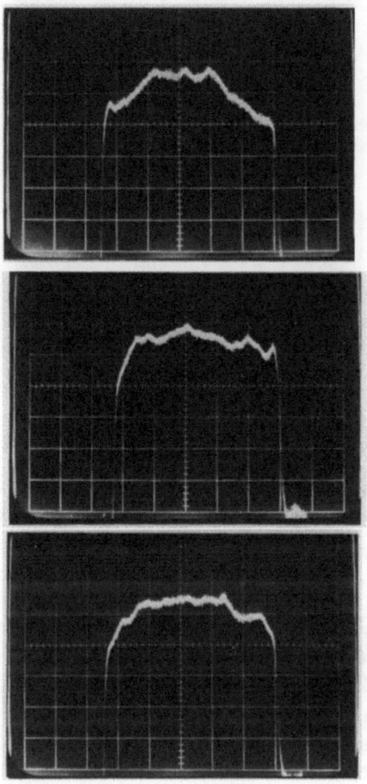

Fig. 3. Symmetric temperature increase over SIJ's.

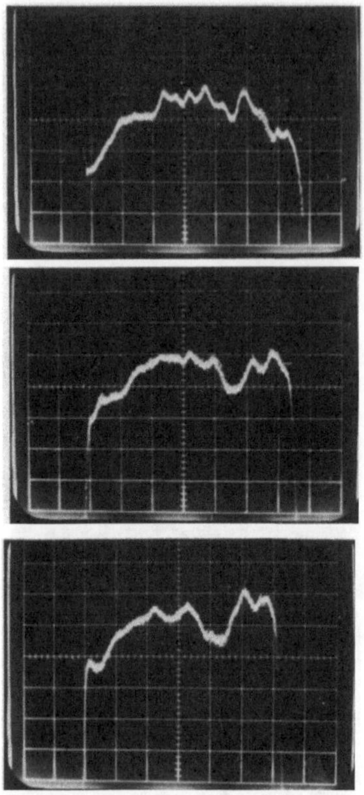

Fig. 4. Asymmetric temperatures over SIJ's.

REFERENCES

1. A. A. Ascarelli, L. Zorzin, M. Castagnoli, F. Taormina, Thermographic assessment of ankylosing spondylitis, Acta Thermographica 2:95 (1980).
2. J. A. Cosh, E. F. J. Ring, G. Evison, B. J. M. Schlecht, Thermography and isotope scanning of the spine in ankylosing spondylitis, Unpublished work, (1978).
3. M. Sadowska-Wroblewska, S. Kruszewski, H. Gawrolinska, A. Filipowicz-Sosnawska, The thermographic examination of sacroiliac joints, Acta Thermographica 2:54 (1976).
4. B. Schlecht, E. F. J. Ring, G. Evison, F. A. Duck, D. G. Scott, An objective view of quantitative thermography in sacroiliitis, XVth International Congress of Rheumatology, Paris (abstract 1007) (1981).
5. D. G. I. Scott, E. F. J. Ring, P. A. Bacon, Problems in the assessment of disease activity in ankylosing spondylitis, Rheumatology and Rehabilitation 2:74 (1981).
6. L. Zorzin, M. Capone, M. G. Maira, Further thermographic study of sacroiliac joint diseases, XVth International Congress of Rheumatology, Paris (abstract 1006) (1981).

THERMOGRAPHY AND RADIO-ISOTOPES IN THE ASSESSMENT OF SACROILIAC INFLAMMATION

D M Grennan, R Bennett, L Taylor, C W Johns, J D Kingsley Brown

**University of Manchester, Rheumatic Diseases Centre
Hope Hospital, Salford, UNITED KINGDOM and
The Departments of Medical Physics and Radiology
Hope Hospital, Salford, UNITED KINGDOM**

This was a cross-sectional study designed to investigate the sensitivity of thermography as compared to radio-nuclide imaging in the detection of human sacroiliac inflammation.

METHODS

Patients

Thirty patients with ankylosing spondylitis and aged between 19 and 55 years were studied. 25 were male. All but 5 satisfied the New York Criteria of definite ankylosing spondylitis.[1] These 5 all gave a history typical of sacroiliitis, all had limitation of lumbar spinal movements noted by a physician and were HLA B27 positive. Clinical sacroiliac disease severity was scored on a 0 - 3 basis on the day of assessment where 0 = no pain, 1 = sacroiliac pain and stiffness on that morning but no tenderness of iliac-crest compression, 2 = definite tenderness on stressing the sacroiliac joints and 3 = severe tenderness on stressing the sacroiliac joints. These clinical assessments were made independently of the results of thermography. Clinical disease activity overall was fairly low. Sacroiliac X-rays were graded according to the New York Criteria on a 0 - 4 basis.[1]

Infrared Thermography

Thermographic measurements were carried out with an AGA 680 camera with THV102 thermal display unit and in a constant temperature

($20^o \pm 0.5^oC$), constant humidity (45% $\pm$ 5% relative humidity) room. An AGA profile adaptor was used to display a profile of temperature across sacroiliac regions and sacrum on a separate oscilloscope. This method has been described in detail in a recent publication.[2] Peak activity of each sacroiliac region after subtraction of the sacral measurement was categorized as abnormal if it lay outside the normal range (mean + 2 SD) calculated in normal volunteers. Photographs were taken of the thermographic pattern and these could be categorized as abnormal in a minority of patients. The abnormalities seen were of a gross increase in the pattern or of loss of the normal demarcation between sacral and sacroiliac representations in a thin person. Categorization of thermographic patterns as abnormal or normal was not possible in obese individuals. Patients were defined as abnormal if either the profile measurements were increased or the thermographic pattern was abnormal.

Radio-isotope Imaging

Imaging of the posterior pelvis was carried out in a sitting position with a gamma camera linked to a computer system. A dose of 550 MBq (15 mCi) of $^{99}Tc^{m}$ - labelled methylene diphosphonate was injected intravenously and imaging carried out approximately 3 hours later. The uptake of radioactivity in each sacroiliac joint was expressed as a sacroiliac index by comparing the total number of counts in each joint with the counts in the trough[3] between sacroiliac joints and sacrum. Forty-eight control subjects were examined. These were patients being investigated for fracture non-union or for suspected bone metastases but without low back pain. In the control group the sacroiliac index declined linearly with age and this decline was more rapid in female subjects. A normal range for sacroiliac index was established at each age for males and females.

RESULTS

The results of thermography and isotope imaging in ankylosing spondylitis patients are shown in Tables 1 and 2. Seven of 30 patients examined were abnormal on imaging and 5 were abnormal thermographically. Both assessments were abnormal in 3 patients. The thermograms of 13 patients with other causes of low back pain were normal.

DISCUSSION

Both thermographic and isotope imaging techniques were insensitive methods of detecting sacroiliac inflammation in this group of patients with overall low disease activity. Seven patients were receiving non-steroidal anti-inflammatory drugs at the time of study possibly lessening the sensitivity of the technique. Neither technique showed abnormalities

Table 1. Comparison of thermographic and radio-nuclide abnormalities versus X-ray grade

XR Grade	Increased isotope Index	Increased thermographic Activity	Increased Isotope or Thermographic Activity
0	0/5	0/5	0/5
1	0/1	0/1	0/1
2	2/4	2/4	3/4
3	5/12	2/12	5/12
4	0/8	1/8	1/8
Total	7/30	5/30	9/30

Table 2. Comparison of thermographic and radio-nuclide abnormalities versus clinical disease activity

Clinical Activity	Increased isotope Index	Increased thermographic Activity	Increased Isotope or Thermographic Activity
0	2/12	1/12	2/12
1	3/11	2/11	4/11
2	2/7	2/7	3/7
Total	7/30	5/30	9/30

in any of the patients with normal X-rays so that neither would seem to be very helpful in the assessment of early, X-ray negative sacroiliitis if the clinical disease activity is low. Radio-nuclide imaging was slightly more sensitive than infrared thermography in this group of patients. Both techniques however were abnormal in only 3 patients so that if used together the two techniques may complement one another. Although these results overall argue against a useful role for either of these objective techniques in the early assessment of sacroiliac inflammation, they do not exclude a role for either technique in the objective, serial, assessment of sacroiliac disease.

SUMMARY

1. 9/30 ankylosing spondylitis patients with generally mild disease activity were abnormal by either thermographic or radio-isotope examination of the sacroiliac regions.

2. None of the ankylosing spondylitis patients with normal X-rays of the sacroiliac joints were abnormal by either assessment so that neither technique is likely to be helpful in the assessment of early X-ray negative sacroiliitis.

3. Radio-nuclide scanning was slightly more sensitive than thermography in the detection of early sacroiliitis.

4. Both techniques were abnormal in only 3 patients as compared to 9 patients with sacroiliac abnormality detectable by either technique so that if used together thermography and radio-nuclide imaging may complement one another.

REFERENCES

1. P. H. Bennet, J. M. Brenner, E. G. C. Bywaters, J. L. Calabro, C. McEwen, W. Mastell, Report from the sub-committee on diagnostic criteria, in: "Population Studies of the Rheumatic Diseases," P. H. Bennett, P. N. H. Wood ed., Excerpta Medica, Amsterdam. 456-457 (1968).
2. D. M. Grennan, L. Caygill, Infrared thermography in the assessment of sacroiliac inflammation, Rheumatology and Rehabilitation (in press) (1982).
3. J. Ayres, A. J. W. Hilson, M. N. Maisey, R. Laurent, G. S. Panayi, A. J. Saunders, An improved method for sacroiliac joint imaging: A study of normal subjects, patients with sacroiliitis and patients with low back pain, Clinical Radiology 32:441-445 (1981).

LUMBAR, THORACIC AND CERVICAL THERMOGRAPHY

C Wexler

Encino
California
USA

The appearance of the normal lumbar thermograph, its variations, and its correlation with the myelogram and surgical findings have previously been described in the literature.[1,2] A method of routinely screening the lower extremities in conjunction with the lumbar thermographic examination has also been described.

The procedure should be objective and be able to stand on its own merit, independent of any other test or finding. As such, it should be reproducible, when properly performed, in the hands of any examiner, assuming no change in the patient's status has occurred. Proper technique is therefore essential. Parameters included in technique are: Use of a proper room, proper cooling techniques, positioning, the taking of sufficient repeat views to determine consistency with time, and proper imaging techniques. As long as the patient is comfortable, not too hot or not too cold, the absolute temperature of the room is not critical. It is extremely important, however, that the patient not be exposed to any direct, or strong indirect, drafts and preferably the room should be draft free. Pre-cooling of the back with either wiping with room temperature water followed by fan blowing, or simply fan blowing to dissipate clothes contact artifact is necessary. The patient should then wait a minimum of 10 minutes in a draft free environment before beginning the thermographic procedure. Forced cooling of the lumbar or thoracic area to enhance visualization of information in these areas should be done as the last thing, before the patient is to get dressed, so as to not unduly influence the remainder of the examination. Positioning should be such that symmetrical portions of opposite sides of the body are presented since interpretation is left-right with the patient being his own control. It is important to establish a routine set of views to be taken with each examination, so that comparable parameters of information are obtained on each patient visit.[3] Although there are exceptions, nerve irritation will usually manifest itself as a relative decrease (it will look colder) in the vascular heat emission pattern along the course of the nerve.

The following are a list of principles of cervical, thoracic and lumbar thermography for nerve root evaluation:

1. The examination should be conducted in a draft-free room. Absolute temperature control is not necessary but the room should be uniform in temperature and the patient comfortable.

2. The patient should be instructed not to smoke for at least one hour before the examination and 3 or 4 hours is preferable.

3. There should be no EMG examination or physiotherapy before thermography.

4. The lumbar area requires qualitative as well as quantitative views; the remainder only quantitative views; however, qualitative views of other areas can be obtained (eg to document scars, varicose veins, etc).

   (a) A one degree (celsius) difference affecting 25% of one surface area as compared to its opposite is the criteria for abnormality in quantitative mode. A one degree change of lesser extent should have supporting evidence contributing to the diagnosis (thermographically or otherwise).

   (b) For the lumbar qualitative patterns, film of superior resolution (celluloid) should be used. Imaging techniques for this area should include multiple views, before and after forced cooling, utilizing several images taken by either varying the camera F stops, or by utilizing "thermal focusing" techniques to ensure proper "seeing" of the information presented.[4] Forced cooling (eg alcohol spray) should be done last after all extremity views are finished.

5. Abnormalities present should be demonstrated to be consistent with time, and therefore, reproducible at any time if there has been no change in the status of the patient. Intermittent abnormalities should be reproducible as such.

   (a) If Polaroid film is used at least a second set 20 minutes following the first should be done, and the studies compared for consistency of information content. If dissimilar, further repeat views at 15 to 20 minute intervals obtained until an equilibrium is reached.

   (b) If 35 mm film is used, three repeat sets at 15 to 20 minute intervals is necessary to ensure sufficient passage of time and repeat views.

6. The examining camera should be at right angles (or close to it) to the parts being examined, and the parts should be parallel with each other, presenting equal surfaces for comparison.

7. Cervical thermogram - examination by definition to include a minimum screening or the posterior neck, posterior shoulders, ventral, dorsal and medial forearms, and ventral and dorsal hands. It is preferable to also include the ventral, lateral and dorsal arms (Table 1).

8. Thoracic thermogram - examination by definition to include as a minimum the posterior thorax. Optional views as necessary are oblique or lateral positions.

9. Lumbar thermogram - examination by definition to include minimum screening of the lumbar area of the back, the buttocks, the anterior, posterior, and lateral thighs, and the dorsal feet and toes. It is preferable to also include the anterior, posterior, lateral and medial legs.

The low back thermogram patterns, by themselves, have been found to be approximately 75% predictive when compared to surgically confirmed herniated lumbar discs. In the same studies, the myelogram was approximately 84% predictive. When the abnormal low back patterns were combined with a study of the extremities, the thermogram accuracy was raised to 90% compared to the myelogram, in surgically confirmed cases.[5]

When the lumbar patterns were combined with a study of the extremities, and the procedure compared to the EMG, and both procedures compared to objective physical examination findings, the thermogram was 93% accurate as opposed to the EMG being 83% accurate (Table 2). There was a 77% direct correlation between the EMG and the thermogram positive or negative. When one examines the sensitivity and specificity of the EMG and the thermogram in this study, one discovers that they

Table 1. Upper extremity anatomic variants

1.	1° centrigrade decrease in temperature affecting the lateral non-dominant forearm (radial aspect) affecting almost 100% (C6).
2.	1° centigrade decrease in triceps in posterior non-dominant arm about 50% of time; 25% in lateral non-dominant arm; rarely in anterior non-dominant arm.
3.	Focal area of cold at T_1 spinous process.

were almost exactly complimentary procedures. It is a moot point as to whether the thermogram "false positive" examinations were not, in fact, clinical "false negative" examinations. The thermogram can detect and graphically demonstrate irritation of, or damage to, a sensory nerve (Fig 1). As such, it provides unique information unobtainable by any other test. This explains why the thermogram may be positive, while all other tests may be negative. For the sake of comparing it to something, such tests as the myelogram and EMG are used, but they are at best only partial standards. Because of the unique information it provides, the thermogram is its own standard, with the patient being his own control.

The question of whether hypothermia along the course of the dermatome actually represents pain is often raised. It may or may not. If the finding is consistent, and has a definitive anatomic correlation, and/or the patient is symptomatic at that derematome level, regardless of which side may or may not be the most prominent clinically, one has to then assume that the finding is a valid one and does represent that symptom. Minimal findings should not be overlooked. The further they

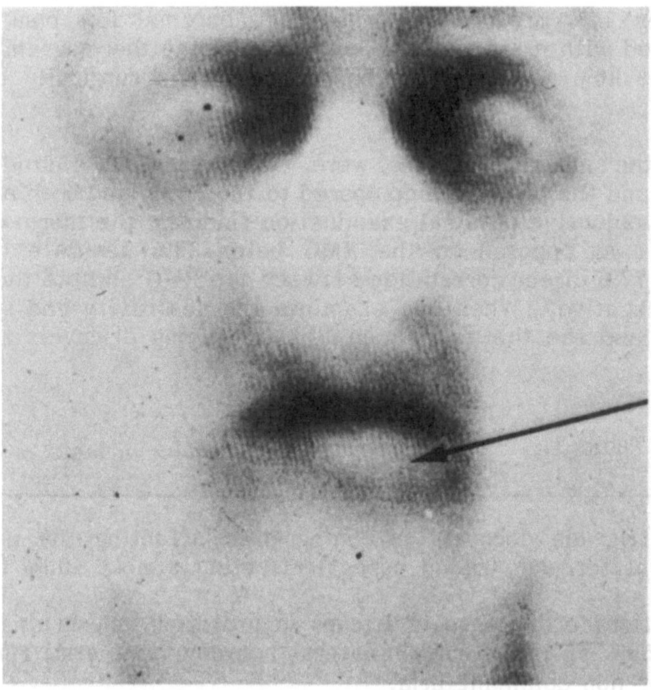

Fig 1. Inferior alveolar nerve injury within the mandible following surgery, with numbness and tingling of the left half of the lower lip. This appears cold on the thermogram.

are from "classic", however, the more one should look for supporting evidence for the abnormality. Proven abnormals with simply an abnormal back pattern and normal extremities or vice versa have been observed (Fig 2). Studies with positive EMG's and the thermogram findings manifested by simply a change in the buttock area and otherwise demonstrate a normal thermogram have also been observed. If there is no corroborating evidence for such minimal findings and they are thought to be, in fact, significant, serial studies would be helpful in evaluating their significance. The thermogram can react to nerve root irritation in only one manner - along the course of distribution of the nerve. The patient, however, may perceive that irritation in a variety of manners depending on his individual pain threshold, crossover pattern, etc. Conversely, if the thermogram is normal, there is probably no organic nerve root irritation or significant musculoligamentous spasm. In discussing "pain" and the thermogram one should confine general comment

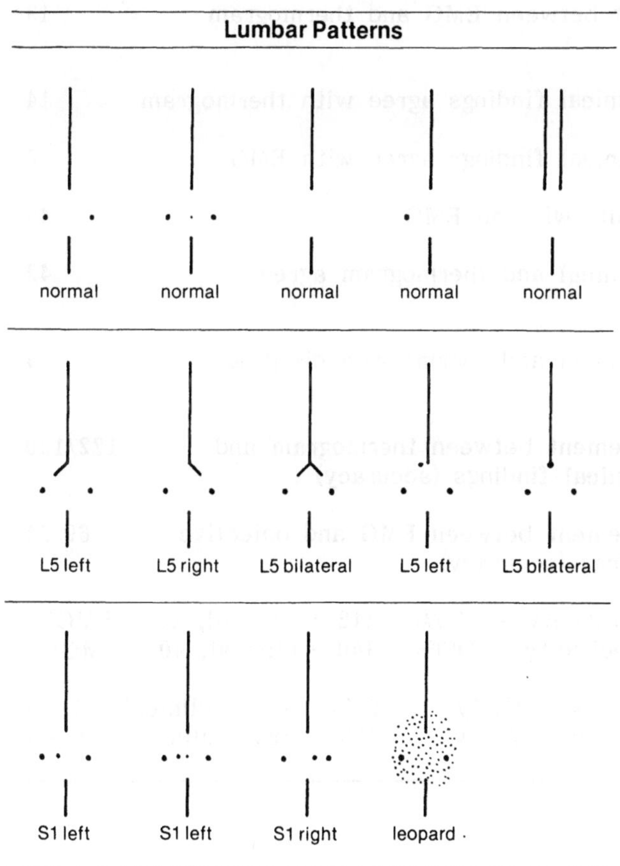

Fig 2. Patterns of heat distribution over the lumbar spine.

Table 2. An evaluation of 130 consecutive patients referred over a 29 month period by a single orthopedist. The electromyelogram (EMG) studies in this series were done by a single neurologist and his associate thereby providing a consistency within the series. A total of 216 cervical and/or thoracic and/or lumbar thermograms were performed.

ANALYSIS OF THERMOGRAM CASES	NO OF CASES	
No. of patients	130	
No. of patients who had EMG (Elkectromyogram)	84	
Agreement between EMG and thermogram	65	(77% - 35 neg,30 pos)
Disagreement between EMG and thermogram	19	(therm +, EMG -)
Objective clinical findings agree with thermogram	14	
Objective clinical findings agree with EMG	5	
No. of patients with no EMG	46	
Objective clinical and thermogram agree	42	(24 neg 18 pos)
Objective clinical and thermogram disagree	4	(therm + cl -)
Overall agreement between thermogram and objective clinical findings (accuracy)	122/130	(93%)
Overall agreement between EMG and objective clinical findings (accuracy)	69/84	(82%)

EMG -	sensitivity -	70%	(42 + clinical, 29 + EMG)	
	specificity -	100%	(40 - clinical, 40 - EMG)	
Thermogram -	sensitivity -	100%	(61 + clinical, 61 - thermogram)	
	specificity -	88%	(69 - clinical, 61 - thermogram)	

to these two entities or otherwise define comments as relating to other types of abnormality. Bone tumors, primary or secondary, for instance, can produce pain, yet may give a normal thermogram.

The thermogram thus provides a graphic complimentary perspective to the EMG and to X-ray evaluation of pain and/or nerve irritation. Its statistical accuracy is comparable with these procedures and it provides a perspective not obtainable by them. Furthermore, it is totally harmless and painless.

REFERENCES

1. J. Edeiken, Thermography and herniated lumbar discs, Am. J. Roentgenol. Radium Ther. Nucl. Med. 102:790 (1968).
2. M. A. Raskin, M. Martinez-Lopez, J. J. Sheldon, Lumbar thermography in discogenic disease, Radiology 119:149-152 (1976).
3. An Overview of Liquid Crystal and Electronic Lumbar, Thoracic and Cervical Thermography. TSI 19135 Rosita Street, Tarzana, CA 91356, USA
4. C. Wexler, Thermographic evaluation of trauma (spine), Acta Thermographica 5;1:3-10 (1980).
5. R. Pochaczevsky, C. E. Wexler, P. Meyers, J. Marc, J. Epstein, Liquid crystal thermography of the spine and extremities - its value in the diagnosis of spinal root syndromes, J Neurosurgery (1982).

LIQUID CRYSTAL THERMOGRAPHY OF THE SPINE AND EXTREMITIES - ITS VALUE IN THE DIAGNOSIS OF SPINAL ROOT SYNDROMES

R Pochaczevsky*, C E Wexler**, P H Meyers+,
J A Epstein*, J A Marc*

* Long Island Jewish-Hillside Medical Center, New York
** Encino, California
+ New Orleans, Louisiana, USA

The liquid crystals utilized for thermography (LCT) are cholesterol derivatives which selectively reflect polarized light in a narrow region of wave lengths. They have strong molecular rotatory power and their molecular arrangement is altered by temperature changes. The latter property results in specific color-temperature responses which are utilized in color thermography.[1,2] In the past, the adaptation of liquid crystals to thermography had been hampered by the necessity of skin preparation with black, water-based paint in the form of a spray, prior to the actual application of liquid crystals to the skin.[1,3] Rigid plastic plates subsequently replaced skin preparation and spraying, but the unyielding plates precluded uniform liquid crystal-skin contact, particularly when applied to the spine or extremities. The resultant thermograms were, therefore, inadequate.

A new thermographic technique for the examination of the breasts utilizing liquid crystals embedded in elastomeric sheaths has been previously reported.[4,5] A modification of this technique has been specifically adapted for contact thermography of the spine and extremities and is described in this paper. The currently-reported adaptation also employs flexible, elastomeric, plastic sheaths containing thermally sensitive liquid crystals.

APPARATUS

The apparatus consists of an hermetically sealed 37 x 28 x 3.5 cm box, one side of which has a transparent plastic wall, while the other is composed of liquid crystals embedded in an elastomeric sheath. Air pumped into it balloons the liquid crystal sheaths, which form a convex, bulging surface. The inflated 'air pillow' is then ready to be used for contact thermography.

Six or more 'air pillow' boxes are available, progressively numbered 26-35 to correspond with tne median Celsius temperature ranges of their incorporated liquid crystal sheaths. Liquid crystals have accurate and reliable color resonses to specific temperature changes (table 1). The lowest temperature is displayed as a dark brown color. This changes with progressive temperature elevation from dark brown to tan, reddish brown, yellow, green, light blue and dark blue. The highest temperature is indicated by a dark blue color.

PROCEDURE

The skin temperature of the back is stabilized by sponging with water and cooling for 10 minutes. A hair dryer set on 'cool' can further expedite skin drying and cooling. Patients should refrain from smoking on the day of the examination since smoking may affect skin temperature.[6]

Routine views of the lumbosacral region and lower extremities consisted of images of the lumbar region, buttocks, both posterior thighs, both lateral thighs, both anterior thighs, the dorsal and plantar aspects of the feet and toes, and the medial, lateral and anterior aspects of both legs and ankles. Routine images of the cervical spine and upper extremities included views of the posterior neck, both shoulders, posterior forearms, anterior forearms, ulnar aspects of the forearms, and the dorsal aspects of the hands and fingers. Abnormal thermographic images were

Table 1. Typical liquid crystal elastomeric temperature calibrations (°C)

No.	1 Dark Brown	2 Tan	3 Reddish Brown	4 Yellow	5 Green	6 Light Blue	7 Dark Blue
32	31.3	31.6	31.9	32.2	32.4	33.2	35.0
33	32.0	32.3	32.7	33.6	33.7	34.6	36.0
34	32.9	33.2	33.4	34.0	34.2	35.3	36.4

These color temperature calibrations permit simultaneous determinations of temperature differentials ($\triangle T^{\circ}$) in comparable regions of the spine and extremities.

repeated at least 3 times in succession to confirm their reliability. Special attention was given to the body dermatomes and myotomes.[7]

THEORETICAL AND ANATOMICAL BASES FOR THERMOGRAPHIC INTERPRETATION

A normal thermogram of the extremities shows symmetrical heat emission. Root lesions of the spine are associated with temperature changes in the corresponding dermatomes and myotomes. The temperature changes may relate to reflex sympathetic vasoconstriction within affected extremity dermatomes and metabolic changes or muscular spasm in corresponding paraspinal myotomes. As a rule, temperature changes appear as zones of hypothermia at the levels of the affected extremity dermatomes. Frequently, however, and particularly in the hands and feet, and for unknown reasons, hyperthermia at the level of the affected dermatomes occurs. Both hypothermic and hyperthermic reactions are abnormal since there should be no significant temperature difference between the extremities in normal individuals. The only exception to this is the occasional increase in temperature of the dominant arm and posterior forearm in very muscular males. In the spine, root compression syndromes usually show lumbo-sacral hyperthermia and cervico-thoracic hypothermia of the ipsilateral adjacent paraspinal myotomes. Musculo-ligamentous injuries of the spine or osseous lesions without root compression appear to have thermographic changes localized to the spine.[7] They are not necessarily associated with thermographic changes in the extremities.

Although there are frequent minor variations and nerve fiber crossovers, temperature fluctuations at particular dermatomes and myotomes reflect a fairly accurate picture of referred pain in root compression syndromes.[4] A normal thermogram of the spine is characterized by a central zone of increased heat emanation in the region of the spinal processes, from the cervical spine down to the level of the lower lumbar spine at the L4-5 level, where it tapers to a relatively cooler, diamond-shaped area - delimited inferiorly by the upper end of the intergluteal fold, and laterally by the sacroiliac joints.[9] A positive or abnormal thermogram will show evidence of increased heat production at myotomes in this zone or lateral to the midline along the spine.

CASE MATERIAL

Our material consisted of 100 patients; 31 were hospitalized prior to myelography for the evaluation of suspected spinal nerve root compression. Fifty examinations were performed in a private practice, simultaneously utilizing liquid crystal contact thermography (LCT) and conventional electronic thermography.

RESULTS

LCT of the spine and extremities showed good correlation with clinical findings. Concomitant myelography and surgery (table 2) were performed on 32 and 21 patients respectively.

COMPARISON WITH MYELOGRAPHY

LCT and myelography were in full agreement in 26 patients or 81% of cases. Surgery showed no false positives or negatives in this group of cases.

Surgery was performed on only one patient who had both a normal LCT and myelography. Although lateral spinal recess stenosis was found at L5, there was no evidence of root compression at surgery. This case is considered a true negative although it may have conceivably caused previous intermittent root compression.

LCT also showed no false negatives in our series while myelography was falsely negative in 4 patients. Myelography, on the other hand, showed no false positives.

In 6 cases in which LCT and myelography disagreed, LCT was fully correct in 4 cases and partially correct in 2 (2 levels involved) while myelography was fully correct in 2 cases and partially correct in 1 case.

Table 2. Root compression syndromes: surgical findings in 21 patients

L5 ROOT COMPRESSION DUE TO HERNIATED DISCS, FACET ENTRAPMENT, OSTEOPHYTE INPINGEMENT AND/OR SPINAL STENOSIS	15 CASES
S1 ROOT COMPRESSION DUE TO HERNIATED DISCS, FACET ENTRAPMENT, OSTEOPHYTE INPINGEMENT AND/OR SPINAL STENOSIS	4 CASES
NEUROFIBROMA - S1 ROOT COMPRESSION	1 CASE
EXTRADURAL METASTASIS - S1 ROOT COMPRESSION	2 CASES
EPENDYMOMA - L3 ROOT COMPRESSION	1 CASE
SYRINGOMYELIA - C3-C7	1 CASE
L4 ROOT COMPRESSION DUE TO SPINAL STENOSIS AND FACET ENTRAPMENT	2 CASES

If only completely correct, surgically proven results were tabulated, the overall accuracy of LCT was 90% and of myelography 81% (table 3).

In 50 patients, both LCT and conventional telethermography were performed with almost identical results.

DISCUSSION

Uematsu and Long[10] reviewed thermograms of 101 cases with chronic pain in a variety of locations of the body. There were 17 abnormal thermograms, none of which included the myotomes of the back. Of these 17 thermograms, all but one demonstrated coldness in the area of pain. Potanin,[11] using liquid crystals, demonstrated coldness of the skin corresponding to the distribution of pain during attacks of experimentally stress-induced angina pectoris.

Tichauer,[12] in his series of patients with backache, found asymmetrical warm areas in the paralumbar area and cold gluteal patches in a high percentage of cases.

Thermography was found to be accurate in predicting disc herniation in 80% of a reported series of 29 surgically proven cases. Its accuracy was comparable to that of myelography. When both thermography and myelography were positive, the probability of negative findings at surgery became much smaller.[8] Accordingly, in our series there were no false positives found at surgery when LCT and myelography were in agreement (table 3).

Raskin et al,[9] in a report based on 24 patients in whom herniated discs were found at surgery, noted that thermography was accurate in 71% and myelography in 88% of the series.

DePalma[13] has estimated the incidence of error in myelography to range from 20% to 30%. This is due to the fact that herniated or extruded disc material may be small in volume and/or may lie far to the side in the lateral recess of the spinal canal without significant contact with the dural sac. Myelography may be unremarkable in cases with lateral recess stenosis or superior facet entrapment. A negative myelogram was noted in 2 of our patients with these conditions (table 3). The fact that myelography may not detect all clinically significant lesions was illustrated in one case in our series in which LCT showed evidence of a left L5 and a left S1 root compression syndrome while myelography solely demonstrated a left L4-L5 defect. Surgery revealed spinal stenosis, an herniated disc at the L4-L5 interspace and an osteophyte impinging on and compressing the left S1 root at the L5-S1 interspace. Myelography may, furthermore, be falsely negative or falsely positive in asymptomatic patients.[13] In our series, there were 4 patients with myelograms which were negative or questionable while LCT on the same patients showed evidence of definite localized root compression syndromes. These LCT

Table 3. LCT and myelography comparison: results in 21 patients

	LCT	Myelography
TRUE POSITIVE	18	16
TRUE NEGATIVE	1	1
FALSE POSITIVE	2	0
FALSE NEGATIVE	0	4
TOTAL	21	21
ACCURACY	90%	81%

findings were surgically confirmed in all 4 patients (table 3). Myelography is, in addition, an invasive examination.

Computerized tomography is highly useful in spinal canal stenosis[14] and appears to show promise in the detection of herniated discs and other abnormalities. However, it may be misleadingly negative, especially in cases of musculo-ligamentous injuries which are positive on thermography.[7] Both myelography and computerized tomography are more expensive examinations which additionally involve radiation exposure.

ROOT SYNDROME PATTERNS

LCT appears to be particularly well suited for the diagnosis of root compression syndromes. Recurrent, typical heat patterns have been observed in this preliminary study which serve to delineate and associate affected myotomes and dermatomes.

Sacral Compression

S1 root involvement is associated with localized increased heat emanation from the affected side of the lumbosacral region. The ipsilateral posterior thigh will, in most cases, show hypothermia along with hypo or hyperthermia along the lateral aspects of the foot and fifth toe. The medial ipsilateral buttock may occasionally show hypothermia.

Lumbar Compression

L5 root involvement is usually characterized by hyperthermia in the corresponding side of the lumbar spine, usually radiating laterally, hypothermia of the ipsilateral midgluteal region and outer aspects of the thigh, and hypothermia or hyperthermia of the dorsum of the foot and first, second, third and fourth toes.

L4 root involvement is indicated by hyperthermia of the lateral aspect of the lumbar spine, hypothermia of the ipsilateral lower anterior thigh and medial aspects of the leg, ankle and foot including the first toe.

Cervical Compression

C7 root involvement is associated with hypothermia of the ipsilateral posterior cervical spine muscles, posterior shoulder and dorsal aspect of the forearm. Hyperthermia or hypothermia of the ipsilateral second and third fingers are also seen in most cases.

C8 root involvement is associated with hypothermia of the ipsilateral posterior cervical spine muscles, posterior shoulder and ulnar aspects of the forearms. Hyperthermia of the ipsilateral fourth and fifth fingers are frequently noted.

It is our opinion that accuracy in diagnosing spinal root compression syndromes was greatly improved by including a study of extremity dermatomes with the thermographic evaluation of the spine.

DIFFERENTIAL DIAGNOSIS

The thermographic differential diagnosis of spinal root compression syndromes includes deep venous thrombosis,[15] ischemic arterial disease, local trauma and arthropathies of extremity joints. They can be distinguished from spinal root compression syndromes in most cases by their distinct clinical pictures and by the failure of thermographic findings to be confined to a definite skin dermatome distribution. Varicose veins can be distinguished by their serpiginous course on the thermograms.[5,6]

In the upper extremities, the differential diagnosis includes carpal tunnel syndrome with median nerve irritation and the ulnar nerve entrapment syndrome. In median nerve involvement, there may be a hot thumb, hot second and third fingers and radial side of the fourth finger. The ulnar nerve entrapment syndrome is usually associated with a cold dorsal forearm and a hot fourth and fifth finger. In leprosy, hypothermic areas, as noted on thermography, are the sites of most marked sensory loss.[16]

CONCLUSION

Despite the fact that the current study is based on a limited number of cases, preliminary results indicate that the accuracy of LCT may be comparable to or better than that of myelography, provided that the extremity dermatomes are simultaneously included in the basic examination (table 3). LCT is a non-invasive technique which appears to have high sensitivity. It could, therefore, be valuable in the selection of patients in whom myelography may be of benefit. It may also be helpful in evaluating patients with negative or questionable myelograms.

The technique additionally has applications in a wide variety of pathologic processes related to the extremities and peripheral joints as well as to the spine.

ACKNOWLEDGMENTS

The technical contribution of Mr Seymour Katz is gratefully acknowledged.

REFERENCES

1. R. Pochaczevsky, P. H. Meyers, Vacuum contoured, liquid crystal, dynamic breast thermoangiography as an aid to mammography in the detection of breast cancer, Clin. Radiol. 30:405-411 (1979).
2. R. Pochaczevsky, P. H. Meyers, The value of vacuum contoured, liquid crystal dynamic breast thermoangiography, Acta Thermographica 4:8-16 (1979).
3. J. Gershon-Cohen, A. G. B. Borden, M. B. Hermal, Thermography of extremities after smoking, Brit. J. Radiol. 42:189-191 (1969).
4. J. J. Keegan, F. D. Garrett, The segmental distribution of the cutaneous nerves in the limbs of man, Anatom. Rec. 102:409-437 (1948).
5. C. Ching, C. E. Wexler, Peripheral thermographic manifestations of lumbar disc disease, Appl. Radiol. pp. 53-58 (1978).
6. C. E. Wexler, Lumbar, thoracic and cervical thermography, J. Neurol. Orthop. Surg. 1:37-41 (1979).
7. H. Karpman, A. Knebel, C. J. Semel, J. Cooper, Clinical studies - application of thermography in evaluating musculo-ligamentous injuries of the spine - a preliminary report, Arch. Environ. Health 20:412-417 (1970).
8. J. Edeiken, J. D. Wallace, R. F. Curley, Thermography and herniated lumbar discs, Am. J. Roent. 102:790-796 (1968).
9. M. R. Raskin, M. Martin-Lopez, J. J. Sheldon, Lumbar thermography in discogenic diseases, Radiol. 119:149-152 (1976).
10. S. Uematsu, D. Long, Thermography in chronic pain, in: "Medical Thermography, Theory and Clinical Applications," Brentwood Publising Corp., Los Angeles, pp. 52-67 (1976).

11. C. Potanin, D. Hunt, L Sheffield, Thermographic patterns of angina pectoris, Circulation 42:199-204 (1970).
12. E. R. Tichauer, The objective corroboration of back pain through thermography, J. Occup. Med. 19:727-731 (1977).
13. A. F. DePalma, R. H. Rothman, in: "The Intervertebral Disc," W. B. Saunders, Philadelphia, pp. 241-247 (1970).
14. I. Ciric, M. A. Mikhael, J. A. Tarkington, N. A. Vick, The lateral recess syndrome: a variant of spinal stenosis, J. Neurosurg. 53:433-443 (1980).
15. W. G. M. Ritchie, M. S. Lapayowker, R. L. Soulen, Thermographic diagnosis of deep thrombosis: anatomically based diagnostic criteria, Radiol. 132:321-329 (1979).
16. T. D. Sobin, Temperature-linked sensory loss: a unique pattern in leprosy, Arch. Neurol. 20:257-262 (1969).

DIFFERENTIAL DIAGNOSIS OF PAIN USING THERMOGRAPHY

W B Hobbins

Wisconsin Breast Cancer Detection Foundation
6818 Seybold Road, Madison
Wisconsin 53719, USA

Before the two subjects can be discussed, certain statements must be made. First, telethermography of the body does not record infrared emission deeper than 0.5 cm having been brought to the surface by vasculature. This vascular bed is under the direct control of the sympathetic nervous system. Therefore, areas of both increased and decreased emissions can be correlated to the nervous system.

Second, "pain" is a perception of a hurt and is a "feeling". Objective evidence of "pain" has always been difficult to demonstrate, but subjective observation of paresthesia in the perceived somatic zone of pain has been recorded.

Having made these statements, we must look at the correlations of the superficial vasculature and its relationship to the pain pathways.

The afferent sympathetic nervous system, which is associated to the smooth muscle of blood vessels, glands, and organs, will be interrelated to thermography. These afferent nerves exist in the extremities in the peripheral nerve, as well as with blood vessels, and thus form the alarm system.

The visceral afferent fibers are, for the most part, carried in to the CNS axis by the vagus, pelvic, splanchnic, and other autonomic nerves. Other autonomic afferents from blood vessels in skeletal muscles and from certain integremental structures are carried in the somatic nerves.

The autonomic afferent fibers are concerned with the mediation of visceral sensation (opposite of "environmental" perception). There are no important physiological differences between visceral and somatic afferent fibers.

Mayer[1] states that there are probably no purely autonomic or somatic centers of integration, and extensive overlap occurs. Somatic responses are always accompanied by visceral responses and vice versa.

The control of the blood vessels of the surface of the body is under the post-ganglionic adrenergic fibers and the post-ganglionic cholingergic fibers. Both of these fibers are carried back to the spinal nerves.

Parasympathetic afferents, such as those in the vagus nerve, do not carry pain. Thus, the autonomic nerves associated with pain are the sympathetic afferents. This is important to keep in mind as well when considering monitoring the infrared emission in relationship to the nervous system and pain.

Three major types of pain can be differentiated and demonstrated by thermography. They are:

1. Visceral afferent
2. Peripheral nerve
3. Spinal cord pain

In appraising the body with thermography, a general statement can be made. Normal physiological observations are usually symmetrical. Examination of the head and neck, thorax, back, abdomen, and extremities should reveal similar patterns contralaterally with equivalent delta Ts. Therefore, before any interpretation of abnormal thermograms can be made, the normal and expected patterns must be understood and recognized.

It should be noted that cutaneous pathology, such as inflammations, tumors, (benign and malignant) and trauma will show characteristic thermographic alterations. As thermograms are measurements of blood flow, as well as vascular thermal content, these conditions influence these observations.

Exceptions to all of these statements thus far is noted where the skin is attached directly to the underlying structures. The knees and phalangeal joints are good examples.

All the discussed influences on the infrared image must be taken into consideration in every study. Changes thought to be due to pain must be subtracted from these changes.

In addition, a further understanding that every area of body surface connects to segments of the spinal cord via the sympathetic nervous system must be known. It is very important, also, to remember that these segmental relationships are different from the cutaneous sensory distribution. This is important when mapping out referred pain or pain distribution as to etiology. Dr LeRoy has given these areas the term "thermatomes".

Thus, we have explained the neurophysiologic basis of the observation of this paper. Since 1972 at the Christian Holistic Health Center, some 500 cases of various pains have been observed and classified.

VISCERAL AFFERENT (REFERRED) PAIN

It is our observation that referred pain shows increased thermal response in the segmental dermatome of referral (thermatome). These responses correspond to the visceral referral area. Examples of commonly understood visceral referred pains are subscapular pain in gall bladder colic to associated dermal area and chest pain in pulmonary embolism. The thermal response in referred pain is that of vasodilation.

RADICULAR PAIN (PERIPHERAL NERVE)

This has been observed as an increased thermal response to the area of distribution and pain area. This corresponds with the peripheral nerve dermatome of the affected nerve. This differentiates it from a visceral (referred) pain. This is an efferent nerve response.

When the peripheral nerve afferent receptor is stimulated, the defensive response through the sympathetic nerves in a large number of these patients is sympathetic dystrophy.

Patients with dystrophic pain were evaluated with thermography pre- and post-sympathectomy. Preoperative patients showed a definite vasoconstrictive pattern with an average of 2 degree decrease in temperature in the affected extremity or area.

CORDAL PAIN

Cord compression and/or transection causes a vasoconstrictive pattern to occur. This is segmental at the area of the cord problem. M Kondo[2] has shown that in the early stages of complete damage of cervical cord, local and finger temperature does not change when immersed into cold water.

Observation at our clinic of cord compression in the upper lumbar area shows there is a segmental cooling at that level and normal thermal recovery after relief of the compression. Lechtape[3] was unable to demonstrate the cordal findings but noted that the reported response of proximally increased heat with caudal temperature reductions at the level of compression. Recovery reverses most of these findings.

FINAL OBSERVATIONS

In all our observations, the most important factor for accurate evaluation of pain is evident; an absolute one is that the pain must be present at the time of the examination. When not present, the resultant study is considered a baseline. Then when the pain is present, the resultant change has greater localizing significance. This is also true of improvement as well as observation of exacerbation of the pain syndrome.

CONCLUSION

The thermal dermal responses (thermatomes) in the presence of pain can be a help in the differential diagnosis of the origin of a pain. Abnormal autonomic nervous system stimulation will change the skin segmental thermal picture. Thermography can then record the pain graphically and, also, any improvement. The pain must be present at the time of the examination, or a false negative response may be recorded.

Visceral dysfunction, no matter the cause, will evidence itself through the autonomic nervous system producing pain. Each organ has its own referred area, and thermography now allows the recording graphically of each referral area.

REFERENCES

1. S. E. Mayer, Neurohumoral transmission of the autonomic nervous system. in: "The Pharmacological Basis of Therapeutics," Goodman Gilman ed., MacMillan, 6th:56-62 (1980).
2. M. Kondo, A cold water immersion test after nerve injuries, Proceedings of the 2nd Cong. Thermo. Euro. 183 (1978).
3. H. Einsiedel-Lechtape, Thermographic studies of the normal back and of spinal lesions, Acta Thermographica 2;2:117-128 (1977).

THERMOGRAPHIC DIAGNOSIS OF LOW BACK PAIN:- RELATIONSHIP OF LOW BACK PAIN TO THERMOGRAPHIC CHANGES IN THE LEG

K Kawai*, S Takehiro*, S Sadamitsu*, A Tanaka**, Y Matsuo**

* The Kanto Teishin Hospital
** Fujitsu Ltd.
Tokyo, JAPAN

INTRODUCTION

We observe many orthopedic patients suffering from low back pain. Though different techniques are employed for diagnosis, many of these cases are labelled "indeterminate", as conventional methods make it difficult to examine the causes of complaints. Cases of low back pain are therefore extremely complicated and it is difficult to clarify the relationship between the cause and the resulting problem.

To solve these problems, we have studied the possible auxiliary application of thermography to ensure adequate diagnostic and therapeutic treatment for low back pain.

TARGET AND METHOD OF RESEARCH

Using an infrared camera unit (Fujitsu INFRA EYE 150), we observed differences in the temperature of the skin of the lower limbs of a total of 651 persons. The subjects included 506 patients complaining principally of low back pain and 145 control subjects (Table 1). Our observations were conducted in the normal environments of orthopedic consultation rooms. The temperature of the lower limbs was measured at the points where the differences were the greatest.

EVALUATION OF TEMPERATURE DIFFERENCES ON LOWER LIMBS

Table 1 shows the temperature differences on the lower limbs of each subject. The following standards of evaluation were established.

Table 1. Relationship between temperature of the skin of the leg and disease

Diseases \ Temperature differences	Average difference	(−)	(±)	(+)	Total (persons)
A Lumbar disc herniation	1.43 ℃	4%	15%	81%	49
B Spondylosis	0.95 ℃	29%	31%	40%	163
C Spondylolysis and-olisthesis	0.83 ℃	33%	28%	39%	34
D Acute low back pain	0.75 ℃	44%	28%	28%	43
E Indeterminate low back pain	0.61 ℃	52%	36%	12%	217
F Sub-total for patient group	0.82 ℃	40%	30%	30%	506
G Control	0.35 ℃	77%	22%	1%	145
H Total (persons)	0.72 ℃	315	184	152	651

Negative Evaluation (-)

The average temperature difference of the control group was 0.35°C (approx 0.4°C), and that of "indeterminate low back pain" patients was 0.61°C (approx 0.6°C). A negative evaluation was therefore given to a temperature difference of 0.5°C or less, where 0.5°C is the average temperature difference between the two groups.

Positive Evaluation (+)

The average temperature difference for patients suffering from lumbar intervertebral disc herniation was as high as 1.43°C and that for spondylosis patients was 0.95°C (approx 1°C). A temperature difference of 1°C or over was considered positive.

Equivocal Evaluation (±)

Positive-negative evaluations were therefore given to temperature differences between 0.6°C to 0.9°C.

RESULTS

Temperature differences vary considerably according to the type of disease. Negative evaluations resulted for 77 percent of the control group and 40 percent of the patients on average, but only 4 percent of those suffering from lumbar intervertebral disc herniation. Positive evaluations resulted for only 1 percent of the control group, 30 percent of the patients on average and 81 percent of herniation patients.

Ninety percent of the patients in the negative group did not suffer from radiated pain, while 86 percent of the patients suffering from radiated pain belong to the positive-negative and positive groups.

When only lumbar interevertebral disc herniation was observed, the correlation was 96.6 percent between the skin temperature of the afflicted and the contralateral side.

In Lasegue's test, only 2 percent of the subjects in the negative group showed a positive reaction below 60° flexion. In the same test, 69 percent of the patients showed negative reactions.

As many as 25 percent of the persons in the positive group showed a reaction below 60°, while 62 percent of all subjects showed a positive reaction.

Those features of the low temperatures on the radiated pain side and the positive side represented in the Lasegue' test indicate reduced blood circulation on these sides.[1]

A CASE STUDY OF LUMBAR INTERVERTEBRAL DISC HERNIATION

The patient suffered from severe low back pain and left sciatica. A myelogram demonstrated a filling defect in the left side at the lumbar 4th-5th level as in Figure 1. A thermogram of both legs showed lower temperatures in the left leg as in Figure 2.

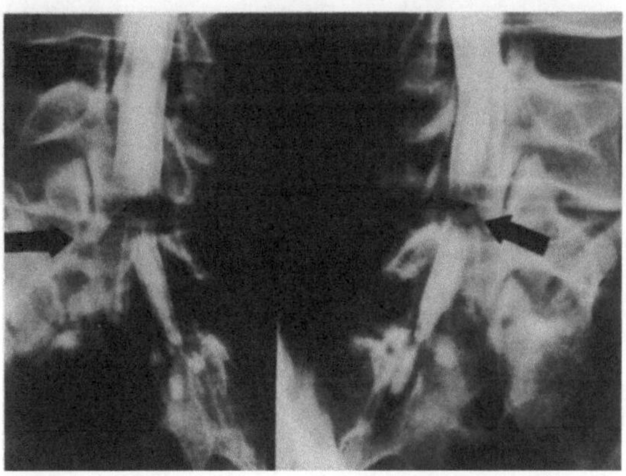

Fig. 1. Myelogram showing a filling defect at the left lumbar 4/5 level.

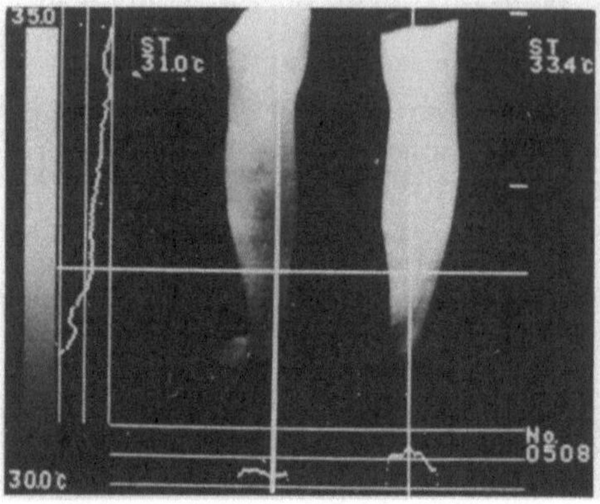

Fig. 2. Thermogram showing lower temperatures on left leg.

Love's removal was performed for L4/5 disc herniation. Two months after the operation, the patient had no complaints and showed no clinical symptoms. Subsequent thermograms do not indicate temperature variations between the legs (Fig 3).

These facts indicate a reduction in temperature differences between the legs as clinical symptoms disappear.

CONCLUSIONS

We studied the possible application of thermography as an auxiliary means to ensure accurate diagnosis and adequate therapeutic treatment

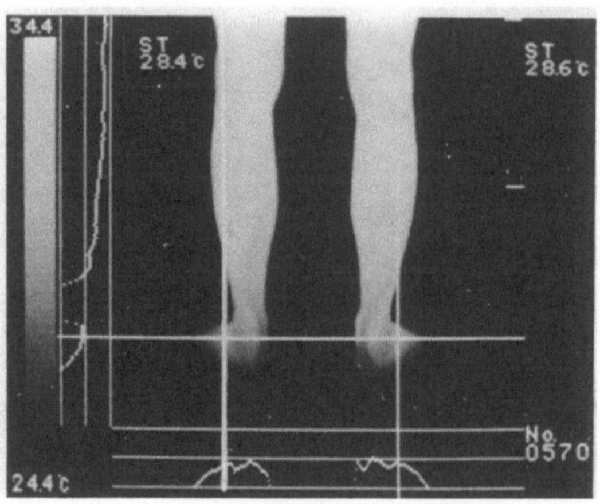

Fig. 3. Post-operative thermogram showing equal lower leg temperatures.

Table 2. Relationship between radiated pain in the leg and temperature difference

Groups	a. (−) group (persons)	b. (±)(+) groups (persons)	c. Sub-total (persons)	b/c
A Without radiated pain	182	170	352	48%
B With radiated pain	21	128	149	86%
C Disagreement between radiated pain side and lower temperature side	—	5	5	100%
D Sub-total	203	303	506	60%
E A/ D	90%	56%	70%	

Table 3. Relationship between temperature differences and Lasegue's test

Temperature difference / Lasègue' Test	(−) group	(±) group	(+) group	Total (persons)
Positive at 30°	1%	1%	12%	22
Positive at 60°	1%	5%	13%	29
Positive at 80° to 90°	29%	41%	37%	177
Sub-total of positive patients	31%	47%	62%	228 (45%)
Sub-total of patients without any reaction	69%	53%	38%	278 (55%)
Total (persons)	203 (100%)	152 (100%)	151 (100%)	506

for patients complaining principally of low back pain. Specifically, we examined the temperature differences of the skin of the lower limbs, and established standard variances in an attempt to obtain the relationship between low back pain and temperature differences on the lower limbs.

The study involved 506 patients and 145 control subjects. This alone is insufficient, however, and the standards for temperature differences still remain hypothetical. As illustrated in Tables 1, 2 and 3, a certain relationship does exist between cases (mainly of low back pain) and temperature differences on the lower limbs.

Our study has revealed that measurements of temperature differences on the skin of the lower limbs can contribute to the diagnosis of lumbar intervertebral disc herniation, and will assist in establishing criteria for choosing conservative treatment or surgery.

Thermographic temperature measurement is a non-invasive method of examination, which can be repeated without discomfort to the patient.[2] It provides an objective means for diagnosing altering conditions[3,4,5] and can be effectively employed in determining therapeutic methods.

REFERENCES

1. I. Fujimasa, Diagnosis of circulatory disturbances and thermography, Jap. J. of Clin. Med. 37:161 (1979).
2. A. H. Barrett, P. C. Myers, Subcutaneous temperature: a method of non-invasive sensing, Science 190:669 (1975).
3. K. Kawai, N. Yamada, Collective medical examination to prevent against cervico-scapulo-brachial syndrome, Teishin Igaku 30:681 (1978).
4. K. Kawai, S. Takehiro, S. Sadamitsu, A. Tanaka, Y. Matsuo, Thermographic changes of the finger on the cervico-brachial-syndrome. Pattern of thermic recovery before and after cold application, Teishin Igaku 33:381 (1981).
5. M. Shiotani, Y. Yuda, B. Wakasugi, Diagnosis and treatment in pain clinic, Teishin Igaku 28:527 (1976).

THERMOGRAPHY OF TENNIS ELBOW

A I Binder, G Parr, P Page Thomas, B L Hazleman

Rheumatology Research Unit
Addenbrooke's Hospital
Cambridge UNITED KINGDOM

SUMMARY

A constant thermographic abnormality in patients with tennis elbow is described. Serial change of this lesion with clinical improvement is considered. Methods of thermographic analysis of a region of interest using a microcomputer revealed gradient analysis to correlate well with clinical severity and accurately reflect recovery. Diurnal variation of temperature but not pattern was found, making temperature-based parameters less useful in analysis.

INTRODUCTION

Tennis elbow or lateral epicondylitis is a painful condition of the lateral elbow. While its clinical diagnosis is usually simple, its exact pathology remains uncertain. A degenerative lesion of the common extensor tendon, near its insertion into the lateral epicondyle of the humerus, is most widely accepted as the site of disease but other local pathologies have been considered. Local tenderness near the lateral epicondyle is an essential feature for diagnosis of this condition and a pilot thermographic study suggested a consistent localized abnormality near the lateral epicondyle, probable reflecting the area of pathology.

Aims of the Study

1. To confirm the presence of the thermographic abnormality previously noted and assess its incidence in patients and normal controls.
2. To investigate its changes with healing.
3. To assess different methods of analysis of the thermographic data

and look for correlations with clinical severity.
4. To note effects of diurnal variation.

Patients

Fifty patients with tennis elbow were included in a therapeutic study which included ultrasound, placebo (mock-insonation) and steroid injection therapy. Thirty three of the patients were female and 17 male. Mean age was 42.5 years. None of the patients had a history of injury and X-ray of the elbow was normal in all cases. Forty of the patients were seen within 3 months of onset of symptoms, the mean duration at presentation being 4.5 months (range 1-18 months).

Fifty normal age- and sex-matched controls (100 elbows) drawn from hospital staff and local industry were also thermographed for comparison purposes.

METHOD

AGA 680M thermograms of the lateral aspect of the elbow were taken after a 15 minute stabilization period in a room kept at a constant temperature of 20°C. All thermograms were taken at a distance of 1 meter using a 0.5°C sensitivity setting. Data were digitized in an OSCAR interface and recorded on tape. Numerical analysis of the gray levels in an area of interest (a constant sized box placed over the lateral epicondyle) was then made on an Apple II microcomputer.

Clinical assessment of severity included 8 parameters on examination, each judged from 0-3 in grades of increasing severity. Local tenderness, pain on resisted dorsiflexion of the wrist and middle finger and pain on resisted pronation and supination were considered. Grip strength with the elbow flexed and extended and a weight test assessing ability to lift weights of ½, 1, 2 and 4 pounds were included - the latter always being performed with the elbow extended and forearm pronated. By summating the score of individual parameters an overall clinical score was obtained.

Patients were reviewed initially and fortnightly during treatment and where possible this was continued until they were asymptomatic on 2 occasions.

For analysis of data the initial visit and the visit at 6 weeks (or on discharge if earlier) have been chosen. Eight of the patients and 4 controls had the thermographic assessments performed diurnally to study the variation during the course of the day.

RESULTS

48 of the 50 patients with tennis elbow lesions had a discrete area of increased temperature near the lateral epicondyle with a temperature gradient of 1-3°C above the background isotherm (Fig 1a). A similar abnormality only occurred in 3 of the control patients, the peak temperature never being more than 1°C above the background. All the other controls, and 3 patients with tennis elbow, showed a gradient across the elbow joint but no specific abnormality near the lateral epicondyle (Fig 1b). Contingency table analysis of patients and controls, with and without the thermographic abnormality, showed a highly significant correlation between patients and the thermographic lesions ($p < 0.001$). With clinical improvement the abnormal area became less clearly demarcated, often increasing in size and merging with muscle heat before returning to normal. The type of therapy given did not appear to substantially alter these changes.

Thermographic analysis on a microcomputer showed the gradient across the area of interest to have a good correlation with clinical score and accurately reflect recovery. Figure 2 shows this relationship between the clinical score and gradient. Temperature-based parameters tended to fluctuate diurnally and were not as useful. Of the 8 patients who were thermographed diurnally, 6 showed a rise in temperature - often as much as 1.5-2°C during the course of the day. This affected the background isotherms and to a lesser extent the abnormal areas. Similar

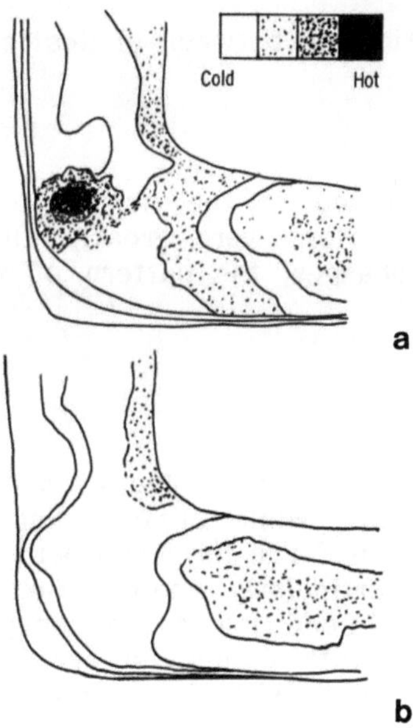

Fig. 1. (a) Temperature increase near epicondyle in tennis elbow.
(b) Normal gradient across the elbow.

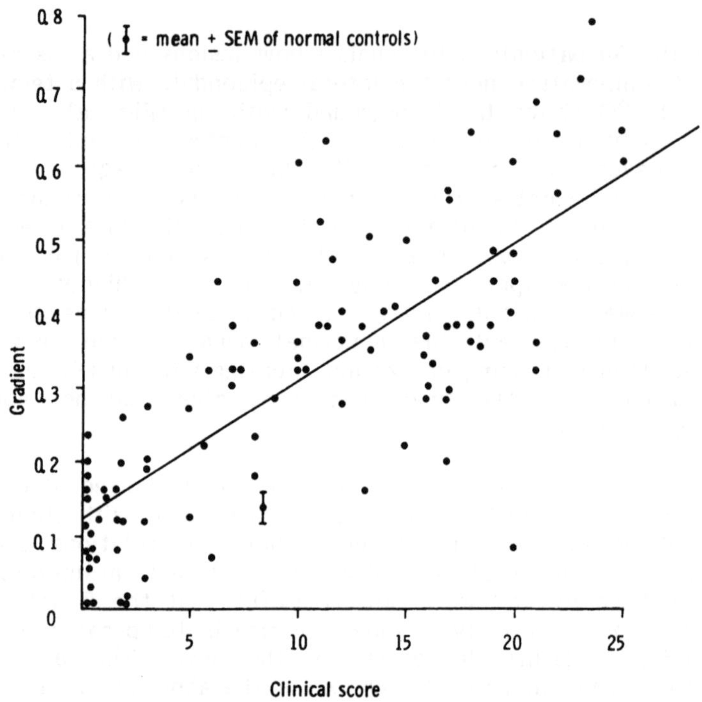

Fig. 2. Relationship between clinical score and gradient.

diurnal variation was noted in the normal controls. Although the temperature-based parameters were greatly altered, gradient analysis showed much smaller changes, the pattern of the abnormality usually staying the same.

DISCUSSION

Pain around the elbow can be due to many causes, tennis elbow being the most common. While clinical diagnosis is usually simple, thermography may prove useful in the small percentage of patients where diagnosis is uncertain, particularly where patients are seen in the early stages of the condition. In addition to confirming the diagnosis, gradient analysis correlated well with clinical severity and reflected recovery, showing much less fluctuation with time, ambient temperature and other variable factors.

Gradient analysis using a microcomputer proved useful in the assessment of this condition and may also be helpful in analysis of thermographic changes in other conditions.

We have confirmed the presence of a consistent thermographic abnormality in tennis elbow and discussed a method of analysis which has closely reflected the clinical state.

A FINGER OSTEOID OSTEOMA: A THERMOGRAPHIC INVESTIGATION

L Zorzin, M Capone, M G Maira

Institute of Rheumatology
University of Rome
ITALY

The osteoid osteoma described for the first time by Jaffe[1] is a benign tumor of the osteogenic connective tissue, which occurs in adolescent or young adults, and which affects in decreasing order, the long bones of the lower limbs, the bones of the upper limbs and the posterior arches of the vertebrae.[2,3]

It may occur in the spongy or the compact cortical section, and, according to the sites involved, the osteosclerotic action is more or less intense. In the case of the cortical site, there is an involvement of the endosteal and periosteal tissue with a fusiform diaphyseal aspect.

Histological examination of the connective tissue shows vascularization with both osteoblast and osteoclast cell elements, and trabeculae with osteoid tissue or bone.

The clinical picture is characterized by a slight bone swelling painful on pressure, but without exterior signs of inflammation and without involvement of the general condition. When the tumor is localized in the juxta-articular site there may also be articular inflammation with swelling and restricted function.

Traditional X-ray examination does not always permit the identification of the lesion (a small round X-ray transparent area, surrounded by osteosclerotic tissue). It may be necessary to perform a tomographic X-ray which clearly demonstrates the lesion.

CASE HISTORY

A case of osteoid osteoma, localized in a proximal phalanx of the finger, which because of its juxta-articular location was difficult to diagnose. A girl of 16 years presented with a painful swelling near the proximal interphalangeal joint of the 3rd finger of the right hand.

The symptoms had been present for 3 months with no history of injury and no response to antibiotic or anti-inflammatory therapy.

An X-ray of the hand (Figure 1) with tomographic examination (Figure 2) revealed a round formation inter-phalangeal with a transparent area surrounded by a tissue adhesion. Xerographic examination excluded involvement of the soft tissues (Figure 3) while thermographic examination (AGA 680 Thermovision) of the hand was normal (Figure 4).

The material for histological examination after biopsy (Figure 5) showed trabeculae of primary bone tissue with osteoblast cell elements (H E x 400), thus confirming an osteoid osteoma.

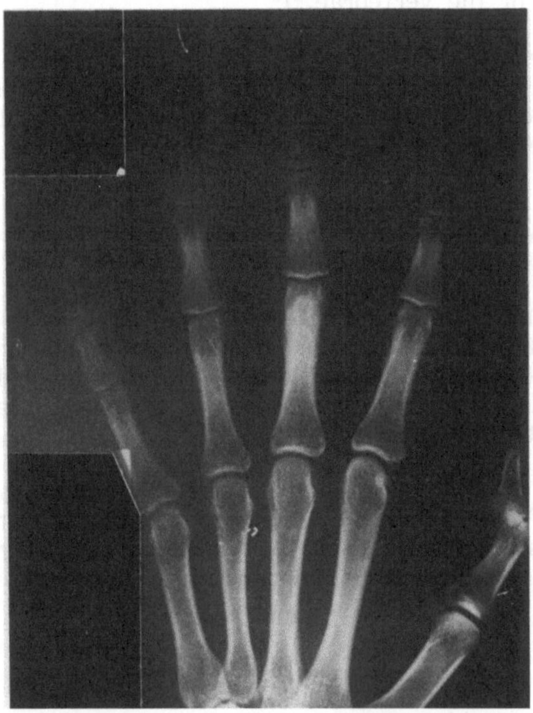

Fig. 1. X-ray of the hand.

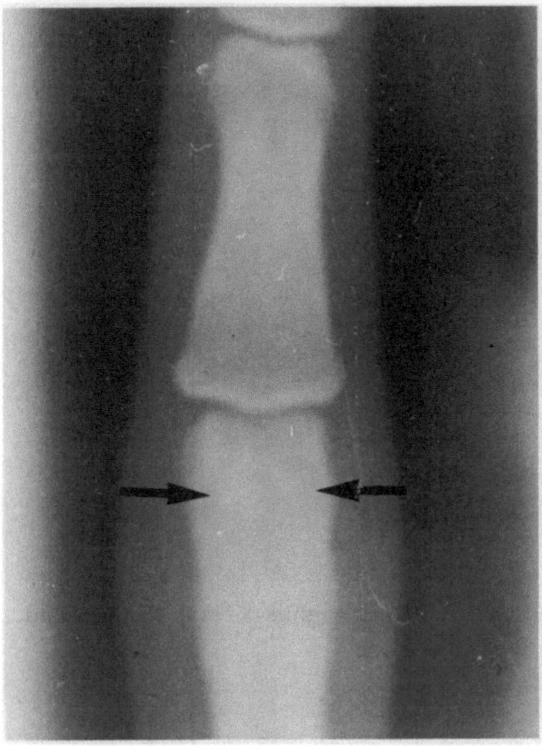

Fig. 2. Tomographic X-ray.

CONCLUSIONS

As already quoted in the literature,[4] the differential diagnosis of osteoid osteoma must be considered with chronic osteomyelitic processes and non suppurative sclerosing osteitis, and excluding inflammatory processes of the joint when the juxta-articular site is involved. The use of X-ray, xerography, and telethermography may exclude an inflammatory process, a malignant neoplastic or even chronic process.[5] Biopsy confirms the clinical diagnosis.

The use of integrated diagnostic techniques in this bone disease is demonstrated in which the thermographic investigation may have a useful role.

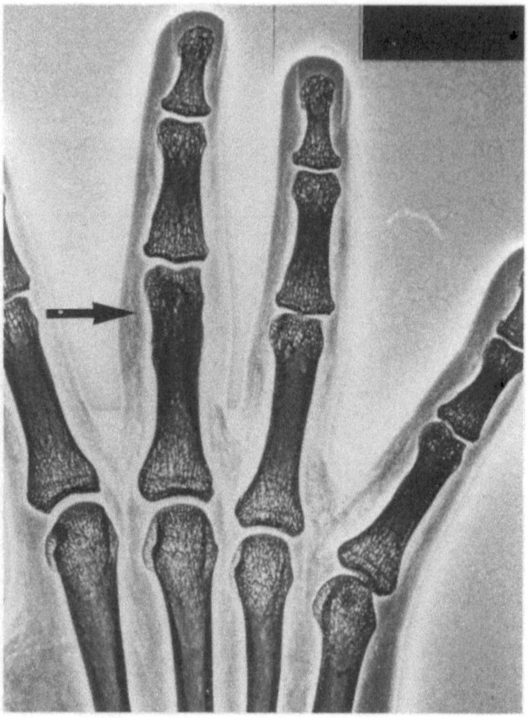

Fig. 3. Xerographic X-ray of the hand.

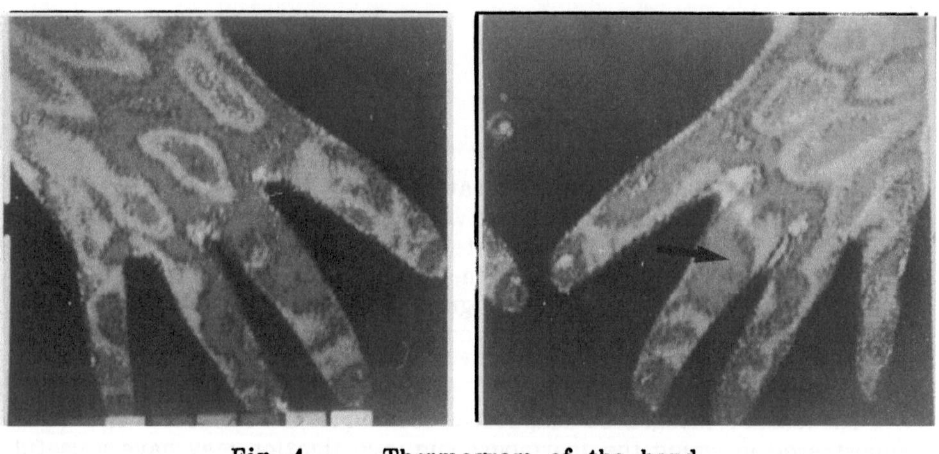

Fig. 4. Thermogram of the hand.

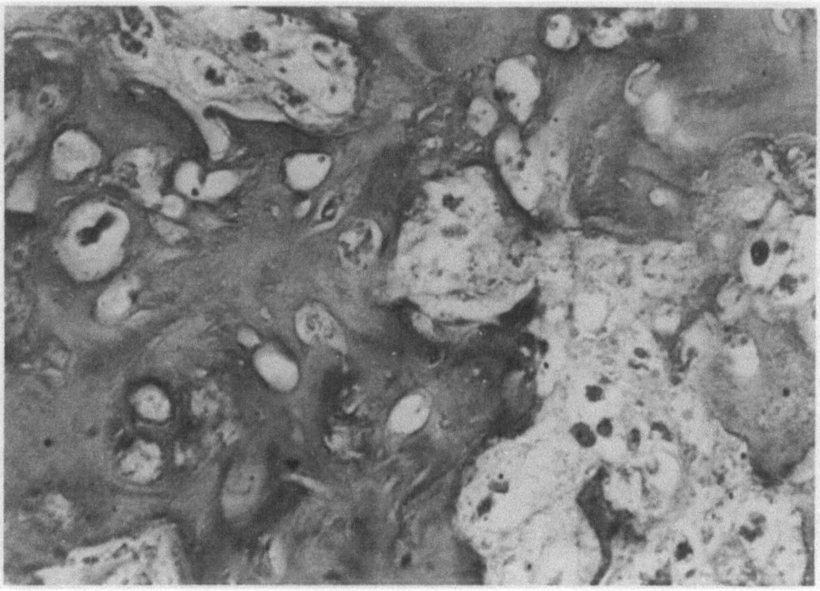

Fig. 5. Histology of the osteoid osteoma.

REFERENCES

1. H. L. Haffe, Osteoid osteoma. A benign osteoblastic tumor composed of osteoid and atypical bone, Arch. Surg. 31:709 (1953).
2. R. H. Freiberger, Osteoid osteoma of the spine, Radiology 75:232 (1960).
3. D. J. McCarty, Lesions of vertebral column, in: "Arthritis and allied conditions," Lea and Febiger ed., Philadelphia (1979).
4. S. DeSeze, A. Ryckewaert, Osteome osteoide, in: "Maladies des os et des Articulations," Flammarion ed., Paris (1954).
5. L. Lovisatti, F. Dalla Palma, G. Gortenuti, F. Marchi, La indagine termografica nella patologia ossea, in: "Termografia Medica," Gaggi ed., Bologna (1974).

INFRARED THERMOGRAPHY IN THE ASSESSMENT OF INTRADERMAL SKIN TESTS IN MAN

L Settas, G M Papadimitriou, D G Macfarlane, M Doherty, P A Dieppe, E F J Ring

Bristol Royal Infirmary, Bristol, and
Royal National Hospital for Rheumatic Diseases
Upper Borough Walls, Bath, UNITED KINGDOM

INTRODUCTION

Skin testing with crystalline material has been used previously to assess anti-inflammatory therapy, and to demonstrate the phlogistic properties of crystals.[1,2,3,4] In this paper the normal response to intradermal injections of urate, pyrophosphate and hydroxyapatite crystals is described, in normal control subjects and in patients with rheumatic diseases. Infrared thermography has been used for the assessment of the response.

METHODS

Crystals

All crystals were prepared in aqueous solution using sterile glassware. They were heated to 180°C for 3 hours prior to use to remove pyrogens, suspended in sterile saline, and briefly sonicated prior to use to reduce aggregation.

Monodosium urate monohydrate crystals were prepared by a modification of the method of Seegmiller et al,[5] and were needle-shaped, ranging from 2-10 µm long. Calcium pyrophosphate dihydrate crystals were made by the method of McCarty et al[6] and had a similar size range. Hydroxyapatite was manufactured by a modification of the method used by Nancollas et al[7]; clumping resulted in particles ranging from 0.01-2 µm in diameter. All preparations were examined by polarized light microscopy, and against standards in a Perkin-Elmer 501 infrared spectrophotometer for chemical purity.

Subjects

Local Ethical Committee approval was obtained for all studies. Normal control subjects included medical students and medical and laboratory staff. Patients included 15 with classical or definite rheumatoid arthritis, 15 with osteoarthritis, and 20 with the typical clinical and radiological features of pyrophosphate arthroapthy, 8 of whom suffered from attacks of 'pseudogout'.

Injections

The injections were made in alcohol-cleaned, normal forearm skin. 0.2 ml of a sterile suspension of crystals in normal saline was injected intra-dermally, using a 23 gauge needle, and a technique analogous to that used for Mantoux testing.

Measurements

In any one experiment, measurements were always made by the same observer. The diameter of erythema was measured in two planes at right angles, and the mean recorded. Skin thickness was measured with calipers; the difference between the readings immediately above and below the lesion, with that at its center was recorded. Skin temperature was recorded with a hand-held radiometer; difference between the temperature of the lesion and that of the surrounding skin was measured. Thermographic images of the lesion were also recorded.

RESULTS

Urate, pyrophosphate and hydroxyapatite crystals all caused an obvious inflammatory response, maximal between 24 and 48 hours after the injections. Before carrying out comparative studies, the dose-response curve and time course of the reactions was assessed in normal control subjects.

Dose Response and Specificity

The maximum erythema response to single injections of 1, 5 or 10 mg of urate, pyrophosphate or hydroxyapatite crystals in one subject are shown in Table 1. On the basis of these data a dose of 5 to 10 mg was considered appropriate, and used in all subsequent experiments.

The table also shows the lack of response to 10 mg doses of uric acid powder and amorphous calcium orthophosphate, suggesting that the response is "crystal-specific" as claimed by Freudweiler in 1899.[8]

Table 1. Dose-response curves for the erythema response to intradermal crystals and powders. All injections were given in the same subject in forearm skin, using the same technique and a suspension volume of 0.2 ml saline

	DOSE	DIAMETER OF ERYTHEMA (mms) (Mean of two recordings at right angles) 24 hrs after injection	48 hrs after injection
Monodosium Urate Monohydrate Crystals	1 mg	3.0	0.5
	5 mg	22.5	6.5
	10 mg	31.0	16.0
Calcium Pyrophosphate Dihydrate Crystals	1 mg	0.5	-
	5 mg	8.5	2.0
	10 mg	14.5	12.5
Hydroxyapatite Crystals	1 mg	1.0	-
	5 mg	5.0	1.0
	10 mg	12.5	13.0
	20 mg	21.5	26.5
Uric Acid Powder	10 mg	2.0	-
Amorphous Calcium Orthophosphate	10 mg	1.5	1.5

Time Course

The time course of the reactions to urate, hydroxyapatite and saline injections is shown in Figure 1. These data were collected from 12 normal control subjects. Individual variation in responses was small. All injections, including saline, caused an initial erythema; this faded, and in the case of the crystals only, a secondary response maximal 24 hours after injection, was seen. On the basis of these data, 24 and 48 hour time points were used in all subsequent experiments comparing responses in different patient groups.

Reproducibility of Measurements

The mean and standard deviation of the three measurements - erythema, induration and temperature - are shown in Table 2. This shows that in normal subjects there is a narrow range of response; the standard deviation was lowest for the diameter of erythema estimation, suggesting that this is as useful as the more sophisticated measurements. No large discrepancies between measurements in individual patients were seen. Two individuals had several injections, with no signfiicant change in the response to repeated stimuli.

Urate and Hydroxyapatite Injections in Normal Osteoarthritis and Rheumatoid Patients

Responses to monosodium urate monohydrate or hydroxyapatite crystal injections in the three groups are shown in Table 3. There was no evidence for reduced responsiveness to urate in rheumatoid arthritis or an enhanced response to hydroxyapatite in osteoarthritis.

Effect of Indomethacin

The effect of anti-inflammatory therapy was evaluated by using the non-steroidal anti-inflammatory drug indomethacin. This was given in a

Table 2. Mean and standard deviation of erythema, induration and temperature responses measured at various times after injection of 5 mg monodosium urate monohydrate crystal (12 normal subjects)

TIME AFTER INJECTION	DIAMETER OF ERYTHEMA (mms)	INDURATION (mms)	TEMPERATURE (°C)
6 hrs	4.8 (1.1)	2.4 (0.8)	0.1 (0.2)
24 hrs	34.8 (4.4)	3.9 (1.4)	1.1 (1.2)
30 hrs	29.6 (4.1)	3.8 (1.6)	0.9 (0.8)
48 hrs	10.8 (5.1)	2.6 (0.8)	0.1 (0.2)

Table 3. Urate and hydroxyapatite injections in control subjects and in patients with rheumatoid arthritis, pyrophosphate arthropathy and osteoarthritis. There were no significant differences in the results

Group	Number	Mean Age	Sex	Mean Disease Duration	Forearm erythema in response to crystals (mean SD)			
					Urate (5 mg)		Hydroxyapatite (10 mg)	
					24 hrs	48 hrs	24 hrs	48 hrs
Normal Controls	12	26.2 yr	8M 4F	-	32.4 (4.4)	10.8 (5.1)	34.2 (3.6)	10.0 (3.8)
Sero-positive Erosive Rheumatoid Arthritis (Mean DAT 1:640)	15	49.1 yr	7M 8F	5.9 yr	29.8 (6.2)	14.7 (8.1)	Not done	Not done
Osteoarthritis	15	63.8 yr	7M 8F	7.1 yr	27.8 (7.3)	12.2 (6.0)	31.6 (7.4)	8.6 (4.3)
Pyrophosphate Arthropathy	20	72.0	6M 14F	8.6 yr	28.8 (5.9)	14.1 (6.6)	32.3 (7.8)	6.4 (3.8)

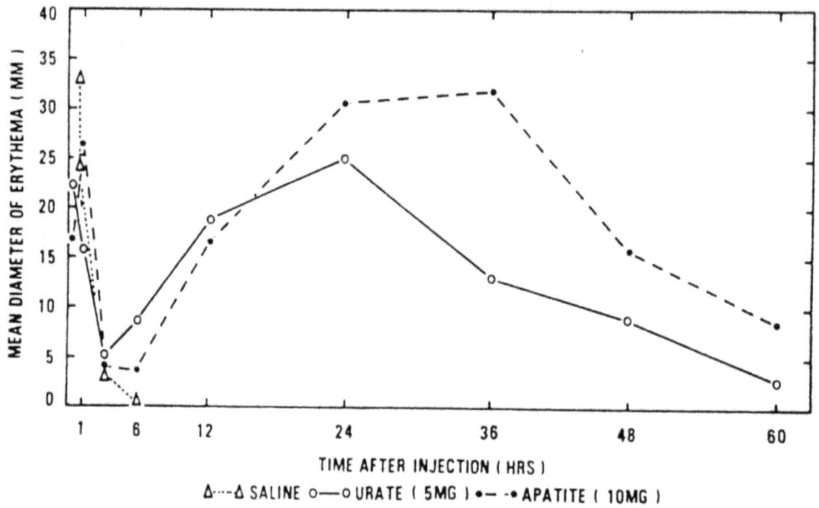

Fig. 1. Time course of the erythema response in 12 male adults.

dose of 50 mg tds in a double-blind cross-over trial with placebo. The trial design is shown in Figure 2. Indomethacin or placebo was administered 24 hours before the injection of crystals and measurements of the response to the injection were performed for the next three days. Indomethacin produced a significant reduction in the inflammatory response in 12 subjects (mean reduction in urate response 21.3%, $p < 0.05$). The mean 1 SD effects of indomethacin are shown in Table 4.

DISCUSSION

Harmless, ethical models of human inflammatory responses have obvious potential in the investigation of inflammatory diseases, and in the development of anti-inflammatory drugs. Freudwiler and Hiss injected crystalline and amorphous material into human skin to investigate the phlogistic potential of crystals.[8] The method has been reproduced in a more sophisticated form by recent workers, and advocated in the assessment of drugs.[1,2,3,4] However, the time course, dose-response, and best ways of measuring these responses are not well documented.

The experiments described here indicate that intra-dermal injections result in an inflammatory response which is dose-dependent and maximal 24 - 36 hours after the injection. The initial erythematous response to all injections, including saline and amorphous material, is probably due to histamine release following the trauma of the injection. Only crystalline material resulted in the secondary response. The time course of each reaction was similar, although urate produced a more rapid reaction than pyrophosphate or hydroxyapatite crystals; this is analogous

Table 4. Mean ± 1 SD Indomethacin data

	ERYTHEMA		INDURATION		TEMPERATURE	
	Placebo	Indomethacin	Placebo	Indomethacin	Placebo	Indomethacin
Urate						
24 hrs	34.38 (4.42)	29.08 (4.86)	38.42 (13.92)	35.17 (13.17)	34.40 (0.55)	34.08 (0.17)
48 hrs	10.83 (5.07)	10.75 (7.18)	26.17 (7.48)	22.5 (7.21)	34.3 (1.17)	33.91 (0.80)
Pyrophosphate						
24 hrs	13.67 (5.04)	11.31 (4.76)	11.3 (4.08)	11.08 (3.32)	33.65 (0.37)	33.65 (0.57)
48 hrs	4.88 (2.04)	4.08 (2.02)	7.58 (3.34)	7.12 (3.42)	33.76 (1.16)	33.31 (1.02)
Apatite						
24 hrs	34.21 (3.61)	27.83 (5.87)	30.08 (7.23)	28.08 (6.08)	34.05 (0.58)	33.5 (0.39)
48 hrs	10.04 (3.80)	10.38 (5.84)	22.83 (6.73)	21.0 (6.37)	34.09 (1.06)	33.78 (1.16)

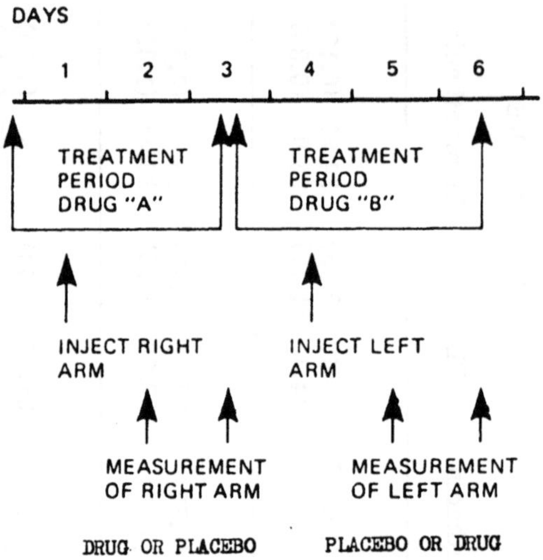

Fig. 2. The trial design.

to the more dramatic clinical course of gout. All of the reactions slowly disappeared, and only minor discomfort was experienced. The procedure can be recommended as a safe, simple and ethical model.

Sterile precautions were taken. Recent work has shown that heating crystals to remove pyrogens also alters their structure,[9,10] and this is a potential source of criticism of the experiments described. However, the data the model provides are probably adequate to test anti-inflammatory drugs and to look for gross difference in patient responses. Infrared thermography was found to be an objective method of measuring the response to crystal injections.

REFERENCES

1. A. D. Steel, D. J. McCarty, An experimental model of acute inflammation in man, Arthritis Rheum. 9:430–442 (1966).
2. W. W. Buchanan, J. R. Klinenberg, J. E. Seegmiller, The inflammatory response to injected microcrystalline monosodium urate in normal, hyperuricemic, gouty and uremic subjects, Arthritis Rheum. 8:361 (1965).
3. U. Ambanelli, G. F. Ferraccioli, P. Mangarelli, G. L. Vaona, The effects of NSAID on monosodium urate (MSU) skin test and on other in vivo skin tests (SK/SD: PHA), Agents and Actions 7 (suppl.):306–309 (1980).
4. P. A. Dieppe, M. Doherty, J. T. Whicher, G. Walter, The treatment of gout with azapropazone: clinical and experimental studies, Eur.

J. Rheumatol. and Inflam. 4;3:392-440 (1981).

5. J. E. Seegmiller, D. R. Howell, S. E. Malawista, The inflammatory reaction to sodium urate, J. Am. Med. Assoc. 180:469-477 (1962).
6. D. J. McCarty, D. W. Palmer, P. B. Halverson, Clearance of calcium pyrophosphate dihydrate crystals in vivo. 1. Studies using ^{169}Yb labelled triclinic crystals, Arthritis Rheum. 22:718-727 (1979).
7. G. Nancollas, The growth of crystals in solution, Adv. In Colloid and Interface Sci. 10:215-252 (1979).
8. D. J. McCarty, "Studies on the nature of gouty tophi" by Max Freudwiler 1899, Ann. Intern. Med. 60:486-505 (1964).
9. N. S. Mandel, Structural changes in sodium urate crystals on heating, Arthritis Rheum. 23:772-776 (1980).
10. P. A. Dieppe, M. Doherty, A. Swan, J. Hornby, N. Luckman, R. Jenkins, A. W. Preece, Changes in monosodium urate monohydrate crystals on heating or grinding, Arthritis Rheum. 24:975-976 (1981).

COMPARATIVE THERMOGRAPHIC ASSESSMENT OF LOWER LEG BATHS IN MEDICINAL MINERAL WATERS (NAUHEIM SPRINGS)

D Rusch, G Kisselbach

Justus-Liebig-Universitat Giessen
D-6350 Bad Nauheim
WEST GERMANY

Natural salt water (≃ 3% NaCl; like sea water) is often said to raise skin blood flow more than fresh water at the same temperature. Skin temperature should be higher after salt water baths as compared with fresh water.[1] CO_2 at higher concentrations (> 1g/kg H_2O) is a direct vasodilator[2] and may have some thermal effect after bathing. The latest measurements[3] compared integral skin temperatures before and after baths in baltic sea water (≃ 1.3% salt) and fresh-water at temperatures of 36 ± 0.5°C. But these and all former results in the literature are derived from whole-body treatments. There is, therefore, a need for direct comparisons of different baths on the same subject.[4] A pilot study[5] based on visual evaluation of isothermograms showed the reported temperature differences if the two baths were kept at 35°C or lower.

METHODS

64 bathing procedures with salt water (brine) and CO_2-salt water were carried out at 32 and 35°C, in each case together with a fresh-water bath of equal temperature. The ambient temperature was 23°C. Each subject bathed twice on two successive days according to the following time-table.

First of all the test subjects had 15 minutes resting in the prone position after which the first thermogram was taken. During the next 15 minutes, while the subjects were sitting in a chair, their legs were immersed. After bathing, the legs were dried by dabbing gently and the subject returned to a prone position for further thermographic examination. The thermograms were documented 1, 5, 10, 15 and 60 minutes after the baths. The same procedure was repeated the following day, but the application of mineral and fresh water was reversed left to right leg. In some experiments the subjects were covered with a blanket for 5 (or 10) minutes after the baths to avoid cooling by evaporation. In these cases thermograms were taken (5), 10, 15, 20 and 60 minutes after the bathing.

Quantitative temperature results were obtained by means of a micro densitometer from enlarged graytone thermograms. Temperature readings from the densitograms were taken on a digitizing tablet and processed by a digital computer. The computer printout was composed of 15 to 16 point measurements and the arithmetic mean temperature of both calves calculated (Fig 1).

RESULTS

Simultaneous baths in CO_2-salt and fresh water of 32°C show better rewarming of the leg bathed in mineral water. The difference however is not significant for 6 experiments in 3 persons ($t=1, 82/P > 5\%$), (Fig 2). Covering the subjects for 10 minutes after bathing reduces the effect. The other comparisons of the study gave similar insignificant results. Therefore, we looked at the temperatures of both legs over the time of one experiment. Thus we are able to compare two day trials with CO_2-salt and salt water at both temperatures. According to this, Figure 3 shows the time-course of temperatures of the 32°C baths and Figure 4 for 35°C. The increase of the CO_2-salt group's temperature on the second day is significant in both cases ($p < 2.5\%$). There is no difference with change of weather for both groups. Especially in the 35°C experiments, we see a swing in the temperature course of the CO_2-salt group. For better understanding we need more temperature information in the first 10 minutes after the baths.

DISCUSSION

Balneologic stimuli are not limited to the skin region bathed in mineral water. Immediately after the bath there is a strong subjective reaction which is different for CO_2-salt and salt water. Unfortunately, we changed the initial experimental scheme by covering some subjects for different times after the baths, so the results are less significant than in the the original study.

From the temperature measurements in Figures 3 and 4 we see that our results do not agree with observations made by former balneologists who measured increasing skin temperatures in a series of salt water baths.[1] In our results, only a longer persisting effect of the CO_2-salt water is indicated. This need not mean that salt water has a less longer lasting (or accumulating) influence on skin temperature (skin blood flow). We are sure we would have seen better signs of what happens to skin temperature if we had had continuous registration of

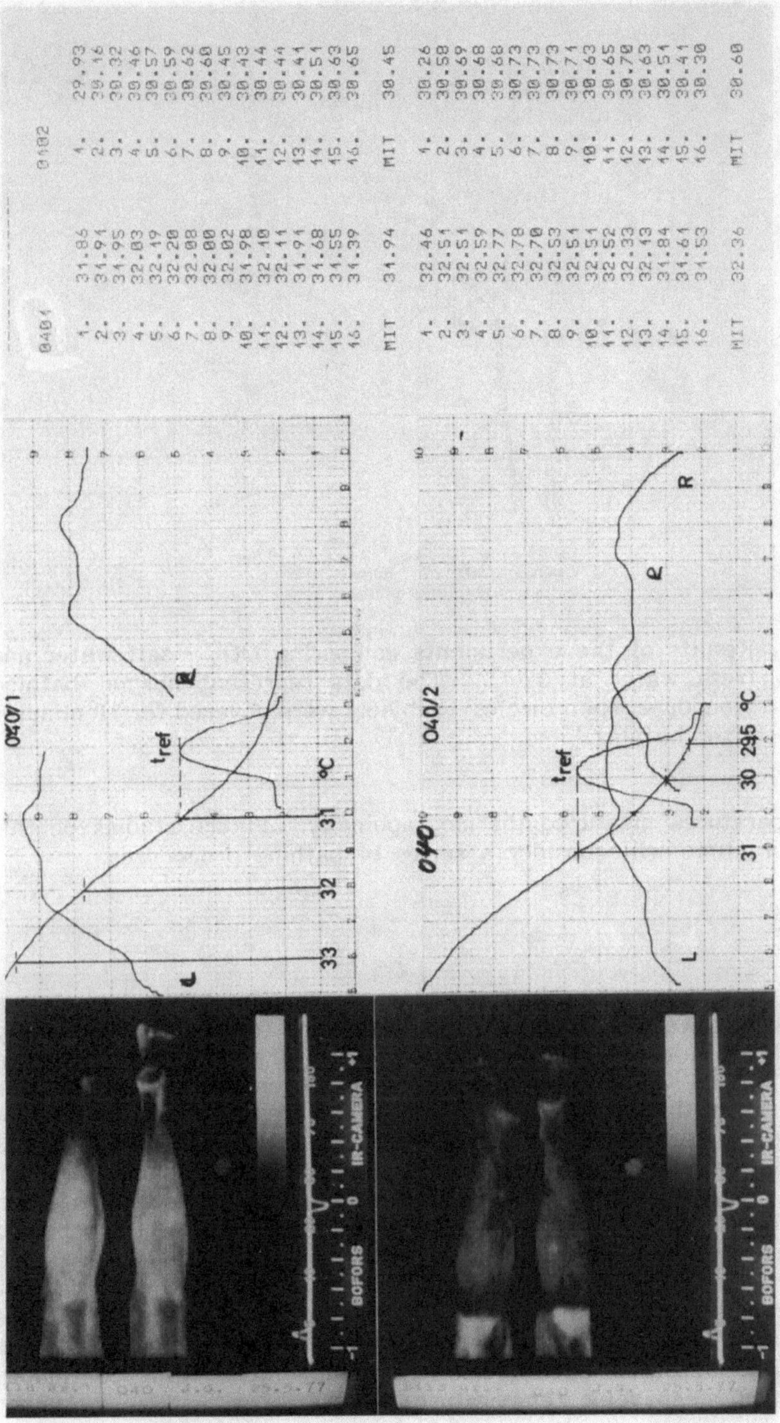

Fig. 1. Densitometry applied to the thermograms to supply point measurements and mean temperature values.

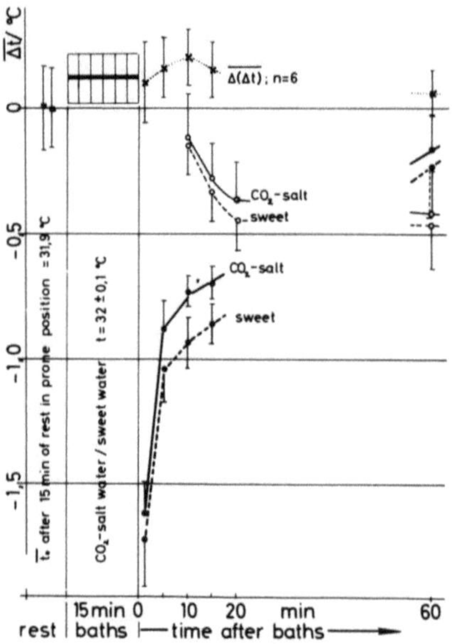

Fig. 2. Results of the experiments comparing CO_2 - salt water and fresh water at 32°C. The dots represent normal bathing procedure; open circles - subjects were covered for 10 minutes after bathing.

mean temperatures measured thermographically. Future studies on this topic in our clinic will consider a series of bathing procedures.

SUMMARY

This study compares skin temperatures over the calves before and after different lower leg baths. All subjects were treated simultaneously with one of two mineral waters and fresh water. Two temperatures were chosen, 32 and 35°C. The subjects bathed twice on two successive days with a change in the side bathed in mineral and fresh water. The results of the original study show a distinct trend towards higher temperatures of the skin bathed in mineral water. But the difference is never significant. The strength of balneologic stimuli was underestimated in planning this study. This is demonstrated by the course of temperature of both legs only one of which was bathed in mineral water. The temperatures are significantly higher after the second bath in CO_2-salt water as compared with salt water. This indicates that future studies require a series of mineral water baths. Direct after effects of different baths could be measured much better with equipment allowing continuous registration of mean value temperatures.

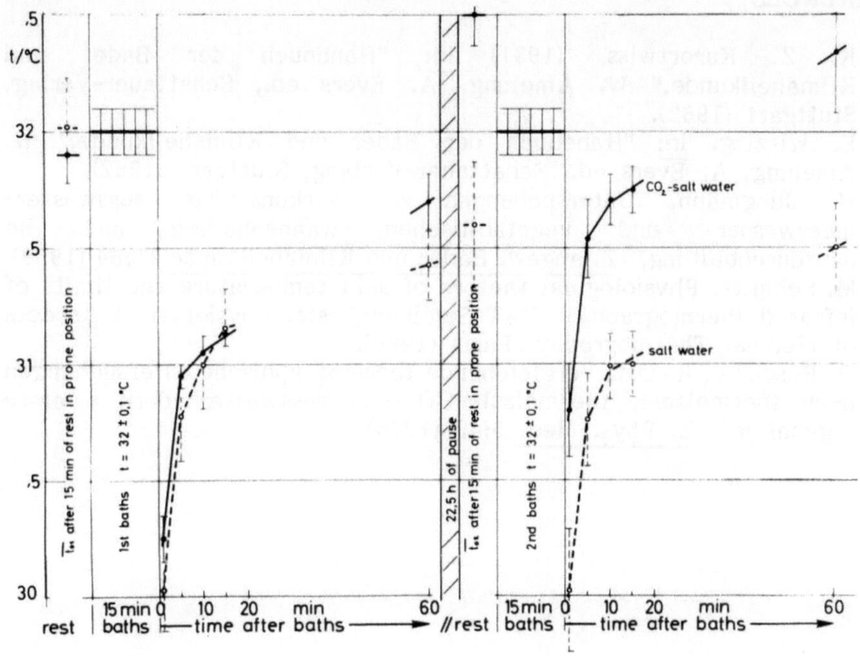

Fig. 3. Results with the salt water and CO_2 salt water groups at 32°C.

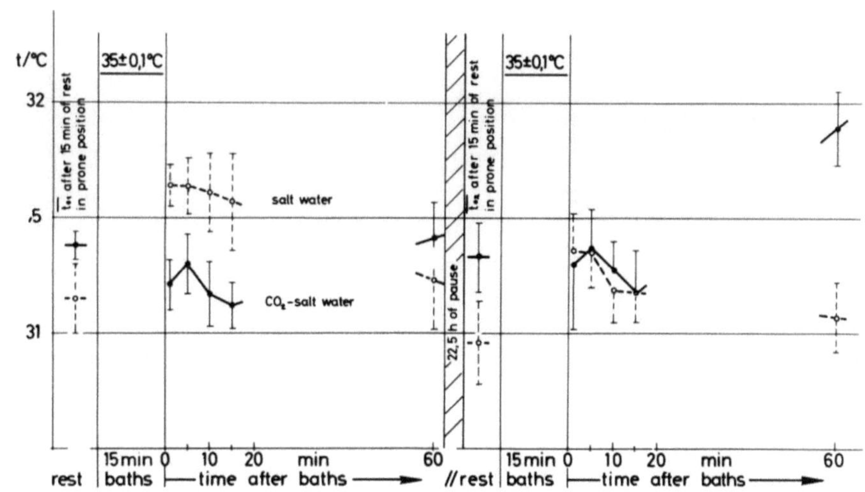

Fig. 4. Results with the salt water and CO_2 salt water groups at 35°C.

REFERENCES

1. K. Z. Kurortwiss, (1931) in: "Handbuch der Bader und Klimaheilkunde," W. Amelung, A. Evers ed., Schattauer-Verlag, Stuttgart (1962).
2. E. Witzleg, in: "Handbuch der Bader und Klimaheilkunde," W. Amelung, A. Evers ed., Schattauer-Verlang, Stuttgart (1962).
3. H. Jungmann, Untersuchungen zur wirkung von susswasser- meerwasser- und vegetabilischen wannenbadern auf die hautdurchblutung, Z. angew. Bader und Klimaheilkunde 16:64 (1969).
4. M. Schmitt, Physiological factors of skin temperature and limits of infrared thermography, EAT Seminar-Abstr. Physiological Aspects of Medical Thermography, Paris (1980).
5. D. Rusch, V. R. Ott, Vergleichende thermographische untersuchungen nach thermalsole, thermalsole-CO_2-und susswasserbadern - erste ergebnisse, Z. Phys. Med. 5:49 (1976).

THERMOGRAPHY OF HEADACHES

P M Ruegsegger
Biotronics Institute
115 East 61st Street
New York, NY 10021, USA

Thermographic study of headaches over the past 12 years in the search for typical heat patterns has shown them to be more complex than peripheral pain.

As we studied a cross-section of muscular, neuralgic and vascular headaches, we were impressed by the fact that differentiating features were absent in all, signs of local hyperthermic tissue reactions were present in all, and most patients did not have headaches typical of one class but often mixed forms which changed in time. All patients, however, showed a remarkable pattern stability of pain which coincided with the thermographic abnormalities.[1]

Thermography demonstrated right from the beginning that the classic headache types did not indicate isolated malfunctions of nerves, blood vessels or muscle tissue but rather that abnormalities of heat emission were widespread and unrelated to the quality of subjectively perceived pain.

We therefore had to dismiss the conventional headache classification as being purely descriptive, unrelated to the observable realities and unrelated to specific pathology which involves more than one type of tissue for each type of headache. In this field, which is pervaded by mysticism and unscientific superstition, it appeared scientifically sound to build on the knowledge secured since antiquity.

Whenever the integrity of our bodily structure and function is disturbed by injury, local tissue reactions are called into play for defense and repair which are clinically manifest by presence of 5 cardinal signs, in form of pain (dolor), swelling (tumor), redness (rubor), disturbed function (functio lesa), and most important heat of inflammation (calor), which can be measured exactly by thermography. Moreover, thermography by quantifying calor as to intensity and extent, permits the measurement of the site of the entire tissue reaction. This approach of thermography in conjunction with minute physical examination quite logically excluded the use of thermography for mass screening and made it mandatory to use thermography and physical examination with exact history in parallel without exception at every examination.[2,3]

Since the historic classification of headaches had been subject to disproof by use of thermography, a new classification more in agreement with our observations, was introduced:

1. Simple self-limiting
2. Acute and chronic headaches secondary to known causes (head trauma, brain tumor, hemorrhage, metastasis, mal-occlusion etc)
3. Refractory headaches: all self-sustaining pain patterns of vascular, neuralgic, muscular or mixed types.

Refractory headaches (class 3) include all others which are resistant to treatment and may last years or a life-time for reasons unknown. On closer study these headaches are set apart by the phenomenon of chronic recurrence with silent intervals and their remarkable pattern stability locating to anatomical sites without changing.

Our clinical experience with thermography and clinical observation on elimination treatment of chronic recurrent headache was first published 10 years ago and more recently with illustrations in analog and isothermic color.[1,3] In this report, emphasis is placed on a more detailed study of the mechanism of these self-sustaining disorders in view of the development of new territories in the future.

From our studies during the past decade, we learned that correlation of thermography with clinical examination allows differentiation of local tissue reaction into three forms:

1. Inflammatory
2. Ischemic injury
3. Regenerative tissue reaction.

Most of the basic research on tissue reactions has been done in the last two decades[5,6] and quite a few questions of interest here have been answered.

MECHANISM OF SELF-PERPETUATING INFLAMMATION AT THE ANATOMICAL LEVEL

Microscopic tissue studies have shown that microblockade in response to cell injury is part of the mechanism for protection and subsequent healing of injured tissue. This local tissue reaction is transitory, because the protective blocks are dissolved by fibrinolysis unless repetitive injury sets in motion a vicious cycle of lysis-blockade-lysis illustrated in Figure 1.

FEEDBACK DERANGEMENT OF SELF-PERPETUATING INFLAMMATION

Figure 2 explains the observation of muscle inflammation made worse by excessive reflex spasm causing compression ischemia instead of stretch-relaxation. This vicious cycle starts itself because of increased irritability of stretch receptors of inflamed muscle spindles causing excessive gamma activity and increased reflex spasm. Cybernetically, increased gamma activity is said to be due to a change in set-point of the system. These foci of chronic injury with increased gamma activity are considered responsible for chronic recurrence of pain patterns. These are also considered as staging sites of headache attacks - if eliminated the attacks should cease.

FEEDBACK REGULATION OF FIBRINOLYSIS

A mechanism for removal of micro-blockade is interacting throughout the entire organism with clotting by dissolving the fibrin

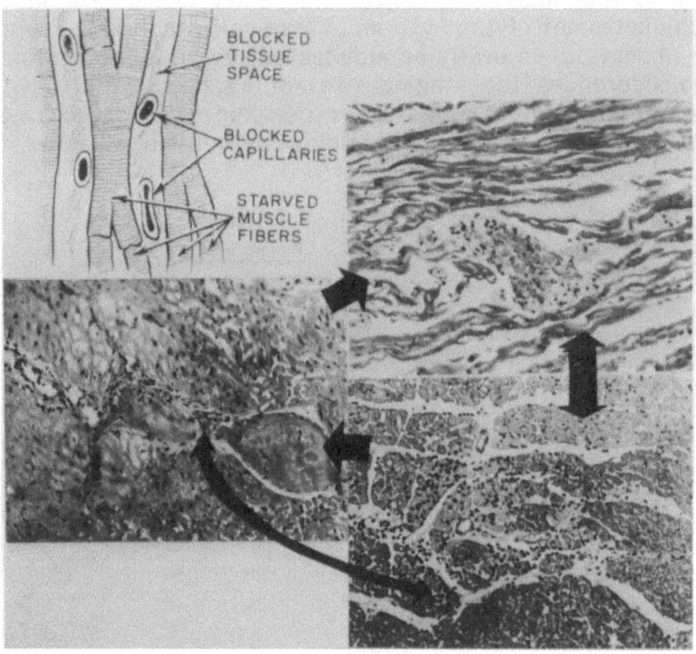

Fig. 1. Anatomy of microblockade in chronic inflammation: cycle of lysis - blockade - lysis.

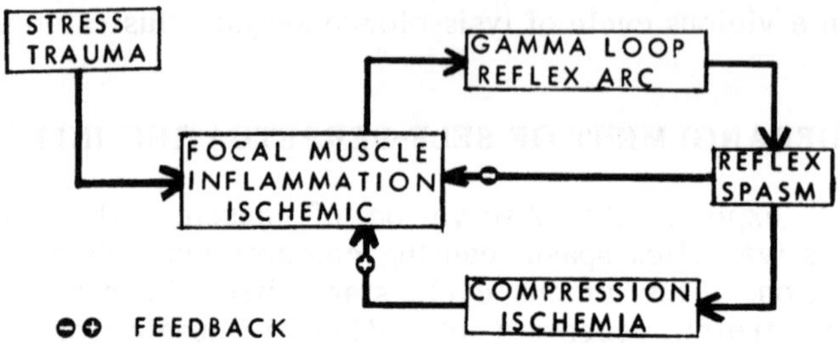

Fig. 2. Feedback derangement in chronic inflammation.

deposits after cessation of injury. The rate of fibrin dissolution depends directly on levels of plasmin and plasminogen at the site of digestion and indirectly on input of circulating activator (eg streptokinase) (Fig 3).

DESCRIPTION OF REFRACTORY HEADACHES (Figs 1–3)

Chronic recurrence is explained by self-exciting and self-perpetuating feedback derangement in focal muscle inflammation. Cybernetically, this increased feedback activity is said to be due to a change in the set-point of the system. These self-stabilizing "penetration-blocks" are impervious to orally administrated systemic medication. These foci are considered as the staging sites of attacks; while active, their increased gamma activity may trigger periodic mild attacks, clinically manifest as unlateral headaches, neuralgia, torticollis or vascular migraine.

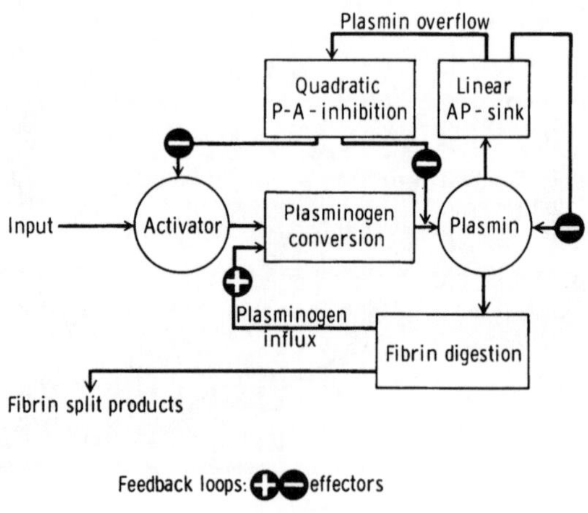

Fig. 3. Cybernetic model of fibrinolysis showing input and output functions.

Fulminating attacks with the "head feeling like bursting" are explained as a runaway to maximum by self-escalating reflex contraction of an over-stretched scalp muscle ring with sharply-increased gamma activity calling for muscle contraction via reflex loop and thereby setting up the positive feedback for head-bursting escalation.

The often quoted serial attacks of cluster migraine are cybernetically defined as relaxation-oscillation due to unbalanced feedback from staging sites usually located in the auriculo-temporal regions; if eliminated by local treatment, the attacks were found to cease in all cases encountered so far.

Therapy of refractory headaches aiming at their elimination must be directed at the interruption of feedback loops by infiltration of self-organizing penentration locks with steroid-procaine thus blocking excess gamma loop activity. Flare-ups may be controlled by suppressing this vicious cycle with psychotropic agents including anti-depressants against stress and muscle relaxers against reflex spasm (Fig 4).

STUDIES ON FIBRINOLYSIS FOR POSSIBLE FUTURE FORMS OF THERAPY

Fibrinolytic System: Regulation by Non-linear Refraction

The lytic curve shows the dependence of the fibrinolytic rate on activator input within the constraints of non-linear inhibition reaching maximum at streptokinase-activity (SK) levels of 3000 units/ml. The fact that activator concentrations up to very high levels are not rate limiting, (in line with penicillin and steroids) shows the potentially powerful fibrinolytic effects obtainable by local injection of SK.[4,5,6,8,9]

The Kinetic Model of the Fibrinolytic Reaction

This shows the possibility of inducing sustained fibrinolytic activity by reducing anti-plasmin (AP) levels by oral medication (eg nicotinic acid, and platelet inhibitors, etc).

The Fibrinolytic Equation

Our mathematical model of enzyme kinetics permits the computer simulation for clinical investigation of the fibrinolytic system. This equation describes the rate of fibrinolysis as a function of 5 components of the fibrinolytic reaction which can be entered as variables.[9,10]

This kinetic system of the fifth order may now be programed into a computer system with greater ease than 20 years ago when we set out to model mathematical blood clots subject to fibrinolytic dissolution.

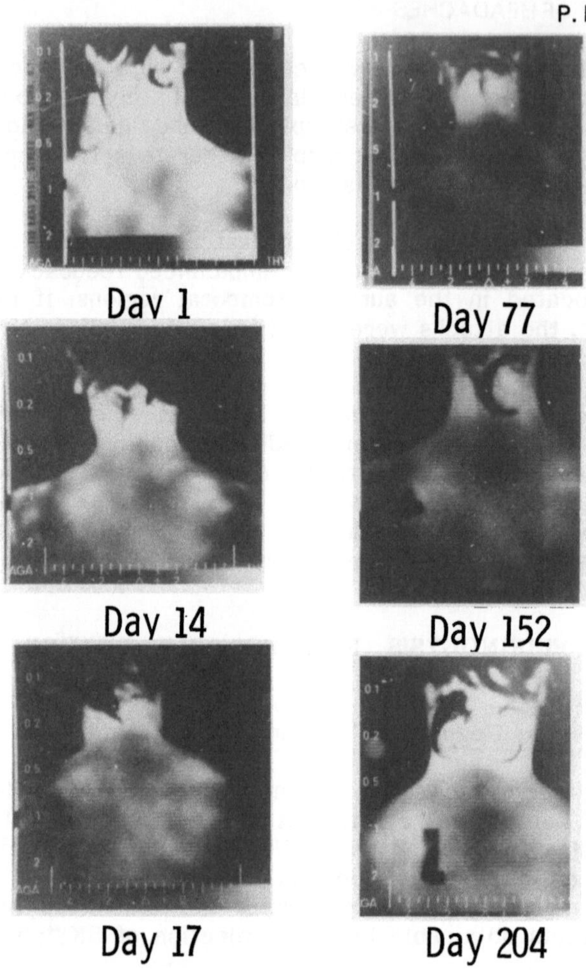

Fig. 4. Treatment resistant penetration blocks: the observation of "instantaneous cessation" (ie within 3 days after chronic recurrent intractable pain of 6 years duration) after direct local treament.

Nevertheless, it was at that time that the surprising effects of local fibrinolytic treatment were discovered by fortunate accident by calculating the extremes of potentially effective therapy. Locally injected streptokinase activator in concentrations a thousand-fold greater than the customary dosage for clinical use exerted very powerful lytic effects.

REFERENCES

1. P. Ruegsegger, Thermographic configuration and distinction of clinical headache varieties, "Proceedings of the international

headache symposium," Sandoz Ltd., Elsinore, Basle, (1971).

2. E. F. J. Ring, A. J. Collins, P. A. Bacon, J. A. Cosh, Quantitation of thermography in arthritis using multi-isothermal analysis, Ann. Rheum. Dis. 33:353-356 (1974).
3. P. Ruegsegger, Use of thermography for evaluation and treatment of chronic pain (mainly recurrent headaches), Acta Thermographica 5:24-31 (1980).
4. I. Nydick, P. Ruegsegger, J. S. Ladue, Salvage of heart muscle by fibrinolytic therapy after experimental coronary occlusion, Am. Heart J. 61:83-100 (1961).
5. P. Ruegsegger, I. Nydick, R. E. Abarquez, F. Reichel, E. E. Cliffton, J. S. Ladue, Effect of fibrinolytic (plasmin) therapy upon the pathophysiology of myocardical infarction, Am. J. Cardiol. 6:519-524 (1960).
6. P. Ruegsegger, Fibrinolytic therapy of myocardial infarction: theory and clinical results, Angiologica 16:1-8 (1968).
7. P. Ruegsegger, Refractive headaches, Headache 9:201-206 (1970).
8. Ching, C. Wexler, Peripheral thermographic manifestations of lumbar disc disease, Appl. Radiol. 7:53 (1978).
9. W. Liniger, P. Ruegsegger, Fibrinolysis: mathematical and experimental studies, Thromb. Diath. Hem. 17:412-417 (1967).
10. W. Liniger, P. Ruegsegger, A mathematical model of fibrinolysis, Math. Bioscienc. 1:263-285 (1967).

MALIGNANT DISEASES - INTRODUCTION

A M Stark

Women's Cancer Detection Society
Queen Elizabeth Hospital, Gateshead
UNITED KINGDOM

The use of thermography in malignant diseases has been confined almost exclusively to breast cancer. Since the early experimental work in 1957 by Lawson of Montreal, followed three years later by that of Lloyd Williams in London, the appreciation of the place of thermography in breast cancer has had many ups and downs. It has been greeted by some with great enthusiasm, while others have considered it to be of no value and many just have not understood its significance at all. Unfortunately, these three factions still exist.

In recent years, however, many experienced thermographers throughout the world have come to appreciate the true value of thermography and to accept its limitations in the assessment of the woman with breast cancer. It is no longer looked upon by many as a diagnostic test for breast cancer but as a non specific test, an abnormal thermogram indicating some abnormal breast activity requiring further investigation. In the event of that investigation being negative, long term follow up is required as it may be an early warning sign of a future cancer. The term 'false positive' should not be applied in this context as a persistent abnormal thermogram is the greatest single risk factor for breast cancer.

Guduric's paper on the use of glucose indicates the possibility of increasing the value of thermography as a method of differential diagnosis in clinical breast conditions.

Thermography has considerable value as a prognostic index in breast cancer - the hotter the cancer, the poorer the prognosis, as indicated by the papers of Lloyd Williams and Spitalier. I wish to sound a word of warning, however, in that I have found 50% of cold cancers to be of poor differentiation and rapid growth, and these have been quickly fatal.

These have all been small tumors which on mammography had a halo due to edema fluid, thought to be a reaction of the surrounding tissues to the activity of the rapidly growing tumor. The edema acts as an insulator and contains the heat evolved by the tumor.

A third value is that serial thermograms, after establishment of a base line, and provided that they are done under identical conditions, may be very useful in repeat screening. A change is significant.

The papers and posters presented in this Congress demonstrate the great advances in thermographic techniques since Lawson's first reports and extend from Haberman's computerized thermography to Karpman's 'do-it-yourself' kit using a form of contact thermography.

Other presentations favorably compare the value of thermography in breast conditions with scintigraphy, ECAT scanning, diaphanography and ultrasound. Isard pointed out that the use of both thermography and diaphanography, which are entirely safe procedures, in conjunction with mammography, improve the overall diagnostic accuracy.

The two papers on melanoma of the skin indicate the continuing controversy as to the value of thermography in this condition.

A THERMOGRAPHIC PROGNOSTIC INDEX

K Lloyd-Williams

Royal United Hospital
Combe Park, Bath
UNITED KINGDOM

Three disturbing facts emerge from any clinical study of breast cancer. The first is that the incidence of the disease increases slightly every year in the Western world, despite improvements in education, in public awareness of the disease, in diagnosis and the publication of 15,000 papers on the subject annually. Of every 16 women alive in Europe today 4 will die of cancer. One out of each four will die of carcinoma of the breast.

The second is that the overall survival rate of 50% at 5 years has remained approximately the same for the last half century in spite of every refinement of surgical technique, hormone therapy, cytotoxic agents and radiotherapy.[1]

The third is the bewildering range and spectrum of disease - cancer cells of the same histological appearance can kill one patient in a few weeks despite every known form of treatment, yet leave another alive and well without any treatment ten years later.

Carcinoma of the breast is a carpet bag diagnosis, the individual cases can differ as chalk from cheese. We need more biological knowledge of the cancer cell in the particular environment of the host.

Earlier diagnosis might improve this depressing picture. Haagensen[2] has shown that a cancer confined to the breast can be treated surgically with 92% 5 year survival rate and 87% 10 year survival. But the classical clinical method of diagnosis depends primarily on the detection of a palpable lump in the breast, with early skin tether, and it is often difficult to detect a small lump, say of 1 cm diameter, even though the breast is easily accessible to palpation. If one believes that a cancer derives from one rogue cell, then a lump of even 1 cm diameter may represent several years of division and the cancer will have been present for a considerable period before it even becomes palpable.[3]

On this hypothesis if you wait until you can feel a lump, treatment may prove ineffective in 50% of cases. We need a diagnostic method more efficient than the palpating hand. Once a breast cancer is palpable and diagnosed, Haagensen's and also Handley's[4] figures have shown that the pathological spread to axillary or internal mammary nodes determines the outcome of the disease. This pathological spread designated by TNM classification and the work of Scarf,[5] Bloom and Richardson[6] looking at cell differentiation in the primary tumor constituted the first glimmer of a prognostic index for breast cancer. This then was the state of the art in 1960. This was an attempt to break down the numbers of breast cancers into differing categories which had a different prognosis and might need different modalities of treatment.

In 1957 Ray Lawson of Montreal[7] described the rise in temperature over breast cancer. In 1960[8] we confirmed his findings from the Middlesex Hospital, London. By 1961 there was interest in a number of centers in the development of infrared scanners, mostly derived from military equipment, and their application in clinical medicine.

The interest and work in America of Bowling Barnes, Gershon Cohen and Joann Haberman culminated in December 1963 with a symposium on Thermography in New York sponsored by the New York Academy of Sciences. At this symposium we presented our experience of the thermal profile of breast cancer.[9]

1. Not all cancers were hot.
2. There was a spectrum of rise in temperature from nothing to 7°C.
3. Some of the very hot cancers were hotter than the body core temperature.
4. The hotter the cancer the worse the prognosis.
5. The hotter the tumor the worse the pathological staging.
6. The hotter the tumor the worse the pathological grading.

The most dramatic of all forms of new growth in the breast is the inflammatory carcinoma which is recognized by its short history and the cardinal signs of inflammation, rubor, calor, tumor and dolor. Often mistaken for an abscess, it was originally described in 1911 by Schuman[10] as a case of acute carcinoma in a lactating breast. We now know that it is not commoner in pregnancy and occurs in any age group. I have personally seen two cases in patients over 90. Haagensen[11] described its incidence as 3% of breast cancers. The skin is red and the temperature raised. The prognosis is poor, the average duration from onset of symptoms to death in Haagensen's series of 56 cases was 20 months.

In our series there was only one case clinically diagnosed as inflammatory - because of redness of the skin - but there were 11% of cases with a temperature rise of 3°C or more who had, despite all forms of treatment, an appalling prognosis - 80% dead and 100% with spread of their disease at 2½ years and not a single survivor at 5 years.

Our measurements were made with a bolometer. Since that time Jones at the Royal Marsden Hospital, London,[12] Gros and Gautherie at Louis Pasteur University, Strasbourg[13] and Amalric and Spitalier from Marseilles have confirmed these findings thermographically with much larger numbers, ie the hotter the tumor the worse the prognosis.

It is interesting that Delarue and others from the Institute Gustav-Roussy have recently (1981)[14] shown that clinically inflammatory tumors tend to have a higher incidence of non-estrogen non-progesterone receptors.

It seems, therefore, that thermography will select a group of cancer cases of rapid growth with undifferentiated cells of poor prognosis which tend to have a high incidence of non-hormone receptors (48% as against 28% of operable cancers). The thermographically hot group covers all age groups and accounts for 11% of all cancers.

This has altered my personal clinical handling of patients. Thermographically inflammatory carcinomas are not now treated surgically but by cytotoxics and supervoltage radiotherapy.

There has been much criticism of thermography in breast cancer because of the number of false positive and false negative thermograms. To deal with the latter first, it is only false in that the cancer is not hot. It is true in that it tells you that the tumor has a good prognosis and is not very active.

In my experience low grade tumors exhibit obvious physical signs of deformity and tethering. So in the clinical situation a negative thermogram is not a disadvantage diagnostically, because there are obvious visual signs or a palpable lump which is amenable to pathological diagnosis by fine needle biopsy. But it is of value prognostically.

Turning to the so-called false positive thermograms from 15% to 25% in various series. These were criticized as being too high for a screening exercise. However, long term follow up of these patients with persistently positive thermograms has shown that they have a very high incidence of developing breast cancer.

Gautherie has published figures showing that out of 1,200 positive thermograms in a 12 year follow up 40% developed a carcinoma. Amalric and Spitalier from Marseilles in a 5 year follow up of persistently positive thermograms, give an incidence of carcinoma in 26% and Stark in Newcastle quotes a 25% incidence of carcinoma in positive thermograms. She says that a persistently positive thermogram is the highest prognostic factor available today in breast cancer.

Bowling Barnes in 1963 referred to a thermographic finger print. This is true in breast screening - women can be recognized by their breast thermograms and changes in their pattern are of ominous significance.

It is 20 years since the first crude thermogram came on the scene. Since then the machinery has improved to a state where it will do all we ask of it. Perhaps we are not asking it the right questions.

Perhaps we are too interested in pictures and not enough in changing state in relation to time, to thermal stress or drugs.

Progress is slow. However, overall I feel we should be optimistic.

The thermal physiology of breast cancer does seem to tell us something about the biological activity of breast cancer - it seems to select a cohort of women of high risk in a non-invasive way.

It gives us a guide to prognosis and helps us to select more appropriate treatment.

REFERENCES

1. M. B. Shimkin, End results in cancer of the breast, Cancer Philad. 20:1039 (1967).
2. C. D. Haagensen, "Disease of the breast," Chapter 27, p. 587 W. B. Saunders (1965).
3. W. H. Bond, The influence of various treatments on survival in breast cancer, Syntex Symposium on the Treatment of Carcinoma of the Breast, Excerpta Med. (1967).
4. R. S. Handley, A. C. Thackray, Invasion of internal mammary lymph nodes in carcinoma of the breast, Br. Med. J. 1:61 (1954).
5. R. W. Scarff, Prognosis in carcinoma of the breast, Br. J. Radiol. 21:594 (1948).
6. H. J. G. Bloom, W. W. Richardson, E. J. Harries, Natural history of untreated breast cancer, Br. Med. J. 213-221 (1962).
7. R. N. Lawson, Thermography, a new tool in the investigation of breast lesions, Can. Ser. Med. J. 13:517 (1957).
8. K. Lloyd-Williams, F. Lloyd Williams, R. S. Handley, Infrared radiation thermometry in clinical practice, Lancet 2:958 (1960).
9. K. Lloyd-Williams, R. B. Barnes, J. Gershon Cohen, J. D. Harberman, Thermography and its clinical applications, Ann. N. Y. Acad. Sci. 121:1-304 (1964).
10. E. A. Schumann, A study of carcinoma mastitoides, Am. Surg. 54:69 (1911).
11. C. D. Haagensen, "Diseases of the breast," Chapter 24, p. 488 W. B. Saunders (1957).
12. C. H. Jones, W. P. Greening, J. B. Davey, V. S. Greeves, Thermography of the female breast. A five year study in relation to the detection and prognosis of cancer, Brit. S. Radiol. 48:532-538 (1975).

13. M. Gautherie, C. M. Gross, Breast thermography and cancer risk prediction, Cancer 45:51-56 (1980).
14. J. C. Delarue, Oestrogen and Progesterone. Cytosolic receptors in clinical inflammatory tumors of the human breast, Brit. J. Cancer 44:911 (1981).

BREAST CANCER DETECTION BY ABSOLUTE TEMPERATURE THERMOGRAPHY AND COMPUTER TECHNIQUES

J Haberman*, J Goin, T Love+, F Ohnsorg++, R Aggarwal++**

*** University of Oklahoma Health Sciences Center, Oklahoma City, Oklahoma, USA**
**** University of Kansas Medical Center, Kansas City, Kansas USA**
+ University of Oklahoma, Norman, Oklahoma, USA
++ Honeywell Systems and Research Center, Honeywell Inc. Minneapolis, Minnesota, USA

ABSTRACT

Absolute temperature thermography (ATT) is the measurement and quantitative representation of thermal radiation emitted at the skin surface. Accurate thermal data are obtained by calibrating a precise radiometric infrared scanner against a set of constant-temperature sources and subsequent digitization. From absolute temperature data variables representing diagnostic features are developed. The variables are utilized in developing classification algorithms which will assign subjects into a normal or cancer class. The ATT approach described here is new and is the basis of a three-phase research program aimed at developing a safe and effective system for mass screening for breast cancer.

During the early months of data collection, when video digitizers providing on-line capability were not available, data collection was limited to recording four basic temperature measurements for each breast. A sample composed of 180 normal and 45 cancer subjects was used to train and test the initial classifier. This preliminary study confirmed that diagnostic information is contained in absolute temperature measurements. The pertinent features and results of this early effort are discussed.

INTRODUCTION

Breast cancer detection was the first and continues today as the most widely utilized application of infrared sensing.[1-11] During the decade following its introduction, there was a great deal of optimism regarding the role of infrared in breast cancer detection, and in 1974 it was included as one of the accepted diagnostic methods in the ACS-NCI sponsored network of Breast Cancer Detection Demonstration Projects. Physicians in these centers were required to use thermography, even though in many instances they lacked prior training and standardized techniques were not available for training technicians. The outcome was predictable, considering that thermography as practised today involves visual interpretation of thermograms by a physician-thermographer. Conventional thermograms permit only a relative indication of temperature differences, and the brightness and range of contrast are dependent upon the film characteristics and any inconsistencies or biases introduced by the technologist. Unavoidably, the interpretation of thermograms is highly subjective, requiring great skill and experience on the part of the physician-thermographer in overcoming the limitations of the visual presentation. Because of these limitations, good results in the United States are limited to a small group of dedicated users, and qualitative thermography is regarded as a disappointment.[12]

During a series of meetings in 1971 and 1972 between scientists at the University of Oklahoma and the Honeywell corporation, a critical assessment was made of thermography's problems, as well as its potential. Numerous instrumentation constraints were identified, including film-human eye gray scale limitations and difficulties in focusing and framing due to the long scan times of existing instruments. It was apparent that the temperature information content of thermograms needed to be standardized in such a way as to minimize the subjectivity inherent in the existing methods.

ABSOLUTE TEMPERATURE THERMOGRAPHY SYSTEM DEVELOPMENT

Since 1974, a system for mass screening for breast cancer by electronic infrared pattern recognition has been under development. The system consists of a unique scanning infrared camera which outputs calibrated analogue absolute temperature signals at a conventional television frame rate. Because the signal is compatible with commercially available television equipment, the infrared image can be viewed real time on a television monitor. The television brightness and contrast levels allow elimination of the film-human eye constraint. Included with the system is a unit which provides a video cursor (adjustable window) which can be viewed on the television monitor display. Various areas can be selected for delineation by the cursor. The temperature of the area so designated can be read from a digital register.

When the electronic instrumentation for this study was initially designed, suitable devices for digitizing each entire video field at the rapid scan rates of the infrared camera were not commercially available. (This capability became available early in phase II, during February 1977). Consequently, hand-recorded data were obtained for each subject examined prior to February 1977, utilizing the cursor unit described above.

INFRARED SCANNER

The Honeywell scanner is an electro-optical device which uses liquid nitrogen-cooled detector arrays (HgCdTe), with optimum response in the body temperature region. Its output is compatible with standard (North American) RS-170 video standards. The scanner includes two internal temperature sources, located on the field stop, which permit stabilization of instrument offset and response.

The scanner subsystem and its associated electronics are mounted in a movable console, shown in figure 1. The scanner uses a refractive optical system optimized for infrared transmission. A system of moving mirrors causes the field of view of a detector cell array to be scanned over the breast area in a raster-scan pattern essentially as in standard television. The detector cells are cooled by liquid nitrogen to provide appropriate sensitivity and low noise in the portion of the infrared spectrum where the skin radiation lies.

The electronic circuits within the scanner serially sum the output of the individual elements of the detector array to enhance the signal-to-noise ratio of the scanner output. In addition, constant temperature sources are provided within the optical system but outside the normal

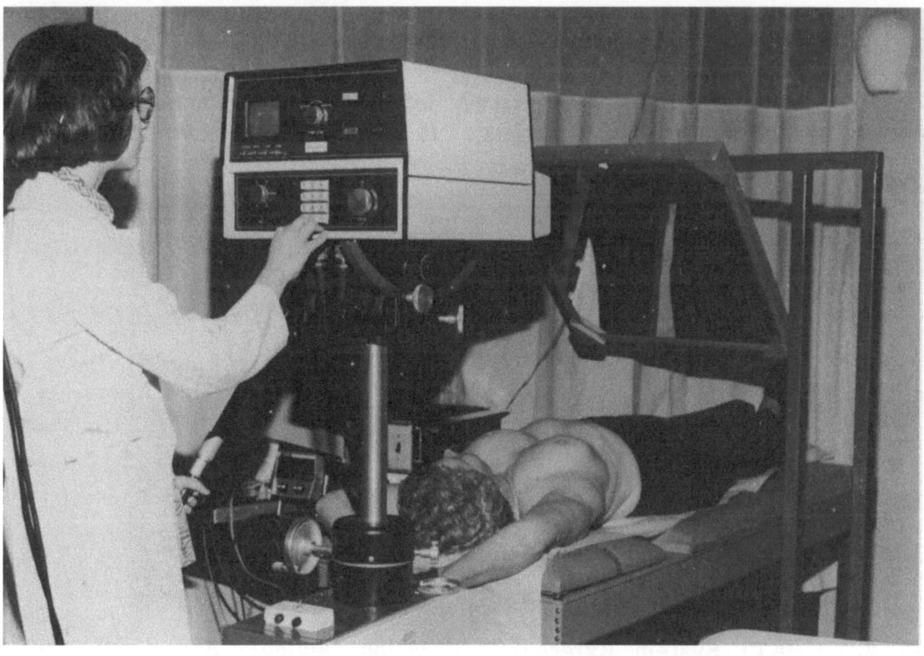

Fig. 1. Absolute Temperature Thermography System in operation.

field of view of the instrument. Electronic circuits use the detector signals, generated when these sources are scanned, to adjust the instrument output to compensate precisely for short-time errors. Accurate calibration is obtained by periodically scanning a set of standard temperature sources (hohlraums), seen in figure 2, and entering the corresponding temperature values by means of a keyboard on the scanner.

The small amount of liquid nitrogen used to cool the detector cell array is contained within an insulated chamber (dewar). The dewar is filled prior to making measurements and periodically, as needed.

The hohlraum calibration measurements were made at the beginning of each day, after the detector array had reached a stable temperature, at the beginning of each new tape and at other times when needed.

METHODS AND PROCEDURES

Environmental Control

The temperature of the skin represents a balance between the heat flow from the tissues to the skin and the heat flow from the skin to the environment. In the use of skin temperatures to represent physiological states, it is very important that the heat flow to the environment be carefully controlled. Two factors influencing the heat transfer to the environment are the room temperature and the air flow over the skin. While these factors are important in qualitative thermography, they become essential in ATT.

The control of room temperature includes both air temperature and temperature of the walls and ceiling. It is necessary to maintain the

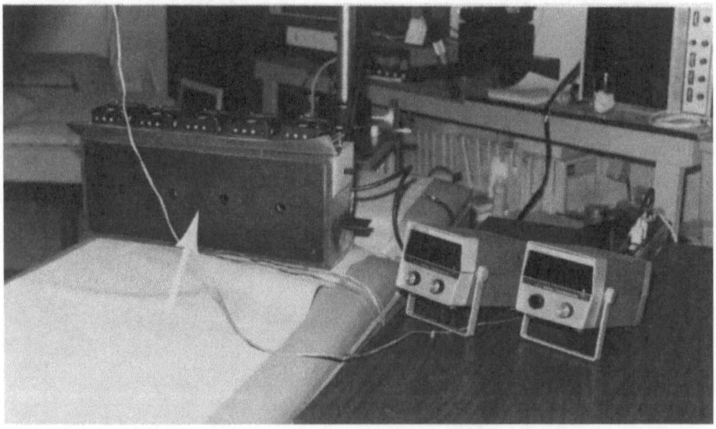

Fig. 2. ATT System External Calibration Device. Arrow indicates the set of four temperature references (hohlraums).

entire room as nearly isothermal as possible, since a signficiant part of the heat loss from the skin is the net thermal radiation exchange with the walls and the ceiling.

The thermography examining room temperature is controlled by the flow of air through the room. The room is located centrally in the screening center and is devoid of windows or outside walls. The air temperature fluctuations are limited to 0.3°C in passing through the clinic. A simple computation indicates that approximately 106 cubic feet of air are required for each BTU of heat gain to equilibrate the room. The normal subject will produce about 600 BTU per hour. Therefore, with 3 patients and 2 technologists, the airflow requirement is 5,300 cubic feet per minute. This forces a minimum requirement of a flow through the ceiling of approximately 15 feet per minute. This flow is evenly distributed over the room by discharging the air from a controlled fan coil unit into a plenum above the ceiling. Air return is provided by floor line registers. This provides a uniform downward flow of air over the patients. Artifacts caused by air currents from air conditioning registers usually present in the thermography examining rooms are eliminated.

The temperature of the air is maintained by a temperature controller, which is activated by a thermistor in the air plenum. The controller has a bandwidth of ± 0.05°C. Chilled water to an individual fan coil unit is thus controlled through a solenoid valve actuated by a relay to the controller. The time constant of the system and the control is such that a uniform temperature of 22°C ± 0.25°C is maintained at the patient's level.

The 22°C temperature was determined to be appropriate during the early part of phase I of this program. This temperature provides an adequate skin temperature contrast and requires less time to approach equilibrium than the 20°C previously recommended by thermographers. It must be emphasized that it is very important that the room temperature be controlled and that uniform airflow be maintained. The usual air conditioning designed for human comfort is totally inadequate for absolute temperature thermography.

Subject Examination Procedure

The subjects equilibrated for 20 minutes in a supine position before thermographic data were recorded. This is an extended period compared to the conventional thermographic procedure. The additional time was required because of the lag time in using the en-face positioning technique. It was also found to be advantageous to increase the clinic temperature from the original 20°C to 22°C. This small increase in room temperature resulted in a more stable breast temperature reading. The 20-minute subject cooling time allocated during the initial months of data collection was reduced to 15 minutes following incorporation of the on-line digitizer.

Following the equilibration period, which was monitored using a stopwatch, the subject was positioned in a semi-oblique position so that the right breast was en-face to the infrared scanning camera. The nipple was located by a position marked in the center of the monitor screen. A 30-second video tape recording was made of the breast, followed by manual recording of temperatures from the digital readout. The manually recorded information was limited to square areas of skin ranging from 0.5 cm^2 to 4 cm^2. Four temperatures were obtained for each breast. They were:

1. Average areola.
2. Average central breast.
3. Warmest vein.
4. Coolest avascular region.

The above procedure was performed for the left breast and immediately repeated for the right and again the left. The two readings taken from the left breast were then averaged to compare with the readings from the right breast. Room temperature and the subject's oral temperature were also obtained. All temperatures were recorded manually on a standard coding sheet and later entered into a general information processing file.

To summarize, the thermal images are obtained under carefully controlled conditions of temperature and airflow. The room temperature is $22^{o}C \pm 0.2^{o}C$, and the patients are equilibrated 20 minutes in a supine position under free convection conditions. This ensures a thermal image free of any transient effects. Cooling also tends to enhance the thermal image.

An en-face presentation of each breast is used in the present program, ie during scanning the subject is supine upon an examining table, with arms extended. The en-face presentation permits the force of gravity to distribute the breast tissue with axial uniformity about the nipple and provides an excellent view of the entire breast. Positioning is adjusted to ensure that the nipple is located approximately in the center of the field of view. Limiting the field of view to a single breast enables the temperature and spatial resolution of the system to be fully utilized. During breast imaging, the operator enters a data sequence which includes a right/left breast code, subject identification and core temperature and the room temperature.

During scanner calibration, the set of reference temperature sources (hohlraums) is placed on the examining table in the position normally occupied by the patient. The operator obtains images of the temperature sources and also enters a data sequence which identifies the temperature reference targets being scanned and their measured temperature values.

Patient Population

As stated earlier, in contrast to later stages of the research program when a digitized array was automatically recorded for each subject, the initial data was obtained by manually recording the absolute temperature data. Patients entered in the first data base are those women examined prior to February 1977 and include asymptomatic screenees from the Breast Cancer Detection Demonstration Project and symptomatic private referrals. Their breast temperature measurements were obtained by the technologist using the adjustable cursor described above. A total of 871 women, including 52 cancer cases, were examined in this manner. Of the 52 cancers, 45 were primary lesions, while the remaining 7 occurred in women who had undergone previous mastectomy of the opposite side. Four standard variables listed above were recorded for each breast.

It should be noted that the data obtained during this time period were handled as a separate study and subjected to statistical analyses similar to those performed on the subsequent larger data base.

Normal Class. Patients participating in the Breast Cancer Detection Project were studied as a normal control group. Subjects included in the normal class satisfied the following criteria:

1. No personal history of breast cancer and no previous mastectomy.
2. No breast symptoms, ie no complaints concerning nipple discharge, lumps or pain.
3. Negative breast physical and mammography examinations at the visit during which the thermographic data were collected.

Cancer Class. Of the 52 consecutive cancers collected during the initial months of the program, 45 were entered into the study. The remaining 7 were excluded because of an earlier mastectomy. Subjects included in the cancer class satisfied the following critiera:

1. No personal history of breast cancer and no previous mastectomy.
2. Confirmation of breast cancer by pathologist's examination of tissue excized following the visit during which the thermographic data were collected.

A copy of the pathology report establishing the diagnosis of cancer was obtained for all subjects belonging to the cancer class. These reports, which may be either the hospital pathologist's and/or the project pathologist's report, have been included in the subjects' permanent files. Biopsy material on all cancers was reviewed by Dr Perry Lambird, the project pathologist.

Variable Development

From the four recorded temperature measurements described above, variables were developed and examined as possible diagnostic indicators. The computed values included a variable which is an estimate of the blood perfusion ratio based on a heat transfer analysis described by Love.[13] Other variables which were computed and tabulated included the value of the right areola temperature minus the left areola temperature (DIFA) and the absolute value of this difference (ADIFA). The difference of the average central temperature was labelled DIFC and the absolute difference ADIFC. Also, the difference between the maximum temperature in each breast and the difference between the minimum temperature in each breast were identified by the labels DIFMX and DIFMN, respectively. The absolute values of these differences were labelled ADIFMX and ADIFMN, respectively. In addition, the maximum observed temperature in either breast was listed. The label MAXSIE was used to identify this variable.

Allocation Rule Development

Allocation rules are algebraic equations that allow subjects to be assigned to one of two classes - a cancer class or a normal class. Various allocation rules were examined, including the Linear Discriminant Function (LDF), the Quadratic Discriminant Function (QDF) and the K-Nearest Neighbor. A discussion of the statistical methodology is beyond the scope of this paper.

Briefly, the sample used in the allocation rule developed from the four limited temperature measurements described above was composed of 180 normal and 45 cancer subjects. Using the four temperature measurements from each breast, variates were defined and examined for discriminating information. The most diagnostic variates were based on temperature differences between the breasts and warmest vein temperature on either side. The pertinent allocation rule applied to this data was the QDF using four variates.

RESULTS

In recent years the scientific community has demonstrated considerable interest in developing methods of performing accuracy analyses of new medical imaging systems to ensure they have been thoroughly tested for diagnostic efficacy before introducing them into clinical use. Presently, the method of receiver operating characteristic (ROC) analysis is the preferred approach.[14,15,16,17] The QDF allocation rule results are presented in the form of a separate ROC curve. The ROC curve is a particularly useful format in which to present the data. Since each allocation rule yields its own ROC curve, a complete analysis of how well the rule performed can be made by examining the generated curve. In other words, the ROC curve is a graphic representation of the results, enabling an examination of sensitivity-specificity relationships throughout the entire spectrum of performance from 0 to 100 per cent.

The point of particular interest on the ROC curve is that point corresponding to the decision criterion which minimizes the maximum error rate, ie that point at which the false positive rate and the false negative rate are the same value. This point and others on the ROC curve will be identified and discussed.

As seen in the ROC curve presented in figure 3, the QDF yields the following estimates:

1. Minimax criterion (equal error rates in both groups) of 72 per cent is estimated.
2. At a 10 per cent false positive rate, a 42 per cent true positive rate is estimated.
3. To achieve a 90 per cent sensitivity requires a 45 per cent false positive rate.

DISCUSSION

Other investigators have reported the use of computerized techniques for the automated diagnosis of breast cancer.[18-20] Generally, the hard

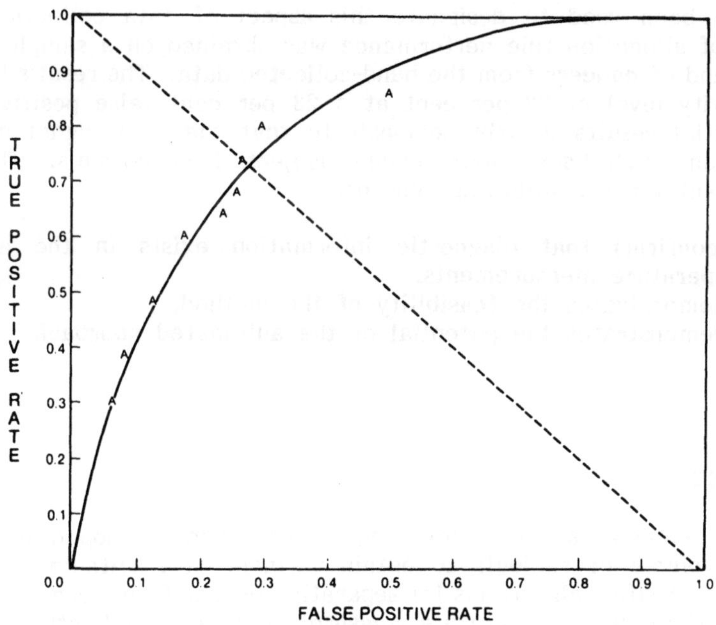

Fig. 3. ROC curve and realized data points resulting from data collected in phase I using the QDF with four variables and based on a sample of 180 normal and 45 cancer subjects.

copy thermogram has been the basis for subsequent digitization. The present program differs by implementing absolute temperatures directly from the infrared scanner.

Shortly after this research program was initiated, a critical issue arose. The issue concerned itself with the ability of thermography to detect breast cancer. Thermography was considered a useful technique when introduced and during the first 15 years of application. Later, due to the very subjective nature of qualitative thermography, lack of interpretation and procedural standards and the absence of training programs, the method developed severe credibility problems during its use in the Breast Cancer Detection Demonstration Projects. The problem was compounded with the publication of an article which concluded that in the area of early cancer detection thermography was no more effective than tossing a coin. This article, published in 1976, was clearly an indictment of the efficacy of thermography.[21]

The first phase of this research program was under way at that time, and the prototype of the absolute temperature thermography system had just been delivered. Following initial testing of the instrument, quality checks and problem solving, data collection commenced. During the first year, data collection was limited to four basic absolute temperature measurements per breast. The absolute temperature values were provided by an analogue-to-digital device and were recorded manually, hence the term "hand-collected data", or limited measurement data, has been used to designate this aspect of data collection. An analysis of allocation rule performance was obtained on a sample of 180 normals and 45 cancers from the hand-collected data. The results indicate a sensitivity level of 72 per cent at a 28 per cent false positive rate. These initial results clearly demonstrate that the ATT signal contains information which discriminates cancer subjects from normals. This fact is important for the following reasons:

1. It confirms that diagnostic information exists in the absolute temperature measurements.
2. It demonstrates the feasibility of the method.
3. It demonstrates the potential of the automated approach.

SUMMARY

An extensive and complex program has been developed to process a large volume of subjects to obtain variable data suitable for use in designing classifier algorithms for separating cancer from normal subjects. Methods of absolute temperature measurements, combined with techniques of pattern recognition and statistical classification (allocation rules), permit a scientific and objective approach. Results indicate the potential of the method.

ACKNOWLEDGEMENTS

This work was supported by research grant N01-CB-43869, National Cancer Institute, National Institutes of Health, Bethseda, Maryland.

The authors are indebted to Martha Linder for assistance in the preparation of this manuscript.

REFERENCES

1. R. Lawson, Implications of surface temperatures in the diagnosis of breast cancer, Canad. Med. Assoc. J. 75:309-310 (1956).
2. K. Lloyd-Williams, Infrared thermometry in the diagnosis of breast disease, Lancet 2:1378 (1961).
3. R. B. Barnes, Thermography of the human body, Amer. Assoc. Advancement of Sci. 140:870 (1963).
4. J. Gershon-Cohen, J. D. Haberman, Thermography, Radiology 82:280-285 (1964).
5. A. M. Lilienfeld, J. M. Barnes, R. B. Barnes, R. Brasfield, J. F. Connell, E. Diamond, J. Gershon-Cohen, J. Haberman, H. J. Isard, W. Z. Lane, R. Lattes, J. Miller, W. Seamon, R. Sherman, An evaluation of thermography in the detection of breast cancer, Cancer 24:1206-1211 (1969).
6. G. D. Dodd, J. D. Wallace, I. M. Freundlich, L. Marsh, A. Zermeno, Thermography and cancer of the breast, Cancer 23:797-803 (1969).
7. H. J. Isard, W. Becker, R. Shilo, B. J. Ostrum, Breast thermography after four years and 10,000 studies, Am. J. Roentgen. 115:811-821 (1972).
8. C. H. Jones, W. P. Greening, J. B. Davey, J. A. McKinna, V. J. Greeves, Thermography of the female breast: a five year study in relation to the detection and prognosis of cancer, Br. J. Radiol. 48:532-538 (1974).
9. O. Melander, Thermography as the primary step in mass screening for breast cancer, Thermography Proceedings, 1st European Congress, Amsterdam. Bibliol. Radiol. 6:91-96, Karger, Basel (1975).
10. J. D. Haberman, T. J. Love, J. E. Francis, Pre-screening a rural population for breast cancer using thermography and physical examination techniques: methods and results - a preliminary report, Ann. N. Y. Acad. Sci. 335:492-500 (1980).
11. M. Gautherie, C. Gros, Breast thermography and cancer risk prediction, Cancer 45:51-56 (1980).
12. B. Threatt, J. M. Norbeck, N. S. Ullman, R. Kummer, P. F. Roselle, Thermography and breast cancer: an analysis of a blind reading, Ann. N. Y. Acad. Sci. 335:501-519 (1980).
13. T. J. Love, R. D. Linsted, Theoretical basis for use of skin temperature as a plethysmographic indicator. Contributions by the Bioengineering Division of ASME, Houston, Texas, December 4 (1975).

14. C. E. Metz, D. J. Goodenough, K. Rossmann, Evaluation of receiver operating characteristic curve data in terms of information theory, with applications in radiography, Radiology 109:297-303 (1973).
15. C. E. Metz, Basic principles of ROC analysis, Seminars in Nuclear Medicine 8:283-298 (1978).
16. L. B. Lusted, General problems in medical decision making with comments on ROC analysis, Seminars in Nuclear Medicine 8:299-306 (1978).
17. J. A. Swets, ROC analysis applied to the evaluation of medical imaging techniques, Investigative Radiology 14:109-121 (1979).
18. J. Winter, M. A. Stein, Computer image processing techniques for automated breast thermogram interpretation, Computers and Biomedical Research 6:522-529 (1973).
19. M. C. Ziskin, M. Negin, C. Piner, M. S. Lapayowker, Computer diagnosis of breast thermograms, Radiology 115:341-347 (1975).
20. P. Friedli, Erkennung das brustkarzinoma mittels computerthermographic dissertation. Eidgenossischen Technischen Hochschule, Zurich, Switzerland (1976).
21. M. Moskowitz, J. Milbrath, P. Gartside, A. Zermeno, D. Mandel, Lack of efficacy of thermography as a screening tool for minimal and stage I breast cancer, N. Engl. J. Med. 295:249-252 (1976).

BREAST THERMOGRAPHY AS AN ADJUNCT TO CLINICAL EVALUATION OF THE BREASTS

I Nyirjesy

Georgetown University School of Medicine
5272 River Road, Bethesda
Maryland 20816, USA

It is estimated that 44% of breast carcinomas and 20% of breast carcinoma deaths occur in women less than 50 years old.[1] Earlier diagnosis of breast carcinomas is rewarded by improved survival rates and by wider choice of therapy.[2]

The two most objective techniques to detect carcinomas are palpation and mammography. Unfortunately, palpation rarely detects tumors smaller than 1 cm in diameter, especially if these are deep seated. Mammography, excellent to detect early carcinomas in older women, must be limited in its use in younger age groups because of the potential carcinogenicity of ionizing radiation.[3]

In the United States, women have been conditioned to obtain annual or semi-annual gynecological examinations and cervical cytosmears. Breast palpation has become a routine part of these periodic check-ups and annual screening by mammography is being recommended to women over 50 years of age.

Since December 1974, infrared breast thermography has been offered to our gynecological patients as an adjunct to their clinical evaluation. Thermograms were taken and classified according to the recommendations of Amalric and Spitalier.[4]

Patients with inconclusive palpatory findings and TH3, TH4 or TH5 thermograms were referred for mammographic studies. Fine needle apsiration biopsies were used liberally and all patients with suspected carcinomas had an open breast biopsy.

The distribution of TH classes in 6,459 thermograms obtained on ·2,799 patients is outlined in table 1.

Table 1. Washington Clinical Thermography Center December 1, 1974 - October 20, 1980

TH Classes	All Examinations (2,799 patients)		Breast Cancer (Prospective)		Breast Cancer (Retrospective)	
	No.	%	No.	%	No.	%
TH-1,2	4,101	63.5	7	20.6	6	17.6
TH-3	1,389	21.5	7	20.6	5	14.7
TH-4	898	13.9	11	32.3	8	23.5
TH-5	71	1.1	9	26.5	15	44.2
Total	6,459	100.0	34	100.0	34	100.0

In prospective reading, about 79% of the carcinomas occurred in 36% of the population with a TH3, 4 or 5 thermogram and 59% of the carcinomas in 15% of the population with TH4 and TH5 thermograms. This distribution tends to confirm that abnormal thermograms are high risk indicators.

The risk of finding a carcinoma in a given patient increased with the thermographic class; the results are outlined in table 2. It is apparent that the great majority of patients with abnormal thermograms had no detectable cancer; consequently the term "positive thermogram" should not be used. Furthermore, it should be emphasized that one of five malignant tumors was detected in women whose thermogram showed no deviation from the normal, and that a negative thermogram, therefore does not rule out the presence of cancer.

Table 2. Risk of breast cancer vs TH class

TH Class	Risk of Breast Cancer	
	Prospective %	Retrospective %
TH1, TH2	0.17	0.15
TH3	0.50	0.36
TH4	1.22	0.89
TH5	12.67	21.13

Table 3. Age of cancer patients*

Age	No.
35 years	4
35-39 years	2
40-49 years	13
50+ years	14
Total	33

* Undeclared - 1

The age distribution of our patients with carcinoma is outlined in table 3. Only 14 of 33 patients were over 50 years of age, and analysis of all the methods used in the detection suggests that there would have been a delay in the detection of 12 of the carcinomas had thermography not been used.

The major events leading to the detection are outlined in table 4 and the correlations between tumor size, node involvement and TH class in table 5.

Among patients without detectable breast cancer, more thermopathological signs were observed when the breasts were anatomically asymmetric, in women with larger breasts and in those who were parous. Twice as many TH4-TH5 thermograms were noted among women with a family history of breast cancer. Acute mastitis was observed to cause marked hyperthermia which disappears promptly following antibiotic therapy. The thermographic manifestation of clinical or radiological fibrocystic disease appears to be variable and needs further study. Visible scars often caused thermographic asymmetry.

Table 4. Detection of 34 carcinomas

Self Detected	6
Mammogram	2
Changed Thermogram	5
Combinations	21
Total	34

Table 5. Thermograms, tumor size, node involvement

TH Class	Tumor Size* 1 cm	1 cm 2 cm	2 cm	Nodes +	Nodes -
TH1, TH2	3	1	1	1	4
TH3	1	2	-	1	2
TH4	3	2	3	1	7
TH5	1	5	10	7	9
Total	8	10	14	10	22

* Unknown - 2 *Unknown - 2

The influence of the menstrual cycle and of exogenous hormones was assessed in 1000 consecutive patients. Breasts influenced by relatively higher estrogen-progesterone concentrations tended to be warmer and slightly more hypervascular. The observed hormonal changes were symmetrical and did not influence significantly the TH class distribution, although high estrogen-progesterone levels make interpretations more difficult. The breasts of postmenopausal patients and of patients using Danazol tended to be avascular. Postmenopausal estrogen increased vascularity but not the asymmetry. Oral contraceptives had no detectable influence on the classification of thermograms. Thermographic changes in pregnant women appear generally early (usually during the fifth menstrual week) and are represented by reticular vascular patterns. Lactational changes are similar to those induced by pregnancy, although in some patients only areolar hyperthermia with radial hyperthermic vessels centered on the nipple could be observed.

Infrared and cholesteric thermograms were compared in 125 patients. Infrared thermograms had two important advantages; they allowed for the observation and comparison of both breasts in a single picture, and they permitted easy temperature measurements. Vascular patterns demonstrated on cholesteric plates were identical with those seen on telethermograms, but imaging of the axillary tails was difficult, and differential temperature measurements were less precise and more time consuming to calculate. The greatest advantage of cholesteric thermography is its relatively low cost.

To conclude, it is believed that breast thermography is a sensitive method of breast imaging which is influenced by numerous anatomical, physiological, pathological and pharmacological variables. Because of the lack of specificity of thermographic images, interpretations without the clinical knowledge of patients may be misleading. In cancer screening, thermography can be used to preselect younger women who are at high risk and who should have a mammogram.

REFERENCES

1. H. P. Leis Jr., Epidemiology of breast cancer: Identification of the high-risk woman, in: "The Breast," H. S, Gallager, H. P. Leis Jr., R. K. Snyderman, J. A. Urban eds., The Mosby Co., Saint Louis, p 38 (1978).
2. H. P. Leis, Selective moderate surgical approach for potentially curable breast cancer, in: "The Breast," H. S. Gallager, H. P. Leis Jr., R. K. Snyderman, J. A. Urban eds., The Mosby Co., Saint Louis, p 232 (1978).
3. C. Henderson, G. P. Canellos, Cancer of the breast, New Eng. J. of Med. 302:21 (1980).
4. R. Amalric, I. Nyirjesy, Analytic classification of breast thermograms and their relationship to breast carcinoma, in: "Breast Disease," D. J. Marchant, I. Nyirjesy eds., Grune and Stratton, New York, p 173-185 (1979).

DETECTION OF SUBCLINICAL BREAST CANCERS BY INFRARED THERMOGRAPHY

R Amalric, D Giraud, L Thomassin, C Altschuler, J M Spitalier

Cancer Institute
Marseille
FRANCE

In our series of 4,000 verified breast cancers we have observed 130 sub-clinical, non-palpable carcinomas of 3 to 5 mm in diameter.

FIRST ALARM

Of these cancers, 10% have been detected by mammography only, 50% by infrared thermography and 40% by both techniques. Thus, there was a thermal alarm in 9 cases out of 10 and this was the first and only sign (Table 1) in 50% of cases.

DIAGNOSIS

These very small, non-palpable cancers, ie belonging to T_0 category of the UICC were located in the following manner (Table 2).

Table 1. First alarm signs of 130 subclinical carcinomas

Radiography only	10%
Radiography + Infrared Thermography	40%
Infrared Thermography only	50%

ie Abnormal radiogram in 50% of cases
Abnormal thermogram in 90% of cases

Table 2. Discovery circumstances of 130 subclinical carcinomas

Unilateral cancers, symptomatic	41%
Bilateral cancers, asymptomatic	59%
- 13% synchronous (at first senological examination)	
- 46% asynchronous (during combined follow-up)	

Bilateral Cancers (59%)

These were identified either through the first senologic examination justified by a contralateral cancer (13% of the cases) or through the combined follow-up of primary cancer treatment in 46% of the cases. These malignancies were all asymptomatic.

Unilateral Cancers (41%)

These were found in women asking for a senologic examination because of a mammary symptom, although a palpable tumor had not been found.

SYMPTOMS OF NON-PALPABLE CARCINOMAS

Mammary pain, mastopathia (54%), nipple abnormalities (32%), superficial mammary and thoracic phlebitis, the so-called Mondor disease (2%), isolated axillary lymph node (2%).

In 10% of the cases, breast examination was suggested only because of a family history of breast cancer (table 3).

It is important that, except in the latter case, these cancers had not been identified by a systemic mass screening.

Table 3. Symptoms of non-palpable carcinomas

Mastopathia, mammary pain	54%
Nipple abnormalities	32%
Family history	10%
Superficial thoracic phlebitis	2%
Isolated axillary lymph node	2%

Table 4. Delay between first alarm sign and appearance of a palpable tumor

Median time	20 months
Average time	24 months
Longest time	60 months

EVALUATION OF THESE NON-PALPABLE CANCERS

In a first retrospective period when the value of thermographic and/or radiological signs were undefined, we waited for the emergence of clinical signs prior to their identification. This occurred in 60% of our cases.

The average time between the first radiothermal signs and the appearance of the smallest palpable tumors was 24 months. The median time was 20 months and the longest is now 60 months (table 4). But, in 12% of the cases an inflammatory tumor appeared, showing thermographic and clinical signs of fast growth, only 1 to 6 months after the first paraclinical alarm. We can therefore say that the fact that the lesion is not palpable is not a guarantee of a good prognosis by itself.

In a second prospective period, we attempted to identify non-palpable cancers through surgery when the thermographic and/or radiographic abnormal signs (suspicious or positive) persisted. The remaining 40% of cases were examined by this method.

TECHNIQUE OF SURGICAL EXPLORATION

This technique involves an exploratory biopsy orientated by the radiothermal pre-operative data: if thermography is the only alarm sign, it is more difficult to identify the precise topography of the carcinoma. The hot spot, or the abnormally hot area is not always centered on the axis of the orthogonal projection of the microcarcinoma on the skin surface. In these cases, the thermal abnormalities are marked out on the skin, under thermoscopic screen. The surgeon then resects the part of the mammary gland surrounding the area determined by the infrared signal. In the case of abnormal mammography with clustered microcalcifications or small stellar images, it is quite easy to orientate the biopsy site for the surgeon. A smaller wedge resection is then possible.

RADIO-SURGICAL CO-OPERATION

The operation for non-palpable cancer must be managed with full radio-surgical co-operation; in every case it is essential to have radiographic control of the excized tissue immediately. The radiological abnormalities are removed and fixed in formalin. Their disappearance is confirmed by means of a new X-ray image. The surgeon closes the skin only when he is sure that the entire abnormality is removed. This technique avoids a false negative caused by an inadequate biopsy. The total duration of this operation is about an hour. The final cosmetic result is satisfactory. The operation is not really mutilating, though a visible scar remains, which is larger than for a simple excisional biopsy done on a small palpable tumor.

Whatever the paraclinical pre-operative data (radiological alone, or thermographic alone) may be, the pre-operative radiological images of the site are finally the sole reliable guide for the pathological specimen. The pathological examination generally requires many serial sections (one hundred) and their reading by the pathologist may take several days. In any case, an immediate histological report is not expected.

RESULTS

Finally, in our experience, a carcinoma has been identified by using this technique 3 times out of 10 in 119 radio-surgical investigations. The other cases correspond more or less to marked mammary dysplasias or border-line lesions or benign infected lesions (table 5).

Infrared thermography and mammography thus enable us to detect an appreciable number of non-palpable breast cancers.

These two methods are complementary: thermography often gives the alarm first, a long time in advance. Radiography is necessary to localize the abnormal site for excision.

Table 5. Histological findings after radio-surgical investigation for thermal and/or radiological alarms

Normal mammary gland	8	7%
Benign anomaly or common dysplasia	66	55%
Severe hyperplasia	12	10%
Non-infiltrating carcinoma	12	10%
Infiltrating carcinoma	21	18%
Total number of cases	119	100%
Total number of carcinomas	33	28%

CONCLUSION

Infrared thermography, combined with mammography, seems to be the best means for early detection of mammary carcinomas, particularly in high risk women.

At the moment, in our Institute, more than 3% of breast cancers are detected and treated at stage T_0 of UICC due to this co-operative approach.

FURTHER READING

1. R. Amalric, D. Giraud, C. Altschuler, F. Amalric, J. M. Spitalier, H. Brandone, Y. Ayme, A. Alvarez-Gardiol, Does infrared thermography truly have a role in present-day breast cancer management? in: "Biomedical Thermology," Alan R Liss, New York (1982).
2. R. Amalric, Y. Ayme, J. M. Spitalier, Infrared thermography in early detection of breast carcinomas, in: "Proceedings of Ninth World Congress of Gynaecology and Obstetrics," Excerpta Medica Publ. Amsterdam, series 512:584 (1980).
3. R. Amalric, M. Gautherie, W. B. Hobbins, A. Stark, R. A. Thierree, Avenir des femmes a thermogrammes infrarouge mammaire anormal isole, La Nouvelle Presse Medicale 10:3153 (1981).
4. M. Gautherie, C. M. Gros, Breast thermography and cancer risk prediction, Cancer 45:1 (1980).
5. M. G. Renzo, del Re Brun, D. Stucki, A. Almendal, J. Torhorst, A new device for isometric localization of non-palpable breast lesions detected by mammography, in: "Proceedings of Ninth World Congress of Gynaecology and Obstetrics," Excerpta Medica Publ. Amsterdam, 512:624 (1980).
6. A. M. Stark, S. Way, The use of thermovision in the detection of early breast cancer, Cancer 33:1664 (1974).
7. A. M. Stark, The results of screening for the early detection of breast cancer, in: "Proceedings of Ninth World Congress of Gynaecology and Obstetrics," Excerpta Medical Publ. Amsterdam, 512:617 (1980).
8. J. M. Spitalier, F. Santamaria, Y. Ayme, D. Giraud, C. Sabbah, R. Amalric, J. Jacquemier, Infrared thermography and non-palpable breast cancers, Acta Thermographica 3:34 (1978).

GLUCOSE TESTS WITH THERMOGRAPHY IN THE DIAGNOSIS OF BREAST CANCERS

B Guduric

Medical Faculty, Novi Sad, Institute for Oncology
Sremska Kamenica
YUGOSLAVIA

Thermography as one of the methods of early detection of breast cancer is well established in many centers all over the world. Opinions on its value vary greatly, from complete negative with only 39% correlation with histology,[1] up to 100% of positive thermograms in cancer patients as shown by Swearingen.[2] In the majority of the papers, however, thermography was positive in 75-85% of carcinoma patients.[3,4,5,6,7] Our own experiences with 10,000 screening examinations and 192 discovered cancers showed 81.64% positive thermograms.[8]

The specificity of thermography had been shown to be rather low with positive findings in benign lesions between 13.8%[9] and 56.4%[10] and between 17%[11] and 50%[10] in normal breasts. We obtained 34.6% positive thermograms in benign lesions and 33.3% in normal persons.[8]

Because of this low specificity the aim of our study was to investigate the possibility of local metabolic elevation of the temperature over breast cancers after administration of glucose and its effects on thermographic examinations.

Based on preliminary investigation of contact temperature measurement over breast cancers after administration of glucose reported by Chiricuta,[12] we started in 1976 to evaluate the effect of the administration of hypertonic glucose on the accuracy of thermography in breast cancer diagnosis.

MATERIALS AND METHODS

Our investigations are divided into two parts:

1. Evaluation of intravenous infusion of 500 ml of 10% hypertonic glucose solution - 127 patients.
2. Oral administration of 100 grams of sugar dissolved in 250 ml of water - 202 patients.

In the first part, all patients admitted for breast biopsy to the Surgical Department of the Institute for Oncology during 1976 and 1977 had a thermographic examination followed by an intravenous infusion of 500 ml of 10% glucose over 30 minutes. Exactly one hour later, under precisely similar conditions, the thermographic examination was repeated. After biopsy, the patients were divided into two groups, according to the histological report - 76 patients with cancer and a benign group of 51 patients. The two groups were analyzed separately.

Intravenous infusion tests proved rather complicated and usually required hospitalization. In the second part of our investigation, glucose was administered orally on an out-patient basis. Patients were given 100 grams of sugar dissolved in 250 ml of water (as practiced in GTT tests). After 90 minutes, thermography was repeated. In this second project, conducted between 1979 and 1981, 202 patients were analysed - 116 cancers and 86 benign conditions in the breast.

An AGA Thermovision type 680B and the modified technique proposed by Gros[13] were used for the thermographic examinations. We modified Gros' criteria for interpretation of thermograms, by lowering temperature values both for delta T_1 and delta T_2, because in our previous investigations we found the original value proposed by Gros to be high.

RESULTS

At the first thermographic examination of 329 patients, positive thermograms were found in 179 patients (56.91%); the result was suspicious in a further 37 patients (11.24%). Thermograms were reported positive if at least 5 of the 8 criteria proposed by Gros were fulfilled and suspicious if only 4 were.

If we considered only carcinoma cases (table 1) then thermography was positive in 119 patients or in 61.97% and suspicious in a further 37 patients (19.18%); so 156 or 81.15% of proven carcinomas of the breast had abnormal thermograms.

After glucose administration in the majority of the carcinoma cases, increased elevation of the temperature (mainly delta T_2 value) was obtained. As shown in Table 1, administration of glucose resulted in an increase of the positive thermograms to 81.25% (156 patients) with suspicious results in a further 16.14% (31 patients), so false-negative results of 18.85% before tests were lowered to 2.61% after administration of glucose. This difference is highly statistically significant with $p < 0.01$.

Table 1. Thermography in cancer patients before and after glucose administration

GLUCOSE TESTS	NUMBER OF PATIENTS	THERMOGRAPHY					
		BEFORE TEST			AFTER TEST		
		Positive	Suspect	Negative	Positive	Suspect	Negative
INTRAVENOUS 10% GLUCOSE INFUSION	76	53 or 69.73%	9 or 11.84%	14 or 18.42%	65 or 83.63%	10 or 13.74%	1 or 2.63%
ORAL GLUCOSE ADMINISTRATION (100 gr OF SUGAR)	116	66 or 56.89%	28 or 24.13%	22 or 18.96%	91 or 78.44%	21 or 18.10%	4 or 3.44%
TOGETHER	192	119 or 61.97%	37 or 19.27%	36 or 18.75%	156 or 81.25%	31 or 16.14%	5 or 2.61%

When we analyzed the two groups separately, ie those who had intravenous glucose and the group who had oral glucose, slightly different results were obtained. In the oral test group, there were 18.96% false negatives at the initial thermogram and 3.44% after the administration of the oral glucose. This is also statistically significant.

In Table 2, elevations of temperature after glucose tests are shown according to different histological types of breast cancer.

Highest elevations were seen in anaplastic and comedo types and lowest in the scirrhous group. This is as expected from the known pathological properties of these tumor types. Tubular and lobular groups, however, also showed high elevation of the temperature, which could not be explained on a pathological basis.

Oral tests produced lower elevations of temperature in almost all histological types but the degree of elevation in both groups according to the pathological behavior of tumors is similar. Patients with carcinoma simplex who received intravenous glucose showed an unexpectedly large elevation in temperature, but in the oral test group the values were much lower and more in accordance with their pathological properties.

Patients with benign breast conditions who received oral or intravenous glucose preoperatively served as a control group for comparison with the cancer patients. Table 3 shows that the elevation of temperature after glucose administration in practically all benign conditions in the breast was very small and insignificant ($p > 0.1$).

DISCUSSION

The origin of the elevated temperature over breast tumors is mainly the result of a higher metabolic rate of tumor cells and increased vascularity in the surrounding tissue.[14] Lawson[15] has also shown that

Table 2. Histological types and mean temperature values before and after administration of glucose

NUMBER OF CASES	HYSTOLOGICAL TYPES	DELTA T_2 ($\bar{x}$)		DELTA T_1 ($\bar{x}$)	
		BEFORE	AFTER	BEFORE	AFTER
88	CA SCIRRHOSUM	1.9°	2.2°	1.0°	1.2°
22	CA MEDULLARE	2.0°	2.8°	1.1°	1.8°
20	CA PAPILLARE	1.6°	2.4°	0.8°	1.4°
18	CA SIMPLEX	2.1°	3.0°	1.4°	1.9°
14	CA COMEDO	1.9°	3.6°	1.5°	1.9°
11	CA GELATINOSUM	1.2°	1.8°	1.2°	1.5°
6	CA ANAPLASTICUM	1.5°	3.8°	0.8°	2.1°
5	CA ADENOIDESCYSTICUM	2.3°	2.8°	1.4°	1.8°
4	CA TUBULARE	1.9°	2.4°	1.0°	1.4°
2	CA LOBULARE	1.6°	2.3°	0.9°	1.2°
2	CA APOCRINES	2.5°	2.9°	1.4°	1.4°
192	TOTAL	1.86°	2.72°	1.13°	1.60°
T° DIFFERENCES		0.86°		0.47°	

the increased metabolism is strictly local and depends on the type of biochemical reaction taking place and the speed at which these reactions occur. Cori and Cori[16] demonstrated experimentally that some rapidly growing tumor cells utilized much greater amounts of glucose than normal cell populations. On the other hand, Bush[17] found that glucose utilization by some tumor cells could be within normal limits, or nearly so. However, one explanation is that the utilization of glucose by cancer cells may depend on environment, the glucose concentration and available oxygen. Finally, Weinhouse and Ono[18] found that rapidly growing tumors have the capacity to utilize more glucose than slow growing tumors or normal tissue, because they have elevated fetal-like enzymes of glycolysis. These enzymes are more active than the usual adult type enzymes. These investigations were confirmed by Ibsen[19] who showed that 40-60% of cell activity during glycolysis in cancer cells was controlled by fetal-like enzymes.

Elevation of the temperature over breast cancers after administration of glucose could be explained on the basis of these experimental data of higher glycolytic properties of cancer cells in general. Several cases of lower elevation of temperature than normally expected on the basis of pathological tumor type can be explained by lower oxygen concentrations in the environment; the majority of these tumors were very large in diameter with poor vascularization of the surrounding tissue.

It is interesting that only two benign lesions gave a measurable elevation of the temperature (0.5° and 0.7°) after administration of glucose. Further histological evaluation showed intraductal papillomatosis and severe adenosis in both cases.

Table 3. Mean temperature values before and after administration of glucose in benign tumors

NUMBER OF PATIENTS	GLUCOSE TESTS	DELTA T_2 ($\bar{x}$)		DELTA T_1 ($\bar{x}$)	
		BEFORE	AFTER	BEFORE	AFTER
51	INTRAVENOUS 10 % GLUCOSE INFUSIONS	1.71°C	1.74°C	0.92°	0.94°
		0.03°		0.02°	
86	ORAL GLUCOSE ADMINISTRATION (100gr OF SUGAR)	1.62°	1.68°	0.84°	0.86°
		0.06°		0.02°	
137	TOGETHER	1.66°	1.71°	0.88°	0.90°
		0.05°		←0.02°	

CONCLUSIONS

The use of hypertonic glucose in our investigations showed good potential for better thermographic evaluation of breast cancers. False-negative thermograms were signficantly lowered from 18.75% before to 2.61% after administration of glucose. The oral glucose test, much simpler to use, showed a slightly decreased potential, but results obtained with this test were also significant.

Although there was relatively good correlation between histological type and the reaction to glucose administration, the latter could not be used to evaluate the malignant potential of any particular tumor because of the many factors which influence tumor behavior.

REFERENCES

1. K. T. Evans, Are physical methods of diagnosis of value? Brit. J. Surg. 56:784-786 (1969).
2. A. G. Swearingen, Thermography: Report of the radiographic and thermographic examinations of the breast of 100 patients, Radiology 85:818-824 (1965).
3. N. J. M. Aarts, Les possibilities de la thermographie dans le depistage des cancer du sein infra-cliniques, Mediterrane Medicale 1:65-68 (1973).
4. R. Amalric, D. Giraud, C. Altschuler, J-M. Spitalier, Value and interest of dynamic telethermography in detection of breast cancers. Acta Thermographica 1:89-97 (1976).
5. C. H. Jones, W. P. Greening, J. B. Davey, J. McKinna, Thermography of the breast: a five year study in relation to the detection and prognosis of cancer, Br. J. Radiol. 48:532-538 (1975).
6. A. M. Stark, The place of absolute temperatures in breast thermography, Acta Thermographica 1:29-33 (1976).
7. A. M. Stark, The significance of an abnormal breast thermogram, Acta Thermographica 1:33-38 (1976).

8. B. Guduric, D. Popovic, B. Sterio, R. Semnic, N. Sljapic, Our experiences with thermography in breast cancer screening on 10,000 examinations, Smene Medicale 34:117-123 (1978).
9. A. P. M. Forrest, I. H. Gravelle, Diagnosis of cancer of the breast with special reference to mammography and thermography, Practitioner 203:137-145 (1969).
10. B. E. Nathan, Value of thermography in breast carcinoma, Br. J. Surg. 2:316-317 (1972).
11. J. Gershon-Cohen, Mammography, thermography and xerography, CA-Cancer J. for Clinicians 17:108-112 (1967).
12. I. Chricuta, M. Bucur, I. Opris, S. Bologa, The correlative value of thermometry and glucose tolerance test in cancer of the breast, Romanian Med. Rev. 15:73-77 (1971).
13. Ch. M. Gros, M. Gautherie, F. Archer, P. Hachnel, C. Colin, Classification thermographique des cancer mammaires, Bull. Cancer (Paris) 58:351-361 (1971).
14. R. N. Lawson, Thermography - a new tool in the investigation of breast lesions, Can. Serv. Med. J. 13:517-519 (1958).
15. R. N. Lawson, J. Gaston, Temperature measurements of localized pathological processes, Ann. New York Acad. Sci. 121:90-98 (1964).
16. C. F. Cori, G. T. Cori, The carbohydrate metabolism of tumors. Change in the sugar, lactic acid and CO_2 combining powers of blood passing through a tumor, J. Biol. Chemi. 65:397-406 (1925).
17. H. Bush in: "An introduction to the biochemistry of the cancer cell," Chapter 10 313-352. Academic Press, New York (1962).
18. S. Weinhouse, T. Ono, Isoenzyme and enzyme regulation in cancer, Cancer Res. Tokyo 13:1-316 (1972).
19. K. H. Ibsen, Interrelationships and functions of the pyruvate kinase isoenzymes and their varian forms, Review Cancer Res. 37:341-353 (1977).

COMBINED DIAGNOSIS: THERMOGRAPHY, MAMMOGRAPHY AND ASPIRATION CYTOLOGY IN BREAST DISEASE

I D Sharma, N C Misra, A Charan, P C Dubey, S Kumar

Breast Cancer Detection Clinic
Department of Surgery, King George's Medical College
Lucknow, INDIA

INTRODUCTION

In India, diseases of the female breast, both benign and malignant, are fairly common. The age-adjusted incidence rate for breast cancer in India per 100,000 population is 25.6. Cancer of the breast is the second greatest killer in the female population after cancer of the cervix uteri. The problem with us is that many patients present with very advanced cancers. Efficient follow-ups were possible only when a detection clinic started screening for both pre-clinical and symptomatic cases. In an effort to improve mortality and morbidity by early detection, a large scale program started in 1979. Self-selected asymptomatic and symptomatic women between the ages of 35 and 75 were eligible to attend and undergo clinical examination, thermography and mammography in addition to answering a questionnaire, and to have the examinations repeated at intervals.

Mammography had proved its value in screening when used in conjunction with clinical examination.[1] Thermography helped to select cases for further investigations such as soft tissue X-rays and aspiration cytology.

The "high risk" group of cases with abnormal thermograms increased the accuracy and the pick up rate. The results of thermography and mammography were compared with histology in a group of patients with definite lumps in the breasts.

The high detection of breast cancer cases in our study is attributed to the fact that the majority of these patients when first seen presented with lumps more than 1 cm in diameter.

MATERIAL AND METHODS

In this study the results, obtained by the combined methods of thermography (T), mammography (M) and aspiration cytology (AC) in 3,000 females with breast symptoms are presented. Cytology was performed under the control of thermography and mammography. The distribution of patients according to age is depicted in figure 1: the majority of patients were in the fifth decade.

Figure 2 shows the relation between TH classification and age. The different surveyed classes progressively increased with age up to the fourth decade, and decreased thereafter, with the exception of the TH5 class.

The majority of our cases were in the fifth decade and therefore they belonged to TH4 and TH5 classes: approximately 300 such cases were operated upon and had histological confirmation. The majority belonged to the "high risk" group.

Thermography was carried out with Thermovision AGA 680 medical equipment, mammography with Mammodiagnostic (Philips) equipment and aspiration cytology using the standard technique. Where aspiration cytology gave an unsatisfactory result at the first attempt, it was sometimes repeated in order to obtain further material.

RESULTS

Table 1 shows the results of the combination of thermography, mammography and clinical examinaation. Breast cancer was detected in 91.8%, and the 8.2% false negatives were attributed to technical faults

Table 1. Thermography and mammography in 300 operated cases

Histological Diagnosis	No. Cases	Thermography and Mammography		
		Detected	False Positives	False Negatives
Cancer	114	105(91.8%)	-	9(8.2%)
Benign tumors	186	160(86.2%)	26(13.8%)	-
Total	300	265(88.3%)	26(8.7%)	9(3.0%)

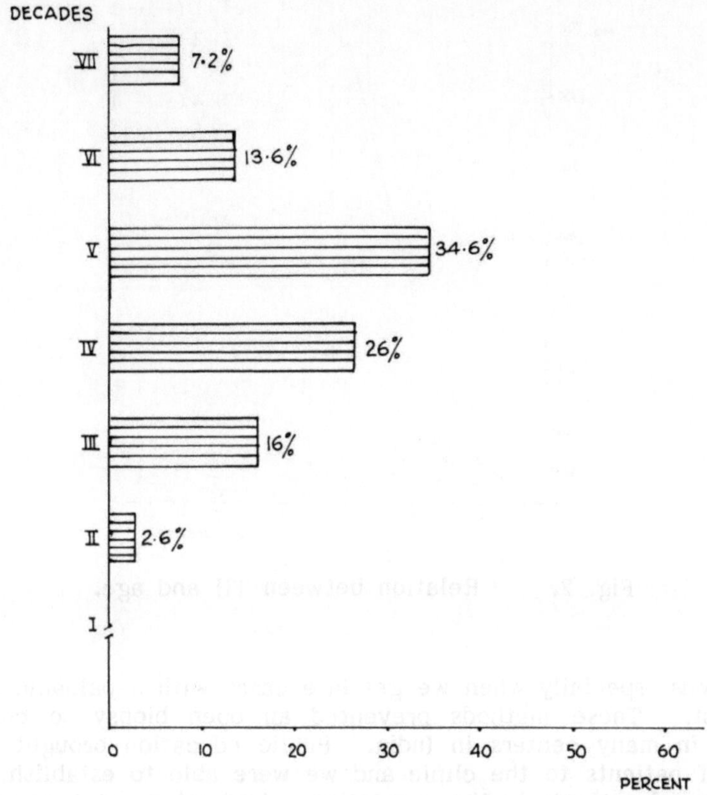

Fig. 1. Distribution of patients according to decades.

and poor vascularity of certain cancers as shown during evaluations.

In the group of benign tumors, fibroadenosis was the commonest condition followed by hard fibroadenoma, reaching a diagnostic accuracy of 86.2% and 13.8% false positive. These 26 cases were borderline on histological examination. Follow-up of such cases continues.

To reduce diagnostic errors, aspiration cytology was added and table 2 shows the results, where AC was included in the pre-operative assessment of 140 cases. There is a sharp decrease in false positives as compared to table 1. Aspiration cytology helped in better detection of cancer and avoided an open biopsy in controversial cases.

CONCLUSION

The combined diagnostic techniques used in our study are best suited

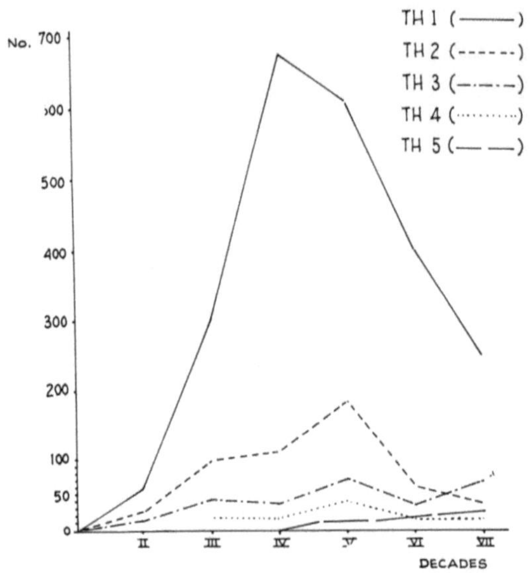

Fig. 2. Relation between TH and age.

to our needs especially when we get late cases with a palpable lump in the breast. These methods prevented an open biopsy so commonly practised in many centers in India. Public education brought a large number of patients to the clinic and we were able to establish a "high risk" group of patients in the population. Aspiration cytology helped

Table 2. The results of combined diagnosis by thermography, mammography and aspiration cytology in operated cases

Histological Diagnosis	No. Cases	Thermography, mammography, aspiration, cytology		
		Detected	False Positive	False Negatives
Cancer	59	55(94%)	-	4(6%)
Benign tumors	81	81(100%)	-	-
Total	140	136(97.1%)	-	4(2.9%)

greatly in achieving diagnostic accuracy, which is in agreement with Donegan.[2] Cases of cancer were detected at an early stage, when the lump in the breast was small and progression of the disease to an advanced stage was prevented.

REFERENCES

1. C. H. Chang, L. Sibala, L. Martin, Breast thermography: identification of differential vascular patterns in breast carcinoma, Acta Thermographica 2:3;138 (1977).
2. W. L. Donegan, "Cancer of the Breast. Diagnosis," W. B. Saunders Company, Philadelphia, London, Toronto (1979).

THERMOGRAPHY IN A BREAST SCREENING UNIT

G V Griffin

Lady McAdden Breast Screening Unit
Southend
UNITED KINGDOM

This unit was formed as a result of discussions at an English society known as the Townswomen's Guild at Southend-on-Sea, UK, in the early 1970's.

A group of women organized a fund to start a local Breast Screening Unit in their town. Approaches were made to the local hospital authorities who referred them to the local Medical Officer of Health. Since the findings of such a unit would immediately call for hospital involvement, two years discussion ensued. During this time the local women continued fund raising. It was finally resolved that in our local situation a breast screening unit was desirable and feasible using palpation and thermography with mammography if palpation or thermography were equivocal.

Our estimate of the cost of setting up such a unit was £80,000. Lady McAdden, wife of our local Member of Parliament, became head of the fund raising campaign and formed a charitable trust to collect the necessary money.

The thermography was to be carried out with a prototype machine devised by physicists at a British military research unit. This system which has a digitalized and computerized output was manufactured by Barr and Stroud, UK.

During fund raising we had to revue our estimates. £100,000 was raised in just over a year.

The equipment was delivered to the unit early in 1976. Nursing and radiography staff were recruited and trained and the unit opened for screening in August 1976.

The unit screens asymptomatic women without charge.

A woman who applies for screening is sent a questionnaire concerning her personal/medical/family/menstrual history. An appointment is arranged at mid-cycle for pre-menopausal women and all are asked not to use perfumes, anti-perspirants or talcum powder on the day of their appointment.

When the patient arrives for screening the data on the questionnaire are checked. She is then invited to strip to the waist and sit in an air conditioned environment at a temperature of $20^oC \pm 1^oC$ with her arms abducted and supported. During this cooling period headphones are available for the patient to listen to music and a message from Lady McAdden explaining the history of this voluntary unit. After 15 minutes cooling a thermographic scan is carried out by a radiographer and a computer print out obtained. Maximum and mean temperatures are computed for the breast areas outlined by the operator who uses a tracker ball to delineate the region of interest. In the post-menopausal an 'A' factor is computed.

$$\frac{(R\ max - L\ max) + (R\ mean - L\ Mean)}{(R\ max + L\ max) - (R\ mean - L\ mean)}$$

and a figure of 350 or above is marked as being suspicious.

With pre-menopausal women a 'vascularity' percentage is computed. Medio-lateral readings of thermal gradients greater than 0.2^oC every 3 mm and extending for no more than 30 mm are automatically noted and the percentage calculated. 26% or over is considered suspicious.

$$\text{"V" L} \quad \frac{\text{Sum L}}{\text{LMO}} \times 100$$

$$\text{"V" R} \quad \frac{\text{Sum R}}{\text{RMO}} \times 100$$

After thermography the woman is then directed to a nurse specifically trained in manual breast examination. At the end of the examination the nurse demonstrates the technique of self-examination of the breast to the woman.

If the results of thermography or manual examination are equivocal either the nurse or radiographer, with the Director's agreement, may ask for a mammogram. This is carried out using GEC equipment and the

Table 1. Lady McAdden Breast Screening Unit - Summary

Date	Number of Screenings 1st time	Recall	Total	Number of Mammograms	No. of women Referred to G.P.	No. of Ca. Breast Confirmed	Ca. Breast Rate/1000 Screenings
Aug 76 - Dec 81	8931	13562	22493	6654	681 (3.0%)	42	1.9

picture is developed by the Xerox process.

The xerographs are read by radiologists in the Southend Hospital Radiology Department. Assuming no abnormality is found, the woman is asked to return to the unit in one year.

Abnormal findings are reported to the patient's own General Practitioner who arranges any necessary clinical action. The screening reports and xerographs are made available to any hospital consultant to whom the patient is referred for further examination.

Table 1 gives a summary of the unit's work to December 1981. 8,961 women have been seen and 22,593 screenings carried out. 681 women were referred to their General Practitioner and 42 carcinomas of breast confirmed. Table 2 shows the age range of women found to have cancer.

We estimate the approximate cost of a single screening visit to be £9. A patient referred for further examination would cost £470 and each diagnosed carcinoma £5,000.

Table 2. Age range of women found to have cancer.

AGE	under 35	36-40	41-45	46-50	51-55	56-60	61-65	66-70	71-80
Aug 76-Dec 77				1	1	1	3		1
Jan - Dec 78					1				
Jan - Dec 79		2	1	2	3	5	2	1	
Jan - Dec 80	1	1	1	1	2	1	1	1	1
Jan - Dec 81			2	2	2			1	1
Aug 76 - Dec 81	1	3	4	6	9	7	6	3	3

CORRELATION OF BREAST IMAGES

H J Isard

Gershon-Cohen Breast Clinic
Albert Einstein Medical Centre
Philadelphia, PA, USA

Early detection is the initial step in the reduction of morbidity and mortality from breast cancer. While breast self-examination is encouraged, the yield for detection of small lesions is disappointing. The expert physician examiner is adept at discovering cancer of the breast, but his ability at detecting lesions under 1.0 cm in diameter or those with only microcalcification is also limited. In the Breast Cancer Detection Demonstration Project conducted in the United States in which 280,000 asymptomatic women were enrolled, it was shown that mammography was more accurate than physical examination, particularly in finding the minimal lesions which accounted for approximately one-third of the cancers detected.

It is well accepted that mammography is the most accurate of the imaging techniques for the detection of breast cancer and therefore should be the standard by which other methods are judged. The anatomy of the breast and the abnormal criteria associated with various diseases are best displayed in mammograms with optimum detail. Microscopic calcifications, circumscribed lesions, tissue distortion, skin thickening and retraction, and differences in radiographic density of the two breasts are important considerations. Currently available film-screen combinations, when used with dedicated mammographic equipment, require a very low radiation dose.

Non-invasive imaging techniques are attractive inasmuch as they are entirely safe and do not require ionizing radiation. These include thermography, diaphanography and ultrasonography. This presentation reflects our experience in correlating mammography, thermography and diaphanography.

Medical thermography is a non-ionizing imaging technique that captures the natural invisible infrared radiation emitted by the skin and translates it to a visible picture, either electronically or by a liquid crystal contact method. The major advantage of thermography is its reflection of biological activity and the associated physiological and pathological processes. Concurrently the configuration of the breast is displayed so that alterations in contour resulting from underlying disease can be seen. Criteria of abnormality have been established and many studies have emphasized the importance of the abnormal thermogram as an indicator of significant risk. In addition, it has been shown that breast thermography can be used as a prognostic tool for estimating survival of patients with breast cancer.

Transillumination of the breast is not a new study and had been used over 50 years ago but it has never been widely accepted. The concept of documenting the visual findings on infrared sensitive color film has certain advantages and the combined technique is referred to as diaphanography. The breast serves as a complex filter to the light and because of the absorption characteristics of human tissue in the near infrared range, the use of infrared sensitive color film is preferable. The normal breast when transilluminated has a reddish-orange color, the vascular structures are undulating and tapered, and their margins are smooth and regular. Marked disparity frequently exists between radiographic and optical density so that breasts with considerable homogeneous radiographic density may be quite translucent. This is, of course, particularly helpful in the evaluation of the radiographically dense breast. Benign cysts which are translucent appear quite pale and yellow while fibroadenomata have a reddish component. Both of these lesions may appear similar on a mammogram but are obviously different on transillumination. Criteria of abnormality reflecting malignancy are being established and consist of both alterations in color and shading as well as distortions and irregularities of the vascular structures.

In symptomatic women mammography approaches 90% accuracy in detecting breast cancer, and during the past decade we have found thermography to be suspicious in approximately 75 to 80% of the proven malignancies. Diaphanography has demonstrated abnormalities in 75% of the first 150 cases of breast cancer that we have studied with this modality. Each of the non-invasive imaging techniques presents a unique dimension with information not otherwise available and, when used in conjunction with mammography, diagnostic accuracy is improved and women can be spared unnecessary biopsies.

BAYESIAN ANALYSIS OF THERMOGRAPHIC DIAGNOSIS OF BREAST CANCER

D Malatin

Institute of Clinical Oncology
Bratislavia
CZECHOSLOVAKIA

In thermographic literature terms like "clinical responsivity of the method", "sensitivity", "false positivity and negativity" are often used. When expressed in percentage form they give a more exact indication of the results achieved. According to the Information Theory and its Bayes Theorem it is obvious that such numerical values can be of different magnitude within the same series of measurements because of their close connection with the phenomenon previously taken to be a manifestation of a positive, negative or equivocal diagnostic result. It is necessary to bear in mind that an increase in one statistical parameter of the method is usually connected with a decrease in one or more other parameters.

In a series of thermographic breast examinations we applied a Bayesian analysis in order to evaluate the accuracy of our findings. From a series of 2,983 thermographic breast examinations a group of 275 women were thus analysed. There were 53 histologically proven cancers and 222 benign tumors. The examinations were made with an AGA Thermovision 680 scanner, using a color monitor and the OSCAR computer system. The examinations were carried out under standardized conditions. The thermograms in both malignant (Fig 1) and benign (Fig 2) cases were evaluated using the 5-grade TH system devised by Amalric.[1] Simultaneous evaluation with the O-system of Gros et al[2] had a validity of 3-6% less and was not used in further analysis.

The main statistical parameters of the thermographic examination were calculated from the four poles contingent table (Table 1a), made up from the distribution of the number, n, of examinations thought to be positive, +ve, or negative, -ve, by the thermographic test, S(thv), with the histological finding D. The false positivity of the test is F +ve and the false negativity F -ve.

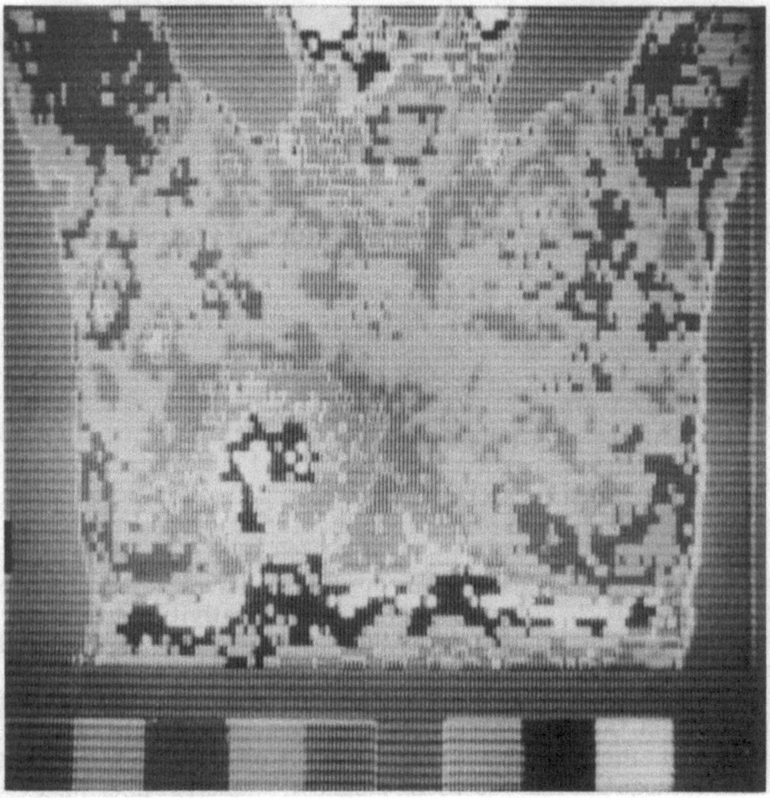

Fig. 1. Thermogram - breast cancer.

The distribution F (%) of the thermographic diagnosis TH in the measured series is shown in Figure 3. For TH5 the probability of malignancy is 0.51 and the probability of benign disease 0.0045 (<1). For TH3 the probability of malignancy is 0.28 and of benign disease 0.21. For TH2 the probability of malignancy is 0.02 and of benign disease 0.55. Therefore, for a thermogram classified as TH5 malignancy is 113 times more frequent than benign disease, while for one classified as TH2 benign disease is 28 times more frequent than malignancy.

If Bayes Theorem is applied to the statistical analysis the results are different (Table 2, a and b). If TH3 - 4 - 5 or TH4 - 5 are called positive and TH 1 - 2 or TH1 2 - 3 negative, sensitivity was 96% (68), specificity 73% (94), positive predictive value 46% (73), negative predictive value 99% (92), repeatability 45% (54) and validity 18% (13), false positivity was 27% (6) and false negativity 4% (32).

From this data it appears that statistical parameters for assessing

Table 1a.

		D_{hist}		
		+ M	− B	Σ
S	+	a +	b_{F+}	a + b
thv	-	c F-	d −	c + d
	Σ	a + b	b + d	a + b + c + d

Table 1b.

P (%)	
$SE = \dfrac{a}{a+c}$	$SP = \dfrac{d}{b+d}$
$PPV = \dfrac{a}{a+b}$	$NPV = \dfrac{d}{c+d}$
$R = \dfrac{a}{a+b+c}$	$V = \dfrac{a}{a+b+c+d}$

Probability data in % (Table 1b) were as follows:-

Sensitivity	SE
Specificity	SP
Positive predictive value	PPV
Negative predictive value	NPV
Repeatability	R
Validity	V

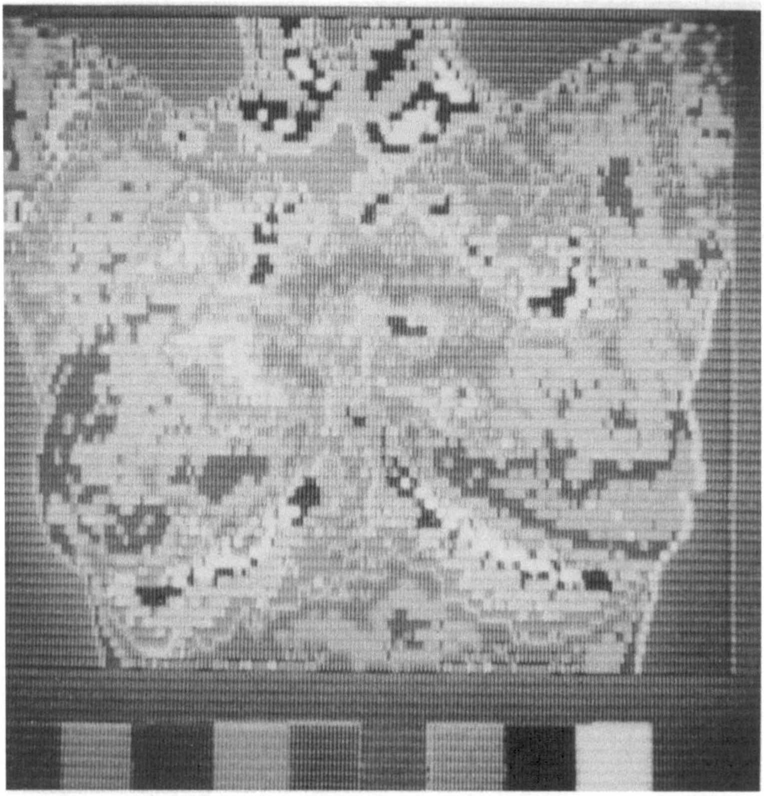

Fig. 2. Thermogram - benign breast disease.

the thermographic findings are to a high degree dependent on which TH grades are designated positive and negative. A shift of the equivocal grade TH3 from positive to negative causes substantial changes in nearly all the statistical parameters. If TH4 - 5 is designated positive there was a very low false positive rate of 6%, but the false negative rate was very high at 32%, which substantially decreases the value of the thermographic examination. In clinical practice it is therefore more advantageous to designate grades TH3 - 4 - 5 positive, in which case the false negative rate is 4% although there is a 21% fall in specificity.

The high negative predictive value of thermography in both variations, as well as the high repeatability and validity are considered to be of practical significance.

Table 2a

n

			D		
		TH	+M	-B	Σ
S thv	+	3 + 4 + 5	51	60	111
	+	4 - 5	36	13	49
	–	1 - 2	2	162	164
	–	1 - 2 - 3	17	209	226
		Σ	53	222	275

Table 2b.

S		
+	TH 3 - 4 - 5	TH 4 - 5
-	TH 1 - 2	TH 1 - 2 - 3
SE	96	68
SP	73	94
PPV	46	73
MPV	99	92
R	45	54
V	18	13

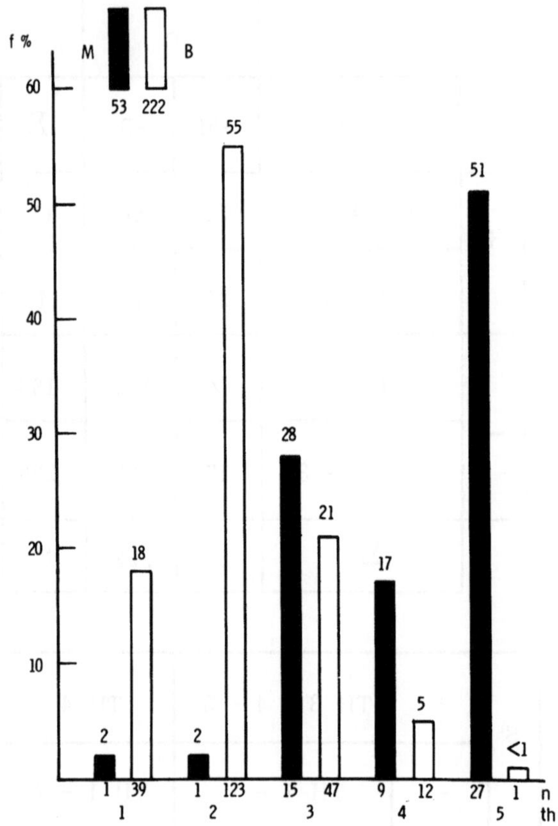

Fig. 3. Distribution of the thermographic diagnosis.

REFERENCES

1. R. Amalric, Mediterranee Medicale 216:13-22 (1972).
2. Ch. Gros, M. Gautherie, F. Archer, Bull. du Cancer 1-58-1:69-90 (1971).

STUDY OF BREAST TUMORS BY THERMOGRAPHY, ^{67}GALLIUM CITRATE AND ECAT USING ^{13}N-AMMONIA

L de Thibault de Boesinghe, K Schelstraete

Department of Radiotherapy and Nuclear Medicine
University Hospital
Ghent, BELGIUM

Thermography is a useful tool for the diagnosis of breast tumors. The thermographic criteria of a malignant tumor are hyperthermia and vascular anarchism. Thermography is a non-invasive technique and no ionising radiation is needed for the examination of the patient. However a number of malignant tumors are "cold" without vascular anarchism and are not diagnosed by this method. For this reason we also investigated other atraumatic methods.

Breast scintigraphy with ^{99m}Tc pertechnetate and ^{67}gallium citrate have been mentioned in the literature.[1,2] ^{99m}Tc pertechnetate demonstrates a good correlation between malignancy and positive scintigraphy; 88% accuracy in 16 cases of breast carcinoma. However, the high false-positive rate (29% of proven benign breast disease) limits the use of ^{99m}Tc as an aid to differential diagnosis. ^{67}gallium citrate is of limited value as a diagnostic adjunct since it localizes in only five out of ten breast malignancies. Since the extent of tumor involvement is a significant prognostic factor, the gallium scan nevertheless merits consideration for the detection of mediastinal metastases.

The ECAT (Emission Computerized Axial Tomography) whole body scanner measures and reconstructs rectilinear and cross-sectional distribution of compounds labelled with positron emitting radionuclides such as ^{15}O, ^{13}N and ^{11}C. In many ways, this system is to in vivo studies what autoradiography is to in vitro studies. It provides quantitative tomographic images of physiological functions (eg blood flow, blood volume, metabolism, membrane transport), depending on the tracer used. We have used ammonia labelled with ^{13}Nitrogen, which has a half life of 10 minutes. The distribution of intravenously injected NH_3 corresponds to the regional capillary blood flow and can also be regarded as a reflection of the local metabolic activity of the tissues.

Table 1.

Patient Number	Thermography	ECAT	Gallium	Tumor Localisation
1	+	+	+	sternal metastasis
2	+	-	+	breast
3	+	-	+	sternal metastasis
4	+	+	(not + so + as ECAT)	breast
5	+	+	-	breast
6	+	+	+	breast
7	±	-	±	breast
8	+	+	no scan	breast
9	+	-	no scan	
10	- -	- +	- -	breast axillary node
11	- +	-	+ -	breast + cutaneous metastasis

<u>Note</u> : patient 3 has thermographically a central cold area due to necrosis and has lung metastases also positive with ECAT

In our study we have made a choice of patients with proven breast cancer and positive thermography and have looked at the results obtained with ^{67}gallium and ECAT using $^{13}NH_3$. Two patients with negative thermography in proven breast cancer were also included in the study. A total number of 11 patients was studied. The results are summarized in table 1.

DISCUSSION

No direct correlation between the three techniques (thermography, ^{67}Ga scan and ECAT with $^{13}NH_3$) was found. Some thermographically "hot" tumors with vascular anarchism were not visualized by gallium or the ECAT examination; on the other hand some tumors that are not diagnosed by thermography can be visualized by ECAT, such as metastases.

Preliminary studies in other tumors such as malignant melanoma give similar results.

The ^{67}Ga scan does not seem to provide additional information.

Further examinations are being performed to determine whether the results obtained by ECAT and $^{13}NH_3$ could be used to predict the response of the tumor to chemotherapy. Our hypothesis is that for a good response the tumor needs a good capillary circulation. This capillary circulation is probably not directly visualized by thermography, but might be expressed by $^{13}NH_3$ uptake.

Further research is required to clarify which are the most significant thermal diagnostic criteria in the diagnosis of malignancies: the temperature gradient or the vascular disturbances which can be visualized by thermography.

REFERENCES

1. S. D. Richman, Breast scintigraphy with ^{99m}Tc pertechnetate and ^{67}Gallium citrate, J. of Nuclear Medicine 16:4 (1975).
2. S. D. Richman, Usefulness of Gallium scintigraphy in primary and metastatic breast carcinoma, J. of Nuclear Medicine 16:11 (1975).

FURTHER READING

1. C. W. Williams, The ortec ECAT. Vth international conference on information processing in medical imaging. Van der Bilt University, Nashville Tenessee (1977).
2. L. Thibault de Boesinghe, La telethermographie infrarouge dans l'etude de la pathologie mammaire monographie : Diagnostic et traitement des affections mammaires, Medical World Press (1980).

THE PROGNOSTIC VALUE OF INFRARED THERMOGRAPHY FOR OPERABLE BREAST CANCER

J M Spitalier, R Amalric, D Giraud, H Brandone, Y Ayme, J F Pollet, F Amalric

Cancer Institute
Marseille
FRANCE

Particular tribute should be given to British research, not only for the discovery of infrared rays by Sir William Herschel in 1800 but also for their application to prognosis in breast cancer during the 1960's by the Congress President, K Lloyd Williams. These studies were pursued further in Germany by D von Fournier and in France by M Gautherie, who showed that the thermogenicity of breast cancer is directly proportional to its doubling time.

But the thermal message emitted by breast cancer still needs to traverse a long and difficult path before it can appear in a form which can be interpreted by the clinician. En route, it encounters numerous potential causes of false negatives due to possible deficiencies of transmission, convection, re-emission and detection. The thermal prognosis of breast cancer is more likely to be falsely favorable rather than falsely unfavorable.

MATERIALS AND METHODS

Patients are submitted to an infrared thermographic examination prior to other diagnostic measures and before any treatment. The thermogram is classified according to our 5 categories of increasing thermographic grade (TH1 - TH5) and into one of the 4 categories estimating the rapidity of growth (PEV_0 - PEV_3). Each patient is then staged using traditional clinical and radiographic tests, treated with conservative or radical therapy according to our criteria and above all, followed regularly for an indefinite time period.

INFRARED THERMOGRAPHY AND OTHER PROGNOSTIC FACTORS

Grade of Carcinoma

Table 1 shows that 62% of the Grade III tumors (according to the system of Scarff and Bloom) are accompanied by a mammary gradient of 3°C or more, whereas only 11% correspond to a 2°C gradient or less.

Axillary Involvement

Among 201 consecutive patients with positive nodes, the mammary gradient was 3°C or greater in 70% and 2°C or less in only 10% (table 2).

INFRARED THERMOGRAPHY AND ESTIMATION OF RAPID GROWTH

As a single investigation, infrared thermography can anticipate the early failures before three years (table 3). Therapeutic success can be shown in patients with more than five years (table 4).

At 3 and 5 years the percentage of patients with a clinical cure is a regular function of the thermographic grading category (table 5).

If one divides the gradients at 3°C or at 4°C, the differences in 3 and 5 year survival are significant (table 6).

INFRARED THERMOGRAPHY AS SOLE PROGNOSTIC TEST

In our experience infrared thermography has prognostic significance in its own right. For a given stage (UICC stage II, that is T1-T2/N1), in a given histological grade (Grade II of Scarff and Bloom), and the absence of any suspicion of rapid growth (PEV_0), the results of axillary staging are very different, depending on whether the thermographic gradient is greater than 3°C or 2°C and less. This is also true when one considers the question of the presence or absence of axillary involvement (table 7).

Table 1. IR thermography in 120 Grade III (SBR) breast cancers

$\Delta >$	3.0°C	62%
=	2.5°C	27%
$\Delta <$	2.0°C	11%

Table 2. Mammary IR thermography in 201 consecutive node + patients

$\Delta \geqslant$	3.0°C	70%
=	2.5°C	20%
$\Delta \leqslant$	2.0°C	10%

Table 3. 3 year mortality in 823 operable (T_{1-3}/N_{0-1}) breast cancers

Thermographic fast-growing type :	60%
Thermographic slow-growing type :	14%

DISCUSSION

We have shown that in breast thermography the greater the thermographic gradient, the shorter the survival of the patient. Infrared thermography thus contributes independently to the results of other investigations (clinical stage, histological grade, axillary staging, hormone receptors) in order to determine whether or not a patient should be considered to have a high risk of treatment failure. This is of great importance in defining sub-groups of patients suitable for adjuvant systemic therapy following curative local regional treatment.

This emphasizes once again that local manifestations of breast cancer should be considered as indicators of more generalized sub-clinical disease, which is much more difficult to control. We request the surgeons not to destroy a breast cancer before first extracting the maximum amount of data. This should include infrared thermography to study the thermal information. We believe that clinicians should note that this is the first time that such an active disease can be examined in terms of energy. This presents an excellent opportunity for the development of a new area of clinical research.

Perhaps the greatest criticism of infrared thermography is that its message is so anticipatory. The information provided has more bearing

Table 4. 5 year disease-free survival in 432 operable (T_{1-3}/N_{0-1}) breast cancers

Thermographic fast-growing type :	19%
Thermographic slow-growing type :	61%

on the patient's future than on the present. This has given rise to many persistent controversies over the use of infrared thermography, reflecting a misunderstanding of its true role, and a lack of on-going studies conducted over adequate time.

Table 5. % of Stage I UICC (T_{1-2}/N_0) patients alive with no evidence of disease

	TH_{1-2}	TH3	TH4	TH5	TH-fast-growing type
3 years	84	81	79	71	42
5 years	64	63	59	53	8

Table 6. % of Stage II UICC (T_3/N_{0-1}) patients alive with no evidence of disease

	$\Delta \leqslant 4^oC$	$\Delta \geqslant 4^oC$
3 years	58	37
5 years	42	21

Table 7. Mammary IR gradient v axillary staging in 188 consecutive Stage II UICC (T_{1-2}/N_1) breast cancers

	$\Delta \leqslant 2.5^oC$	$\Delta \geqslant 3^oC$
N + Total	36%	61%
N + > 4/Total	25%	49%
N + > 4/N+	9%	30%

RESULTS OF CONTACT THERMOGRAPHY USING THE BTD SYSTEM IN 1500 PATIENTS WITH 100 MAMMARY CARCINOMAS

M Bauer, V D Fournier, R Schulz-Wendtland, F Kubli, U Spah
Universitats-Frauenklinik
Heidelberg
WEST GERMANY

A prospective study on contact thermography with 1500 patients was carried out at the Universitats-Frauenklinik and Universitats-Strahlenklinik between August 1981 and January 1982. This was done at the clinic for breast diseases in Heidelberg, where about 7500 patients are examined each year. A total of 232 breast carcinomas were diagnosed and operated upon during 1981.

Within this clinical routine we have tested a new liquid crystal thermography system, developed by Rohm-Pharma. The apparatus is called BTD, the letters standing for Breast Thermography Detector.

This system claims to offer two advantages:

1. For the thermographic diagnosis of breast conditions one single contact thermography plate should be sufficient.
2. A flexible frame system should enable a simultaneous critical assessment of both breasts.

In the first case, the test plate has an exact thermographic image within the temperature range 31°C to 33°C. The image is definite and the thermographic sensitivity good.

On colder breasts the amplitude of the temperature, however, is not sufficient on this plate; only the increased heat over the tumor is revealed. In this case an additional plate with a temperature range of 29° to 31°C must be used. In the case of hot tumors, the thermographic image was frequently uniform within the dark blue range of the spectrum. Only examination with other plates (33.5° - 36°C) gave sufficient information for diagnosis.

Thus, in our opinion, at least three thermographic plates with the following grades: 29.5 to 31°, 31 to 33.5° and 33.5 to 36°C are required.

Simultaneous examination of both breasts proved to be valuable. However, there were considerable problems at first with photographic recordings due to high reflection of the foil. To overcome this, fixation of the image had to be made with a specially constructed plate for each separate breast.

RESULTS

Tables 1-5 show the results of the 1500 thermographed patients. On principle both a static and a dynamic thermogram were assessed. The dynamic, ie rate of heating of the plate, proved more useful than the static examination. The classification of results was made using Tricoire's criteria.[1]

In Table 1 thermographic findings are shown in 1278 cases investigated clinically and radiographically and in 158 cases histologically verified. False positive thermograms (TH IV/V) were the same in both groups, just under 5%.

Table 2 shows the thermographic findings in 40 preinvasive carcinomas and high-grade proliferations (III) with atypical cells. 32.5% results were suspicious and 27.5% required further investigation, ie when both groups were combined, 60% of the results were abnormal.

Table 3 reproduces the thermographic findings of 100 verified carcinomas, in the left column without knowledge of the clinical findings. A correct positive diagnosis was made in 58%. When clinical findings were known, as in the right hand column, the accuracy was 77%. That means that in 23%, only mammography was suspicious. Also interesting was the vascularization pattern (table 4). In three out of four cases type C was found.

Table 1. Normal or benign findings : n = 1278

With histological verification	Without histological verification
n = 158	n = 1120
Th I/II = 74.7%	Th I/II = 84.5%
Th III = 20.9%	Th III = 10.6%
Th IV/V = 4.4%	Th IV/V = 4.9%

Table 2. Preinvasive Ca-in-situ
Proliferations grade III according to Bassler : n = 40

Th I/III	= 40.0%
Th III	= 27.5%
Th IV/V	= 32.5%

Table 3. Carcinoma : n = 100
Thermographic findings - Carcinoma = 100%

	Thermographic Findings Without Clinical findings	Thermographic Findings With Clinical findings
Th I/II	33.0%	13.0%
Th III	9.0%	10.0%
Th IV/V	58.0%	77.0% ie in 23% of further cases mammography only was suspect.

Table 4. Carcinoma : n = 100

Vascularization	Frequency
A	4.0%
B	11.0%
C	75.0%
D	8.0%
E	2.0%

More important was the certainty of diagnosis in relation to the tumor size (table 5).

The most essential requirement for early diagnosis, however, is the ability to detect pre-clinical tumors. Here we found the correct positive rate by this method is reduced from 38 to 20%.

Table 5. Carcinoma - tumor size - reliability of diagnosis: n = 100

Tumor size	Number of cases	Reliability of diagnosis = "correct positive" (of the actually found carcinoma)	
T_0	15	3/15 = 20.0%	5/15 = 33.3%
T_1	24	9/24 = 37.5%	17/24= 70.8%
T_2	42	28/42 = 66.7%	36/42 = 85.7%
$T_{3/4}$	19	18/19 = 94.7%	19/19 = 100.0%
T_{0-4}	100	58/100 = 58.0%	77.0% ie: in 23% of all further cases mammography suspect only.

False Positive Thermograms

There are an unknown number of correct positive thermograms which have not been verified histologically because of negative clinical and mammographic findings. This is because during the natural history of any tumor, there is a period when it is undetectable clinically or radiologically.

SUMMARY

The results using this method of contact thermography are as follows:

1. Within the temperature range 31° to 33.5°C the sensitivity of the plate is equal to that of other contact systems for thermographic diagnosis.

2. However, two additional plates are required - one in the lower and one in the higher range of temperature. The comparison of both breasts obtained by using two plates simultaneously seems to have potential. Further studies are required to confirm the value of this technique.

REFERENCES

1. J. L. Tricoire, J. P. Marier, G. Poirot, J. Lacour, S. Fajbisowicz, Thermographie en plaque, Press Med. 78:2483 (1970).
2. J. R. Bissler, Pathologie der Brustdruse, in: Dorr Seifert Uelinger: Spezielle pathologische Anatomie, Band II," Springer, Basel, Heidelberg, New York (1978).

BREAST CANCER SCREENING INDICATOR (BCSI): A HOME-USE DEVICE FOR THE DETECTION OF BREAST PATHOLOGY

H L Karpman*, B Hamilton, S Cutler****

*** School of Medicine, Los Angeles, California, USA**
**** Georgetown University, School of Medicine, Washington DC, USA**

Breast cancer is the major malignant lesion[1] and leading cause of death from cancer in the United States. It rates second only to heart disease as the leading killer of American women,[2] and is the leading disease-related cause of death in women between the ages of 40 and 65. The disease is diagnosed every 7 minutes and, in addition, a death occurs every 15 minutes due to its effects in the United States. Approximately 106,000 new cases of breast cancer in the United States are diagnosed each year and 1 out of every 11 women will develop breast cancer during her lifetime. Until breast cancer can be prevented, the greatest hope for its control is early detection, diagnosis and treatment; if the disease can be detected prior to the appearance of signs and symptoms, marked improvement is possible in the survival rate. Intensive research has been and continues to be conducted in each of these areas, but to date it is generally accepted that the most effective means of control lies in the early detection of the lesion.[1,3,4,5]

Self examination of the breast currently remains a valuable method for the "early" detection of breast malignancies since 90% of cancers are detected by the patient herself.[6] Women who practice self examination have been reported to detect malignancies at a signficantly earlier stage[7] and have experienced a higher survival rate.[8] However, despite the fact that cancer cells have been demonstrated to double in size as rapidly as within 23 days,[4] the average doubling time for breast cancers has been demonstrated to extend for as long as 4 months[9]; therefore, before a breast mass is large enough to be palpated by the patient or the physician, it has, on average, been slowly enlarging in the breast for at least 2 years. Curability of breast cancer obviously depends upon the detection of the malignancy at its incipiency, before it has grown sufficiently to metastasize locally or distally and, therefore, it follows that the malignancy should ideally be detected before it is palpable.

Regular screening programs with resultant early detection of malignancies have reduced the incidence of nodal involvement by approximately 30%.[10] Detection of the malignancy before its spread is an important factor in prognosis, especially for women over the age of 50.[11]

A thermogram is a pictorial representation of the infrared energy emission from the body. Newly formed malignant tumors are thought to emit more heat per unit size and, therefore, because of these metabolic abnormalities combined with an increased regional blood supply, emit more heat than the surrounding normal breast tissue.

The Breast Cancer Screening Indicator (BCSI) device[12,13] consists of a pair of pads made of spun fibre material with wafer-thin, pliant thermosensitive units attached to the inner surface. The number of units depends upon the breast size, with cup sizes A and B using three units on a pad approximately six inches in diameter, and cup sizes C and D utilizing four units on a larger pad.

The BCSI device has been tested clinically at several locations. Validation studies[12,13] to determine the efficacy of the chemical heat sensor in detecting breast cancer in women scheduled for biopsies of the breast were performed at Memorial Sloan-Kettering Hospital, New York, MD Anderson Memorial Hospital, Houston, Texas, and at Brotman Memorial Hospital and UCLA, Los Angeles, California. The subjects for the study were women of all age groups who were admitted to hospital for a definitive biopsy procedure. Studies were performed with the BCSI device prior to the scheduled surgical biopsy, and the results were subsequently evaluated statistically by Dr Sidney Cutler of the Division of Biostatistics and Epidemiology, Georgetown University School of Medicine, Washington DC.

Of the 133 women who participated in the study to date, 79 were found to have cancer. The BCSI device was found to be more accurate in women subjected to unilateral breast biopsies than in patients who were subjected to bilateral biopsies; bilateral cancers may produce similar heat emission patterns and therefore differences of over 1.5^{o}F may not be present in a large percentage of patients with bilateral pathology.

Unilateral biopsies were performed in 101 patients and the device was positive in 87% of the 53 cancers detected. One patient had an axillary tail cancer which was outside of the range of the BCSI device; if we exclude this patient from the results, the device was actually positive in 89% of the unilateral breast cancers. The sensitivity of the BCSI device in symptomatic patients therefore appears extremely high and at least equally as sensitive as other non-invasive methods for breast cancer detection in the symptomatic patient.

A second major study was performed in asymptomatic women who were being screened at the Guttman Breast Diagnostic Institute, New

York. This study was designed to test the efficacy of the BCSI device when compared to standard clinic screening in detecting a significant breast abnormality. A "significant breast abnormality" is defined as a condition detected via clinic screening which warranted a recommendation for further investigation for the possibility of cancer of the breast, eg either biopsy or aspiration. Standard clinic screening consisted of obtaining a history, performing a physical examination, and then correlating these results with mammographic and/or thermographic studies which were performed on the majority of the patients. The basic questions investigated in this study were:

1. How often does the device produce a positive reading in women with significant breast abnormalities; and/or
2. How often does the device produce a negative reading in women with no significant breast abnormalities.

The BCSI tests were performed under the supervision of the clinic assistant before the standard screening procedures were performed and the results were correlated with the screening findings.

The study demonstrated that, of the 1,350 women screened, 1,312 were not recommended for biopsies or aspirations and of these, 1,116 had negative BCSI readings for a specificity index of 85%. The BCSI reading was positive in 43 of the 48 women recommended for biopsy or aspiration, for a sensitivity index of 90%. Those women with a "suspicious" clinic screening were recommended for a three-month recall. If these three-month recall women were added to the total group, the total group number would increase to 63 women, 55 of whom tested positively with the BCSI device, for a sensitivity index of 87%. False negatives in this group consisted of five out of 48 women, for an incidence of 10.4%. However, when this value is corrected statistically it is important to note that, based upon the experience of the Guttman Institute during 1978 and 1979, the percentage of true false negatives with respect to breast cancer dropped to as low as 1.3%, but probably not more than 2.3% of the total.

The false positive rate in the asymptomatic group was 14.9%; obviously, a so-called false positive errs on the side of being conservative. Previous thermographic studies[14] have clearly demonstrated that more than one-third of patients with "suspicious" thermographic studies will develop breast cancer within a period of five years, therefore, a so-called "false positive" finding, in addition to being caused by thermal variations induced by the effects of environment, stress, menses and/or other factors, may actually be an indication of a high-risk patient.

The BCSI device is one of a growing class of devices which are intended for home use. It is designed to complement, not replace, other non-invasive methods for the evaluation of breast pathology, including, but not limited to self examination of the breast. The end result of its use will be to bring a growing group (ie true and "false" positive patients) of women to the physician's office for a more thorough examination including physical examination, mammography, thermography, ultrasonography, and possibly even biopsy. If adequate utilization occurs, a population of women with a higher frequency of breast pathology, including cancer, will present themselves to their physicians for these expanded diagnostic studies.

Large-scale studies are now being conducted on asymptomatic women who are wearing the BCSI device on a monthly basis in an attempt to detect the thermal changes in the breasts at the earliest possible time. Women will wear the device for 15 minutes each month and, in addition to comparing corresponding areas of each breast on a monthly basis, each section of each breast will become its own control and will be compared to itself on a continuing basis for months or years. In this way, any thermal alteration which occurs as a result of breast pathology and/or cancer which is thermally active will make itself immediately known to the patient, who will then consult her physician for advanced diagnostic studies. Bringing the patient into the diagnostic cycle at this early phase will undoubtedly improve the morbidity and mortality statistics in breast cancer.

REFERENCES

1. National Cancer Institute, Cancer patient survival report number 5, 1976. D. H. E. W. Publications No (NIH) 77-992, Bethesda MD. National Cancer Institute 316 (1977).
2. U. S. Department of Health, Education and Welfare. National Center for Health Statistics (1977) Advanced report, final monthly statistics 1975. Monthly vital statistics report 25, no 11 (supplement). D. H. E. W. Publication No. (CHRA-77-1120). Rockville MD, February 11 (1977).
3. R. L. Egan, Mammography, xeroradiography and thermography, Clinical Obstetrics and Gynaecology 18:197-209 (1975).
4. L. V. Ackerman, J. A. del Regato, "Cancer: diagnoses, treatment and prognosis," 4th edition St Louis C. V. Mosby:1049 (1975).
5. R. S. Paffenbarger Jr, Value in the early diagnosis of cancer, Cancer suppl. 33:1712-1719 (1970).
6. American Cancer Society, "1978 Cancer Facts and Figures, New York," The American Cancer Society 31 (1977).
7. R. L. Brown, Breast cancer in Georgia: observations on 1,100 cases surveyed by the Georgia Cancer Management Network, Inc. J. Med. Assoc. Ga. 67:422-424 (1978).
8. R. S. Foster, S. P. Lang, Breast self-examination practices and breast cancer stage, New Eng. J. Med. 299:265-270 (1978).
9. R. L. A. Kirch, M. Klein, Prospective evaluation of periodic breast examination programs, Cancer 42:728-736 (1978).
10. R. L. A. Kirch, A. Klein, Prospective evaluation of periodic breast examination programs. Cancer 38:265-272
11. National Cancer Institute Final reports of National Cancer Institute

ad hoc groups on mammography screening for breast cancer and a summary report of their joint findings and recommendations. D. H. E. W. Publication No (NIH) 77-1400, Bethesda, MD. National Cancer Institute, March 124 (1977).

12. H. L. Karpman, B. Hamilton, Self-applied breast pathology detecting sensor device: description and preliminary results, The Fourth International Symposium on Prevention and Detection of Cancer. July, 1980, London, England (1980).

13. H. L. Karpman, B. Hamilton, S. Cutler, R. Snyder, Breast cancer screening indicator, a device for home use: description and report of preliminary results in evaluation of symptomatic patients, Presented at 19th National Conference on Breast Cancer, March, 1981, San Diego, CA (1981).

14. M. Gautherie, C. M. Gros, Breast thermography and cancer risk prediction, Cancer 45:51-56 (1980).

VALUE OF INFRARED THERMOGRAPHY IN THE ASSESSMENT OF MALIGNANT MELANOMAS OF THE SKIN

R Amalric, C Altschuler, D Giraud, L Thomassin, J M Spitalier

Cancer Institute
Marseille
FRANCE

Our material comprises 98 histologically confirmed cases of cutaneous malignant melanoma which we have been able to study since 1970 by means of infrared (IR) thermography. Attention to the application of this technique was previously drawn by Gros[1] and Maillard and Hessler.[2] From ten years experience we have found infrared thermography to be of practical use in four ways:

1. For positive diagnosis.
2. To determine the spread of the disease.
3. For prognosis.
4. For monitoring the effects of treatment.

POSITIVE DIAGNOSIS

To facilitate thermographic interpretation of these skin lesions, in 1972 we established a classification similar to that used in mammary thermography. The classification is a function of two parameters:

1. The thermal gradient of local heat emission (1-4°C).
2. The area of the hot zone.
   (a) strictly limited to the clinically visible lesion.
   (b) exceeding the visible lesion to a greater or lesser extent.

Five categories have been defined:

1. TH1 Normal thermogram.
2. TH2 Indicates a thermogram of benign type with a local rise in temperature of not more than 1°C.
3. TH3 Corresponds to an equivocal thermogram with limited

temperature elevation between 1 and 2°C.

4. TH4 Corresponds to a thermogram of malignant type with a temperature elevation of 2-3°C limited to the clinically visible lesion or extending beyond it by a small amount.
5. TH5 Corresponds to a very malignant type with a temperature elevation of 2-4°C and extension of the hot area beyond the apparent limits of the lesion. Sometimes this hyperthermia extends in an elliptical gradient of at least 2°C towards the proximal end of the limb or towards the probable zone of lymphatic drainage. This we have called "thermal flare".

To simplify this study we have defined melanomas as "cold" if the maximum temperature elevation is 1°C, corresponding to categories TH1 and TH2. "Hot" melanomas are those with a temperature elevation equal to or greater than 1.5°C ie categories TH3, 4 and 5.

In our series 76% of malignant melanomas have shown an objective thermal abnormality and fall into categories TH3, 4 and 5. Table 1 shows that in seven European series, from 60 to 92% of malignant melanomas are "hot" (mean = 83%).

False negatives occur in from 8 to 27% of cases (mean 17%). In our series the false negative rate was 24%. In half of these a false negative result was due to ulceration, exudation or necrosis. Others appear to be explained by slow growing tumors, as discussed under "Prognosis".

False positives are less common. They represent 7% of cases in one series of 100 benign cutaneous pigmented lesions. These were usually due to infection or angiomatous abnormalities. The diagnosis is easily made by clinical examination, and confirmed by biopsy taken with suitable precautions to avoid the spread of malignant cells.

In practice a positive thermogram has great predictive value (93%) and is a non-invasive means for identifying equivocal nevi, which require histological examination.

A negative thermogram has no diagnostic value (as with cytology) when taken in isolation. Clinical examination is essential and the thermogram should be considered in the light of the clinical findings and other ancillary investigations.

DETERMINING THE EXTENT OF THE MELANOMA

The IR thermogram may show evidence of a hot area extending beyond the clinical lesion and thus provide an objective means of measuring the size of area involved. In all, 18% of our cases have shown evidence of a "thermal flare" to a greater or lesser degree. Thermography can also locate regional spread by visualizing node metastases which are not

Table 1. Percentage of "hot" and "cold" cutaneous malignant melanomas in seven European series

	Number Of Cases	"Hot"		"Cold"	
Maillard & Hessler, 1973	44	32	73%	12	27%
Pierson & Schmitt, 1973	35	31	88%	4	12%
Rabbi, 1973	25	23	92%	2	8%
Cristofolini, 1976	30	27	90%	3	10%
Hartmann, 1981	15	9	60%	6	40%
Gautherie, 1981	136	121	89%	15	11%
Amalric & Spitalier, 1982	98	74	76%	24	24%
TOTAL	383	317	83%	66	17%

always apparent clinically. In our series of 40 cases with metastatic adenopathy there was significant hyperthermia in 28 (70%). Finally, by scanning the rest of the body, one can examine for distant spread by identifying metastases in transit in the cutaneous lymphatics, and also skeletal, hepatic or ocular secondaries.

In practice this information improves the selection of therapy whether by surgery, radiotherapy or a combination of the two. It is possible to define more precisely the area to be excized or irradiated, depending on whether or not there is evidence of spread. Chemotherapy and/or immunotherapy is indicated where there is evidence of generalized disease or of a "thermal flare".

PROGNOSIS

With regard to sex ratio one knows that the prognosis for malignant melanomas is more unfavorable in men. Infrared thermography shows us that two-thirds of nevo-carcinomas in men are "hot", while in women there are equal numbers of "hot" and "cold" (Table 2).

Level of Histological Invasion

We have compared the results of thermography with the levels of invasion described by Clarke[3] for which we know the prognostic significance. There is a very clear parallel between the local rise of temperature and the degree of invasion. At the second level of invasion 22% of melanomas were "hot", while at the fifth level 86% were "hot" (Table 3). It should be noted that the percentage of "hot" melanomas doubles between the second and third levels of invasion and quadruples between the second and fifth. The more a melanoma infiltrates the greater the chances of its being "hot" and of poor prognosis. Superficial lesions may remain "cold" in which case the prognosis is better.

Node Metastases

We know the prognostic value of node involvement by malignant melanoma. When there is node involvement the primary tumor is "hot" in eight cases out of ten (Table 4).

Length of Survival

A further thermographic prognostic factor is the degree of local hyperthermia when related to the length of survival. At three years, two thirds of "cold" melanoma patients had survived and were symptom-free, as compared to one third of those with "hot" melanomas (Table 5). Of patients with melanomas showing a "thermal flare" only 18% survived three years and none had survived at five years. Mean survival in these cases was only nine months. At five years the difference in survival rate between patients with "hot" and "cold" tumors is even more apparent with a ratio of three to one (Table 6). The same relationship has been found by Hessler and Battaglino[4] in a series of 22 cases. 69% of "cold"

Table 2 Sex ratio and hyperthermia in 60 patients with cutaneous malignant melanomas

	"Hot"		"Cold"	
Male	17/26	65%	9/26	35%
Female	18/34	53%	16/34	47%

Table 3. Clark's level of invasion and hyperthermia in 55 patients with cutaneous malignant melanomas

Level	Number of "Hot" lesions	%
1	0/1	0
2	2/9	22
3	8/18	44
4	9/12	75
5	13/15	86

melanoma patients survived to five years as compared to 22% with "hot" tumors. The same correlation between the degree of local hyperthermia and prognosis has also been confirmed by Grosshans and Gautherie.[5]

In practice it is necessary to take these new factors into account when deciding on treatment. For example, "hot" melanomas and especially those showing a "thermal flare" may benefit from adjuvant medical treatment. This may not be required for "cold" tumors, which do not have the same adverse prognostic factors.

Table 4. Node involvement and hyperthermia in 36 cases of primary malignant melanomas

	"Hot"	"Cold"
Node positive	14/17 82%	3/17 18%
Node negative	4/19 21%	15/19 79%

Table 5. Symptom-free 3-year survival rate of 50 cases of malignant melanomas in relation to their thermograms

	Number	%
"Cold"	8/12	67
"Hot"	11/38	29
Thermal flare	2/11	18

SURVEILLANCE

Regular thermographic examination of the primary site, the regional nodes and even of the whole body provides a non-invasive method for follow-up after treatment. After excision the scar becomes cold and isothermic within a month and should remain so. After radiotherapy there is hyperthermia of the irradiated area for two to three months followed by progressive cooling until finally the area becomes isothermic. Any subsequent rise of temperature is suspicious and is an indication for biopsy. Node involvement and even distant metastases may be apparent thermographically before there is any clinical evidence of spread.

CONCLUSIONS

Thermography can supply additional information in the following applications.

Table 6. Symptom-free 5-year survival rate of 33 malignant melanomas in relation to their thermograms

	Number	%
"Cold"	3/7	43
"Hot"	4/26	15
Thermal flare	0/9	0

1. Diagnosis: As an adjunct to clinical examination it may reveal hidden forms of tumor, unexpected extension and unsuspected metastases.
2. Prognosis: This is favorable if the malignant melanoma is "cold", other sources of error having been eliminated. It is unfavorable if the melanoma is "hot" and very poor if the lesion shows a "thermal flare".
3. Choice of Treatment: Thermography is useful when delineating an area for surgical excision or irradiation. The findings may indicate the need for adjuvant chemotherapy.

Finally, thermography has a practical role in follow-up after treatment. In some instances, it can detect local, regional and distant recurrences before these become clinically apparent.

REFERENCES

1. C. M. Gros, A. Basset, J. Alt, C. Vrousos, La thermographie des tumeurs pigmentees, Bull. Soc. Franc. Dermat. Syphil. 72:726-729 (1966).
2. G. F. Maillard, C. Hessler, La thermographie des melanomes malins cutanes, Dermatologica 139:353-358 (1969).
3. W. H. Clark, L. From, E. A. Bernardino, M. C. Mihm, The histogenesis and biologic behavior of primary human malignant melanoma of the skin, Cancer Research 29:705-726 (1969).
4. C. Hessler, G. Battaglino, Prognostic thermographique des melenomes malins cutanes, Mediterranee Medicale 142:115-120 (1977).
5. E. M. Grosshans, M. Gautherie, The value of infrared thermography for diagnosis, prognosis and surveillance of malignant melanomas of the skin, Symposium International de Thermologie Biomedicale, Strasbourg (1981).

VALUE OF THERMOGRAPHY IN THE DIAGNOSIS OF MALIGNANT MELANOMAS OF THE SKIN

M Cristofolini, S Boi, B Perani, G Recchia, G Zumiani

Ospedale S Chiara di Trento
ITALY

INTRODUCTION

Thermographic examination of cutaneous tumors shows that most malignant lesions are thermally positive. Benign tumors are cold, with the exception of inflammatory and hypervascular lesions.[1] Epithelial tumors are easily diagnosed clinically and on thermographic examination show numerous false positives and negatives[1]. In these cases thermography is of no practical use. Cutaneous malignant melanoma (CMM), on the other hand, may be difficult to diagnose clinically, especially in the early stages, but it is the hottest of the cutaneous tumors.[2,3,4,5,6,7,8] Thermographic examination is useful in these cases in so far as it may:

1. Confirm clinical suspicion.
2. Show a possible nevus-CMM association.
3. Facilitate the differential diagnosis between CMM and other pigmented lesions.

MATERIALS AND METHODS

143 patients with CMM and 289 patients with pigmented lesions simulating CMM were examined thermographically. The thermograms were made with the Bofors IR Camera, Mark III and the AGA Thermovision 680, after the patient had rested for 20 minutes at $20^{o}C \pm 1^{o}C$. All the lesions were biopsied and examined histologically.

Table 1. Primary cutaneous malignant melanomas

	Thermography			
	No.	+	-	%-
Total	143	103	40	27.9
Intra-epidermal CMM	26	10	16	61.5

RESULTS

Of the 143 CMM, 40 (27.9%) were thermographically cold. Taking into consideration the 26 CMM of less than 0.76 mm thickness, the number of false negatives was 16 (61.5%) (Table 1).

Of the 289 pigmented lesions simulating CMM, 64 (22%) were hot (Table 2).

DISCUSSION

It has been suggested that thermography is useful in the diagnosis of cutaneous melanoma because it is non-invasive and there is a fear that biopsy may spread neoplastic cells.

In our series the high false positive rate for thermography in pigmented lesions simulating CMM makes it of little diagnostic value.

Table 2. Cutaneous lesions simulating melanomas

Thermography			
No.	-	+	%+
289	225	64	22

In our cases the false positive rate for clinical examination (5%) was lower than for thermography (21.6%). Therefore the technique was not found to be helpful in clinically doubtful cases: of the 7 clinical false positives, 4 also had false positive thermograms.

In 143 cases of CMM, false negative thermograms (27.9%) were much more numerous than clinical false negatives (3%). Moreover two clinical false negatives also had a false negative thermogram.

Thermography could be useful in the early diagnosis of tumors less than 0.76 mm thick. At this stage the prognosis is favorable, 5-year survival being 100%, but unfortunately in this group, the false negative rate for thermography is very high at 61.5%. If pigmented lesions are clinically suspicious, biopsy and histological examination are necessary regardless of the thermographic result.

Histological examination provides the diagnosis and also precise information about the thickness of melanomas. Thickness is the most important prognostic factor.[9] If biopsy is followed by tumor excision within 20 days, it does not affect the prognosis or influence survival rates of patients with cutaneous malignant melanoma.[10,11]

SUMMARY

Thermography is useless for the diagnosis of cutaneous malignant melanoma. In fact false negative thermograms (27.9%) are much more numerous than clinically suspicious cases.

False negative thermograms are extremely frequent (61.5%) when the malignant melanoma is less than 0.76 mm thick. False positive thermograms are frequent (22%) in pigmented lesions simulating cutaneous malignant melanoma.

REFERENCES

1. M. Cristofolini, C. Valdagni, D. Cattoi, B. Perani, La teletermografia nei tumori cutanei, Giornale Italiano di Dermatologia Min. Derm. 110:10;523-534 (1975).
2. C. Altschuler, La telethermovision des melanomes malins, These Med. 1-21 Marseille (1972).
3. P. Bourjat, M. Gautherie, E. Grosshans, Diagnosis, follow-up and prognosis of malignant melanomas by thermography, Radiol. 6:115-127 (1975).
4. M. Cristofolini, F. Piscioli, C. Valdagni, A. Della Selva, B. Perani, Corrrelations between thermography and morphology of primary cutaneous malignant melanomas, Acta Thermographica 1:3-11 (1976).

5. M. Gautherie, E. Grosshans, J. Juillard, The value of infrared thermography for diagnosis, prognosis and surveillance of malignant melanomas, Philips 1-18 (1978).
6. Ch. Gros, A. Basset, J. Alt, C. Vrousos, La thermographie de tumeurs pigmentees, Bull. Soc. Franc. Derm. Syphil. 73:726-728 (1966).
7. C. Hessler, G. F. Maillard, Apport de la thermographie dans le diagnostic et le traitement du melanome malin, Schweiz. Med. Wschr. 100:972-975 (1970).
8. J. M. Spitalier, R. Amalric, J. Pollet, J. D'Estienne d'Orves, La thermovision des melanomes malins cutanes, Corse Medit. Med. 216:28-31 (1972).
9. A. Breslow, Problems in the measurement of tumour thickness and level of invasion in cutaneous melanoma, Human Pathology 8:1-2 (1977).
10. A. W. Kopf, R. S. Bart, R. S. Rodriguez-Sains, Malignant melanoma: a review, J. Dermatol. Surg. Oncol. 3:41-125 (1977).
11. U. Veronesi, E. Bayetta, N. Cascinelli, C. Clemente, F. Rilke, New trends in the treatment of malignant melanoma, Surgical Oncology 1:113-130 (1978).

HYPERTHERMIA - INTRODUCTION

P Carnochan

Royal Marsden Hospital, Fulham Road
London SW3 6JJ
UNITED KINGDOM

There is accumulating experimental and early clinical evidence to support the view that hyperthermia may be a useful modality for the treatment of solid tumors.

The rationale for such treatment is based largely upon experimental observations of the time dependant survival of cells exposed to temperatures in the appropriate range of 42-45°C. Further studies have highlighted the selective thermosensitivity of poorly oxygenated and hyperacidic cell populations; such environmental conditions may be found in the central regions of large tumors. With regard to the relative radioresistance of such regions it has been suggested that an improved therapeutic strategy would result from combined radiotherapy and localized hyperthermia treatments.

The successful clinical use of localized hyperthermia is dependent critically upon the ability to deliver an adequate thermal 'dose' to tumor tissues without enhancing normal tissue toxicity. Various heating techniques, commonly based upon the use of electromagnetic fields and waves or ultrasound, are being investigated to enable these aims to be met. Novel techniques, based upon microwave radiometry, are being developed to aid thermal dosimetry which at the present time may only be achieved by invasive means.

Presented here are papers describing typical examples of such techniques currently under development or being employed clinically. Further contributions included in this section are examples of experimental work directed towards a closer understanding of the biological and physiological effects of hyperthermia, and clinical studies aimed at the evaluation of hyperthermia treatment given either alone or in combination with radiotherapy or chemotherapy.

INFRARED THERMOGRAPHY AND HYPERTHERMIA

P Carnochan, C H Jones

Royal Marsden Hospital, Fulham Road
London SW3
UNITED KINGDOM

The treatment of cancer by hyperthermia requires careful control of the temperature within the tumor as well as in overlying healthy tissues. For hyperthermia to be effective clinically it is necessary to maintain the tumor at a higher temperature than normal tissues.[1] Tumor cell kill and normal tissue tolerance are dependent critically upon temperature magnitude and duration of heating, as summarized by Overgaard.[2] The length of a treatment will therefore depend upon the temperature attained within the tumor and also the temperature differential between healthy and abnormal cells. Clinical studies indicate the effectiveness of achieving tumor temperatures in the range 42-43°C for periods of 45 minutes when incorporated into a treatment schedule involving the use of radiotherapy, chemotherapy or radiation sensitizers.[3,4] The effective use of hyperthermia alone however, may require higher tumor temperatures owing to the influence of vascular cooling as discussed by Lagendijk.[5] Ideally the hyperthermia technique should generate adequate heating throughout the tumor without causing similar temperature increases in surrounding normal tissues, and to achieve this end it is necessary to investigate the optimization of various heating modalities with respect to each particular clinical site and tumor type to be treated. Popular heating techniques currently under investigation utilize electromagnetic fields, microwaves and ultrasound to transfer energy to the tissues. The relative merits of these methods have been described by Hand and ter Haar.[6] This paper is concerned primarily with the use of RF (27 MHz) electromagnetic field techniques for the induction of localized hyperthermia, and the ways in which IR thermography may be used as an aid to temperature measurement. In principle, the thermographic methods described may be applied when other heating modalities are employed.

There are five areas in which infrared thermography measurements are useful for the study of hyperthermia:

SELECTION OF THE MOST APPROPRIATE FORM OF HEATING TECHNIQUE

The distribution of temperature induced by a particular heating modality depends largely upon the physiological, thermal and electrical properties of the heated tissues, combined with the configuration of the electromagnetic field 'applicator' coupled to the source of RF power. Applicators may be classed as either capacitive or inductive, depending upon whether energy is transferred to tissues by electric fields or magnetic fields respectively. Table 1 summarizes the inductive method and three capacitive methods which have been developed for clinical hyperthermia. Some of the physical advantages and disadvantages of the respective RF techniques are tabulated: reference may be made to Field and Bleehen[7] for information about the merits of hyperthermia as a treatment modality, the interactions between hyperthermia and other treatment modalities and for details of the biological mechanisms involved.

The intracavitary capacitive technique is particularly useful for

Table 1. RF electromagnetic field techniques used for localized clinical hyperthermia.

METHOD	INTRACAVITARY	INTERSTITIAL	EXTERNAL APPLICATION	
			CAPACITIVE	INDUCTIVE
FREQUENCIES USED (MHz)	0.3-0.5	0.1-1	3-30	10-30
EXPERIMENTAL STUDIES	Harzmann et al, (1979,1981) A.	Doss and McCabe, (1976) P. Cetas and Connor, (1978) A,P. Manning et al, (1982) P.	Strang (1978) M. Armitage (1980) M. Manning et al, (1982) P.	Lehmann et al, (1969) H. Guy et al, (1974) M. Armitage (1980) M. Hand (1982) M.
CLINICAL STUDIES	Harzmann et al, (1979,1981) D.	Cosset (1981) S. Manning et al, (1982) S.	LeVeen et al, (1976) D. Storm et al, (1979,1981) S. Manning et al, (1982) S.	Kim et al, (1978) S. Storm et al, (1979,1980,1981) D.
ADVANTAGES	Good localisation.	Good localisation. May be combined with Radiotherapy afterloading techniques.	Superficial heating area determined by electrode shape. IR thermography may be useful.	Good coupling to tissues medial to subcutaneous fat.
DISADVANTAGES	Limited access to cavities. Thermometry difficult.	Invasive.	Can be difficult to heat tissues medial to subcutaneous fat.	Non uniform heating area of many practical systems.

A-Animals. H-Healthy volunteers. M-Mathematical models. P-Phantoms. D-Deep visceral tumours. S-Superficial tumours.

some visceral tumor sites where access is a practical possibility, for example the bladder or rectum, and insufficient heat penetration by external methods is likely.

Interstitial methods, although invasive, have the desirable feature of minimizing the unwanted heating of superficial tissues without compromize with respect to the localization of heating. These methods may also be conveniently combined with low dose rate radiotherapy afterloading techniques wherever such treatment is indicated.

Most clinical experience has been gained using external techniques, which are inherently more versatile and generally present fewer problems with respect to thermal dosimetry. Typical examples of capacitive and inductive techniques are outlined below.

The thermal characteristics associated with particular applicators may be assessed experimentally by different methods. Mathematical modelling has been employed to predict the distribution of power absorbed within tissues having dielectric discontinuities of various geometries.[8,9]

The present lack of pertinent data however, coupled with the excessive requirements in terms of computing resources, has precluded the extension of these studies to clinically representative predictions of temperature distributions. Presently it seems unlikely that the development of mathematical models will sufficiently represent the important anisotropic characteristics of blood flow, and therefore be of limited further clinical value. Alternatively, distributions of absorbed power may be studied using phantoms having electrical characteristics similar to those of tissues. Heating the phantoms for intervals short enough to preclude significant thermal conduction allows the distribution of absorbed power to be deduced from the temperature distribution within the phantom. Such temperature distributions may be visualized instantaneously by IR thermography using tissue equivalent phantoms which may be sectioned after heating along planes of interest. Heterogeneities can be incorporated into the phantom structure so that the effects of, for example, fat, muscle and bone may be examined quantitatively. Typical examples of phantom studies using various heating methods at 27 MHz are shown in Figure 1. Some characteristic features of inductive heating are illustrated in Figure 1(a). Heating was achieved by passing a RF current of 20A (peak) through the three turn coil shown for 10 minutes, immediately followed by thermographic observation of the phantom across a plane halfway along its length. The heating pattern has approximate cylindrical symmetry about the axis of the coil and can be seen to have relatively little heating along and proximal to the applicator axis. The heating pattern may be described by an annulus of diameter approximately equal to that of the applicator.

An example of the external capacitive technique is illustrated in Figure 1(b). A non-uniform electric field is produced between two electrodes of unequal size placed across a three layered cylindrical

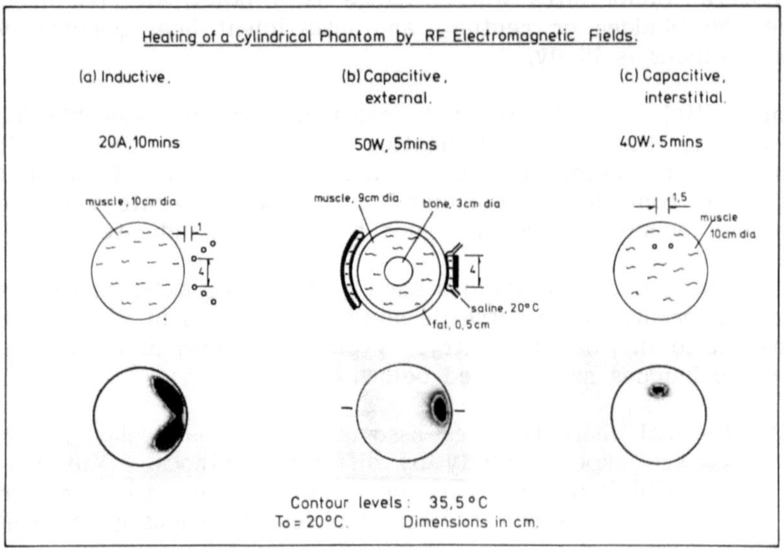

Fig. 1. Heating patterns produced by electromagnetic fields in cylindrical phantoms.

phantom, good electrical contact being made between each electrode and the phantom by a saline 'pillow'. Heating is concentrated in the region of high electric field magnitude close to the small electrode, and circulating saline of constant temperature is used to further control the heating of the superficial area of the phantom in this region.

Figure 1(c) illustrates an example of interstitial heating. Two nylon tubes of the type used for the afterloading of ^{192}Iridium wires were inserted into a muscle equivalent phantom, into which electrode wires were placed. The tubes were perfused with constant temperature saline, thereby cooling the electrodes and surrounding region in order to enhance the uniformity of the heating pattern. The uniformity of heating in the direction of the electrode wires is largely determined by their separation and therefore the accuracy of implantation of the nylon tubes.

An example of the optimization of applicator design is illustrated in Figure 2. The externally applied non-uniform electric field technique requires careful selection of the appropriate electrode size to correspond to the required heating area, and also to minimize the heating of regions close to the larger of the two electrodes. It can be deduced from the examples shown that the electrical potential boundary is defined by the size of the intervening saline pillow rather than the metal electrode, and minimal relative heating near the larger electrode occurs in the intermediate example. Sensitivity of the heating pattern to a further increase in the size of the larger electrode is not seen in the examples shown; however, care should be taken to preclude excessive heating near

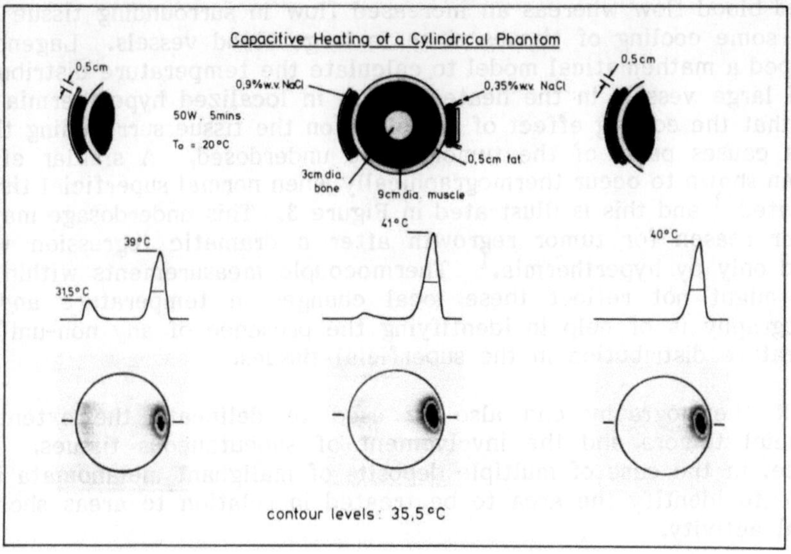

Fig. 2. **The optimization of applicator design for capacitive diathermy.**

electrode edges due to the increasing non-uniformity of the electric fields produced.

It may be concluded that IR thermographic studies of phantoms are of value for the selection of the appropriate configuration of electromagnetic field applicator, but it must be understood that the in vivo temperature distributions will be modified by vascularity and thermal conduction, the effects of which cannot be predicted with certainty and for this reason it is useful to investigate the uses of IR thermography in vivo prior to and throughout hyperthermia therapy.

EXAMINATION OF THE SKIN SURFACE OVERLYING THE TUMOR PRIOR TO TREATMENT

Many tumors have the ability to induce angiogenesis in the tissues in which they are located.[10] After the vasculature has been established there are frequently signs of abnormal morphology and physiology; there are sometimes giant capillaries around the bulk of the tumor, the presence of arterio-venous shunts and the veins draining the tumor become varicose. When the tumor is relatively superficial there will be considerable likelihood that vasculature between tumor and skin will be pronounced and there is much thermographic evidence to support this view. IR thermography is a useful means of locating and monitoring the effect of treatment on this vasculature. In the case of large tumors heat may be accumulated in the central regions of the tumor as a result of reduced blood flow whereas an increased flow in surrounding tissue may induce some cooling of tissues close to large blood vessels. Lagendijk[5] developed a mathematical model to calculate the temperature distribution around large vessels in the heated region in localized hyperthermia and found that the cooling effect of the blood on the tissue surrounding these vessels causes parts of the tumor to be underdosed. A similar effect has been shown to occur thermographically when normal superficial tissues are heated[11] and this is illustrated in Figure 3. This underdosage may be a major reason for tumor regrowth after a dramatic regression when treated only by hyperthermia.[1] Thermocouple measurements within the tumor might not reflect these local changes in temperature and IR thermography is of help in identifying the presence of any non-uniform temperature distribution in the superficial tissues.

IR thermography can also be used to delineate the extent of superficial tumors and the involvement of subcutaneous tissues. For example, in the case of multiple deposits of malignant melanomata it is possible to identify the area to be treated in relation to areas showing thermal activity.

The prognosis of some tumors is also reflected by their thermal activity[12,13] and it is appropriate to measure the temperature over the

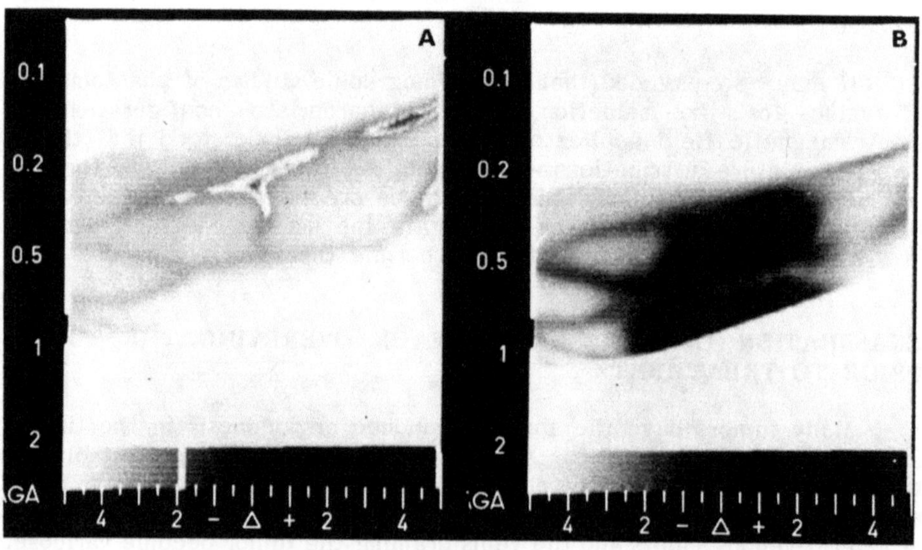

Fig. 3. Thermographic visualization of the inductive heating of a normal forearm, showing modification of the heating patterns by venous blood flow, (a) before heating, (b) after 5 mins heating with three term 'pancake' type coil. MLT = 35°C, Maximum temperature after heating = 39°C. (after Carnochan et al[11]).

tumor in advance of therapy as a baseline for future assessment purposes.

If previous radiotherapy or chemotherapy has been given or if it is intended to give such therapy concurrently, IR thermography may be used to monitor the skin reaction over the treated area. Chemotherapy can cause accentuation of a radiation skin reaction and thermography is an effective means of monitoring its magnitude so that further injury by hyperthermia can be avoided or at least minimized. Figure 4 shows the effect of adjuvant chemotherapy on a typical radiation skin reaction.

ASSESSMENT OF THE SKIN SENSITIVITY TO HEAT AND THE EFFECT OF VASCULATURE ON THERMAL LOAD

The degree of hyperthermia that can be achieved clinically depends upon the ability of the patient to withstand local temperatures above about 42°C. Some patients are unable to tolerate this temperature level and before hyperthermia treatment it is useful to assess the patients' reactions to various levels of temperature. The rate of loss of heat from the heated area is also of importance and estimates of this, as well as the maximum surface temperature that the patient can withstand without discomfort, may be assessed thermographically with the aid of a specially designed temperature reference block.

THE USE OF IR THERMOGRAPHY IN MONITORING SURFACE HEATING PATTERNS DURING HYPERTHERMIA THERAPY

There are various heating methods used in hyperthermia therapy, and the way in which IR thermography may be used will depend upon the technique employed. For example, the area heated by the external capacitive technique described above may be visualized by temporary removal of the electrode from the skin surface. This procedure would serve to check the correct optimization of electrode size with respect to the desired heating area.

THE MEASUREMENT OF POST THERAPY CHANGES

Amalric et al[14] have shown that IR thermography is an effective way of monitoring tumor response to radiotherapy. It is likely that IR thermography could provide useful data for monitoring post-therapy changes after hyperthermia or after hyperthermia combined with other treatment methods.

The skin reactions associated with multi-modal therapy are difficult to predict and assess: when hyperthermia is used with radiotherapy and other modalities such reactions can be monitored quantitatively and provide data from which it might be possible to predict and maybe optimize the effects of a certain treatment regime. For example it is

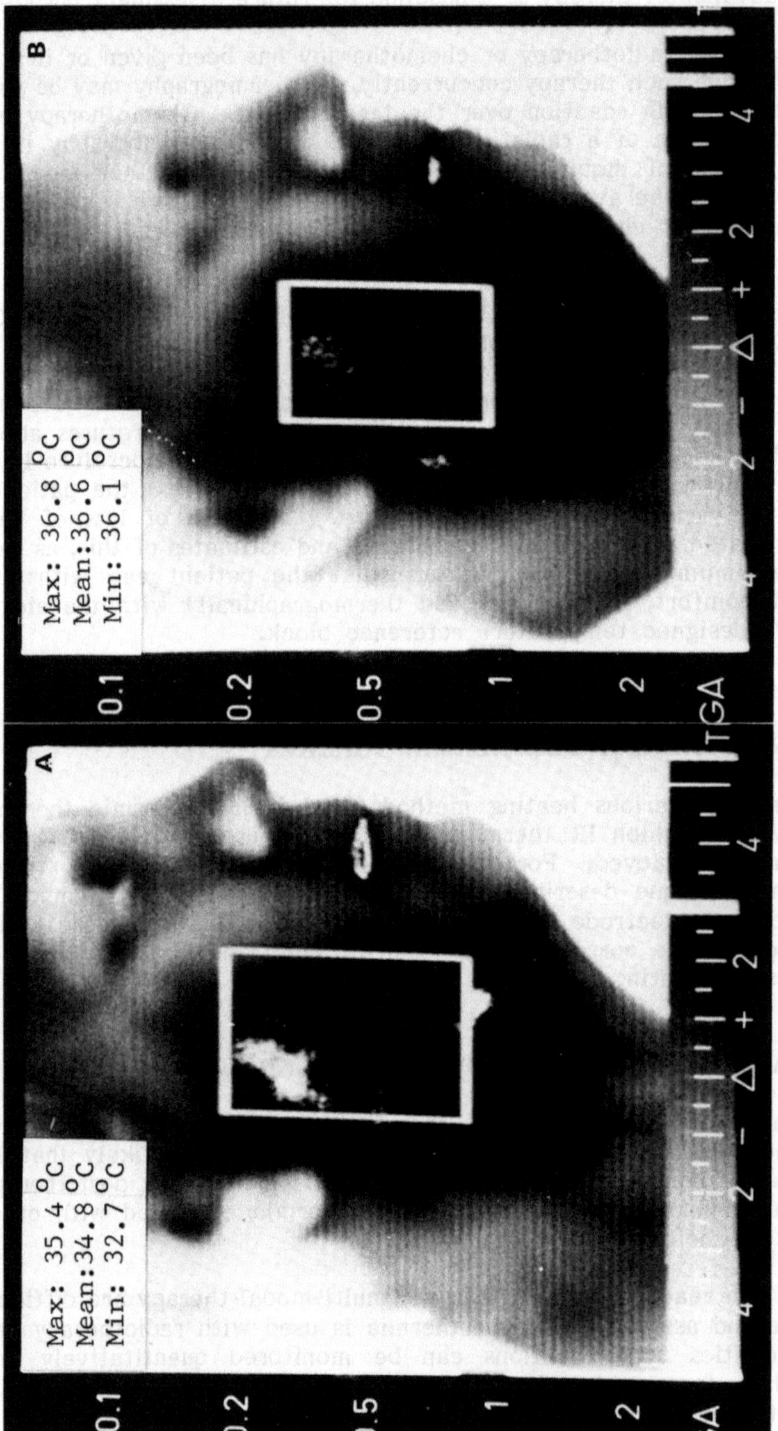

Fig. 4. Thermogram showing skin reaction after Cobalt 60 radiotherapy, (a) before a second dose of chemotherapy, (b) 24 hours after the second dose of chemotherapy.

known that chemotherapy influences the sensitivity of tissues to ionizing radiation so when radiotherapy, chemotherapy and hyperthermia are used conjointly, care must be exercised to ensure that damage to normal tissues is maintained within acceptable limits. Although the measurement of temperature at depth must be achieved by using either implanted thermocouples or surface contact microwave radiometers, infrared thermography provides means by which skin damage may be assessed quantitatively throughout a course of treatment.

REFERENCES

1. M. R. Manning, T. C. Cetas, R. C. Miller, J. R. Oleson, W. G. Connor, E. W. Gerner, Clinical hyperthermia: Results of a phase 1 trial employing hyperthermia alone or in combination with external beam or interstitial radiotherapy, Cancer 49:205 (1982).
2. J. Overgaard, The effect of local hyperthermia alone, and in combination with radiation, on solid tumors, in: "Cancer Therapy by Hyperthermia and Radiation," C. Stoffer ed., Urban and Schwarzenberg, Baltimore, Munich (1978).
3. G. Arcangeli, A. Cividalli, G. Lovisolo, F. Mauro, G. Creton, C. Nervi, G. Pavin, Effectiveness of local hyperthermia in association with radiotherapy or chemotherapy: comparison of multimodality treatments on multiple neck node metastases, in: "Proceedings 1st Meeting of European Group of Hyperthermia in Radiation Oncology," Cambridge, Masson (1979).
4. G. Arcangeli, A. Barocas, F. Mauro, C. Nervi, M. Spano, A. Tabocchini, Multiple daily fractionation (MDF) radiotherapy in association with hyperthermia and/or misonidazole: experimental and clinical results, Cancer 45:2707 (1980).
5. J. J. W. Lagendijk, The influence of bloodflow in large vessels on the temperature distribution in hyperthermia, Phys. Med. Biol. 27:17 (1982).
6. J. W. Hand, G. R. ter Haar, Heating techniques in hyperthermia, Brit. J. Radiol. 54:443 (1981).
7. S. B. Field, N. M. Bleehen, Hyperthermia in the treatment of cancer, Cancer Treatment Reviews 6:63 (1979).
8. J. W. Hand, J. L. Ledda, N. T. S. Evans, Considerations of radiofrequency induction heating for localized hyperthermia, Phys. Med. Biol. 27:1 (1982).
9. D. W. Armitage, Three dimensional simulation of power absorbed by the human body from radiofrequency applicators (abstract). 2nd Meeting of European Group of Hyperthermia in Radiation Oncology, Rome (1980).
10. J. Folkman, Tumor angiogenesis, Adv. Cancer Res. 19:331 (1974).
11. P. Carnochan, M. Jancar, C. H. Jones, The assessment of RF inductive applicators suitable for clinical hyperthermia, Brit. J. Cancer 45 (suppl. V):25 (1982).
12. K. Lloyd-Williams, Thermography in the prognosis of cancer, in: "Medical Thermography," Proceedings of a Boerhaave Course for

Postgraduate Medical Education. Leiden 1968, Bibl. Radiol. 5:62 Karger, Basel/New York (1969).

13. P. Bourjat, M. Gautherie, E. Grosshaus, Diagnosis, follow-up and prognosis of malignant melanomas by thermography, in: "Thermography," Proceedings 1st European Congress Amsterdam 1974. Bibl. Radiol. 6:115 Karger, Basel (1975).
14. F. Amalric, D. Giraud, C. Altschuler, R. Amalric, J. M. Spitalier, Infrared thermographic follow-up after breast cancer curative radiotherapy, Acta Thermographica 4:54 (1979).

PROGRESS IN MICROWAVE AND RADIOFREQUENCY HYPERTHERMIA CONTROLLED BY MICROWAVE THERMOGRAPHY

M Chive*, M Plancot*, Y Leroy*, G Giaux, B Prevost***

*** Centre Hyperfrequences et Semiconducteurs, LA CNRS No 287, Universite de Lille I - Bat P4, 59655 Villeneuve D'Ascq Cedex, FRANCE**

**** Centre Anti Cancer "Oscar Lambret", Rue Frederic Combemale, 59020 Lille Cedex, FRANCE**

INTRODUCTION

Previous feasibility experiments have indicated the possibility of achieving microwave local heating and microwave thermography with the same system.[1,2] This combined process is now available for measurement and control of the local increase of temperature in hyperthermia therapy. Several systems have been built according to this principle[3,4,5] which are intended for biomedical applications.

The first microwave system we present here works according to the alternating process[2] with the heating frequency at 2.45 GHz and a radiometric receiver operating around this central frequency with a 1 GHz bandwidth.

The second microwave system associates the previous 2-4 GHz radiometer with an heating generator which operates at 1 GHz.

Finally a 2-4 microwave thermograph (or radiometer) has been associated with the 13.56 MHz hyperthermia apparatus as a complementary process for temperature control in clinical hyperthermia tumor treatment.

In this paper we present the different systems and the results obtained in clinical applications.

THE 2.45 GHz PROTOTYPE FOR CONTROLLED LOCAL HYPERTHERMIA

The system which combines the two functions (heating and radiometric temperature measurement) operates in the frequency range (2.450 MHz) allowed for ISM applications. It offers several new advantages: firstly it allows the use of atraumatic and easily made probe applicators (such as dielectric filled waveguide antennas); secondly this is the condition in which the detected noise signal is emitted in the very volume in which the microwave power is dissipated. The latter, which is a consequence of the antenna reciprocity principle, holds even if the tissues are heterogeneous.

The block diagram of this system given in Figure 1 shows that it is based on an alternating process: the heating is practically achieved at any time (switch in position 1), the temperature measurement (switch in position 0) being made during short intervals (5 seconds every minute).

The microwave power for heating is provided by a cavity oscillator followed by a 2-4 GHz TWT amplifier (maximum output power 16 watts).

The microwave radiometer includes a heterodyne receiver involving a low noise preamplifier (40 dB gain), a double balanced mixer and an IF amplifier (30 dB gain, 500 MHz bandwidth). The comparator or modulator (as in Dicke configuration) is a pin diode switch. The temperature reference is obtained by means of a noise source which emits its signal through a circulator. The detection and signal processing are based on a numerical filtering, including a microprocessor.[6] We note

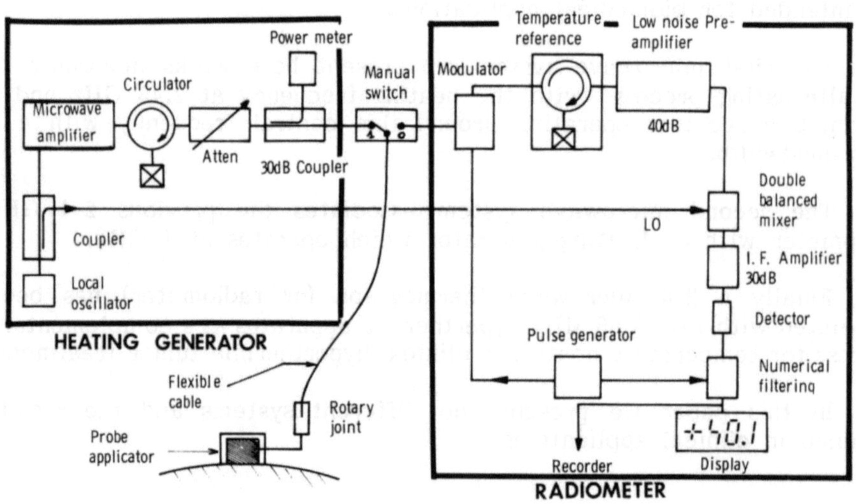

Fig. 1. Microwave system for local hyperthermia monitored by microwave thermography based on the alternate method.

that the local oscillator input for the double balanced mixer uses a small part of the heating power.
With this process, which helps to make the radiometer insensitive to the heating signal, the spectral sensitivity pattern of the radiometer is automatically locked to the generator frequency.[2] The noise factor of this microwave radiometer is 5 dB and its temperature sensitivity is about 0.1°C.[5]

The probe applicator is a rectangular waveguide filled with a low loss solid dielectric of high permittivity (ε = 25) in order to reduce the aperture in contact with tissues (2.2 cm x 1.1 cm) and also to reach an almost perfect matching at the applicator-tissue interface; the power reflection coefficient of the applicator in direct contact with skin is lower than 0.1 in the bandwidth of the radiometer.

Numerous experiments on phantoms and animals have been performed with this system[7] before its clinical use on patients and they have shown that it is possible to achieve atraumatic control of the temperature in a microwave hyperthermia process by means of a microwave radiometer.

An example of a clinical result is given in Figure 2. It concerns the heating of a neck tumor previously treated by curie-therapy. A thermocouple was inserted in the center of the tumor at 1.4 cm below the probe applicator (it had been located by X-ray) and compared with the radiometer. To avoid any interference with the microwave heating power, during radiometric and thermocouple temperature measurements (switch position 0) the TWT amplifier was switched off. In this condition the results obtained, given in Figure 2, show a good correlation between the radiometric temperature measurements and the thermocouple readings.

THE 1 GHz HEATING SYSTEM

This prototype is also based on the alternating process and it associates a 1 GHz heating generator with the previous 2-4 GHz radiometer.

The heating generator consists of a 1 GHz cavity oscillator followed by a TWT amplifier (maximum output power 12 watts).

The probe applicator is a rectangular waveguide filled with a low loss dielectric (ε = 16); the aperture dimensions are 5 cm x 2.5 cm and this applicator operates in the TE_{10} mode for the heating process (switch position 1) and in the TE_{30} mode when working in the radiometric process (switch position 0).

With this system clinical treatments have been performed during the last six months, particularly for recurrent tumors in previously irradiated areas. An example of temperature profile versus time in the case of a recurrent breast tumor is given in Figure 3.

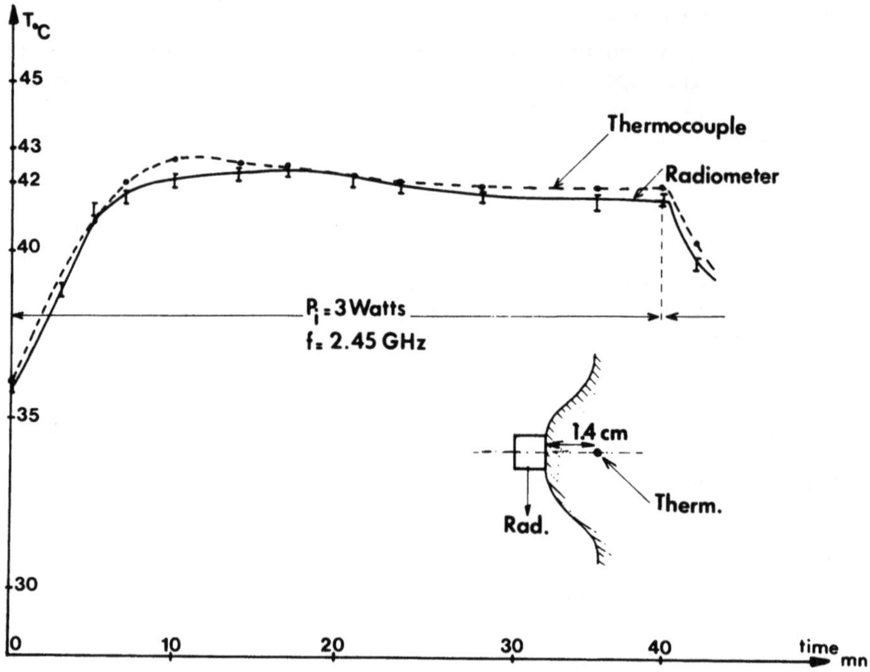

Fig. 2. Temperature profile versus time during a heating session of a cervical lymph node. (f = 2.45 GH_Z; P_i = 3 watts).

THE 13.56 MHz HYPERTHERMIA APPARATUS

Clinical experiments at the Centre Anti Cancereux Oscar Lambret (Lille, France) have been undertaken by Professor Giaux and his team using the 13.56 MHz hyperthermia SAT system.[8,9] This apparatus is based on the well known capacitive heating process and the temperature measurement is achieved by thermoprobes inserted at different depths in tissues and tumor. As this temperature control is an invasive technique, we have associated with the SAT system the previous 2-4 GHz radiometer as a complementary and atraumatic process for temperature control in tissues. To avoid any intermodulation from the generator to the microwave radiometer, the alternate method has been used - the radiofrequency heating generator was switched off during a short time for radiometric temperature measurement. Feasibility experiments have been performed on living anesthetized animals[9] and a comparison between thermoprobe temperature measurements and radiometric readings have shown good correlation between the two types of temperature measurement. This was confirmed in clinical experiments on patients and as an example we give in Figure 4 the results obtained during heating of a cervical lymph node metastasis.

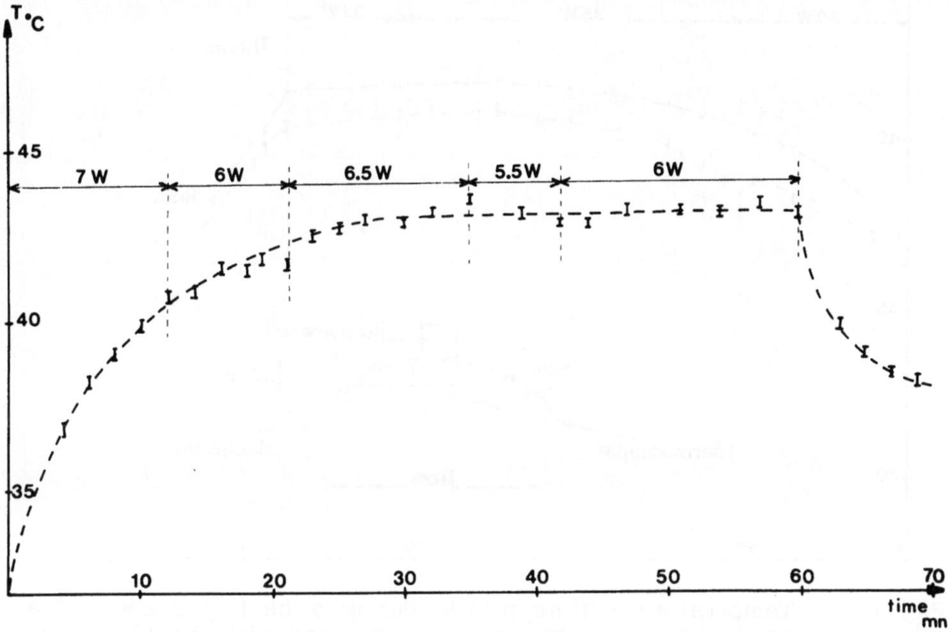

Fig. 3. Temperature - time profile of a recurrent breast tumor during a heating session. (f = 1 GH_z).

CONCLUSION

The clinical results show that it is possible to achieve non-invasive control of temperature during hyperthermia by means of a microwave thermograph associated with a heating system (microwave or radiofrequency). This is of benefit in the case of repeated heating sessions because it is now possible to substitute non-invasive control by microwave thermography for traumatic temperature measurement with inserted thermoprobes.

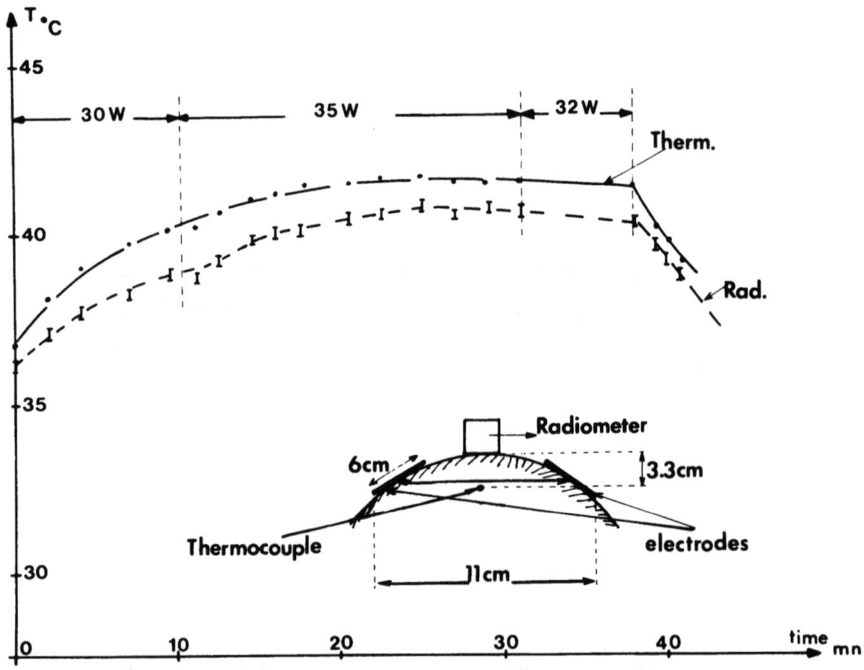

Fig. 4. Temperature - time profile during a heating session of a cervical lymph node metastasis obtained with the 13.56 MH_Z apparatus.

REFERENCES

1. D. D. Nguyen, A. Mamouni, Y. Leroy, E. Constant, Workshop on diagnosis and therapy using microwaves, Eur. Micr. Conf. Paris, (1978). Also published in J. M. P. 14:2 (1979).
2. D. D. Nguyen, M. Chive, Y. Leroy, E. Constant, IEEE Transactions on instrumentation and measurements, IM-29;2:143-144 (1980).
3. D. D. Nguyen, M. Chive, Y. Leroy, URSI Conf.: OEMB Jouy en Josas (France) (1980).
4. K. L. Carr, A. M. Elmahdi, J. Shaeffer, IEEE MTT Symposium, Washington DC (USA) (1980).
5. D. D. Nguyen, Thesis: 3e cycle, Univ Sciences et Techniques de Lille I - France (1980).
6. J. Baudet, J. L. Van de Velde, Electr. Applic. Indust. 282 (1980).
7. M. Chive, D. D. Nguyen, Y. Leroy, Onde Electrique 62:2 (1982).
8. M. Chive, M. Robillard, J. J. Fabre, G. Giaux, B. Prevost, Symposium International de Thermologie Biomedicale, Strasbourg (1981).
9. S. Lacroix, Advances in Electronics and Electron physics (1980).

EXPERIMENTAL AND CLINICAL STUDIES OF HYPERTHERMIA IN ONCOLOGY USING DIFFERENT ELECTROMAGNETIC SYSTEMS

J L Guerquin-Kern*, H-M Lauche, P Pillement**, G Methlin**, M Gautherie***

*** Laboratoire de Thermologie Biomedicale, Faculte de Medicine, 11 rue Humann, 67085 Strasbourg Cedex FRANCE**

**** Centre Regional de Lutte contre le Cancer "Paul Strauss" 3 rue de la Porte de l'Hopital, 67005 Strasbourg, Cedex FRANCE**

INTRODUCTION

For more than a century, heating tissues to slightly higher temperatures than physiologically normal has been envisaged as a means of antitumoral therapy.[1]

Various techniques have been tried for artificially producing an increase in temperature of suitable intensity, duration and localization. Local hyperthermia in particular, can be induced at depth in the tissues by the conversion of electromagnetic energy into thermal energy.

To obtain sufficiently high temperatures at depth poses many problems, largely stemming from the technology of the system rather than the actual heat exchange inside the tissues. Problems arise from the variation in target volume, and the difficulty of controlling temperature in the presence of an electromagnetic field. The degree of hyperthermia achieved with radiofrequency and hyperfrequency electromagnetic waves must therefore first be established on several different models, prior to clinical investigations. For some years we have tested different experimental systems to evaluate the influence of the various parameters of treatment.

MATERIALS

To obtain the frequency necessary to achieve a heating effect we have used two types of generating systems, one capacitive, the other radiative. The capacitive systems function at 0.5 MHz and 13.56 MHz in the field of radiofrequencies. They use either flexible electrodes applied directly to the skin or implantable electrodes in the form of needle-probes. In these systems thermal energy is carried by electric currents circulating between the electrodes.

At microwave frequencies of 434 MHz and 2450 MHz, the radiative type of system was used. This depends upon the absorption of electromagnetic waves by the tissues for a heating effect. These systems have external applicators applied to the tissues, which permit recooling of the surface during treatment to prevent superficial lesions of healthy tissue.

PHYSICAL AND BIOLOGICAL EXPERIMENTS

Simulation of Physical Models

During these experiments two types of physical model were used:-

1. Excized animal tissue: homogenized samples of adipose or muscle tissue in which the thermal distribution after electromagnetic radiation was analysed by IRT following an identical method to that proposed by A W Guy.[3]. This technique of simulation was found to be practical, but it only provides information after treatment.

2. Original phantoms to simulate the tissues were created in order to visualize the distribution of temperatures in continuity. They were made from transparent gels with thermal and dielectric properties corresponding to those of the tissues. Inside the model a film of thermosensitive liquid crystals reveals the temperature differences by changes in color (Fig 1).

These simulation techniques have enabled us to compare different systems of treatment and study the influence of certain parameters on hyperthermia.[4] Thus at 2450 MHz on adipose tissue, tests with a charge between two applicators placed perpendicularly, one upon the other (Fig 2) showed that such a system would prove interesting and practical, providing a means of control of the heating depth depending on the phase difference between the two applicators (Fig 3).

Influence of Blood Flow

In these different models the dielectric and thermal characteristics remain more or less constant whatever the temperature, but these characteristics vary considerably with the thermal state of the tissues, notably thermal conductivity, which is linked to circulatory and thermoregulatory functions. In order to complete these studies it seemed

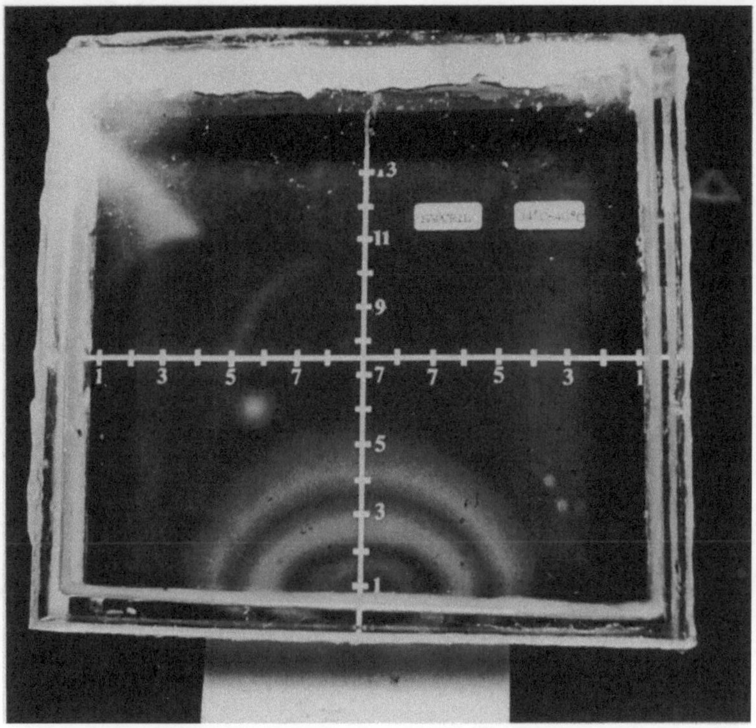

Fig. 1. Cross-fire irradiation of fat by means of two microwave applicators (2450 MHz).

essential to take such physiological variations into account. We therefore attempted to improve the gel phantom by perfusing it with physiological serum circulating in coils of tubing.[5]

The influence of circulatory phenomena of thermoregulatory origin is also evidenced by measurements in animals. These are made with the double aim of verifying the results obtained with the model, and to set up a preliminary phase prior to clinical investigation. An experiment was made on the hind leg of a dog (Fig 4) and carried out in two stages:-

1. In vivo, on an anesthetized animal.
2. Post mortem, the animal having been killed at the start of the return to the initial thermal state.

In both cases the experimental conditions of irradiation and temperature measurement were identical. The treatment was applied in sequence, temperature measurements being carried out before and after each period of hyperfrequency irradiation. Needle-probes with thermistors were implanted at different depths to monitor the temperature of the area.

Fig. 2. Analysis of the temperature distribution on the medial surface of the block of fat using an infrared thermograph.

Measurements carried out on the living animal during treatment have shown that the temperature at depth is considerably higher than in the areas close to the skin surface. The role of the thermoregulatory effect of the circulation is particularly evident in Figure 5. A comparison of thermal profiles obtained just after irradiation of the living animal (Fig 6a) compared with the post mortem specimen (Fig 6b) shows that in vivo:-

1. Hyperthermia is much less intense for an identical period of treatment.
2. The hottest zone is the deepest.

Moreover, we found that the pattern of recooling differs in vivo from post mortem. This is clearly in keeping with the convective role of blood heatflow. Furthermore, a comparison of the post mortem results with those obtained from the perfused phantom showed similar curves. This indicates that the simulation of vascular phenomena made on the phantom is not significant. This is probably due to the complexity of circulatory and thermoregulatory phenomena.

As the physical models were not able to allow for the full range of physiological variations, it was necessary to further the tests on the animal in vivo. The results seemed favorable in the production of a

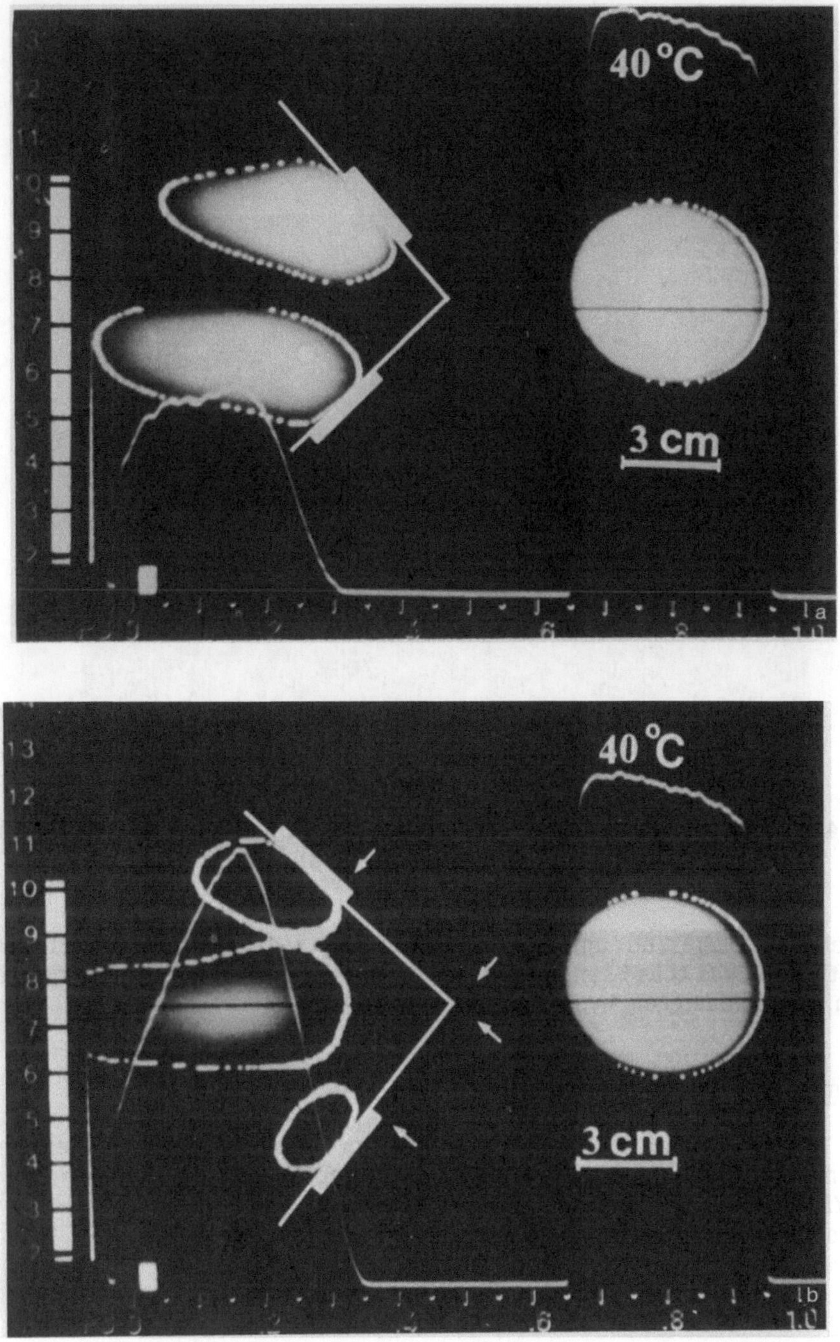

Fig. 3. Infrared thermograms recorded after cross-fire irradiation: (a) relative superficial heating with a 180° phase shift between the two applicators. (b) deeper heating with a 0° phase shift.

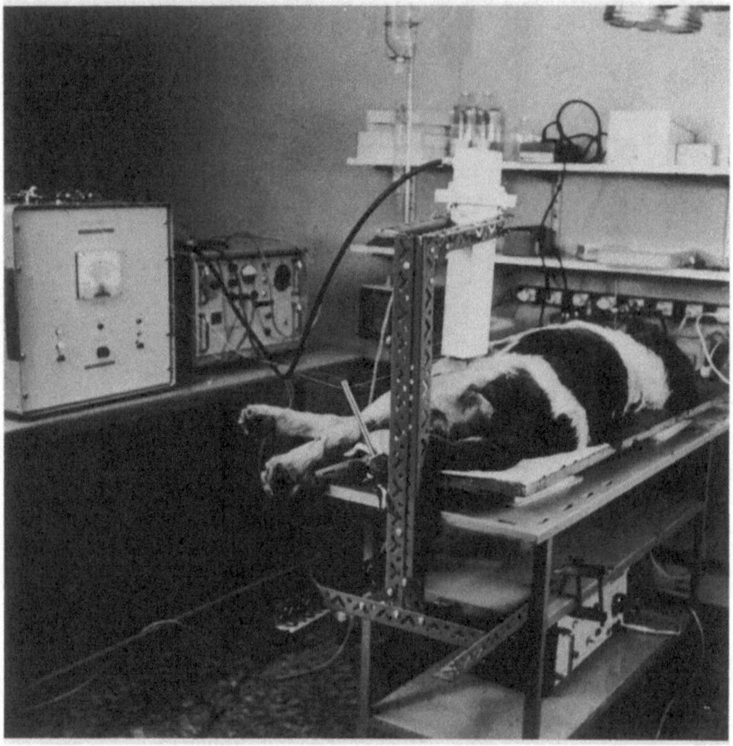

Fig. 4. Experimental hyperthermia studies: microwave heating of the thigh muscles of a dog (2450 MHz).

sufficiently deep and uniform hyperthermia. A protocol for clinical investigation has thus been established to continue this research in greater depth, particularly on hyper- or hypo-vascular tumors.

Preliminary Clinical Investigations

Having established some technical progress with the generating systems at 2450 MHz and 13.56 MHz and the production of hyperthermia, the first series of clinical trials commenced with strict technical and ethical precautions.

Recruitment of Patients

In these tests patients presenting with various superficial tumors (cutaneous, subcutaneous or in the mucosa) were recruited. Such tumors are considered to be susceptible to superficial hyperthermia and were

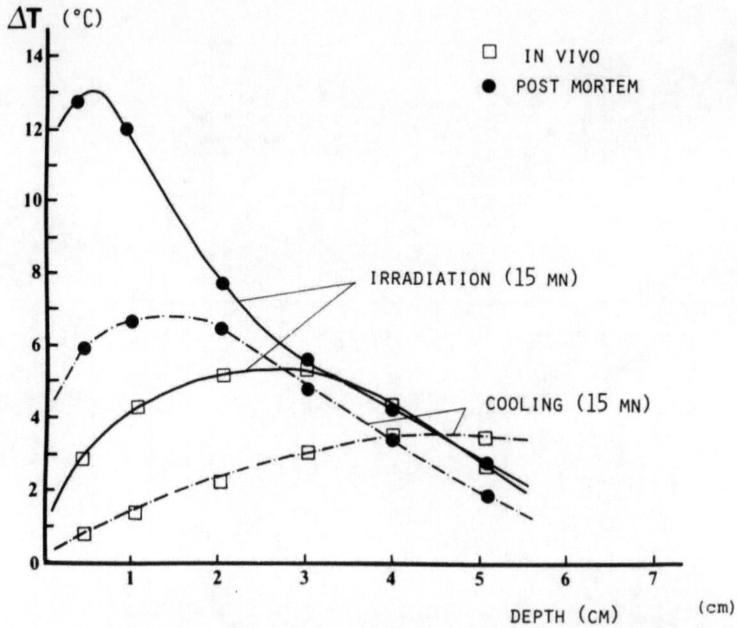

Fig. 5. Subcutaneous thermal profiles measured using fine needle probes just after stopping microwave irradiation (reference to Figure 4) and 15 min later after cooling: (a) in vivo on the anesthetized dog, (b) post mortem.

easy to evaluate; they included nodules beneath cutaneous metastases confirmed by biopsy, nodules arising from breast cancer, superficial adenopathies and recurrences of vulval carcinomas. This purely local treatment does not interfere with the basic treatment of the primary tumor (chemotherapy, hormonotherapy).

Research Methods

The purpose of the research was both to analyze the production of heat by electromagnetic waves in the region of the tumor and to show its efficiency in the eradication of superficial tumor nodules. It was possible to take measurements from four nodules of approximately equal size using different therapeutic methods in each case:-

1. Hyperthermia and radiotherapy.
2. Hyperthermia alone.
3. Radiotherapy alone.
4. No treatment.

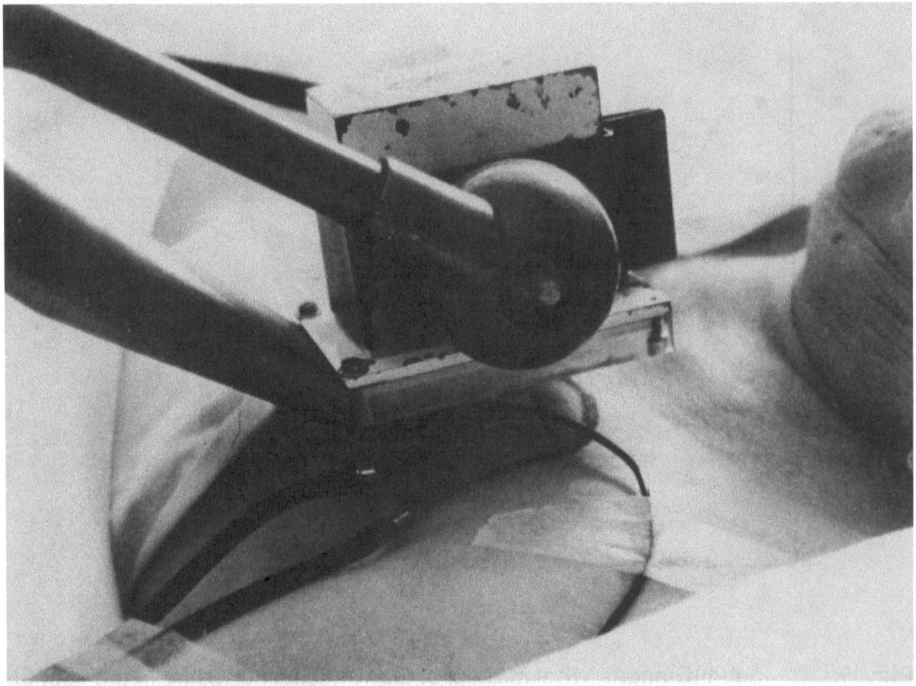

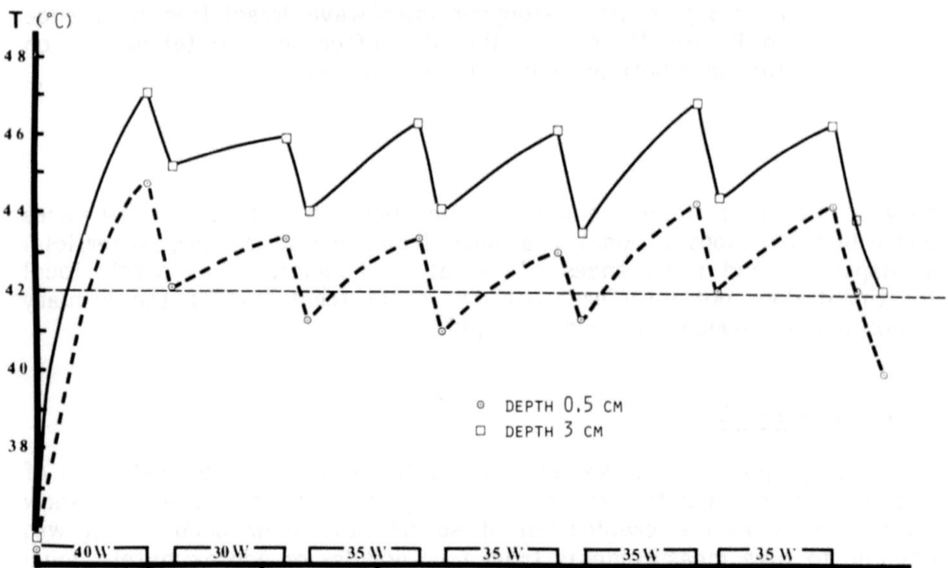

Fig. 6. Clinical hyperthermia studies: microwave irradiation of metastatic cutaneous nodes (2450 MHz).

The choice of heating system and frequency depended on the function of the area and depth of the zones to be treated. For superficial localized lesions the radiative 2450 MHz system was used, while for more extensive or deeper lesions preference was given to a capacitive system at 13.56 MHz.

Temperature measurements were made with two thermocouples, one placed on the superficial surface of the lesion and the other at depth after blocking electromagnetic radiation to eliminate interference. In practice these measurements were taken in sequence every five minutes.

The hyperthermic session lasted thirty minutes, starting when the threshold of efficiency of 42°C was attained, and immediately preceded the session of radiotherapy. The frequency of treatment was based on that of the radiotherapy, ie two to three sessions a week.

CRITERIA FOR EVALUATION

Clinical

Subjective: effective analgesia.
Objective: measurement of nodules before and after treatment.

Thermographic

A thermographic examination with an IR camera was made before treatment, then weekly, and at the end of treatment.

Fluvography

A direct measure of thermal conduction and vascular flow in the treated region was made at the start and finish of treatment.

Histology

Histological proof was required before treatment and further histological examination was made at the end of treatment to assess any modifications in histo-cytology.

CONCLUSION

The clinical research program is currently still in its early stages with few patients. These are principally cases of multiple subcutaneous metastases arising from carcinoma of the breast, a uterine leomyosarcoma and an adenosarcoma of the lung treated by the 2450 MHz generator. A 13.56 MHz system has been used for larger lesions such as glandular secondaries from the upper respiratory and digestive tracts or recurrences of vulval carcinomas.

While it is still too early to present definitive results, the combination of hyperthermia and radiotherapy seem to be promising, particularly in the treatment of subcutaneous nodules.

ACKNOWLEDGEMENTS

This work has been carried out with the assistance of La Delegation Generale a La Recherche Scientifique et Technique of France DGRST number 78-72271 and 79-70812. We also acknowledge support from La Caisse Nationale de l'Assurance Maladie des Travailleurs Salaries (INSERM-CNAMTS 1978)

REFERENCES

1. J. Dutreix, J. P. Le Bourgeois, M. Salama, L'hyperthermie dans le traitement des cancers, J. Radio/Electrol. 59:5;323-334 (1978).
2. A. Priou, Ch. Fournet Fayas, E. Gimonet, J. L. Guerquin-Kern, L. Palas, M. Gautherie, A 434 MHz computer program simulation and reduced size direct contact applicators for therapeutic heating of tissues, in: "Proceedings International Microwave Symposium (IMPI)," Iowa City (1980).
3. A. W. Guy, Analysis of electromagnetic field induced in biological tissues by thermographic studies on equivalent phantom models, IEEE, Trans. on MTT 19:2;205-214 (1971).
4. J. L. Guerquin-Kern, L. Palas, A. Priou, M. Gautherie, Local hyperthermia using microwaves for therapeutic purposes; experimental studies of various applicators, J. Microwave Power 16:3;305-311 (1981).
5. J. L. Guerquin-Kern, L. Palas, M. Samsel, M. Gautherie, Hyperthermie micro-ondes: Influence du flux sanguin et des phenomenes thermoregulateurs, Bull. Cancer (Paris) 68:3;273-280 (1981).

KALIOMIMETIC ION TRANSPORT BY INJURED MEMBRANE DURING IN VITRO HYPERTHERMIA OF TUMOR CELLS

J Robert, C Marchal, F Brunotte, L Anghileri, P Thouvenot

Laboratoire de Biophysique
Faculte de Medecine, 18 rue Lionnois
54000 Nancy, FRANCE

INTRODUCTION

Despite the fact that the mechanisms responsible for tumor cell death during hyperthermia are not well known, there is evidence that cell membrane changes related to small temperature variations are the possible cause of the phenomenon. Hyperthermia changes the physical and chemical characteristics of the lipid double layer at the plasma membrane level with consequent modification of permeability to big molecules.[1,2] Early changes of cell membrane function may be assessed by means of radiotracers. Smigielski and Janiak[3] have demonstrated hyperthermia-induced early damage to the cell membrane by alteration of ^{86}Rb transmembrane transport.

Thallium (^{201}Tl), which has physical characteristics suitable for use in vivo, has been thought to be a useful tool for studying the effects of hyperthermia on cell membrane transport.

MATERIALS AND METHODS

Erhlich ascites carcinoma cells kept in Swiss mice, and DS carcinosarcoma, also in ascitic form, kept in Sprague Dawley rats, and human red blood cells have been used for these studies. The cells were washed three times with $MgCl^2$ (0.112 M) and centrifuged. In order to have a standardized cell concentration, a cell count using trypan blue was performed. Series of tubes containing 800 µl of cell suspension (in the range of 3.5 to 5 x 10^7 cells/ml for the ascites cells, and 10^9 cells/ml for the erythrocytes) plus 100 µl of Krebs-Ringer solution (with 0.25 mM KCl) and 0.25 µCi of ^{201}TlCl (Byk-Mallinchrodt) were incubated at different temperatures in a water-bath. The temperatures used were

37°C, 39°C, 41°C, 43°C, 44°C and 45°C and the incubation period was in the range 8 minutes to 2 hours. A ± 0.05°C precision thermometer was used to control the water-bath temperature. After incubation the cells were washed three times with normal saline at 4°C. A viability test and radio-activity count were performed on the centrifuged cells.

RESULTS

Radioactivity uptake of red blood cells as a time function (at 4°C, 37°C and 44°C) shows a significant increase of ^{201}Tl incorporation for the short incubation time. Under these conditions no changes in cell viability were observed (Fig 1). At 4°C, the low radioactivity incorporation may be related to enzymatic inhibition, as well as to a change in cell membrane physical and chemical characteristics (lipid double layer fluidity). On the contrary, after further incubation the ^{201}Tl uptake differences at various temperatures seem to be related to cell mortality. In the case of radioactivity uptake by ascitic tumor cells the observed behavior as a temperature function is the opposite. As shown in Figure 2, a representative curve of all the series of experiments, a heating temperature of 44°C presents an increase of 30% in ^{201}Tl uptake when compared to that at 37°C. This temperature-related difference is more significant at shorter times (first 15 minutes) when the cell viability is the same in each case. Later on, the higher the temperature the lower the radioactivity uptake. Two main possibilities could explain this phenomenon; a cell mortality increase or a cell permeability augmentation. Comparative studies of ^{201}Tl and ^{42}K incorporation by red blood cells and ascites tumor cells (Fig 3) corroborate the existence of a thermic inhibition of radioactivity uptake only for the tumor cells.

DISCUSSION

These experimental results indicate that hyperthermia has an inhibitory effect on ^{201}Tl uptake by tumor cells. If we consider red blood cells as a model for normal cells, this thermo-dependent inhibition appears to be specific for tumor cells, and this effect is detectable before a positive trypan blue inclusion is observed. This effect begins at 40°C and increases with higher temperatures. The ^{201}Tl and ^{42}K uptake experiments appear to indicate that hyperthermia acts partially on the transmembrane potassium transport systems.

Additional experiments done in the presence of ouabain (0.334 μM), an inhibitor of the Na-K ATPase system, show an equal percentage of inhibition by ouabain action or by hyperthermia over 42°C.

This similarity seems to point to an early thermic effect on the enzymatic systems responsible for potassium transport.

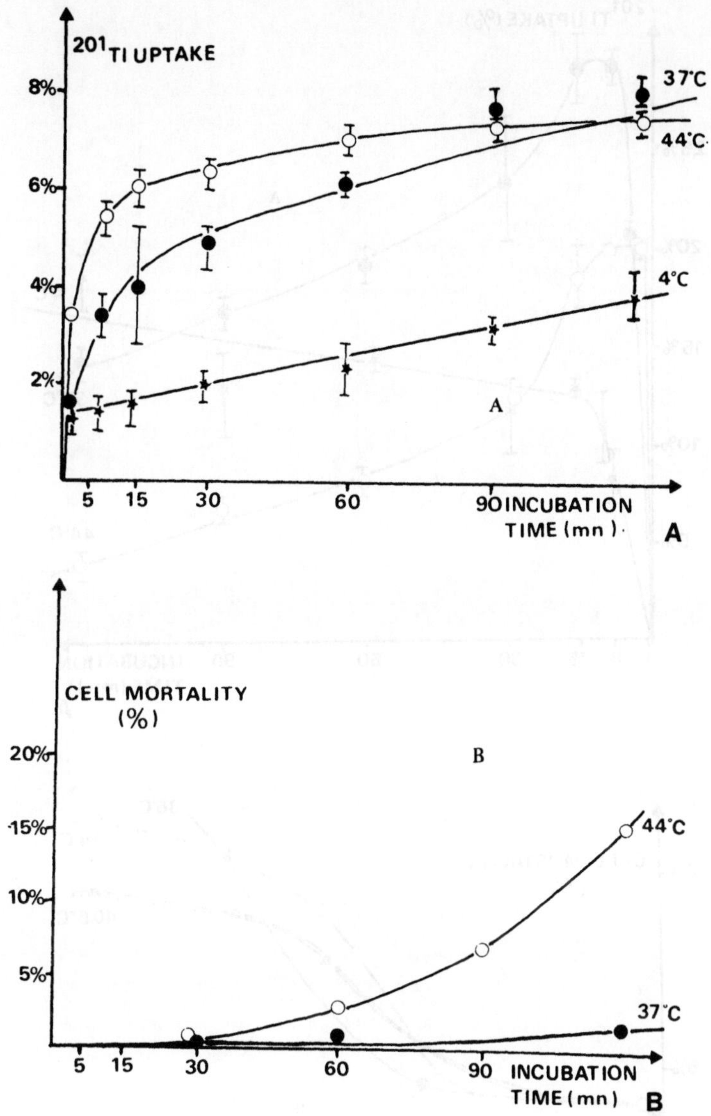

Fig. 1. ^{201}Tl uptake (A) and cell viability (B) for erythrocytes as a function of time and incubation temperature (● 37°C, O 44°C, * 4°C).

ACKNOWLEDGEMENTS

Gratitude is expressed to Mrs S Boyet and Mr P Debas for their valuable technical assistance.

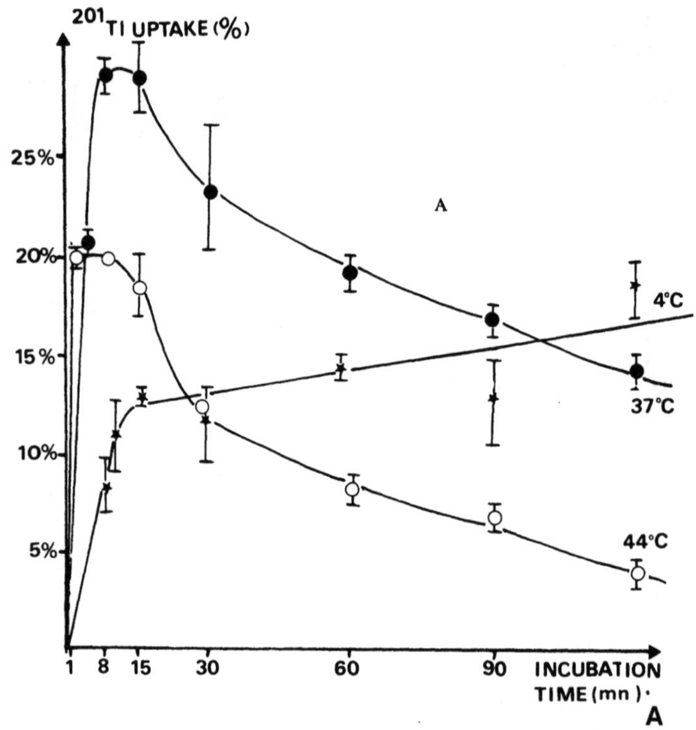

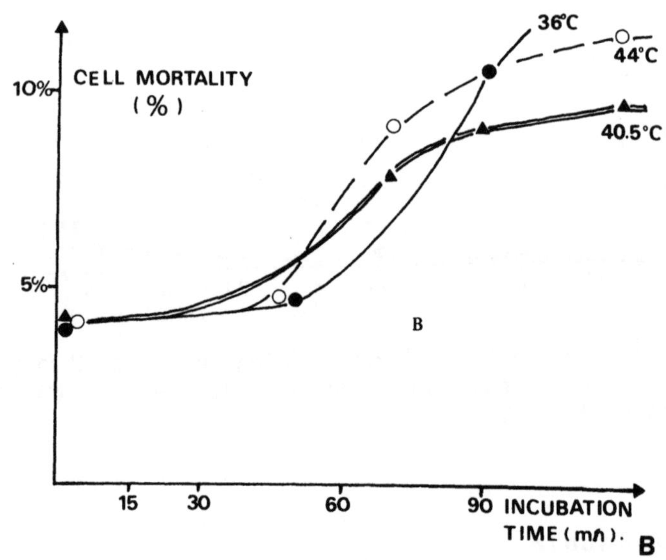

Fig. 2. Kinetics of ^{201}Tl uptake (A) and cell viability (B) of Erhlich ascites Carcinoma cells. (● 37°C, O 44°C, * 4°C).

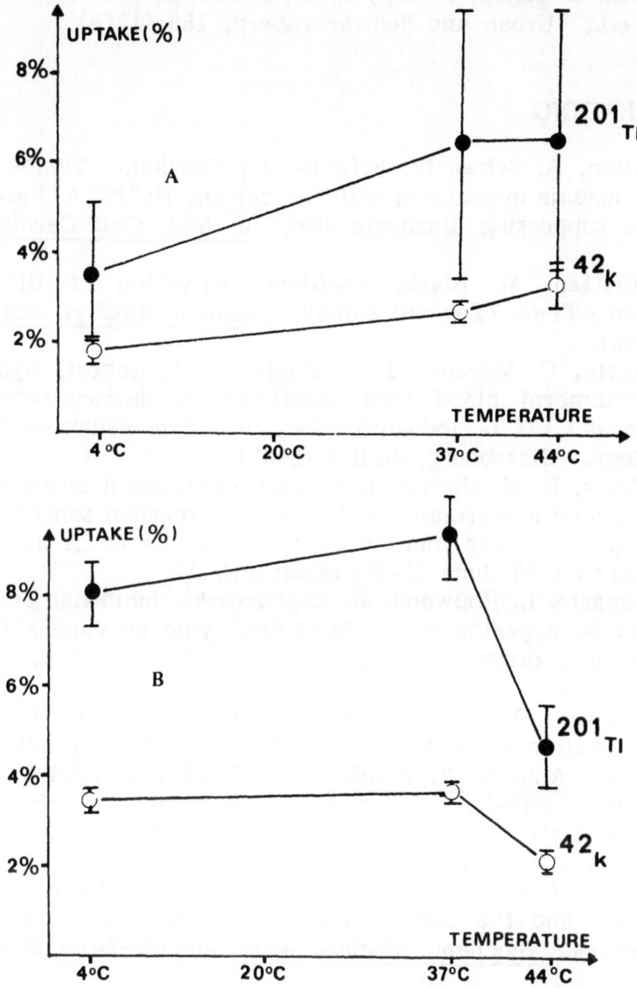

Fig. 3. Comparative study of ^{201}Tl and ^{42}K uptake by red blood cells and Erhlich ascites cells as a function of temperature after 1 hour incubation.

REFERENCES

1. G. M. Hahn, J. Braun, I. Har-Kedar, Thermochemotherapy: synergism between hyperthermia (42-43°) and adriamycin (or Bleomycin) in mammalian cell inactivation, Prov. Nat. Ac. Sc. USA 72:937 (1975).
2. G. M. Hahn, G. C. Li, E. Shiu, Interaction of amphotericin B and 43° hyperthermia, Cancer Res. 37:761 (1977).
3. S. Szmigielski, M. Janiak, Membrane injury in cells exposed in vitro to 43°C hyperthermia, in: "Proceedings of the 2nd International

Symposium on cancer therapy by hyperthermia and radiation," Essen, Strefer ed., Urban and Schwarzenberg, 169 (1978).

FURTHER READING

1. J. C. Allen, A. Schwartz, Effects of potassium. Temperature and time on ouabain interaction with the cardiac Na^+K^+ ATPase : further evidence supporting allosteric site, J. Mal. Cell Cardiol. 1:39-45 (1970).
2. J. S. Britten, M. Blank, Thallium activation of the (Na^+K^+). Activated ATPase of rabbit kidney, Biochem. Biophys. Acta 159:160-166 (1968).
3. F. Brunotte, C. Marchal, L. J. Anghileri, J. Robert, Hyperthermia induced tumoral blood flow modifications demonstrated by 201 Thallium and 131 I. Iodoantipyrine, Int. Symposium on Biomedical thermology. Strasbourg, Juillet (1981).
4. A. E. Cress, E. W. Gerner, Decreased cholesterol levels in cultured mammalian cells correlates with increased thermal sensitivity. Third Int. Symp on Cancer therapy by hyperthermia drugs and radiation, Fort Collins USA June (1980) (abstract), 17.
5. H. S. Dimario, L. Hopwood, M. Kapiszewska, Inhibition of membrane transport by hyperthermia, Third Int. Symp on cancer therapy by hyperthermia. drugs and radiation, Fort Collins USA, June (1980) (abstract) 18.
6. E. W. Gerner, A. E. Cress, D. G. Stickney, D. K. Holmes, P. S. Culver, Factors regulating membrane permeability alter thermal resistance, Ann. N. Y. Acad. Sci. 242:215-233 (1974).
7. Y. Ito, A. Muranaka, T. Harada, A. Matsudo, T. Yokobayaski, H. Terashima, Experimental study of tumour affinity of 201 T1-chloride, Eur. J. Nucl. Med. 3:81-86 (1978).
8. I. A. Sluskii, V. Manninen, J. Jarnefelt, Factors affecting the relative magnitudes and the ouabain insensitive fluxes of thallium ion in erythrocytes, Biochem. Biophys. Acta 506:233-241 (1978).

EFFECT OF LOCALIZED HYPERTHERMIA ON TUMOR BLOOD FLOW AND OXYGENATION

W Mueller-Klieser*, R Manz, J Otte*, P Vaupel***

*** Department of Physiology, University of Mainz, D-6500 Mainz, WEST GERMANY**

**** Department of Physiology, University of Regensburg, D-8400 Regensburg, WEST GERMANY**

INTRODUCTION

Hyperthermia has been shown to have a lethal effect in tumor cell cultures, particularly at temperatures above 42°C.[1] Moreover, it has been demonstrated that hyperthermia retards the growth rate of certain types of malignant tumors and can be effective in completely eradicating some tumors without any major damage to normal tissue.[2,3,4] The differential response of solid tumors in vivo to heat treatment in comparison to normal tissue probably results from several factors. Among these the in vivo micro-environment seems to play a decisive role. In general, the interstitial milieu of solid tumors is characterized by tissue hypoxia and anoxia[5,6] and by severe tissue acidosis.[7] These unfavorable micro-environmental conditions are mainly caused by a deterioration of nutritive blood flow and by inadequate removal of metabolic waste products. Since both pH and O_2 tensions are known to be critical determinants of the thermal sensitivity of tumor cells,[8,9] the effectiveness of hyperthermia in solid tumors may vary according to the efficiency of tumor blood supply. On the other hand, a possible impact of elevated tissue temperatures on tumor blood flow and on cellular metabolism can induce changes in tissue pH and/or tissue pO_2 values thus modifying the direct cell-killing effect of hyperthermia.

Since there is only little information on the interdependence of hyperthermia and tumor blood flow, the goals of the present series of experiments were (i) to measure total tumor blood flow as a function of tumor temperature and (ii) to determine the tumor oxygenation during and immediately after heat treatment at various tissue temperature levels and different heat application times. The results of these investigations should yield a better understanding of the mechanisms involved in tumor therapy by hyperthermia.

MATERIALS AND METHODS

For evaluating the response of tumor perfusion to hyperthermia a series of experiments was carried out on Sprague-Dawley rats with 'tissue-isolated' preparations of DS-Carcinosarcoma implanted into the kidney. In this model, the tumor is connected to the host by only two vessels, ie the former renal artery and vein.[6] Total tumor blood flow can then be assessed by measuring the tumor-venous outflow. Mean arterial blood pressure was continuously registered throughout the experiments, and the relevant parameters of respiratory gas exchange were determined in the arterial and the tumor-venous blood. From this data the O_2 consumption rate of solid tumor tissue as a function of tissue temperature could be calculated taking into account tumor blood flow and arterio-venous O_2 concentration differences. Local hyperthermia was induced by 2.45 GHz microwaves using a special applicator. Mean tumor tissue temperature was monitored with miniaturized thermocouples (shaft diameter 250 μm). Tumor temperature was kept at four levels: 33-36°C (control), 39.5°C, 42.0°C and 44.0°C. Tumor blood flow was measured after a period of 20-30 minutes of constant temperature maintenance. The rectal temperature of the animals was continuously measured with small thermistor probes. There were no substantial changes in the core temperature during heat application to the tumor.

In a preliminary series of experiments the temporal behavior of tumor blood flow during hyperthermia was investigated. Experiments were performed in DS-Carcinosarcoma implanted into the dorsum of a hind foot of Sprague-Dawley rats. Hyperthermia was induced by ultrasound employing a feed-back control system.[10] Tumor temperature was monitored with a thermocouple, the thermal voltage of which was used to control the electrical power of the sonicator in such a way that tumor temperature was maintained constant. This ultrasonic power was continuously recorded through the entire period of heating and served as a relative measure of tumor perfusion rate.

Two microtechniques were employed to assess tumor oxygenation during heat treatment in two different experimental tumor types.

In a first series of experiments, the oxyhemoglobin saturation (HbO_2) in tumor microvessels with diameters of 3-10 μm was determined using a cryophotometric micromethod. This method has been proved in previous investigations to be valid for quantifying the oxygen supply to solid tumors.[5,11] For the determination of HbO_2 values in tumor microvessels, tissue cryobiopsies were taken from tumors with special tongs pre-cooled in liquid nitrogen. Thus, the in vivo HbO_2 saturation is fixed by rapid freezing. The biopsies were then sectioned into 15 μm thin slices, and the HbO_2 saturations of single red blood cells in tumor microvessels were determined using a special photometer (Zeiss UMSP 1). The method was applied to solid tumors of DS-Carcinosarcoma subcutaneously implanted into the dorsum of a hind foot of Sprague-Dawley rats. Tumors were used for experiments 9 to 20 days after implantation, a period of time in which the tumors reached volumes of 1.2 to 5.0 ml. Heating and temperature control were performed in the same way as in the investigations on tumor blood flow described above. The tumor sites used for temperature monitoring were not included in the HbO_2 measurements. Thus, possible tissue damage by the thermoprobes could not affect the experimental results. Tumor temperature was kept at four levels: 35°C (control), 40°C, 43°C and 45°C. The application time was 30 or 60 minutes. Mean arterial blood pressure and the relevant respiratory gas parameters were monitored throughout the experiments.

In a second series of experiments concerning the influence of hyperthermia on tumor oxygenation, tissue oxygen partial pressures (pO_2 values) were obtained by means of O_2-sensitive gold microelectrodes with tip diameters of 1-2 µm. This technique, described in a previous investigation,[7] was applied to C3H mammary adenocarcinoma 10-20 days after implantation of the tumor cells into the thighs of mice. Elevated tumor temperatures were maintained and monitored in the same way as described above. Tissue temperature levels in the present series of experiments were 35°C (control), 37°C, 40°C, 42°C and 45°C.

In order to get a better insight into the effect of hyperthermia on tumor cell metabolism, the O_2 consumption rate of isolated tumor ascites cells of DS-Carcinosarcoma was determined as a function of temperature in the range of 35-45°C. The tumor cells were suspended in native ascitic fluid, and then exposed to elevated temperatures for 30 minutes. The O_2 consumption rate was measured with a recently developed technique.[12] Hereby the decline of the O_2 concentration in a gas-tight syringe containing a defined volume of a cell suspension with a known cell concentration is determined as a function of time. For this determination, defined volumes are injected from the gas-tight syringe into a measuring system at certain time intervals. The measuring system which is based on the autoanalyzer principle is described elsewhere.[13] Oxygen concentration is determined photometrically utilizing the sensitive color reaction of pyrocatechol with O_2.

RESULTS

The relevant results concerning the influence of hyperthermia on tumor blood flow and on tumor oxygenation are displayed in Figure 1. The exposure times were about 30 minutes in all experiments. Figure 1(a) shows the average tumor blood flow of 'tissue-isolated' DS-Carcinosarcoma as a function of tissue temperature at comparable mean arterial blood pressures. It is obvious that moderate hyperthermia induces a signficant increase ($p < 0.005$) in tumor blood flow to levels about 20% above the flow value at control temperatures. A further rise of tumor temperature up to 42°C and 44°C leads to a drop in the perfusion rate to values even lower than those at control temperatures.

The mean HbO_2 saturations as a measure of the efficiency of oxygen supply to tumor tissue at four different temperature levels are plotted in Figure 1(b). The curve essentially exhibits the same characteristic features as the tumor blood flow in Figure 1(a): The mean HbO_2 saturation increased from 51% at control temperatures to 66% at 40°C. Heating up to 43°C caused a decrease in mean HbO_2 to 43%, and during 45°C hyperthermia a drastic decline in the mean HbO_2 values to 23% occurred. This behavior of tissue oxygenation can also be found in the temperature-dependence of the mean tissue pO_2 values recorded in C3H mouse mammary tumors (Fig 1(c)).

During heating from control temperatures up to 40°C, the mean pO_2 increased from 14 to 20 mm Hg, and then decreased to 13.5 and 11 mm Hg at 42°C and 45°C respectively. The mean O_2 consumption rate of 'tissue-isolated' solid tumors of DS-Carcinosarcoma as a function of tissue temperature is shown in Figure 1(d). Again, a characteristic maximum is found at moderate hyperthermia (39.5°C) with an O_2 consumption of 10 µl O_2/g_{tissue}/min. At control levels the O_2 consumption was 6 $\mu l_{O_2}/g_{tissue}$/min. The O_2 consumption rate dropped at 42°C and 44°C, yet this drop is less pronounced than the decline in blood flow or oxygenation at corresponding temperatures. In contrast to this, the O_2 consumption rate of ascites cells of DS-Carcinosarcoma exhibits a distinctly different dependence on temperature than the parameters considered so far (Fig 1(e)). Upon heating, ascites tumor cells increase their O_2 consumption to maximum values at 42°C. Here, the consumption rate is increased to 42 $\mu l_{O_2}/g_{cells}$/min in comparison to 25 $\mu l_{O_2}/g_{cells}$/min at 35°C. Heating up to 43.5°C decreased the O_2 consumption to values of 31 and 24 $\mu l_{O_2}/g_{cells}$/min, respectively.

For investigating the influence of the exposure time on tumor oxygenation during heat application, HbO_2 saturations in Ds-Carcinosarcoma were measured after microwave-induced hyperthermia maintained for 60 minutes. These experiments led to results that differ considerably from those obtained after a heating period of 30 minutes. A marked drop of the mean HbO_2 values can be registered already at 40°C. Elevating the tumor temperature to 43°C and 45°C is followed by a further deterioration of the O_2 supply to the tumors with mean HbO_2 saturations of 17.5% and 19%, respectively.

The critical role of exposure time during heat application to solid tumors is also demonstrated in Figure 2. It shows an original registration representing preliminary results from experiments with ultrasound-induced hyperthermia in solid tumors of DS-Carcinosarcoma. During constant tumor temperature and constant mean arterial blood pressure, the power required for maintaining the tumor temperature at 44°C shows a maximum within the first half of the heating period and then drops to about 50% of the starting level towards the end of the experiment. Even though no

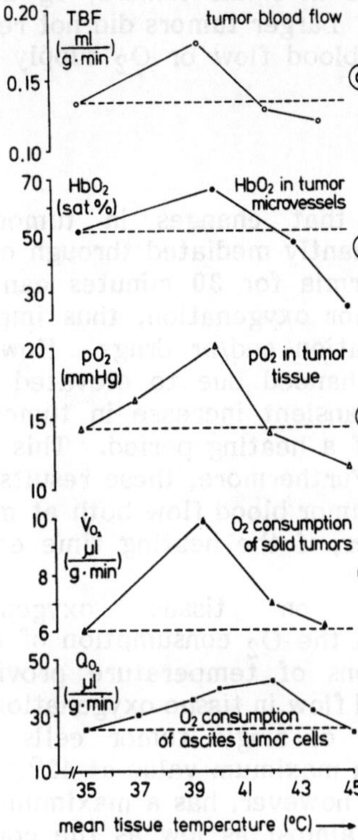

Fig. 1. Compilation of measured parameters determining tumor oxygenation during localized hyperthermia.

(a) Tumor blood flow (TBF) through "tissue-isolated" DS-Carcinosarcoma implanted into the kidney

(b) HbO_2 saturation of single red blood cells in microvessels of DS-Carcinosarcoma implanted into the hind foot dorsum of rats

(c) Tissue pO_2 values in mammary tumors implanted into the thigh of C3H mice

(d) In vivo oxygen consumption rate V_{O_2} of "tissue-isolated" DS-Carcinosarcoma

(e) In vitro oxygen consumption rate (Q_{O_2}) of ascites cells of DS-Carcinosarcoma.

quantitative correlation between the acoustic power and tumor blood flow can be given at present, the time-dependent changes described indicate a qualitative rise and decline of tumor blood flow during 44°C hyperthermia.

It should be emphasized that the latter variations, as well as the characteristic variations in tumor blood flow and oxygenation shown in

Figure 1, were only found in small tumors, eg in DS-Carcinosarcomas with wet weights of 3-5 g. Larger tumors did not respond to hyperthermia by significant changes in blood flow or O_2 supply to the tumor tissue.

DISCUSSION

The data indicate that changes in tumor oxygenation during hyperthermia are predominantly mediated through changes in tumor blood flow. Moderate hyperthermia for 30 minutes can cause an increase in tumor blood flow and tumor oxygenation, thus improving the efficiency of tumor therapy by radiation and/or drugs. However, the propagation of metastases may be enhanced due to elevated perfusion rates. At higher temperatures, a transient increase in tumor blood flow is often present at the beginning of a heating period. This can be seen from the registration in Figure 2. Furthermore, these results indicate that there is a substantial decrease in tumor blood flow both at moderate hyperthermia and at higher temperatures, if the heating time exceeds 30 minutes.

The investigations on tissue oxygen consumption of DS-Carcinosarcoma and on the O_2 consumption of ascites cells from the same cell line as functions of temperature provide evidence for the critical role of tumor blood flow in tissue oxygenation during hyperthermia. The O_2 consumption rate of single tumor cells increases with rising temperature, and reaches a maximum value at 42°C. The O_2 consumption rate of solid tumor tissue, however, has a maximum at 39.5°C. At 42°C there is a drop to values almost as low as the consumption rate of the tumors at control temperatures. It can be concluded from these differences that the respiration rate of solid tumors is mainly determined

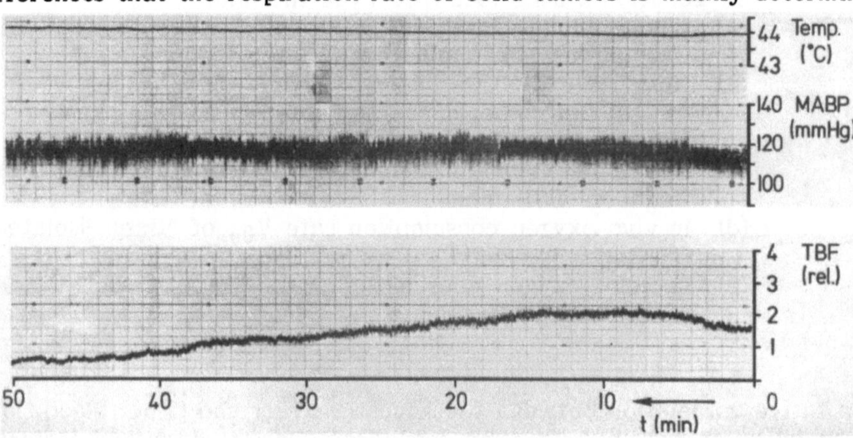

Fig. 2. Mean arterial blood pressure (MABP) and total tumor blood flow (TBF) as a function of exposure time during 44°C - hyperthermia. TBF is given in arbitrary units derived from the acoustic power necessary to maintain a constant tumor tissue temperature.

by the efficiency of O_2 delivery to the cells. Oxygen consumption of solid tumors at 42°C is lowered by a restricted oxygen supply to the tumor cells as a consequence of a reduced tumor blood flow. In single cell suspensions where O_2 supply is not impeded, heating from 39.5°C to 42°C results in an increase in the O_2 consumption rate.

It is obvious from Figure 1(c) and Figure 1(d) that the tissue pO_2 and the O_2 consumption rate in solid tumors run parallel with increasing temperature. Both curves suggest that, at moderate hyperthermia, the amount of oxygen delivered to the tumor tissue is larger than that removed by an increase in the respiration rate. Thus, an increase in tissue pO_2 is induced. Corresponding considerations also apply to the changes in pO_2 and V_{O_2} at higher temperatures. This again demonstrates the predominant role of tumor blood flow for in situ tissue oxygenation during hyperthermia.

Presumably several mechanisms are involved in the flow changes described. Previous investigations[14] have shown that vasoreactions seem to be partially responsible for the increase in blood flow during moderate hyperthermia. Changes in flow properties of blood may also be important in this regard, yet play a more decisive part in reducing tumor perfusion at higher temperatures. Tissue acidosis and hyperthermia can induce a loss in red blood cell flexibility, which in combination with other alterations in the terminal vascular bed can cause severe disturbances of the tumor microcirculation. Changes in the permeability of the tumor microvessels may also account for a decline in nutritive tumor blood flow[15] since it has been observed that tumors can increase considerably in volume during hyperthermia, probably due to congestion, extravasation and/or edema.

SUMMARY

Investigations were carried out on various tumor models using different techniques to assess the impact of hyperthermia on tumor blood flow and on tissue oxygenation. Experiments using cryophotometry or O_2-sensitive microelectrodes both demonstrated an improvement in tumor tissue oxygenation during moderate hyperthermia at 39-40°C with exposure times of 30 minutes. In contrast to this, a drop in oxygenation was registered either at temperatures of 44-45°C maintained for 30 minutes or at moderate hyperthermia for 60 mintues. This implies that tissue temperature levels and the exposure time play decisive roles in heat treatment. Furthermore, it could be shown that these characteristic changes are predominantly mediated through parallel modulations of tumor blood flow.

REFERENCES

1. W. C. Dewey, L. E. Hopwood, S. A. Sapareto, L. E. Gerweck, Cellular responses to combinations of hyperthermia and radiation,

Radiology 123:463 (1977).
2. J. A. Dickson, The effects of hyperthermia in animal tumor systems, Rec. Res. Cancer Res. 59:43 (1977).
3. J. Overgaard, Effect of hyperthermia on malignant cells in vivo, Cancer 39:2637 (1977).
4. H. D. Suit, Hyperthermic effects on animal tissues, Radiology 123:483 (1977).
5. P. Vaupel, Hypoxia in neoplastic tissue, Microvasc. Res. 13:399 (1977).
6. P. Vaupel, R. Manz, W. Mueller-Klieser, W. A. Grunewald, Intracapillary HbO_2 saturation in malignant tumors during normoxia and hyperoxia, Microvasc. Res. 17:181 (1979).
7. P. Vaupel, S. Frinak, H. I. Bicher, Heterogeneous pO_2 and pH distribution in C3H mouse mammary carcinoma, Cancer Res. 41:2008 (1981).
8. L. E. Gerweck, E. L. Gillette, W. C. Dewey, Killing of Chinese hamster cells in vitro by heating under hypoxic or aerobic conditions, Eur. J. Cancer 10:691 (1974).
9. L. E. Gerweck, E. Rottinger, Enhancement of mammalian cell sensitivity to hyperthermia by pH alteration, Radiat. Res. 67:508 (1976).
10. W. Mueller-Klieser, E. G. Lierke, P. Vaupel, A feedback control system for localized ultrasonic hyperthermia in tumors. II. Application and first experiments in DS-Carcinosarcoma. IVth Meeting European Co-operative Hyperthermia Group, London (1982).
11. W. Mueller-Klieser, P. Vauple, R. Manz, R. Schmidseder, Intracapillary oxyhemoglobin saturation of malignant tumors in humans, Int. J. Radiol. Biol. 7:1397 (1981).
12. W. Mueller-Klieser, R. Zander, P. Vauple, Oxygen consumption of tumor cells suspended in native ascitic fluid at 1-42°C, Pflugers Arch. 377;R:17 (1978).
13. W. Lang, H. U. Wolf, R. Zander, A sensitive continuous and discontinuous photometric determination of oxygen, carbon dioxide and carbon monoxide in gases and fluids, Anal. Biochem. 92:255 (1979).
14. P. Vaupel, K. Ostheimer, W. Mueller-Klieser, Circulatory and metabolic responses of malignant tumors during localized hyperthermia, J. Cancer Res. Clin. Oncol. 98:15 (1980).
15. B. Emami, G. H. Nussbaum, N. Hahn, A. J. Piro, A. Dritschilo, F. Quimby, Histopathological study of the effects of hyperthermia on microvasculature, Int. J. Radiat. Oncol. Biol. Phys. 7:343 (1981).

IMMUNOLOGICAL RESPONSE TO WHOLE BODY MICROWAVE HYPERTHERMIA IN NORMAL AND TUMOR-BEARING ANIMALS

S Szmigielski

Department of Biological Effects of Non-ionizing Radiation
Center for Radiobiology and Radioprotection, 128 Szaserow
00-909 Warsaw, POLAND

Whole body microwave hyperthermia (MwH) results in inhibition of tumor growth in experimental animals, due to the high sensitivity of neoplastic cells to increased temperatures and direct thermal injury to the overheated tumor tissues. However, some other mechanisms, notably those originating from the cell-mediated immune system and non-specific natural antineoplastic resistance are also present.

In a large series of experiments we tested the response of cell-mediated immunity (function of macrophages in vitro and in vivo, delayed hypersensitivity to oxazolone, reactivity of spleen lymphocytes to mitogens in vitro, natural killer cell activity) and non-specific antineoplastic (lung cancer colonies assay) and antiviral (course of acute herpes and vaccinia virus infections) resistance in normal and transplantable sarcoma bearing BALB/c mice. The animals were exposed to daily 2 hour sessions of MwH (2450 MHz, anechoic chamber with controlled environmental conditions, field power density 40 mW/cm^2, specific absorption rate 18 mW/g, rectal temperature of 41.5 $\pm$ 0.5°C) for 4, 7, 10 or 14 days. The above listed methods were carried out in all the animals; in tumor-bearing mice the growth of sarcomas (size and mass) and survival were also registered.

It was found that MwH repeated daily results in temporary regression of the tumors; however, total cure of the neoplasms is not possible unless the tumor-inhibiting effect of MwH is enhanced by additional treatment with immunomodulators (P granulosum, killed lyophylized bacteria with established immunomodulating and anti-neoplastic properties). The response from cell-mediated immunity to MwH was found to be identical both in normal and tumor-bearing mice and was characterized by a biphasic course: stimulation of immune reactivity during the first 4 days of treatment, followed by rapid, but transient and reversible, inhibition of cell-mediated immunity and lowering of natural resistance after 10-14 days of MwH.

TREATMENT OF MALIGNANT NEOPLASMS WITH RADIOFREQUENCY HYPERTHERMIA - CLINICAL EXPERIENCE

Z Dreznik, R E Falk, U Ambus, V Howard, D Venturi, L Makowka

Goldie Rotman Oncology Unit and the Department of Surgery Toronto General Hospital, Faculty of Medicine, University of Toronto, Toronto, Ontario, CANADA

The use of 'heat' for the treatment of malignant neoplasms was mentioned in the medical literature as early as the 19th century.[1] This subject remained unnoticed and it was only in the last decade when better and safer equipment became available, that this treatment modality gained a renewed interest.

Neoplastic tissue reacts differently from normal tissue to thermotherapy. It was observed that tumor cells show an increased sensitivity to heat, as compared to normal cells, at temperatures of 42°C - 44°C.[2,3] This difference is probably related to the blood supply within the neoplasm as compared to normal tissue.[4] The selective heating allows a core tumor temperature to be recorded at 45°C - 50°C, producing substantial tumor necrosis and sparing the normal surrounding tissue from excessive heating.[5]

The treatment of malignant tumors with hyperthermia includes the whole electromagnetic spectrum. The most common techniques in clinical use are: whole body, regional and local hyperthermia. Clinical experience, using thermo-therapy alone, or combined with other modalities of antitumor therapy, has been limited to relatively superficial tumors, as effective deep heating could not be easily or safely achieved.[6] Limited experimental and clinical reports of combining hyperthermia with chemotherapy are available.[7,8]

The present report reviews our experience using the combination of local radiofrequency hyperthermia with chemotherapy for the treatment of deep malignant neoplasms.

PATIENTS AND METHOD OF TREATMENT

Since August 1981, 88 patients with malignancy have been treated with radiofrequency hyperthermia. All patients had previously failed to respond to other therapeutic modalities. The heat was generated by the International Institute of Medical Sciences (IIMS) capacitive type of machine with a radiowave frequency of 13.56 MHz.

Gastrointestinal tumors were the most frequent malignancy treated and the primary tumor was colorectal in 45 patients (Fig 1). Most of the patients have been treated for deep seated tumors (89.7%) (Table 1) with a combination of RF hyperthermia and chemotherapy (90.8%) (Table 2). In 14 of 28 patients with liver metastases, chemotherapy was administered through a percutaneous catheter placed in the hepatic artery, using a continuous pump infusion for three days. Concomitantly, local thermo-therapy was directed toward the liver. The systemic chemotherapy was given intravenously during the last 15 minutes of the hyperthermia treatment.

Local RF hyperthermia was applied for 90-120 minutes each day for 5 consecutive days. Tumor core temperatures were measured in the liver and palpable recurrent abdominal cancers, provided coagulation studies were normal. One to three thermocouples were inserted through plastic angiocatheters into the tumor mass under fluoroscopy and local anesthesia. The thermocouples remained in the same position for the 5 days of treatment and were attached to a clinical thermometer providing constant

Table 1. Treatment of malignant neoplasms with radiofrequency hyperthermia - clinical experience

AREA TREATED	
DEEP SEATED TUMORS - 89.7%	
Liver	28
Inop. or Rec. Abd.	27
Pelvic	15
Lung	7
Mediastinum	1
Brain	1
SUPERFICIAL TUMORS - 10.3%	
Neck	3
Chest Wall	3
Other Superficial Lesions	3

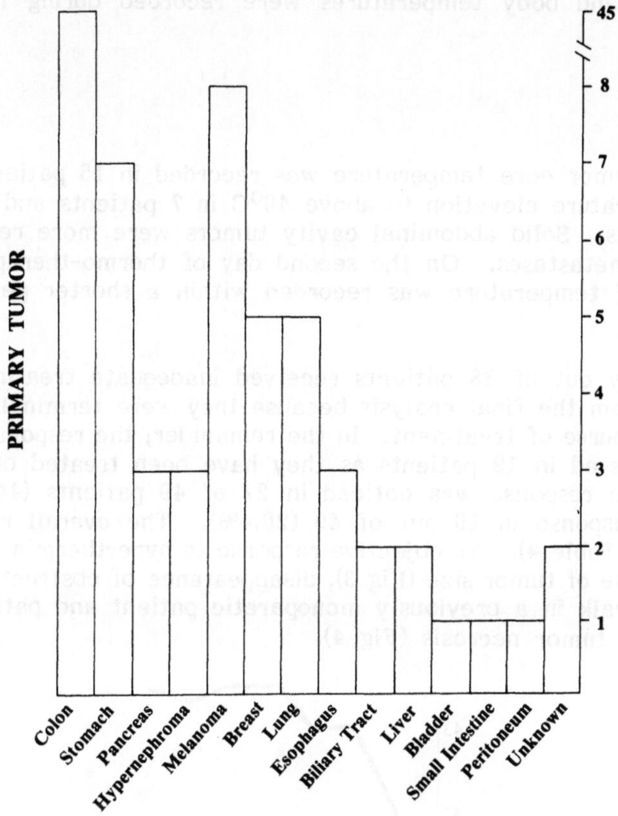

Fig. 1. Location of primary tumor.

recording of the tumor core temperature. Clinical evaluation, radiographic assessment and routine laboratory testing were performed before and after each course of treatment. The patients were sedated but conscious;

Table 2. Combination of hyperthermia with chemotherapy

METASTATIC LIVER (28 PTS)		OTHER (60 PTS)	
Hepatic Infusion	14 pts		
Systemic	12 pts	54 pts	
Total	26 pts	54 pts	90.9%

vital signs and body temperatures were recorded during hyperthermia treatment.

RESULTS

The tumor core temperature was recorded in 15 patients (Table 3) with temperature elevation to above 40°C in 7 patients and above 42°C in 6 patients. Solid abdominal cavity tumors were more readily heated than liver metastases. On the second day of thermo-therapy a greater elevation of temperature was recorded within a shorter period of time (Fig 2).

Twenty out of 88 patients received inadequate treatment and are excluded from the final analysis because they were terminally ill or had a limited course of treatment. In the remainder, the response rate could not be assessed in 19 patients as they have been treated only recently. A subjective response was noticed in 22 of 49 patients (44.9%) and an objective response in 10 out of 49 (20.4%). The overall response rate was 65.3% (Table 4). An objective response to hyperthermia is considered as a decrease of tumor size (Fig 3), disappearance of obstructive jaundice, ability to walk in a previously monoparetic patient and pathological evidence of tumor necrosis (Fig 4).

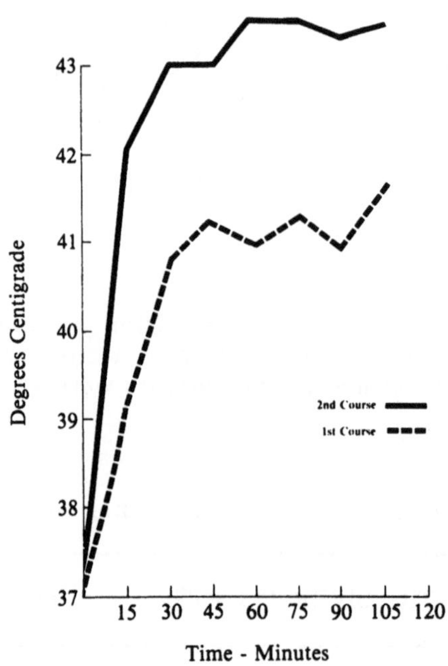

Fig. 2. The difference in tumor temperature on the first and second day of treatment.

Table 3. Treatment of malignant neoplasms with radiofrequency hyperthermia - clinical experience

TUMOR CORE TEMPERATURES IN 15 PATIENTS		
Core temp	Liver : 12	Local recurrence : 3
40°C	4 of 12	3 of 3
42°C	3 of 4	3 of 3
max temp	42.7°C	44.1°C

A decrease in alkaline phosphatase in all the patients with metastatic liver disease was noted. The level of alkaline phosphatase decreased after hyperthermia to about 60% of the pre-treatment values (Table 5 and Fig 5).

The combination of hyperthermia and chemotherapy had no additional toxicity and small superficial burns in 8 patients were the only recorded complication.

Changes of pulse rate, blood pressure and body temperature as a result of hyperthermia were within physiological ranges.

Table 4. Data analysis (88 patients)

Inadequate treatment : 20 terminally ill, limited course, no follow-up

Adequate treatment : 68

Improved 32 (47%)	No response 5 (7.3%)
Undetermined 19 (27.9%)	Progression 4 (5.8%)
Stable 8 (11.7%)	

Overall response in 32 of 49 patients (65.3%)

Subjective	Objective
22 of 49 (44.9%)	10 of 49 (20.4%)

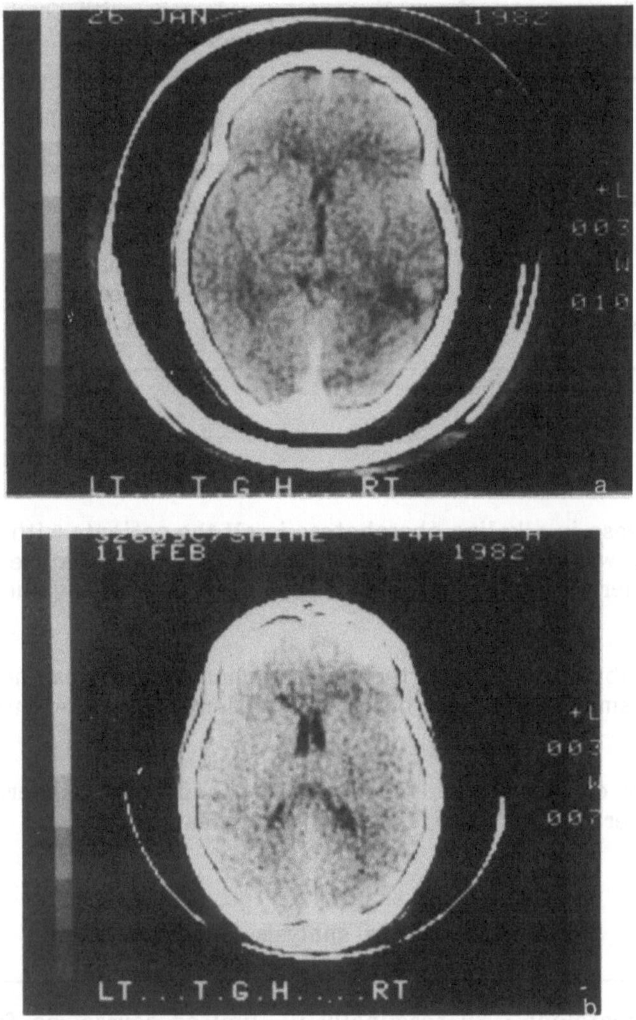

Fig. 3. (a) Brain CT scan - metastatic lesion. (b) The lesion disappeared following thermo-therapy.

DISCUSSION

There is an increasing body of evidence which demonstrates that hyperthermia has a destructive effect upon malignant tumors, which is dependent on the tumor mass, intensity of heating and rate of cellular division.[9] Since effective heating may be accomplished by raising the tumor temperature above 42°C, continuous recording during hyperthermia is preferred. However, the use of thermocouples in deep seated tumors is sometimes hazardous and may lead to complications. Since there is no significant damage to normal tissue, as a result of the present

Table 5. Biochemical Changes

MEAN ALKALINE PHOSPHATASE IN 17 PATIENTS WITH LIVER METASTASES	
Before Thermotherapy	After Thermotherapy
493 μ/l	310 μ/l

As a result of hyperthermia and chemotherapy a decrease of about 40% in alkaline phosphatase was noted

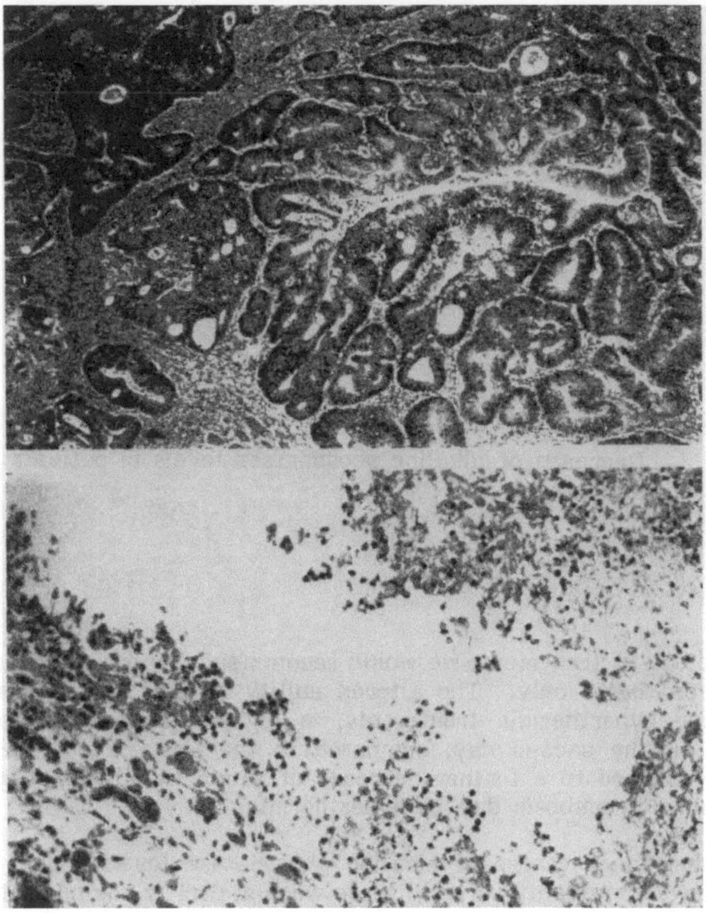

Fig. 4. Adenocarcinoma of the colon before and after thermo-therapy.

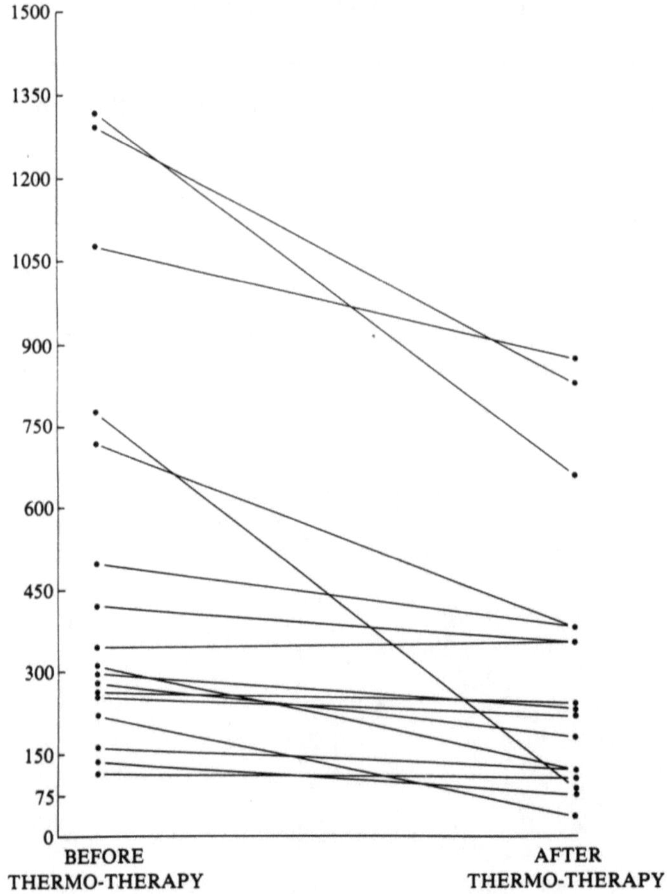

Fig. 5. Decrease of alkaline phosphatase levels in patients with liver metastases.

methodology of treatment, we would recommend the use of thermocouples in selected cases only. The altered ability to heat tumor tissue during successive hyperthermia treatments, ie higher temperatures and faster heating on the second day, compared to the first day of treatment is probably related to a further decrease of tumor blood supply as a result of vascular thrombosis due to hyperthermia.

Since the decrease of alkaline phosphatase levels found in all our patients with metastatic liver disease was parallel to the patient's clinical condition, it probably represents a resultant tumor mass necrosis due to hyperthermia.

The high response rate (65.3%), no toxicity and only minor complications, make the combination of local RF hyperthermia with chemotherapy a safe and potentially useful method of tumor therapy. Technical improvement to increase energy output, and non-invasive monitoring of the tumor temperature are the major requirements for further improvement in results. Local hyperthermia should be considered for the treatment of malignant neoplasms as an adjunct to chemotherapy, radiotherapy or selective immune stimulation, as it increases tumor cell destruction with essentially no significant addition to toxicity.

SUMMARY

Eighty-eight patients with malignant tumors, who failed to respond to previous antitumor therapy, have been treated with local RF hyperthermia and chemotherapy since August 1981. The origin of the tumors was mainly gastrointestinal and most of them were located in 'deep seated' areas. Tumor core temperature was recorded in selected patients. An overall response rate of 65.3% was noticed and a decrease in alkaline phosphatase levels of about 40% is reported in patients with metastatic liver disease. Detailed clinical and laboratory analyses, including pathological evidence of tumor necrosis as a result of hyperthermia, are reported.

REFERENCES

1. W. B. Coley, The treatment of malignant tumors by repeated innoculations of erysipelas with a report of original cases, Am. J. Med. 105:487 (1893).
2. G. Jr. Crile, Heat as an adjunct to the treatment of cancer: experimental studies, Cleveland Clin. Quart. 28:75 (1961).
3. D. S. Muckle, J. A. Dickson, The selective inhibitory effect of hyperthermia on the metabolism and growth of malignant cells, Brit. J. Cancer 15:771 (1978).
4. P. M. Gullino, F. H. Grantham, Studies on the exchange of fluids between host and tumor. II. The blood flow of hepatomas and other tumors in rats and mice, J. Natl. Cancer Inst. 27:1465 (1961).
5. J. A. Dickson, S. A. Shah, Technology for the hyperthermia treatment of large solid tumors at 50°C, Clin. Oncol. 3:301 (1977).
6. E. Bender, T. Schramm, Untersuchungen sur thermosensibilitat von tumor and normalzellen in vitro, Acta Biol. Med. Germ. 17:527 (1966).
7. J. B. Marmor, Interaction of hyperthermia and chemotherapy in animals, Cancer Res. 39:2269 (1979).
8. G. M. Hahn, J. Braun, I. Har-Kedar, Thermochemotherapy: synergism between hyperthermia (42°C - 43°C) and Adriamycin (or Bleomycin) in mammalian cell inactivation, Proc. Natl. Acad. Sci. USA 72:937 (1975).

9. J. A. Dickson, H. A. Ellis, The influence of tumor volume and the degree of heating on the response of the solid Yoshida sarcoma to hyperthermia (40°C - 42°C), Cancer Res. 36:1188 (1976).

RADIOFREQUENCY HYPERTHERMIA (13.56 MHz) ASSOCIATED WITH RADIOTHERAPY OR CHEMOTHERAPY IN THE TREATMENT OF CANCER, WITH ATRAUMATIC CONTROL BY MICROWAVE THERMOMETRY (2.4 GHz)

G Giaux*, B Prevost*, Y Leroy, M Chive****

*** Centre de Lutte contre le Cancer, Lille, FRANCE**
**** Centre Hyperfrequences et Semi-conducteurs, Villeneuve d'Ascq, FRANCE**

This study has been carried out to determine the value of thermometry and the tolerance of local radiofrequency hyperthermia associated with radiotherapy or chemotherapy of advanced cancers and post-radiation recurrences.

A 13.56 MHz Generator (HLI 500 od SEM, France) is connected to two electrodes (by means of a coaxial cable) making a capacitor, which delivers a maximum power of 500 watts. These two electrodes, which are in contact with the biological tissues by a paste with ground quartz, are made of conductive tissue on metallic contact. They are kept in place by a permeable resin mould, which allows cooling of the superficial tissues by means of blown fresh air. We are testing a method using water to cool the edge of the electrodes where slight burns may occur.

At the beginning, the temperature was controlled by metallic thermocouples inserted into the tissues at different depths and connected with a filter to eliminate interference.

To avoid traumatic temperature measurement with inserted thermocouples, we tested a microwave radiometer (2.4 GHz) which was developed in the Microwaves and Semi-conductors Center of the University of Lille and made by Odam-Bruker-Wissembourg, France. The probe of this apparatus is placed on the skin surface of the heated area either between the electrodes or through a small hole made in the middle of one electrode. We have found good correlation between the radiometric temperature measurements and the thermocouple values between 2 and 3 cm.

MATERIALS AND METHOD

For the clinical study, we have used two protocols:

1. Local hyperthermia for one hour during a course of chemotherapy with Cis-platyl 100 mg/m^2 once every three weeks, and Bleomycine 15 mg IV twice a week, during the other two weeks.
2. Local hyperthermia for one hour, four hours after radiotherapy two or three times a week.

Among the 11 patients selected for this study, we had:

1. Seven recurrences of tumors of the head and neck after radiotherapy (squamous cell carcinoma)
   (a) Two cases of metastatic neck nodes treated by hyperthermia and chemotherapy
   (b) Five cases of local recurrences treated by hyperthermia and radiotherapy (30 Gy)
2. Two initial treatments for tumors of the pyriform sinus with very large metastatic neck nodes (11 cm diameter)
3. One tumor of the posterior wall of the hypopharynx, which occurred in a previously irradiated area.
4. One inoperable tumor of the ovary (psamno-carcinoma)

RESULTS

1. Three regressions of more than 50% during one year
   (a) Two metastatic neck nodes which occurred in a previously irradiated area, treated by hyperthermia and chemotherapy
   (b) One initial treatment of a carcinoma of the pyriform sinus with a large metastatic neck node (11 cm diameter) with combined radiotherapy and hyperthermia.
2. Three regressions of less than 50%
3. Three cases with short functional improvement
4. Two cases with no positive result

CONCLUSION

Local hyperthermia (42° - $43^{o}C$) for tumors of 5 or 6 cm diameter under the skin, for example cervical lymph node metastases, is well tolerated, even after previous treatments.

A temperature of 42° - $43^{o}C$ is obtained in these tumors during one hour with an average power of 40 $watts/dm^3$ of treated tissue.

After comparison between the measurements obtained from inserted thermocouples and by microwave radiometer, regular control of the treatments is usually carried out with the atraumatic radiometer with good accuracy.

Finally, it is necessary to cool the electrodes, particularly at the edges, by air flow or with peripheral circulation of water. By thus protecting the skin, the average applied power can be increased.

FURTHER READING

1. M. Chive, D. D. Nguyen, Y. Leroy, Localized hyperthermia. Towards an atraumatic method of heat control based on microwave thermography, Bull. Cancer (Paris) 68;3:293-294 (1980).
2. M. Chive, Y. Leroy, G. Giaux, B. Prevost, Microwave thermography for controlled local hyperthermia at 2.5 GHz. Digest Microwave Power Symposium, Toronto, Canada (1981).
3. M. Chive, M. Robillard, J. J. Fabre, G. Giaux, B. Prevost, Radiofrequency and microwave hyperthermia controlled by microwave thermography. International Symposium of Biological Thermology, Strasbourg, France (1981).
4. L. Dorbec, Y. Chabble, P. Rault, P. Lavallee, G. Legeay, J. L. Playe, A. Grassat, A 13.56 MHz hyperthermia apparatus for local treatment of deep tumors, J. Eur. Radiother. T2;3:157-167 (1981).
5. L. Israel, M. Besenval, Phase I and phase II trials of the use of localized radiofrequency hyperthermia in 49 cases of deep-seated cancer, Bull. Cancer (Paris) 68;3:296 (1981).
6. S. Lacroix, Hyperthermie par utilisation des radiofrequences a 13.56 MHz, Contrat D.G.R.S.T. no DIF/GBM/79.7.0544.
7. D. D. Nguyen, M. Chive, Y. Leroy, Hyperthermie locale controle par thermographie microonde a 2.5 GHz. Symposium International URSI. Ondes Electromagnetiques et Biologie. Jouy en Josas, France (1980).
8. R. Pagione, F. Sterzer, J. Mendecki, E. Friedenthal, C. Botstein, Microwave Journal 71-80 (1981).

HYPERTHERMIA COMBINED WITH IMMUNE-MODULATION IN CANCER THERAPY

P Pontiggia, G M Pigliucci, N Gentiloni, G Follini, R Cervellin

Department of Hematology and Oncology, Centro Sanrocco Como, ITALY
Department of Surgery and Department of Medical Pathology Universita S Cuore-Roma, ITALY

INTRODUCTION

Today's strategies for treatment of neoplastic disease, including extensive surgery, chemotherapy and radiotherapy, alone or in combination, have given acceptable results in a limited number of tumors. However, the results obtained in over thirty years with these methods cannot be considered satisfactory in the large majority of malignant tumors.

An immunological reactivity, clearly demonstrated, directed towards hindering the growth of a tumor, may be enhanced by methods of immune stimulation (specific or non-specific), of immune modulation (thymus extracts), or of passive transfer of natural, immune active substances, able to hinder the expansion of a tumor (transfer factor, interferon, interleukins).

We have shown that the level of the total T cells (and of the percentage of them) in our neoplastic patients is signficiantly low in comparison with the normal population. This immune depression is the reason for our attempt to improve the immune reactivity in our patients.

In this case the use, besides the usual treatments of immune stimulation or immune modulation, of an oncolitic method that does not have toxic side effects and above all does not have a suppressive effect on the immune system, seems a rational choice. Hyperthermia, and in particular localized hyperthermia, seems to fill these requirements.

From a study of the literature it is possible to draw the following conclusions:

1. Hyperthermia strengthens immunocompetence.
2. The immune system is an important intermediary of a thermic effect able to amplify the toxic action of it on neoplastic tissues.
3. Immune stimulation by heat seems to mainly involve all the T lymphocytes and the macrophages.

MATERIAL AND METHODS

For this reason we have started a pilot study using thermotherapy and a stimulating treatment with thymic hormones (active on T-cell population) and a Coryne bacterium Parvum (which stimulates macrophage activity).

From RF heating units we have chosen the LeVeen machine, at 13.56 MHz. In the light of the results obtained, this reliable machine has been a good choice, especially for deep-seated tumors (chest and abdominal tumors). It acts with three pairs of paddles activated one after the other. The schedule of therapy adopted is as follows:

1. Localized thermotherapy	six applications, at intervals of at least 48 hours, for the duration of 180'.
2. Collateral therapies	strengthening chemotherapy in concomitance with the cycles of hyperthermia (usually Adriamycin or Bleomycin), or radiotherapy in cases which are likely to be more sensitive to a combination of hyperthermia and radiotherapy.
3. Immune stimulation	active aspecific immunotherapy with Thymostimulin (Tp1 - 0.5 mg/kg/2 times a week) and/or Corybacterium Parvum.

RESULTS

In Table 1 we summarize the results obtained in 127 cases of advanced malignant disease, treated during the first 18 months of utilization of the LeVeen's machine.

We must emphasize that these results have been obtained in advanced cases, where previous treatments (radio and/or chemotherapy) had failed. If we put ourselves in a position to treat patients at the onset of the disease, the response might be much better.

The use of chemotherapy at low doses was planned with the aim of obtaining a satisfactory cytotoxic effect, while avoiding the resulting immune depression frequent in patients affected by neoplasia when treated aggressively.

Table 1. Patients treated with localized thermotherapy (LeVeen method)

	Cases	Cases not assessed	CR	PR	CR + PR assessed cases
Pulmonary Ca	72				
Adeno	27	3	2	8	10/24
Epider.	37	8	2	12	14/29
Oat Cell	8	-	1	2	3/8
Breast Ca	10	3	4	3	7/7
Colon Ca	6	5	-	-	0/1
Oropharynx Ca	6	2	-	3	3/4
Gastric Ca	5	1	1	3	4/4
Osteosarcoma	5	2	-	2	2/3
Bladder Ca	4	2	-	1	1/2
Ca of the Pancreas	4	1	-	2	2/3
Ca of the Ovaries	3	2		1	1/1
Connective Tissue (Sarc.)	3	1		1	1/2
Melanoma	3	-	3	-	3/3
Miscellaneous	6	-	-	2	2/6
TOTALS	127	30	13	40	53/97

CR - Complete remission

PR - Partial remission

In the following case (Fig 1a and b) of a lung microcytoma we have achieved remarkable resolution of the clinical and radiological picture with progressive reventilation of the whole lung. Dyspnea has disappeared, and there has been improvement in the patient's general condition.

An example of the change in a localized lesion during thermic treatment is demonstrated in Figure 2a and b. The patient, a man of 45 with carcinoma of the lung, has shown a radiologically evident response to the treatment. The neoplastic mass has been almost completely resolved and now, after 15 months, his general condition remains good.

In some cases we have noticed only partial regression (or in some cases no change) in the radiological appearance. Despite the unsatisfactory morphological evolution, the clinical course seems favorable.

Our impression is that a balance has been created between host and tumor with persistence of the initial lesion, but without signs of distant dissemination.

CONCLUSION

On the basis of our results and those of other researchers, we can say that today thermotherapy is indicated in those tumors in which surgery does not offer reasonable prospects of radical cure and in which chemo and/or radiotherapy have been ineffective.

The eventual effect of immune stimulation seems extremely interesting and it can be strengthened and modulated by the rational utilization of drugs which are employed more and more in the therapy of malignant tumors.

In the future we hope to be able, not only to improve the length of survival, but above all to improve the quality of this survival, using techniques and substances less toxic towards normal tissues and selectively active on neoplastic tissues.

We are trying to change from treatments based on chemotherapy, radiotherapy and extensive surgery (all of which are destructive of healthy as well as neoplastic tissues, including the bone marrow and the immune system) to methods which respect the integrity of the organism and the immune system. In this program thermotherapy can play an important role.

We think that the results already obtained indicate the effectiveness of this treatment, even if we are not yet able to furnish valid statistics on the length of survival.

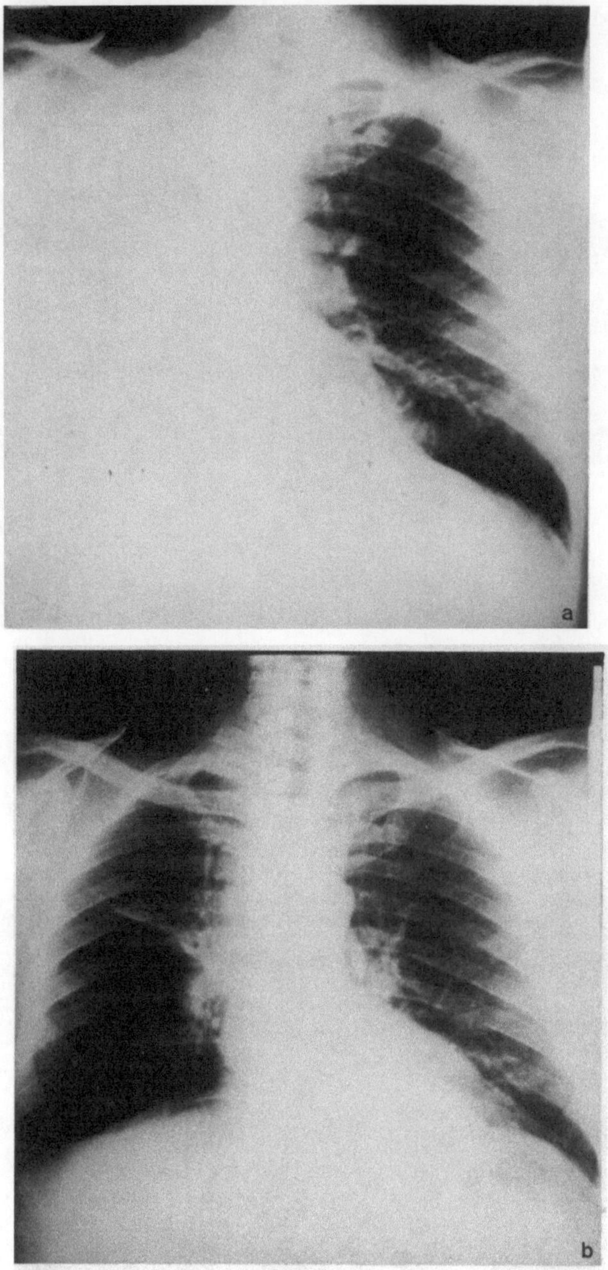

Fig. 1. (a) Microcytoma of lung before treatment. (b) Microcytoma of lung after treatment.

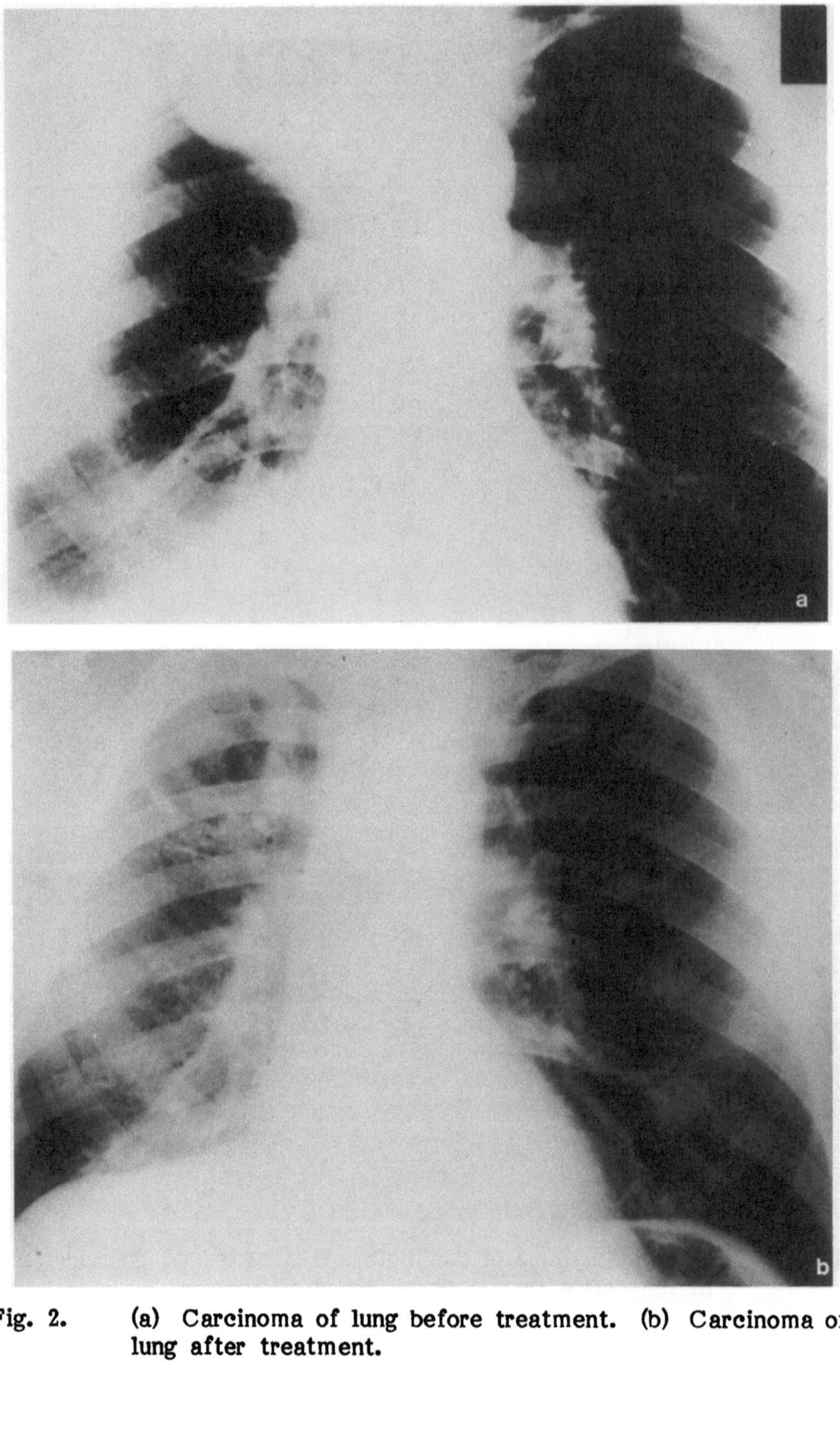

Fig. 2. (a) Carcinoma of lung before treatment. (b) Carcinoma of lung after treatment.

Heat sensitizers and metabolic handling which allow one to obtain selective acidosis of tumors and hypoxic conditions in neoplastic tissue are still at the experimental stage. When these become available for clinical use, we can expect further developments in thermotherapy.

FURTHER READING

1. US National Centre for Health Statistics and US Bureau of the Census.
2. L. A. Shafer, M. L. Washington, A. L. Golstein, Thymosin immunotherapy. A phase I study, Am. Soc. Clin. Oncol. 16:233 (1975).
3. M. W. Dewhirst, W. G. Connor, T. E. Moon, Evaluation of the heat, radiation and heat and radiation response of spontaneous large animal tumors: preliminary results of a phase III randomized trial, The Third International Symposium Cancer Therapy by Hyperthermia, Drugs and Radiation, Colorado (1980).
4. H. H. LeVeen, S. Wapnick, V. A. Piccone, G. Falk, N. Ahmed, Tumor eradication by radiofrequency therapy, J. Am. Med. Assoc. 235;20:2198 (1976).
5. H. H. LeVeen, "Radiofrequency in the treatment of malignant tumors. Radiotherapy and Oncology," T. J. Deeley ed., Volume II, Wright PSG, Bristol (1982).
6. P. Pontiggia, H. H. LeVeen, G. L. Ganzetti, La termoterapia localizzata nella cura dei tumori maligni. Valutazione clinica preliminare, Rec. Progr. Med. 70:5 (1981).

AUTHOR INDEX

SUBJECT INDEX

MIX
Papier aus verantwortungsvollen Quellen
Paper from responsible sources
FSC® C105338

If you have any concerns about our products,
you can contact us on
ProductSafety@springernature.com

In case Publisher is established outside the EU,
the EU authorized representative is:
Springer Nature Customer Service Center GmbH
Europaplatz 3, 69115 Heidelberg, Germany

Printed by Libri Plureos GmbH
in Hamburg, Germany